CONCEPTUAL PHYSICS

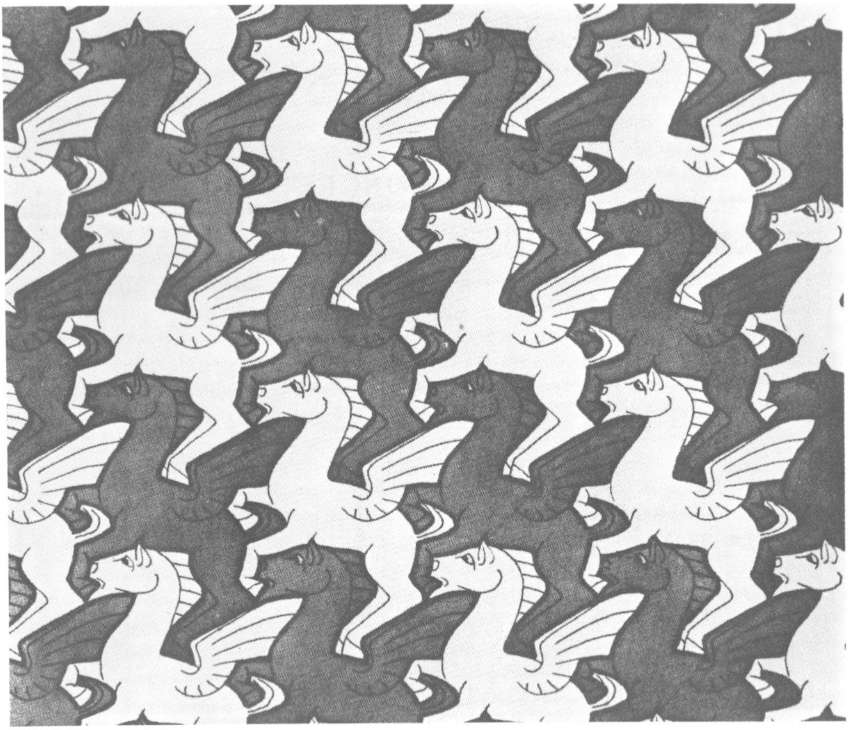

Reproduced from *Symmetry Aspects of M. C. Escher's Periodic Drawings* by Caroline H. MacGillavry with the permission of the International Union of Crystallography.

JAE R. BALLIF
WILLIAM E. DIBBLE

CONCEPTUAL PHYSICS

MATTER IN MOTION

JOHN WILEY & SONS, INC.
New York · London · Sydney · Toronto

10 9 8 7 6 5 4 3 2 1
Library of Congress Catalog Card Number: 69-13676
SBN 471 04640 X
Printed in the United States of America

PREFACE

In this book we have tried to present and explain the fundamental principles of physics in a way that non-scientists will find comprehensible and interesting. Selected applications of these principles are also discussed, mainly to illustrate and clarify the principles. Although we begin with some topics that are usually regarded as part of classical physics, we have presented them from a modern point of view and have included those which we believe are necessary for an understanding of later developments. In our opinion, the basic elements of modern physics include the fundamental particles, quantities, constants, and interactions; the principle of relativity and other symmetry principles; the conservation laws; quantum physics, including the uncertainty principle; entropy; and the laws of motion. Although our understanding of the second law of motion has been changed by modern developments, we believe that the concept of force and its relation to motion is still very useful in teaching the fundamental principles of physics, even from a modern point of view. Of course in a short text we cannot discuss all of the ramifications of these fundamentals in great detail.

Our approach to the principles of physics is essentially conceptual rather than mathematical. Although simple equations appear throughout the text, they have been kept to a minimum except in some optional sections and are used mainly for the precise statement of important principles. Often the equations follow a qualitative discussion of the principle being considered. The concepts we discuss are not trivial, and the mere lack of mathematics does not necessarily make them easy. Nevertheless we have found that our students can understand these concepts and that they very often find them exciting.

The student for whom this book is written is a successful high school graduate who has probably had some algebra, but is not skillful at algebraic manipulations. We have included questions at the end of each chapter that can be used as exercises to test and enlarge the student's understanding of the subject. We have also included problems and worked examples but these are strictly optional. Our students are capable of solving simple problems if the instructor will take the time and effort to guide him through examples. One of us regularly uses the questions provided at the end of each chapter and not the problems as exercises for the student. The other regularly uses problems in one or two areas to impress the students with the precision of certain physical laws. Too much time spent with problems, however, can lead to omission of too many important principles, especially in a short course.

This book is designed for a one semester course. Some of the sections of the text are marked with an asterisk and may be omitted. If most of this optional material is omitted the book may be used for a one-quarter course.

The optional sections are of two kinds. Some of them are no more difficult than the rest of the text, and may be used or omitted as time allows. On the other hand, some of the optional sections are more difficult, usually more quantitative, than the rest of the text, and can be used by students with better preparation or greater ability. For example, the section on the New Philosophy in Chapter One is not difficult, and we always use it. Chapter Five can be omitted in its entirety, although the first half of the Chapter is not difficult. The optional sections in Chapter Ten, including the section on general relativity, although not necessarily more mathematical, are more challenging and are considered optional for that reason.

At the end of each chapter we have included a chapter summary. This summary is not an outline of the chapter, but is a list of the things introduced in the chapter which we think are important.

The sections in the appendix which discuss the use of powers of ten and proportion might be useful for all students, since we express quantities in the text using powers of ten, and speak of proportion a number of times. Appendix II contains the necessary mathematics for even the more difficult problems and can be used by those who prefer a more quantitative treatment. Appendix III contains an algebraic derivation of the Lorentz Transformation and other equations of the special theory of relativity. Appendix IV contains a discussion of instantaneous velocity, with a simple introduction to the calculus.

These two appendices (III and IV) are more difficult and succinct than the other parts of the book.

We have used the meter-kilogram-second-coulomb system of units throughout this book. For reference, however, Appendix I gives the units of two other systems in terms of the units we use.

We appreciate the assistance of the staff of John Wiley & Sons, and those associated with them. We are also indebted to our colleagues and friends, particularly Dr. B. Kent Harrison, and to our students.

Jae R. Ballif
William E. Dibble

Provo, Utah
August, 1968

CONTENTS

CONCEPTUAL PHYSICS

1

THE FIRST LAW OF MOTION AND THE SPECIAL PRINCIPLE OF RELATIVITY

If we believe in the existence of specifically modern attitudes to the world, distinct in quality from those held by men at earlier periods, the elements which go into the making of "modernity" may be seen to make their first historical appearance in the sixteenth and seventeenth centuries. Some historians attribute the change to the liberation of men's minds during the Renaissance and the Reformation. Today, many historians would be more likely to stress the conservatism of these two movements rather than their radicalism. Their emphasis tends instead to fall on the effects of the change in men's attitude to the universe, which has been summed up in the phrase "the Scientific Revolution."

By this is meant above all the imaginative achievement associated with the names of Copernicus, Galileo, and Newton. . . . The new-world picture worked far more with mathematical abstractions, upon which general laws were erected. The result was that mathematics in 1700 enjoyed the prestige which in 1400 had been held by Aristotelian logic and philosophy.

. . . what is not clear is how it all came about.[1]

GALILEO GALILEI (1564–1642), THE ITALIAN MATHEMATICIAN, philosopher, and scientist, will figure prominently in our discussion of the beginnings of modern science. Galileo helped to develop and establish new attitudes toward nature and toward truth, attitudes that are still part of our scientific outlook. A complete historical study of the beginnings of modern science would require a discussion of Galileo's contemporaries and of his renaissance and medieval prede-

[1]Hugh Kearney, *Origins of the Scientific Revolution*, London: Longmans, Green and Co., 1964, p. xi.

cessors as well. It would also require a discussion of a number of ancient Greek philosophers and mathematicians, such as the "divine" Archimedes[2] and members of the mysterious Pythagorean brotherhood.

In their study of the ancients, Galileo and his fellow scientists went back beyond the Greek philosopher Aristotle (384–322 B.C.) to Plato and others. However, their work was not just a revival of ancient learning, nor was it simply an attack against the opinions common in their day. Galileo represents those who developed a new philosophy, the philosophy that we now call "science."

This new philosophy has drawn together many of the activities and ideas that developed from ancient and medieval medicine, philosophy, astrology, alchemy, civil and military engineering, arts, skills, and even theology. This new and changing philosophy has been characterized as "the search to discover unity in the wild variety of nature" and "the creation of concepts and their exploration in the facts. . . . Truth is the drive at the center of science; it [science] must have the habit of truth, not as a dogma, but as a process."[3] This "truth" is something that can be applied in operations and actions in the actual world, not something that exists only in some logically

[2]Galileo referred to Archimedes as "that divine man" in his paper "The Little Balance." See Galileo Galilei, *Dialogue Concerning the Two Chief World Systems, Ptolemic and Copernican,* translated by Stillman Drake, Berkeley and Los Angeles: University of California Press, 1953, p. 483.

[3]J. Bronowski, *Science and Human Values*, rev. ed., New York: Harper Torchbooks, Harper & Row, 1965, pp. 16, 60.

consistent intellectual world of its own. Thus we consider it no accident that the rise of the new philosophy, science, has been associated with increased power and wealth in the societies in which it took place. The Italian Galileo, his German contemporary Johannes Kepler (1571–1630), and the Englishman Isaac Newton (1642–1727) were prominent among those who helped to construct this new philosophy and use it effectively.

Most philosophers have believed that there was some order, sense, or unity in their world of experience. Therefore, they have been interested in the fundamental substances of which everything is made and the basic concepts in terms of which the world of experience can be understood. Physicists in particular have been concerned with this effort as it applies to space, time, and matter. Indeed, physics may be considered as the study of the properties of matter, space, and time.

Most physicists have believed that matter is made up of certain fundamental particles, and that much or all of our experience can be explained by the motion and interaction of these particles as they stick together or bounce apart while moving through otherwise empty space. Others, unwilling to believe in the existence of nothing—that is, empty space (the void)—preferred to believe that some smooth, possibly flowing spirit, or "ether," is spread throughout all space, and that whirls or eddies in that ethereal substance manifest themselves to us as ordinary matter. Others believed that matter itself is a smooth and continuous substance, not made up of particles. Still others, usually not scientists, emphasizing the importance of observation, contended that mind is the fundamental substance and that matter exists only as it is observed or thought of by a mind. All these ideas have had an influence on science.

In keeping with this philosophical and scientific tradition, we shall be very much concerned with certain fundamental entities and principles, the elements that make up our world and the basic ideas in terms of which we describe and explain the behavior of these elements. These are the particles, principles, forces, laws, and quantities that we consider basic, from which others may be constructed or derived. To some extent, of course, we have our choice as to what things we call fundamental.

In this chapter we discuss two fundamental quantities, distance and time duration, which are obtained by measuring space and time. We also discuss the standards and units for these quantities. We next describe the derived quantities speed and velocity. Next we present two of the fundamental principles of physics that arose in part from Galileo's work: the first law of motion, and the special principle of relativity. The special principle of relativity is our first example of a symmetry principle in nature. We also discuss five other fundamental symmetry principles. Finally, we offer a historical perspective on the "new philosophy."

QUANTITIES AND MEASUREMENT

A *quantity* may be defined, roughly, as an observable property or process in nature with which a number may be associated. This number is obtained by the operation of *measurement*, which in simple cases may be merely the process of counting. The number may be obtained directly by a single measurement, or indirectly, as, for example, by multiplying together two numbers obtained in separate operations of measurement. In this section we will first be concerned with the quantities *distance* and *time duration*. We obtain distances by making measurements of space. We obtain time durations by making measurements of time. Almost everyone has had practical experience making measurements of distance using rulers and making measurements of time duration using clocks. Deeper concepts of space and time will be discussed later.

Fundamental quantities are those which we choose *not* to define in terms of other quantities. For example, we choose distance and time duration to be fundamental; space and time seem very basic to us. Other quantities, which we can then define in terms of the fundamental quantities, are called *derived* quantities.

Scalar quantities can be specified using a single number. For example, the length of a rope is a scalar quantity. Some quantities cannot be specified completely without a direction as well as a number. For example, we cannot completely describe the motion of an object without giving the direction of its motion. A quantity that has a direction as well as a number is called a *vector* quantity. We must distinguish between scalar and vector quantities because they obey different rules of arithmetic, as we shall see in Chapter Two.

The Unit and Standard of Distance

We measure a distance in terms of some other distance. We say, for example, that the distance from Chicago to New York is 713 miles. The number 713 alone is not enough; we need the whole term, *713 miles*, in order to indicate the distance correctly. The *mile*, which is itself a distance, is the unit in terms of which we usually measure the distance from Chicago to New York. We may thus define a *unit* of distance as that distance in terms of which we measure other distances. We could, of course, choose any distance we liked for our unit: a mile, a foot, an inch. In fact, we shall choose a unit commonly used in scientific work—the *meter*.

In order to define the meter (or any unit of distance), we need to have some sample distance we can rely on. We need some material object that will give us a distance to which we can refer. This material object, properly used, is the *standard* of distance. For a long time the standard of distance was a rigid bar, made of a corrosion-resistant metal alloy, which was kept in a controlled environment at a constant temperature and pressure (see Fig. 1-1). This bar had a thin scratch near each end. The distance between those scratches was defined to be one meter. Rulers one meter long were constructed by

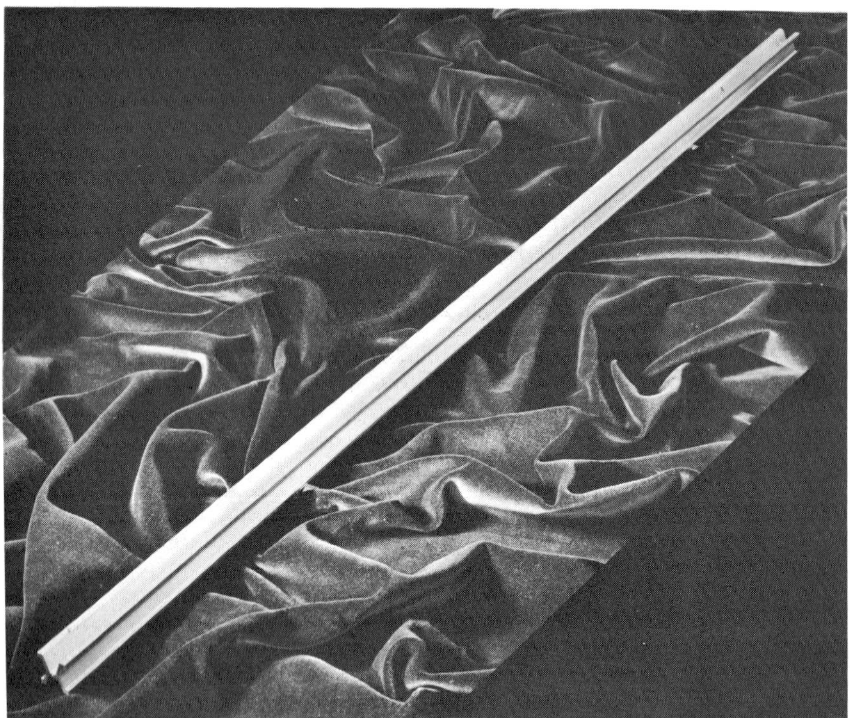

Fig. 1-1 *Meter No. 27 at the National Bureau of Standards. This bar is similar to the original standard bar in Paris. Courtesy of the National Bureau of Standards.*

comparing them with the bar, and other distances were measured in meters using these rulers. A meter is exactly equal to 39.37 inches.

The rigid metal bar has been supplanted by another standard of distance. However, it illustrates the use of standards and units in physics: quantities are measured in terms of units. These units can be defined by using suitable material objects or processes as standards.

The Unit and Standard of Time Duration

For our unit of time duration we shall choose the familiar *second.* There is, of course, no direct way to compare a second of time that elapsed twenty years ago with a second of time today. Thus there is a problem in deciding just what should be called a second of time. The solution is to consider some process which happens over and over again in essentially the same way, such as the swinging of a pendulum or the rotation of the earth. We then agree that each seemingly identical process takes the same time. For our purposes here, we may consider that the earth takes the same time for each rotation (which is very nearly true) and obtain the *second* as our unit of time duration as follows: the second is defined as one 86,400th part of the earth's daily rotation period. In other words, there are 86,400 seconds in a day. (There are 60 seconds in a minute, 60 minutes in an hour, and 24 hours in a day. $60 \times 60 \times 24 = 86,400$.)

Once we have some understanding of the laws of physics we can make more precise and more convenient choices for our standards of distance and time.

Counting and Measurement: The Discrete and the Smooth

Most of the instruments used to obtain information about the physical world are equipped either to *measure a distance* or to *count.* An example of an important counting measurement is shown schematically in Figure 1-2. The dust particles are counted separately, as they strike the detector on the spacecraft. An example of the other kind of instrument is the speedometer of an automobile (see Fig. 1-3). Actually, it is the distance that the indicator has traveled along the dial that tells us the speed. Similarly, an ordinary clock indicates elapsed time by the distance the hands have traveled around the clock face. In these two examples we measure a quantity—speed or elapsed time—by measuring the distance the end of a pointer moves on a dial.

A collection of separate objects that can be counted is called a

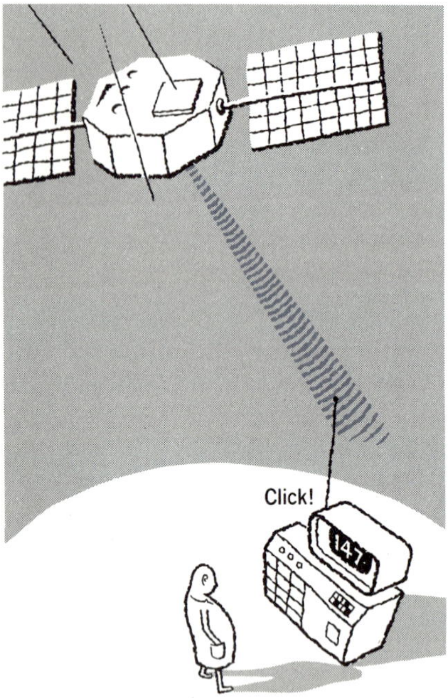

Fig. 1-2

discrete collection. A basket of eggs, for example, gives us a discrete collection of eggs. We can divide this collection into smaller and smaller collections. However, we finally reach the point where we can no longer divide these collections and still have whole eggs. Half an egg is very different from an egg, and "all the king's horses and all the king's men" can't put the two halves of an egg together and get the original egg. Furthermore, the two halves could never be mistaken for two small eggs. We say that the collection of eggs is a discrete collection, made up of individual units or particles called eggs. A collection made up of separate, countable bits is called discrete.

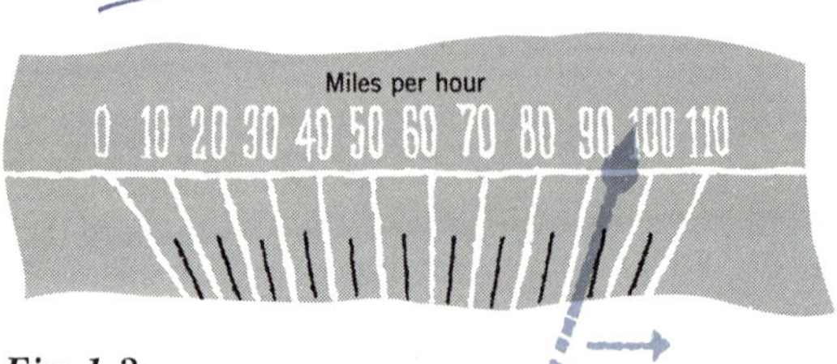

Fig. 1-3

A distance, however, is different in nature from an egg. If we divide a distance in half, we obtain two other distances that are very much like the original distance except that they are smaller. It appears that we can continue to divide distances in half forever, getting smaller and smaller distances, without ever stopping or getting something that could not be called a distance. Since distances, unlike eggs, have this property, we conclude that space is not discrete; rather, we call it smooth, or *continuous*. The measurement of distance is not the simple process of counting but a different process. One consequence of this difference is that whereas counting can result in an exact whole number, the measurement of distance always involves some error or approximation.

We shall not discuss all the properties of continuous quantities; we are content to say that they are smooth and can be divided indefinitely without changing their nature. Space and time are generally regarded as continuous; thus distance and time duration are considered to be continuous quantities.

Scientists and philosophers have long discussed whether matter is smooth (continuous) or made up of particles (discrete). We shall discuss this question later; for the time being we shall simply assume that matter is made up of small, invisible particles. True, such material as water certainly looks smooth, but we believe that it is really made up of particles and simply gives the illusion of smoothness because the particles are so small. Even the sand of the seashore appears smooth when viewed from afar. Up close, the individual grains of sand become visible, giving the sand a discrete, or grainy, appearance (see Fig. 1-4).

Derived Quantities

In our considerations of matter in motion, we will need to define many other quantities. However, we do not consider most of these quantities fundamental since they can be derived from other quantities. Area, speed, and velocity are examples of such *derived quantities*.

AREA. If we have a rectangular region six meters long and three meters wide, we obtain the area of the region by multiplying these two distances together:

$$\begin{aligned} 6 \text{ meter} \times 3 \text{ meter} &= 18 \times (\text{meter}) \times (\text{meter}) = 8 \text{ meter}^2 \\ &= 18 \text{ square meters} \end{aligned}$$

Note that we can multiply units together just as if they were numbers. We multiply meter times meter and get meter squared, or, as we call it, "square meters." The unit of area is the square meter, or $meter^2$.

Just as we obtained an area by multiplying two distances together, so we can obtain a volume by multiplying three distances together. To find the volume of a rectangular object—for example, a brick—we multiply together its length, width, and height. The unit of volume is the cubic meter, or $meter^3$.

SPEED AND VELOCITY. A useful quantity associated with the motion of an object is its *speed*. The speed is determined by the amount of distance traveled by the object in a given time duration. In order to derive the quantity speed, we divide the fundamental quantity distance by the fundamental quantity time duration. For example, suppose that an object travels 10 meters in two seconds. We divide 10 meters by two seconds and obtain a speed of five meters per second:

$$\frac{10 \text{ meter}}{2 \text{ second}} = 5 \frac{\text{meter}}{\text{second}} = 5 \text{ meters per second}$$

(In this example, we are assuming that the speed does not change in these two seconds.) Thus, in general we obtain the speed of a

Fig. 1-4 Continuous and discrete quantities.

moving object by dividing the distance traveled by the time duration. The unit of speed is

$$\text{meter per second} = \frac{\text{meter}}{\text{second}}$$

This unit is obtained by dividing the unit of distance by the unit of time duration.

Speed alone does not completely describe the motion of an object. We also need to know the direction in which the object is moving. Therefore, it is convenient to define a new quantity, which includes both the speed and the direction of a moving object. This quantity is called *velocity*. In order to specify the velocity of an object, we must give both the speed of the object and the direction in which it is traveling. Because it includes the direction, velocity is a vector quantity. Speed, on the other hand, is a scalar quantity and it omits the direction.

THE FIRST LAW OF MOTION

Galileo was particularly interested in the motion of objects. Some of his contemporaries explained motion, as the Greeks had done, with one set of rules for celestial objects and another for objects near the earth. The methods, insight, and personality of Galileo brought him quickly into conflict with the entrenched principles of the ancients. From his studies, Galileo came very close to understanding what we now call the first law of motion.

Statement of the First Law of Motion

Sir Isaac Newton stated the first law of motion as follows:

> Every object continues in its state of rest, or of uniform motion in a straight line, unless it is compelled to change that state of motion by forces impressed upon it.[4]

Newton gives credit to Galileo for using this law; however, Galileo evidently did not understand it completely.

Let us consider the first part of the law: if an object is at rest it will remain at rest unless something starts it moving. In order to start the

[4]Isaac Newton, *Mathematical Principles of Natural Philosophy and His System of the World*, translated by Andrew Motte, 1729; rev. by Florian Cajori, Berkeley: University of California Press, 1934, p. 13. We have changed the word "body" to "object" and the word "right" to "straight."

object moving, it is necessary to use force—that is to say, to push it or pull it in some way. (Force will be discussed more fully in the next chapter.) Without using some force it is impossible to start the object moving. This is true even if the object is not fastened or held in any way. The force may be very small, but *some* force is required.

The second part of the law appears to be simple also, but it was unknown for thousands of years. A moving object will keep moving at the same speed and in the same direction unless some force acts on it. For example, imagine a moving object far out in space, so far from any planet or star that we can ignore their influence on it. Then, according to the first law of motion, this object would continue to move forever, never changing its speed or its direction—in other words, never changing its velocity. At one time philosophers (such as Aristotle) thought otherwise. They thought that force was required to keep a body moving. This is false. Actually, a force applied to the object will change its motion by changing its speed, its direction, or both.

Galileo ridiculed those followers of Aristotle who claimed that an arrow, after it leaves the bowstring, continues to move forward only because the air stirred up by the bowstring keeps pushing the arrow along. Galileo pointed out that if this were true, the arrow would point feathers first, as does an arrow blown by the wind.[5] In fact, the air exerts a retarding force, called friction.

Friction

There are nearly always forces that cause moving objects to slow down and stop. This is why people did not discover the first law of motion sooner. Objects that are sliding, swimming, or flying are rubbing against something—a solid surface, the water, or the air. The rubbing causes a force which tends to slow the objects down and eventually stops them. Such retarding forces are called friction.

Illustrations of the First Law of Motion

Our experience with automobiles furnishes us with many illustrations of the first law of motion. Once a car is rolling along, a force is required to stop it. If we wait long enough, friction will stop it; if we are unwilling to wait, we increase the frictional force by stepping on the brakes. If we jam on the brakes, the frictional forces stop the car

[5]Galileo Galilei, *op. cit.*, pp. 152–153.

quickly. But the brakes do not act directly on the passengers inside the car. Therefore, they continue their forward motion in accordance with the first law—right on through the windshield if nothing prevents it. Hence the use of seat belts. If the car turns suddenly, the occupants continue in the original direction and, therefore, slide across the seat unless friction prevents it.

The first law of motion also helps us to understand what produces "whiplash" injury in an automobile collision. When a car is struck from the rear, the seat, which is firmly attached to the car itself, pushes the occupant's body forward. But his head, which is not rigidly fastened to his body or supported by the car seat, tends to remain where it was. The motorist therefore suffers an injury at the top of the spine, where the head and body meet. The situation is illustrated in Figure 1-5, for one common posture. The unsuspecting driver nearly loses his head, so to speak.

A bomb dropped from an airplane provides another illustration of Newton's first law. The bomb does not fall straight down; it has the same forward speed as the plane when it is released, and it preserves that forward motion—somewhat retarded by friction—as it falls. Therefore, it keeps moving forward at the same time as it moves downward (see Fig. 1-6). The bombardier allows for this forward motion in deciding when to release the bomb.

There are various ways of reducing friction that make the first law more evident. Figure 1-7 illustrates one way. The gas escaping from

Fig. 1-5

the hole in the bottom of the container forces a thin layer of air to flow between the container and the flat surface over which it moves. Thus the container has no direct contact with the surface; the only friction is the slight amount between the two surfaces and the escaping gas, which only slightly retards the device. The modern hovercraft works on the same principle: it is kept above the surface of the water or ground by a strong jet of air blown downward.

The difficulty of stopping when one is sliding on ice or on a banana peel also illustrates the first law of motion. Recently, chemicals such as those used in "hydrolube" (see Fig. 1-8) have been developed that can greatly reduce the friction on a surface. We see from these examples that the forces of friction are very important in controlling our bodily motion. Without them we should keep sliding forever.

Galileo's Reasoning[6]

For most physical principles, the process of discovery is either unknown or mathematically inacessible to the general reader. Gali-

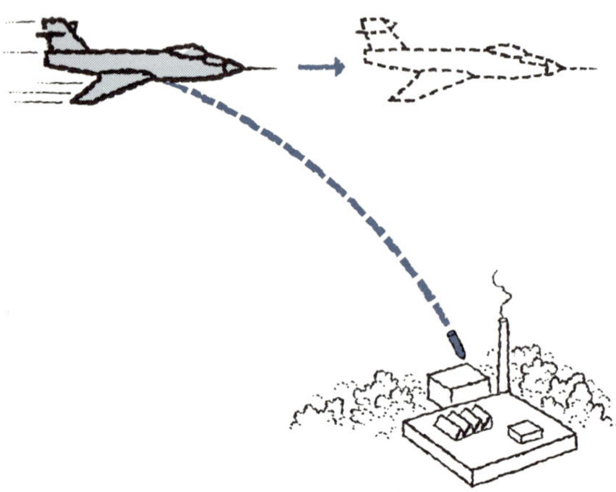

Fig. 1-6

[6]Galileo Galilei, *Discourses and Mathematical Demonstrations Concerning Two New Sciences*; see *Dialogues Concerning Two New Sciences*, translated by Henry Crew and Alfonso de Salvio, New York: McGraw Hill, 1963, pp. 164–166. (We have adopted the alternate title suggested by I. B. Cohen, *The Birth of a New Physics*, New York: Doubleday Anchor Books, 1960, p. 193.) We have not been able to find this argument in Galileo, but it is ascribed to him by Ernst Mach, in *The Science of Mechanics*, LaSalle, Ill.: Open Court, 1960, pp. 168–169; see also Galileo Galilei, *Dialogue Concerning the Two Chief World Systems, op. cit.*, pp. 146–147.

leo, however, often presented the development of his ideas in ways that were "elegant" and simple.

Consider the inclined planes shown in Figure 1-9*a*. We allow a ball to roll up and down the planes. You will notice that we have rounded off the corners where the planes meet the level; this is done so that the ball will roll smoothly. If a ball struck a corner, it might bounce.

We release a ball from the left-hand plane, letting it roll down that plane, across the level, and then up the other inclined plane. How far up the other plane will it go? Intuitively we answer: as high above the level as it was when it started on the left-hand plane. Then we think more carefully; if there is some friction, the ball will probably not go quite so far up as it was when it started. So we idealize the experiment we are thinking about. In practice we cannot remove friction entirely. However, the smoother and harder the surface and the ball, the less friction there will be and the more nearly the ball will come back to the same height.

Now we lower the plane on the right, still repeating the same experiment (see Fig. 1-9*b*). The ball should still come up to the same height, but the same height will put the ball farther and farther to the right as the plane is made less steep. In the limiting case, when the plane on the right is horizontal, the ball should keep going forever. Thus we have arrived at the first law of motion: the ball, being neither retarded nor pushed ahead by any force, keeps moving uniformly forever.

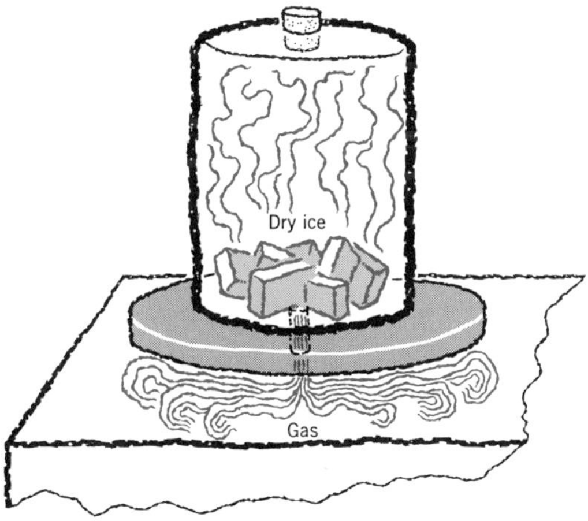

Fig. 1-7

In the preceding paragraphs we performed what is sometimes called a "thought experiment." We "talked" an experiment without actually doing it. (There are those who accuse Galileo of doing too much of this.) We assumed certain ideal conditions that would not exist in actuality—for example, the absence of all friction. This type of thinking focuses our attention on certain ideas and helps us to work out their consequences. We were in this way led to the first law of motion, which gives the behavior expected of objects with no forces acting on them. Actually, of course, there do not appear to be any such force-free objects—there always appear to be some forces acting on an object. Thus the first law of motion describes an idealized condition that never occurs. Yet, after making this idealization, we are able to understand real processes better than if we merely made detailed observations of the real processes themselves. Experimental science abstracts and idealizes from experience to arrive at principles,

***Fig. 1-8** Hydrolube used by permission of the Western Company Research Division, Richardson, Texas. Courtesy of the Western Company, Research Division.*

which form a sort of model of reality. It is amazing how precisely this abstract model of reality can predict the behavior of real objects.

Of course principles such as the first law of motion are not established by thought experiments alone, just by thinking, without actual observations. Since we only talked about the experiment, we need to check its conclusions by actual experiment, as Galileo clearly pointed out.[7] Also, some of our experience went into the discussion of the thought experiment. What really convinced us that the ball could rise no higher than it started was our general experience that balls do not run uphill by themselves. Similar notions were used by other early scientists in developing theories.

Later we shall fasten our attention again on the interesting fact that the ball returns to the same height it started from. It suggests that something remains the same, or is conserved, as the ball moves on the inclined planes.

THE SPECIAL PRINCIPLE OF RELATIVITY

Reference Frames

Anyone who has followed a map to find his way from one place to another has used what we call a *reference frame*. You cannot begin such a trip until you first "get oriented." This orientation amounts to establishing a reference frame. You must know your present position

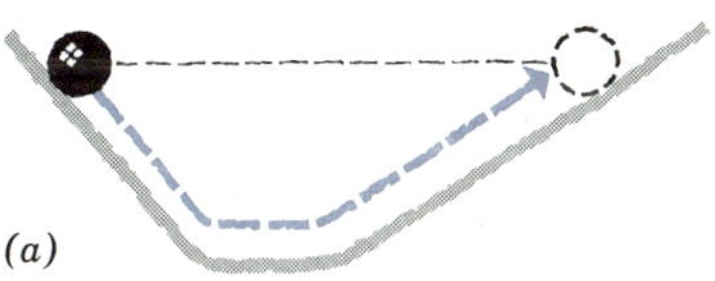

(a)

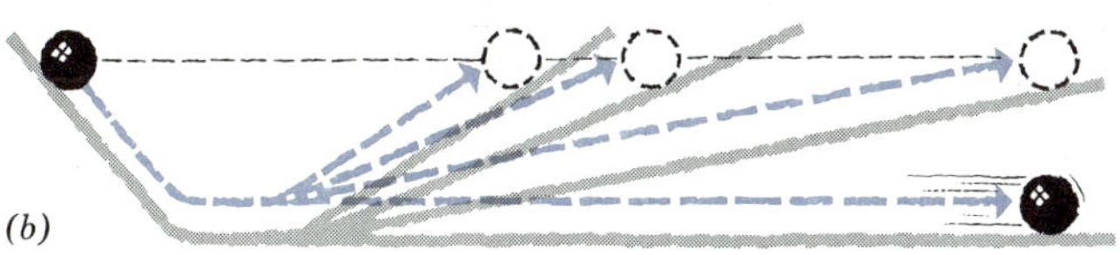

(b)

Fig. 1-9

[7]Galileo Galilei, *Dialogues Concerning Two New Sciences, op. cit.*, pp. 163–166.

with respect to some object (such as a tree, a rock, or a city), and you must also be able to establish the direction of your goal. To establish direction on a map, we ordinarily imagine two intersecting lines, one pointing north and the other east. The first can be established with a compass needle, and the other is at right angles to it. These imaginary lines give us an example of a reference frame. Now, travel instructions have some meaning. For example, we can go 8000 meters north from the rock, 10,000 meters east, and another 16,000 meters north and find the treasure (see Fig. 1-10*a*).

The two imaginary lines that intersect at right angles to make up the reference frame are called *axes*. The point at which these axes intersect is called the *origin*. In our example, the axes, intersect at the rock. The rock, then, is our origin.

Many different sets of instructions could be made up that would eventually take us from the rock to the treasure. Once we have set up the reference frame, the simplest instructions would require just two statements. The only instructions actually needed to go to the treasure from the rock (origin) are the following ones:

(a) Go 10,000 meters east, parallel to the east-west axis.

(b) Go 24,000 meters north, parallel to the north-south axis.

The order in which we follow these instructions does not matter; either way, we will arrive at the same spot (see Fig. 1-10*b*).

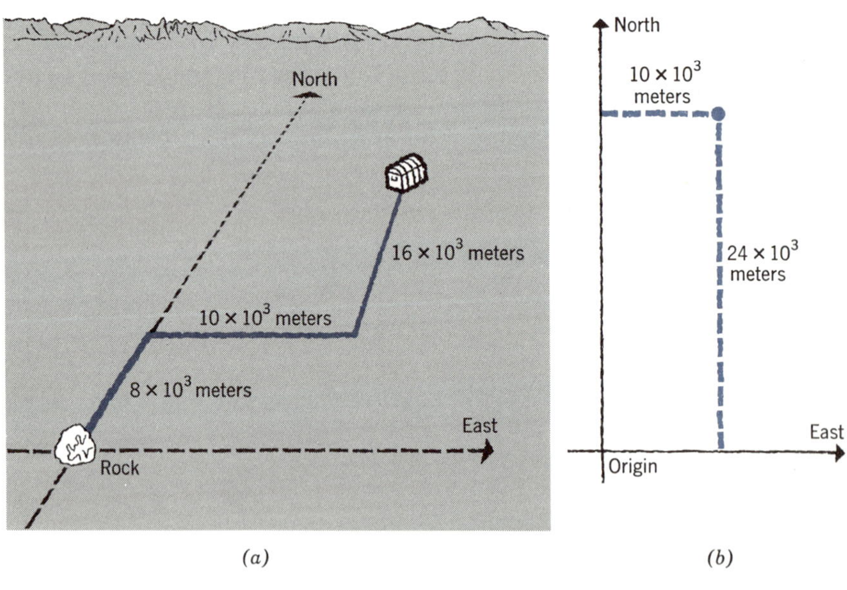

(*a*) (*b*)

Fig. 1-10

In using just two lines in our reference frame, we have assumed that we were on a surface. If the treasure were buried underground or hidden in an old oak tree, we would need another instruction in order to locate it. This instruction would tell us how far above or below the surface we must go to get to the position of the treasure. As this example suggests, to locate objects in ordinary space we need three reference lines or axes in our reference frame, all intersecting at the same point (the origin) in such a way that each is perpendicular to the other two (Fig. 1-11). To locate any point in ordinary space, three instructions, or items of information, are necessary. For this reason we refer to ordinary space as three-dimensional.

We have seen that it is necessary to have a reference frame if we want to locate a position in space from a set of instructions. It is also necessary to use a reference frame in describing the motion of an object. The velocity (speed and direction) of an object should be given in reference to something else. Most often, the reference frame is similar to that described above for the location of a point or position. For example, the train in Figure 1-12*a* is moving to the northeast at 30 meters/second.

We are so accustomed to using compass directions fixed to the earth for our reference frame that we may not realize that reference frames attached to other objects and oriented in different directions are just as valid and sometimes more useful. Figure 1-12*b* shows the

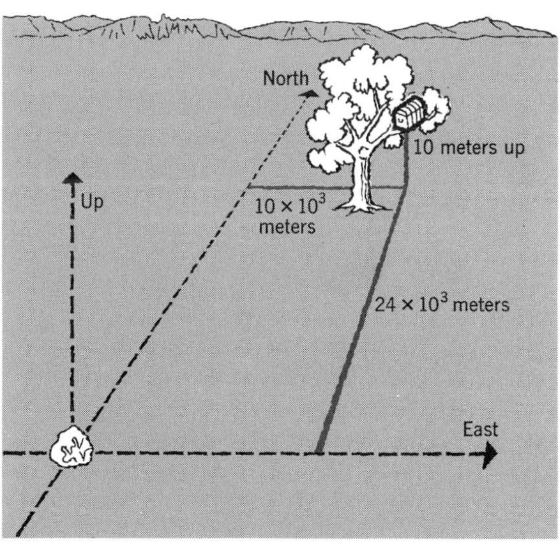

Fig. 1-11

same train moving on the surface of the earth from a different point of view. The earth with the train on it is itself rotating and moving through space around the sun. What, we might ask, is the "real" motion of the train—its motion across the earth's surface, or its motion that includes the motion of the earth around the sun? Actually, either motion is real—the difference is in the reference frame with respect to which the motion is considered. In the former case we use the conventional reference frame fixed to the earth and describe the motion of the train as observed from the earth. In the latter case we view the motion as though we were located on the sun and refer it to a reference frame centered in the sun, with axes pointed outward toward selected stars. Although the motion looks very different from different reference frames, it is the same physical situation that is being observed and described.

Another example of the fact that the same motion can be described from different reference frames with equal validity is afforded by the motion of a stewardess in an airplane. Although the airplane may be traveling 600 miles per hour, to the passengers the stewardess appears to move around normally. That is, relative to a reference frame attached to the airplane, her motion appears no faster than it would be if she were walking on the ground. Relative to a reference frame fixed to the earth, however, she is rushing by at about 600 miles per hour. Since the same motion appears different from different reference frames, we cannot completely describe speed or velocity unless we tell what reference frame we are using.

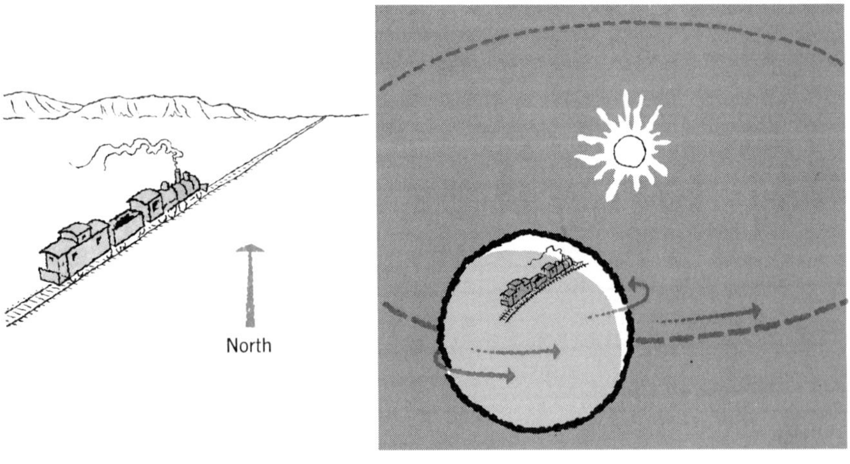

(a) (b)

Fig. 1-12

The Special Principle of Relativity

Most people in Galileo's day thought that the earth was motionless in the center of the universe and that the sun and other heavenly bodies revolved around it. Galileo, on the other hand, was convinced that the earth itself moved around the sun, even though the motion was not apparent to people on the earth. Lacking any direct evidence that the earth was actually rushing through space at a speed of some nineteen miles per second, he had to persuade his opponents by explaining how the earth could appear to be stationary when in fact it was not. Such rapid motion of the earth, his antagonists claimed, would leave the air behind. Even a bird perched on a branch would be left behind in space the instant it let go of the branch, as the earth rushed on—or so it was argued (see Fig. 1-13). Galileo countered by pointing out that the bird was moving *with* the earth while it was on the branch and retained that motion when it let go. Thus the relative motion of the bird with respect to the tree would be the same as if the earth were at rest. According to Galileo, it is the relative motion that is important. This assertion was an early approach to the *special principle of relativity*. We shall state this principle in the slightly modified words of Albert Einstein (1879–1955), who emphasized it about three hundred years after Galileo:

All laws of nature are the same in all reference frames moving uniformly.[8]

Fig. 1-13

[8]Albert Einstein and Leopold Infeld, *The Evolution of Physics*, New York: Simon and Schuster, 1938, p. 186.

Objects are said to be moving uniformly, or to be in uniform motion, when they are moving in a straight line at a constant speed.

The special principle of relativity tells us that under a certain restricted condition, the laws of physics will appear to be the same in different reference frames. The restriction is that these reference frames must be either at rest or moving in a straight line with constant speed, and not rotating. In other words, they must be moving in the same way that a non-rotating object moves when there are no forces applied to it, according to the first law of motion. Under these conditions each reference frame, when viewed from another, will appear to be moving in a straight line with constant speed, and not rotating. An example will help us to understand this principle.

Suppose that we are in a railway car, but that the train has not yet started to move. It is sitting in the station (see Fig. 1-14*a*. If we drop a ball, it will fall straight down. Now suppose that we have left the station, and that the train is going eighty miles an hour along a straight smooth track (Fig. 1-14*b*). Once again, we drop the ball. What happens? To us in the train it looks as though the ball drops straight down. In other words, the result looks to us in the train exactly as it did when the train was at rest with respect to the ground. To an observer outside the train, however, the ball moves forward with the train as it drops.

We can account for this phenomenon by using the first law of motion: since the ball was moving with the train *before* it was dropped, it continues to move forward with the train *as* it drops. But an observer in the train is also moving forward with the train. Thus it looks to him as if the ball falls straight toward the earth. In fact, if we were to set up a scientific laboratory on the train and test all the known laws of physics, they would be just exactly the same as if we tested them in a laboratory that was at rest. The fact that the laws are tested in a moving reference frame fixed to the train makes no

(*a*) (*b*)

Fig. 1-14

difference whatsoever. We say that the laws of physics are *invariant*, regardless of the straight uniform motion of the reference frame. This is another way of stating the special principle of relativity. Invariant means unchanged. The laws are not changed by using another reference frame that is moving uniformly (at a constant speed in a constant direction).

If all the laws are the same in the moving system as in a stationary system, then there is no way of telling that the system is moving without looking outside the system at something else. From the train we must look outside and see the trees and houses going by; then we know that we are in motion relative to them. There is no way to tell which is "really" moving, the train or the trees. Uniform motion is relative. That is, uniform motion is meaningful only as a relation between one frame of reference and another, or one object and another. Because all the laws of physics are unaffected by uniform motion, there is no way of finding any "real" or "absolute" uniform motion that a particular reference frame may have. We can measure relative motion only with respect to other objects or frames.

In all of this, of course, we are talking only about reference frames moving uniformly. If a frame is in nonuniform motion, we can detect it. If we are spinning, speeding up, or otherwise changing motion, we can feel or detect it without looking outside our system. If the train is vibrating, or if it turns a corner, we can tell it without looking out the window. In fact, of course, the train is moving on the earth, which is rotating and revolving. The effects of this nonuniform motion are too small for us to feel, but they can be detected by means of instruments. We can sometimes feel the vibrations of a moving train or automobile, but we do not feel its uniform motion.

Another demonstration of the special principle of relativity is provided by the standard laboratory toy illustrated in Figure 1-15. The "cannon" fires a ball straight up while the cannon itself is in motion to the right. Where will the ball fall? Assuming that we can disregard friction and that the cannon does not change its speed or direction, the ball will fall back into the mouth of the cannon. This, of course, is just where the ball would fall if the cannon were not moving. As far as the view from the cannon is concerned, the uniform motion of the cannon makes no difference.

An amusing example of this fact is given in Galileo's *Dialogue*. The *Dialogue* describes an imaginary conversation among three people, one of whom, Simplicio by name, opposes the idea that the earth moves. In the course of the argument, Simplicio claims that sailors measure the speed of a boat by dropping a rock from the mast.

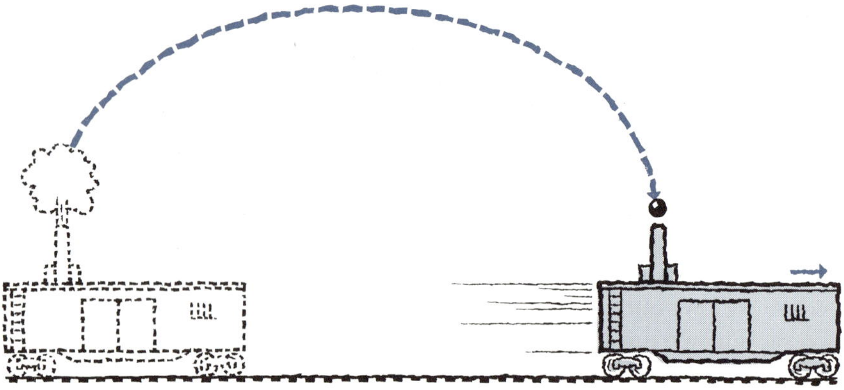

Fig. 1-15

If the boat is in motion, he erroneously states, it will move forward (or backward) past the rock as the rock falls, and therefore the rock will strike the deck at some distance from the bottom of the mast. The faster the boat is moving, he claims, the greater this distance. Actually, this method will not work (see Fig. 1-16). Galileo criticizes Simplicio, and the "authorities" who made such claims, for merely quoting each other and for putting:

> . . . faith in their predecessors, right on back without ever arriving at anyone who had ever performed it. For anyone who does will find that the experiment shows exactly the opposite of what is written; that is, it will show that the stone always falls in the same place on the ship, whether the ship is standing still or moving with any speed you please.[9]

Thus, if the ship is in uniform motion, the experiment will come out the same, regardless of the speed of the ship.

Fig. 1-16

[9]Galileo Galilei, *Dialogue Concerning the Two Chief World Systems*, *op. cit.*, p. 144.

Care must be taken in using the principle of relativity. In Figure 1-17 we illustrate two ways in which a badminton game can be played on a moving train. If the car is enclosed, as in Figure 1-17*a*, the game will proceed just as if the train were at rest. If the game is played on an open flatcar, however, the "wind" generated by the movement of the train might spoil the game. The principle of relativity does not guarantee that the physical situation will be the same in a moving frame of reference as in a stationary frame. It *does* guarantee that the fundamental laws and principles of physics will be the same in two uniformly moving reference frames.

The Motion of the Earth

THE FOUCAULT PENDULUM. We have said that the uniform straight-line motion of a system cannot be determined by measurements made within the system but must be determined by looking outside the system. The earth, however, is not moving in a straight line; it is spinning and revolving in near circles. These circles are large, so that the circular paths are very close to straight lines for short time periods. Still, there should be and are small effects. There are indeed ways to detect the earth's motion without looking out to see the stars go by.

One such method involves the use of the Foucault pendulum, which detects the rotation of the earth. In Figure 1-18 we show such a pendulum set up at the North Pole. The pendulum bob is not in rigid contact with the earth, since it has connection with the earth only through the cord and the framework that supports it. When the pendulum is set to swinging back and forth, it tends to go in as straight a line as possible, in accordance with the first law of motion.

Fig. 1-17

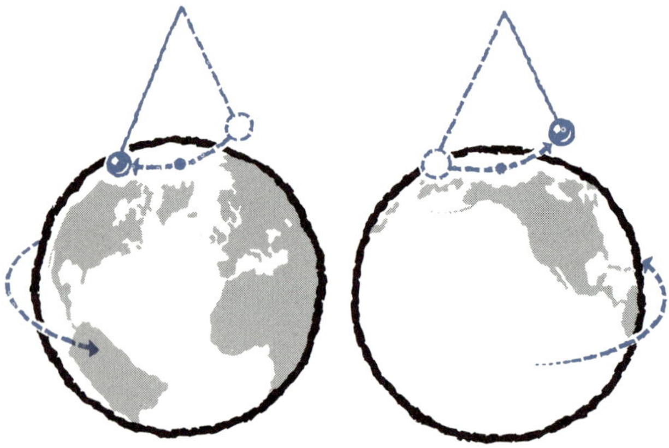

Fig. 1-18

As the earth turns under it, the pendulum bob tends to keep swinging back and forth in the same direction in space—that is, to keep swinging in the general direction of the same stars. To a person on earth, the pendulum appears to be mysteriously changing its direction of swing. Actually it is the earth that is moving, rotating underneath the pendulum. If the pendulum is not at the North Pole, its behavior is less simple to explain. What happens is that the direction of swing changes more slowly if a Foucault pendulum is moved to a location closer to the equator. (This pendulum effect was observed by J. L. Foucault in 1851.)

Even if the earth were covered with clouds so that we could not see the sky, we could still measure the rotation of the earth by means of such a pendulum. There are also other mechanical effects of the earth's rotation, small effects in the motion of bodies on the earth, but we shall not consider them here.

THE ABERRATION OF STARLIGHT. One phenomenon that is used to demonstrate the motion of the earth by observing the stars is called the aberration of starlight. This was first observed in 1725 by James Bradley with Samuel Molyneux, and later correctly interpreted. To understand this phenomenon, suppose that we are in a train, with rain falling straight down outside. If the rain makes streaks on the window under these circumstances, these streaks are also straight down, as shown in Figure 1-19*a*. If the train is moving, however, the streaks will be slanted as in Figure 1-19*b* since, as the rain moves down the window, the window itself is moving forward. Thus the rain will appear to be coming down in a different direction if the train is moving. If the train were to move back and forth, the direc-

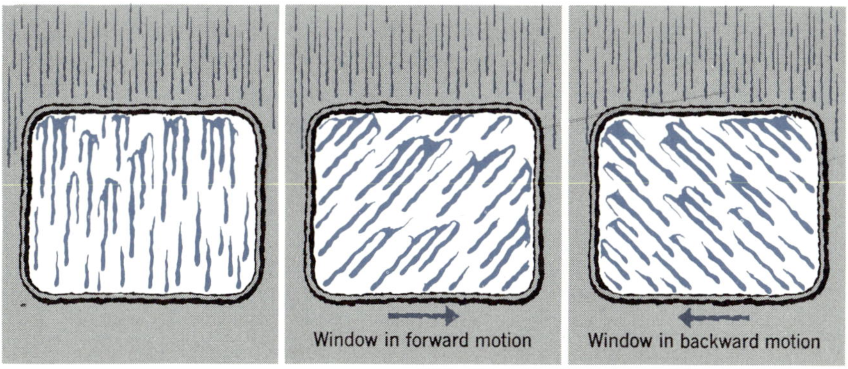

Fig. 1-19

tion of the rain would appear to change periodically, coming first from the right and then from the left, as illustrated.

Similarly, as the earth moves around the sun in its orbit, the apparent direction of incoming starlight changes, and thus the star positions appear to change (see Fig. 1-20). The change is very small, since the speed of light coming from the stars is so great compared to the speed of the earth, but instruments were sensitive enough to detect this small change by Bradley's time.

STELLAR PARALLAX. In 1838 F. W. Bessel succeeded in obtaining a long-expected piece of evidence for the motion of the earth: the

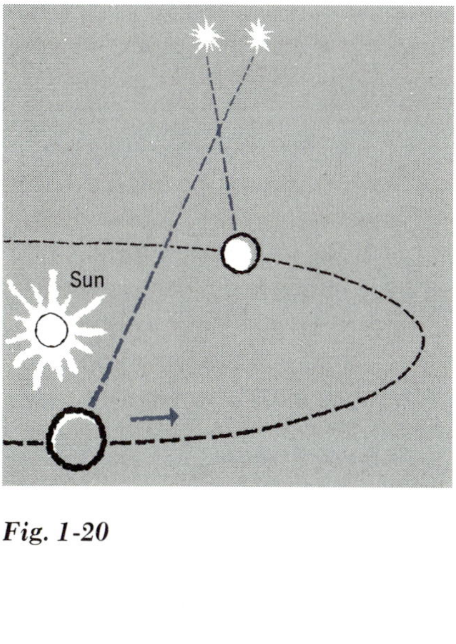

Fig. 1-20

parallax of a star. Figure 1-21 illustrates what is meant by parallax. Since the earth is in different places in space at opposite points of its orbit around the sun, the relative position of stars should be slightly different as observed from the earth at these two points. This effect is extremely small because of the great distance of the stars, but Bessel finally observed it.

Parallax, like stellar aberration, is observed by looking out at the stars from the earth. One could, of course, argue against the evidence of the earth's motion supplied by stellar parallax and the aberration of starlight, by saying that the stars were actually moving just enough to cause these small regular changes in their positions. Actually, the motion of the stars would have to be even more complex, because the earth, in addition to its daily rotation and its yearly revolution about the sun, also wobbles somewhat as it rotates. If we did not admit the motion of the earth, this wobble would have to be ascribed to the sun, the moon, the planets, and the extremely distant stars. It is vastly simpler to ascribe the motion to the earth.

OTHER SYMMETRY PRINCIPLES

The special principle of relativity tells us that the same laws of physics are obtained in all reference frames in uniform straight-line motion. That is to say, the laws of physics are invariant under a change from one of these reference frames to another. There are many other principles that tell us that we can change the reference frame without changing the way in which we express the laws of

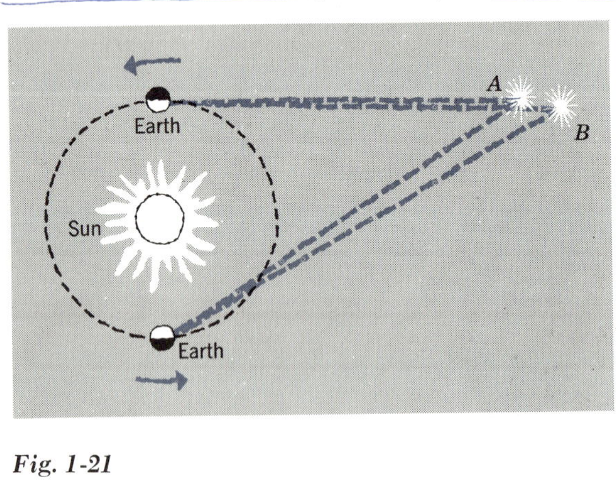

Fig. 1-21

physics. These principles are called *symmetry principles*. Before describing them, let us first explain in greater detail why they are called symmetry principles.

Symmetry

Figure 1-22*a* shows a square, with the corners labeled for convenience. We pretend that the letters are not there in reality but that we have imagined them there in order to keep track of the corners. Now suppose that the square has been rotated counterclockwise through one quarter of a full revolution so that it appears as in Figure 1-22*b*. Without looking at the letters, you cannot tell that it has been moved. It looks exactly the same in its new position as it did in the old one. The square is a symmetric figure—there are certain things we can do to it without changing the way it looks. Something is said to be symmetric if it can be changed in such a way that its new condition is not detectibly different from its original condition. Any operation which makes such a change is called a symmetry operation. The operation that rotates the square through one quarter of a revolution is one of the symmetry operations of the square. There are, of course, others.

We can study the symmetry of the square by changing the position of the observer instead of rotating the square. In Figure 1-23, for example, we have observers viewing the square from four different positions, yet the square looks the same to all of them. Thus, there are certain different ways of looking at the square for which the square still looks the same.

We are interested in the laws of physics. We look at them by using some reference frame in which we study the universe. There are certain ways of changing to new reference frames in which the laws of physics still look the same. That is, there are certain symmetry operations that will change our reference frame without changing the laws of physics. For example, according to the special principle of

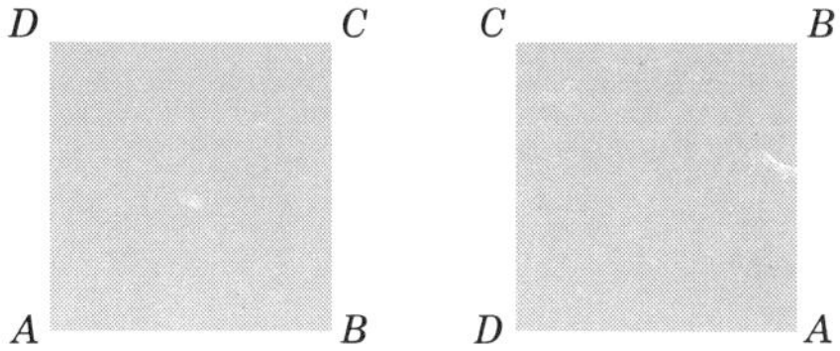

(*a*) (*b*)

Fig. 1-22

relativity, we can change to a frame in uniform motion without changing the laws of physics which we obtain. Therefore, we consider the special principle of relativity to be a symmetry principle.

From a practical point of view, symmetry principles are often useful in analyzing situations when we do not know the precise laws that apply. Besides, they have a certain elegance and simplicity about them, which is aesthetically appealing, and some of them also seem philosophically satisfying. They help us to express our faith that the universe is the same for all observers even though each may see it from a different vantage point. A law that can be expressed in the same way in different frames of reference seems likely to be an expression of some actual outside reality, something that is not completely dependent upon the situation of the individual observing mind.

Position Symmetry

The laws of physics do not depend upon the position of the origin of the reference frame. This means that the laws do not depend on the position in space at which we perform our experiments. The laws should turn out to be the same whether we make our experiments in New York, London, Moscow, or on the moon (see Fig. 1-24). In Galileo's day, the laws were supposed to be different on earth from those on the moon or above it. The heavens were thought to be different from the earth, and to obey different laws. Galileo's work helped to change that idea.

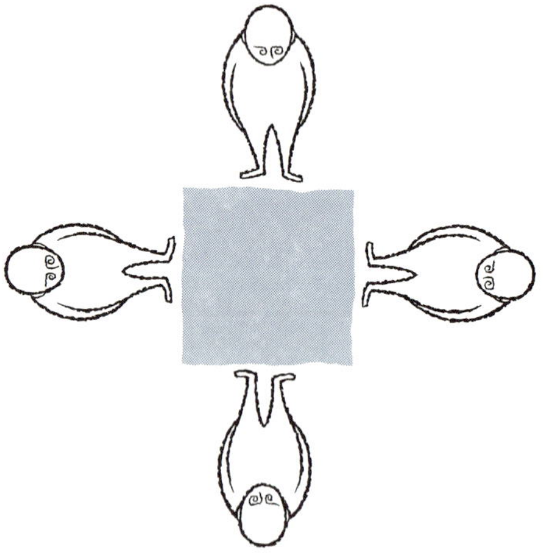

Fig. 1-23

Directional Symmetry

The laws of physics do not depend upon the direction of the axes of the reference frame. (We continue to assume, however, that the axes are perpendicular to each other.) It does not matter how the whole reference frame is oriented in space.) As far as the laws are concerned, our axis might point toward either Chicago or Miami with no effect on the results (see Fig. 1-25). In Galileo's day, however, the direction toward the center of the earth was regarded as very different from the other directions because heavy bodies would naturally fall in that direction. When it was realized that the earth was in motion, spinning and going around the sun, this idea had to change. What is "up" or "down" at one time may not be "up" or "down" later on, because the earth turns. The center of the earth also

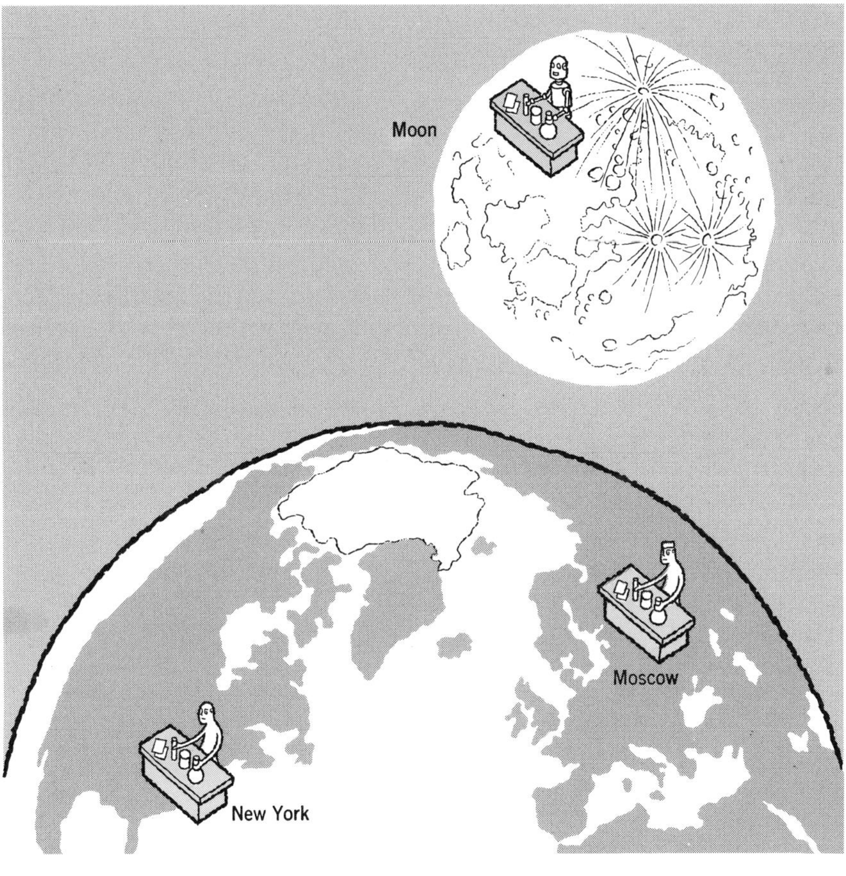

Fig. 1-24

moves; therefore the direction to that center is not always the same.

We can appreciate the philosophical shock Galileo's contemporaries experienced when the "new physics" which implicitly contained these principles became widely known. The earth, which they had believed to be steady, was now flying through space. "Up to the heavens" and "down to earth" were now merely relative terms. The heavens and the earth obeyed the same laws. The earth was an arbitrary planet, moving through arbitrary space, in an arbitrary direction. We cannot blame them for feeling "lost in space."

Time Symmetry

We can set up a reference frame for time somewhat as we do for space. A point in time corresponding to some event is taken as the reference point or origin in time, and events are dated forward and backward with respect to this point of reference. Our ordinary habit of referring to events as A.D. or B.C. uses the birth of Christ as such a reference point in time. As far as we know, it does not matter which instant is chosen. This gives us another symmetry principle: the laws of physics do not depend upon our chosen point of reference in time. Therefore, the laws alone can give us no reason for using one such reference point rather than another. It does not matter whether our experiments are done today, yesterday, or tomorrow; if we have taken everything into account, the laws should be the same (see Fig. 1-26). We may, of course, find out that we have been wrong, that our present laws are incorrect, but we believe that valid laws should be independent of time.

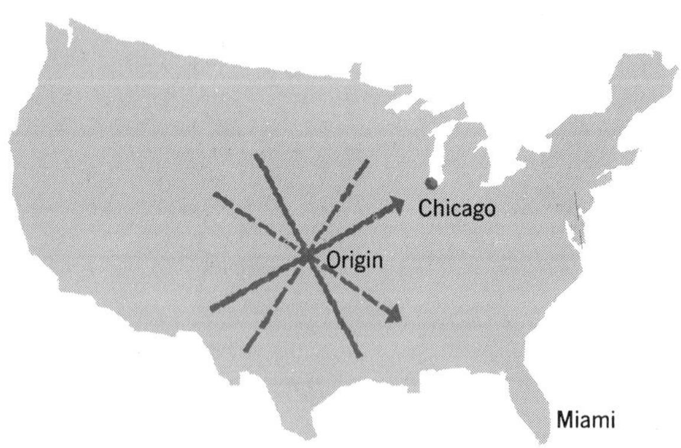

Fig. 1-25

Time-Reversal Symmetry

The laws of physics should be the same whether time is running forward or is reversed to run backward. This may sound strange, since we cannot actually make time go backward. The easiest way to understand this principle is to suppose that we take motion pictures of all our scientific experiments. Then we determine the laws of physics from what we see when the film is run backward. If the laws turn out to be the same, then we say that the laws of physics are time-reversible, or that they have *time-reversal symmetry*. Although many of the laws of physics are reversible in this way, it is not known at present whether or not they all are. People, for example, are *not* time-reversible. If we take a movie of a man diving into a swimming pool and run it backward, we see the man come out of the water back onto the diving board (see Fig. 1-27). This reversed process does not happen in real life. The growth of a child is another example of a process that never occurs in reverse. Some processes in this world, therefore, do not appear to be time-reversible. We shall discuss such apparent violations of this symmetry principle in Chapter Seven.

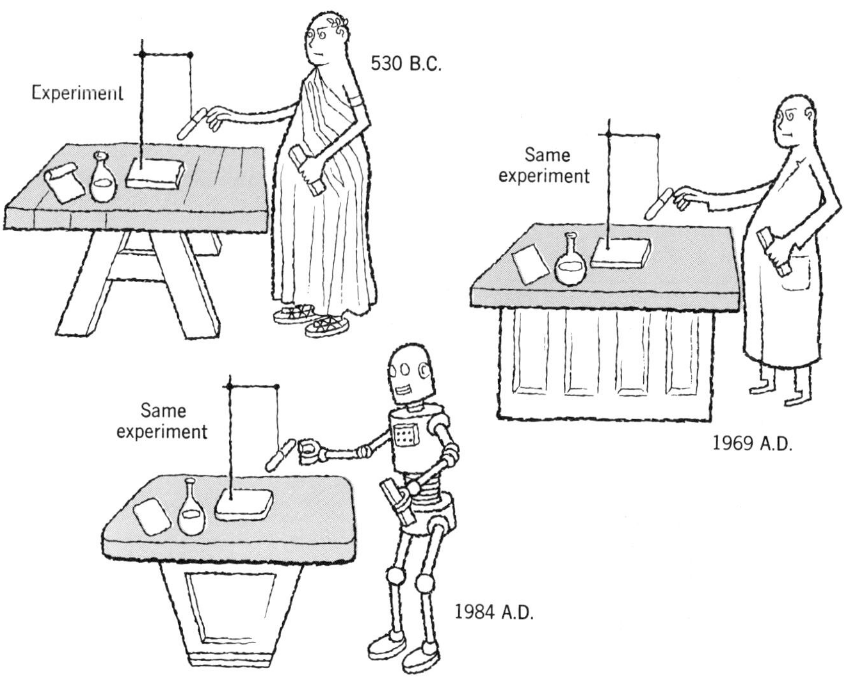

Fig. 1-26

Fig. 1-27

Mirror Symmetry

We might expect the laws of physics to be the same even if they were obtained by looking at our experiments in a mirror. This means that the laws we would discover by regarding the universe through a looking glass would still be the same (see Fig. 1-28). Thus we would have no way of telling the real universe from its mirror image as far as the laws of physics are concerned. This also means that if any process can occur, a mirror image of the process can also occur. This symmetry principle fails for certain situations, as far as we know at present. We shall describe one such situation in Chapter Twelve.

Fig. 1-28

* The new philosophy

And new Philosophy calls all in doubt,
The Element of fire is quite put out;
The Sun is lost, and th' earth, and no mans wit
Can well direct him where to looke for it. . . .
. . .
'Tis all in peeces, all cohaerence gone;
All just supply, and all Relation.
(John Donne)[10]

The New Astronomy

At the time of Galileo, tradition favored the physics of Aristotle and the astronomy of Ptolemy (2nd century A.D.). This old astronomy and physics had not gone unchallenged, but it had been reconciled with Christian theology in the so-called Scholastic philosophy of Thomas Acquinas (1225–1274). Thus it had great authority for both philosophical and theological reasons.

The old physics and astronomy divided space into two regions. Both the nature of objects and their motion were different in the two realms. All objects higher than the moon were celestial. They were thought to be light, unchanging, perfectly smooth and spotless spheres. Many thought that the moon also was perfectly smooth and spherical. Furthermore, these celestial spheres were assumed to move in perfectly circular paths. In the terrestrial realm below the moon, heavy objects were imperfect, changeable, and had a natural motion toward the center of the earth. Objects such as arrows and cannonballs could also have an unnatural motion in other directions if they were continually pushed.

According to Aristotle and Ptolemy, the earth itself was motionless in the center of the universe. The heavens revolved about it once a day, causing sunrise, sunset, and the motion of the stars seen in the night sky. The revolution of objects in the solar system was troublesome in this scheme since these objects appear to change their position in the sphere of stars.

The planets in particular presented a formidable problem. When viewed against the starry background at the same hour on many consecutive nights, their motion is seen to be eastward, with occasional temporary reversals that for a time have a westward

[10]John Donne, as quoted in I. B. Cohen, *The Birth of a New Physics, op. cit.*, p. 89.

direction (see Fig. 1-29). It was considered essential for all the heavenly bodies to move in circular motion, for this was supposed to be the natural motion of perfect celestial objects. Since the planets did not appear to revolve in this simple way, it was necessary to invent a scheme or trick to save the theory. Ptolemy's scheme (and that of others) was to use epicycles. To explain what an epicycle is, let us imagine a certain point moving around the earth in a perfect circle (see Fig. 1-30). Instead of letting that point represent a planet, however, we let it be the moving center of another circle. It is around this moving center that the planet is supposed to revolve. In a sense, then, an epicycle is a circle on another circle. If necessary, other epicycles could be added, until the theoretical model fit the observed movement—that is, until the calculated motion agreed with the recorded observations and thus "saved the appearances." Actually, dozens of epicycles were necessary, as well as other tricks, to make the whole system agree with the observed appearances.

Copernicus (1473–1543) suggested that the sun is the center of the universe, and that the planets, including the earth, revolve around the sun in perfect order (see Fig. 1-31). He further suggested that the earth also rotates on its axis once a day, making the heavens appear to go around the earth. He could then account for the temporary reversal in the motion of the planets. For example, when the fast-moving earth passes closest to the slower-moving Mars, Mars appears to go backward in its orbit (see Fig. 1-31*b*). Since the planets do not go around the sun in perfect circles, it was necessary for Copernicus also to use epicycles to account exactly for the observed plenetary motion and thus save the appearances.

Copernicus knew that he was reviving the doctrine of Aristarchus and certain other ancient Greek philosophers and astronomers when he suggested that the earth moved. These doctrines, however, had long been out of favor. Osiander, in an unsigned preface to Copernicus's main treatise, made the doctrine more palatable to some by

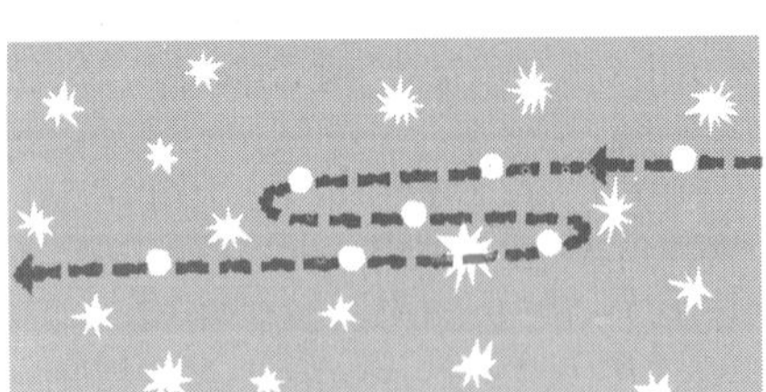

Fig. 1-29

emphasizing that the suggestion was merely a mathematical technique to save appearances. Johannes Kepler, convinced of the reality of the sun-centered system, said that this preface was "written by a jackass for the use of other jackasses."[11]

Tycho Brahe (1586–1601), the Danish astronomer famous for his very accurate measurements of star and planet positions, constructed a compromise system known as the Tychonic system. He felt that it was contrary to the laws of physics for the earth to move. He therefore suggested a modified system in which all the planets except the earth go around the sun, and the whole system, including the sun, goes around the earth. Thus all observations could be exactly the same for the Copernican system and a correctly designed Tychonic system. Appearances would be saved. Conventional physics would also be saved. Unfortunately, conventional physics was wrong.

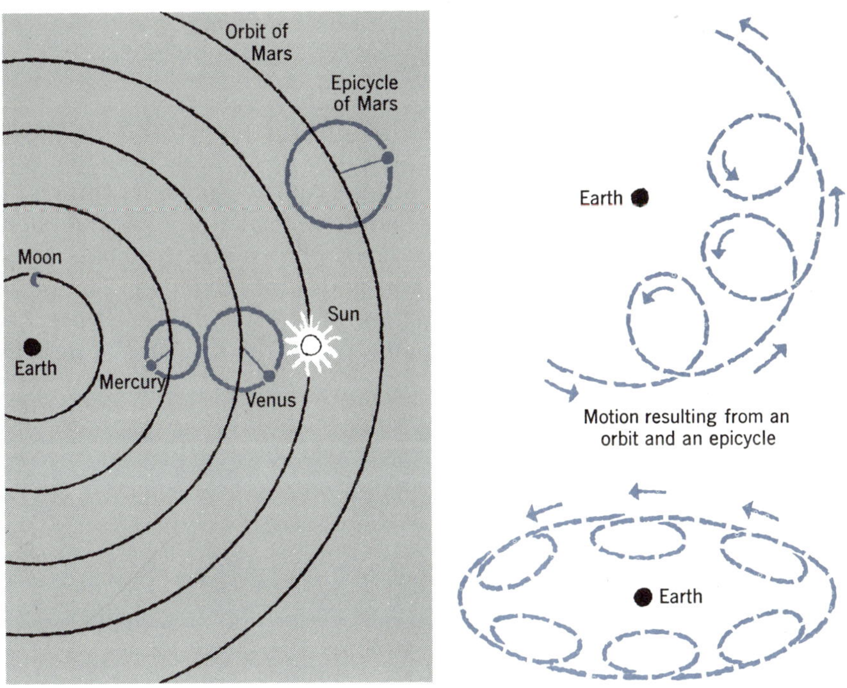

Planets in orbit with epicycles

Fig. 1-30

[11]Quoted in Giorgio de Santillana, *The Crime of Galileo*, Chicago: University of Chicago Press, 1955, p. 101.

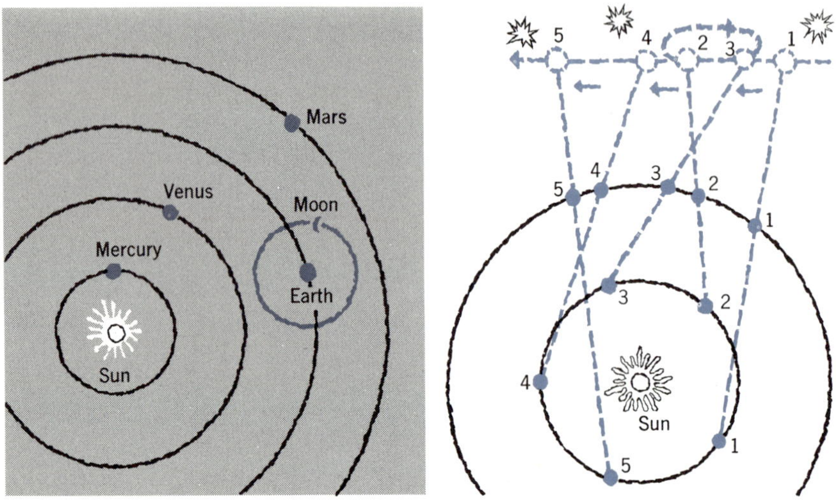

(a) (b)

Fig. 1-31

It was Kepler who finally discovered that the planets did not travel in perfect circles but in *ellipses*. The ellipse is an oval curve that can be more or less like a circle. An ellipse can be drawn as shown in Figure 1-32 by looping a string around two nails or pins and then pulling a pencil around the loop. The ellipse can be made very similar to a circle by putting the two nails close together. The position of the two nails gives the position of the foci of the ellipse. Of course, if the two foci are at the same point, the ellipse is a perfect

Fig. 1-32

circle. The planets go around the sun in ellipses that are very close to circles.

Kepler worked laboriously for a long time, using the accurate data on the orbit of Mars which Tycho Brahe had obtained. No matter what scheme he used in putting together an orbit from perfect circles, he could never quite match the observed orbit of Mars. He was always left with a very small error in angles of observation. However, he found that a single ellipse matched the observations to their limit of accuracy.

> But for us, who by divine kindness were given an accurate observer such as Tycho Brahe, for us it is fitting that we should acknowledge this divine gift and put it to use. . . . For if I had believed that we could ignore these eight minutes [a very small angle] I would have patched up my hypothesis accordingly. But since it was not permissible to ignore them, those eight minutes point the road to a complete reformation of astronomy. . . .[12]

And to much else besides, we might add.

Kepler arrived at three famous laws of planetary motion from his analysis of Tycho Brahe's data.

1. The planet moves in an ellipse about the sun, with the sun at one focus (Fig. 1-33).
2. The line between the sun and the planet sweeps across equal amounts of the area of the ellipse in equal amounts of time (Fig. 1-34).
3. The square of the period of revolution of the planet around the sun is proportional to the cube of its average distance from the sun (Fig. 1-35).

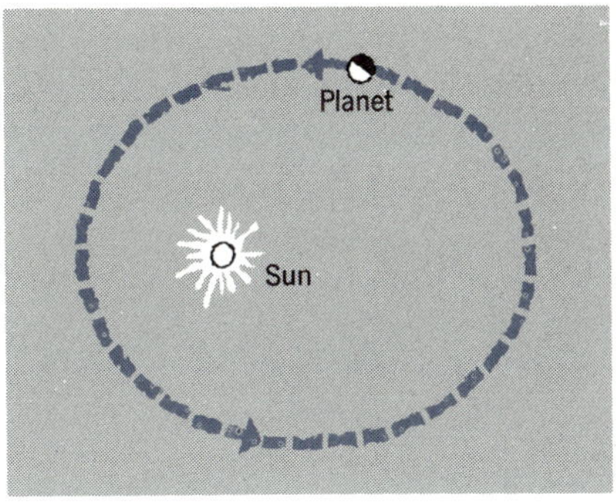

Fig. 1-33

[12]Johannes Kepler, *Astronomia Nova, II*, Chap. 19, as quoted in Arthur Koestler, *The Sleepwalkers*, New York: Macmillan, 1959, p. 322.

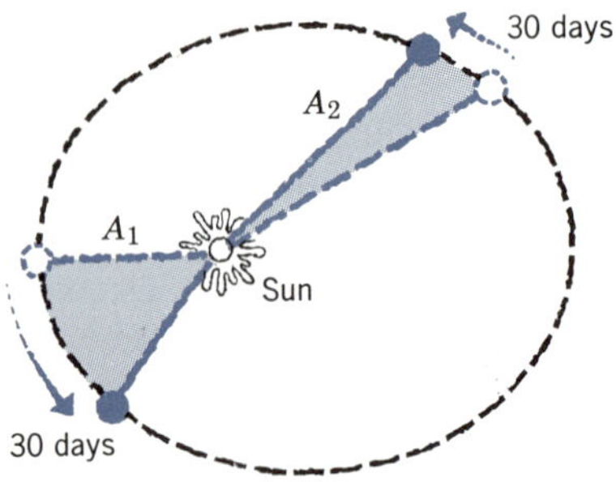

Fig. 1-34 Kepler's second law—planets sweep out equal areas in equal time intervals: $A_1 = A_2$ *(the planet travels slower when it is further away).*

We do not wish to enter into a detailed explanation of these laws; it should be noted, however, that they gave a precise mathematical description of planetary motion. If Galileo had known and appreciated Kepler's laws he would have had an easier time defending the motion of the earth. Kepler's book, *Astronomia Nova* (The New Astronomy), containing his first two laws, may even have been sitting on Galileo's shelf unheeded.

Galileo improved the telescope, which had just been invented, and made many astronomical discoveries that upset the old ideas. He saw the planet Jupiter with its moons going back and forth in the vicinity of the planet, and concluded that the moons were revolving around the planet. Here was a solar system in miniature, for all to see. He observed that the planet Venus showed changes like the

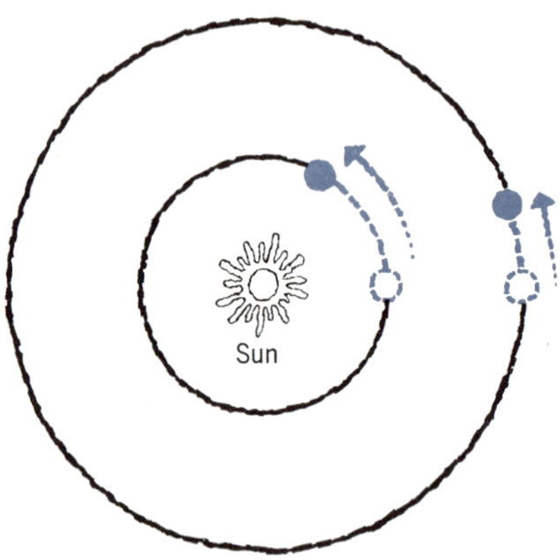

Fig. 1-35 Kepler's third law implies that the further the planet is from the sun, the slower it moves in its orbit.

moon, appearing at various times as full, half full, and crescent, in a way that would not occur in the Ptolemaic system. Then he saw mountains and valleys on the moon—a supposedly smooth celestial body. He saw spots on the sun, a supposedly spotless celestial body. These were not the first observational assaults on the old astronomy. Even before this time, in 1572 and 1604, two new stars, the very rare supernovae, had been seen in the supposedly unchangeable heavens. Tycho Brahe showed that they were much further away than the moon because they showed no measurable parallax.

The Conflict with Theology

Traditional philosophers and conservative theologians were disturbed by the new astronomy. At least one Aristotelian philosopher refused to look through the telescope. Another claimed that "geometry is of the Devil" and that "mathematicians should be banished as the authors of all heresies."[13] Still another charged that "to see the satellites of Jupiter, men had to make an instrument that would create them."[14]

Even before Galileo's time, responsible religious leaders were concerned because they thought the astronomy of Copernicus conflicted with the sacred scriptures. John Calvin (1509–1564) opposed it: "Who will venture to place the authority of Copernicus above that of the Holy Spirit?"[15] Martin Luther (1483–1546) wrote:

> People gave ear to an upstart astrologer who strove to show that the earth revolves, not the heavens or the firmament, the sun and the moon. . . . Whoever wishes to appear clever must devise some new system, which of all systems is of course the very best. This fool wishes to reverse the entire science of astronomy; but sacred scriptures tell us that Joshua commanded the sun to stand still and not the earth.[16]

St. Robert Bellarmine, in Galileo's day, wrote:

> . . . To say that on the supposition of the Earth's movement and the Sun's quiescence all the celestial appearances are explained better than by the theory of eccentrics and epicycles is to speak with excellent good sense and to run no risk whatever. Such a manner of speaking is enough for a mathematician. But to want to affirm that the Sun, in very truth, is at the center of the universe and only rotates on its axis without going from east to west, is a very

[13]A. D. White, *A History of the Warfare of Science with Theology in Christendom*, Vol. I, New York: Dover, 1960, p. 133.

[14]Quoted in *ibid.*, p. 132.

[15]*Ibid.*, p. 127.

[16]*Ibid.*, p. 126.

dangerous attitude and one calculated not only to arouse all Scholastic philosophers and theologians but also to injure our holy faith by contradicting the Scriptures. . . .

If there were a real proof that the Sun is in the center of the universe, that the Earth is in the third heaven, and that the Sun does not go around the Earth but the Earth around the Sun, then we should have to proceed with great circumspection in explaining passages of Scripture which appear to teach the contrary, and rather admit that we did not understand them than declare an opinion to be false which is proved to be true. . . .[17]

Galileo could not prove that the earth moved but only that the arguments against its motion were invalid. In 1623, after conferring with his apparent admirer,[18] the newly elected Pope Urban VIII, Galileo published his *Dialogue Concerning the Two Chief World Systems*, which presented an apparently strong argument for the Copernican system. Galileo's preface, addressed "To the Discerning Reader," artfully reveals his purpose in writing the book.

Several years ago there was published in Rome a salutary edict which, in order to obviate the dangerous tendencies of our present age, imposed a seasonable silence upon the Pythagorean opinion that the earth moves. There were those who impudently asserted that this decree had its origin not in judicious inquiry, but in passion none too well informed. Complaints were to be heard that advisors who were totally unskilled at astronomical observations ought not to clip the wings of reflective intellects by means of rash prohibitions.

Upon hearing such carping insolence, my zeal could not be contained. . . .

To this end I have taken the Copernican side in the discourse, proceeding as with a pure mathematical hypothesis and striving by every artifice to represent it as superior to supposing the earth motionless—not, indeed, absolutely, but as against the arguments of some professed Peripatetics [Aristotelian philosophers]. These men indeed deserve not even that name, for they do not walk about; they are content to adore the shadows, philosophizing not with due circumspection but merely from having memorized a few ill-understood principles. . . .[19]

In the book Galileo discussed only the Ptolemaic and Copernican systems, omitting mention of the Tychonic compromise. He attempted to clinch his argument with a discussion of the tides which was erroneous.

Apparently Galileo's argument was stronger than the Pope had

[17]As quoted in Santillana, *op. cit.*, pp. 99–100.
[18]*Ibid.*, pp. 156, 157, and Chap. VIII.
[19]Galileo Galilei, *Dialogue. . .* , *op. cit.*, pp. 5, 6.

expected, and the book, though very popular, was banned. Galileo was tried, convicted of heresy, and forced to renounce his theory. Although three cardinals refused to sign the condemnation, and the doctrine of Copernicus was never formally declared heretical, Galileo was sentenced to imprisonment by the Inquisition.[20] The sentence includes the following charge.

> The proposition that the Sun is the center of the world and does not move from its place is absurd and false philosophically and formally heretical, because it is expressly contrary to the Holy Scripture.
>
> The proposition that the Earth is not the center of the world and immovable but that it moves, and also with a diurnal motion, is equally absurd and false philosophically and theologically considered at least erroneous in faith.[21]

What were some of the Scriptures which the new philosophy supposedly contradicted?

> Then spake Joshua to the Lord in the day when the Lord delivered up the Amorites before the children of Israel, and he said in the sight of Israel, Sun, stand thou still upon Gibeon; and thou, Moon, in the valley of Ajalon. (Joshua 10:12)
>
> . . . the world also is stablished, that it cannot be moved. (Psalm 93:1)

Galileo himself felt that there was no conflict when the Scriptures were properly understood, but he feared that ignorant and anti-scientific theologians would bring the scriptures into disrepute.[22] In a famous letter to the Grand Duchess Christina (1615),[23] he wrote that if the Bible had spoken openly of the earth's motion no one would have understood, pointing out that even Copernicans speak of sunrise and sunset without deception. He urged that the so-called scientific contradictions of the Scriptures be given the same cautious treatment given to instances in which the Scriptures appear to contradict themselves. He quoted Proverbs 8:26, which referred to the "hinges of the terrestrial orb," and suggested that "hinges" might be translated as "poles."[24] He also stated:

> . . . I think that in discussions of physical problems we ought to begin not from the authority of scriptural passages, but from sense-experiences and

[20]For a detailed discussion of the political and theological motives for this action against Galileo and his book, see Santillana, *op. cit.*, Chaps. VIII, XV.

[21]As quoted in *ibid.*, p. 307.

[22]From *Discoveries and Opinions of Galileo*, by Stillman Drake. Copyright ©, 1957 by Stillman Drake. Reprinted by permission of Doubleday & Company, Inc.

[23]*Ibid.*, pp. 175–216.

[24]*Ibid.*, p. 216. See also Proverbs 8:26, Douay version.

necessary demonstrations; for the holy Bible and the phenomena of nature proceed alike from the divine Word. . . . It is necessary for the Bible, in order to be accomodated to the understanding of every man, to speak many things which appear to differ from the absolute truth so far as the bare meaning of the words is concerned. But Nature, on the other hand, is inexorable and immutable; she never transgresses the laws imposed upon her, or cares a whit whether her abstruse reasons and methods of operation are understandable to men. For that reason it appears that nothing physical which sense-experience sets before our eyes, or which necessary demonstrations prove to us, ought to be called in question (much less condemned) upon the testimony of biblical passages which may have some different meaning beneath their words.[25]

With regard to this opinion, and others which are not directly matters of faith, certainly no one doubts that the Supreme Pontiff has always an absolute power to approve or condemn; but it is not in the power of any created being to make things true or false, for this belongs to their own nature and to the fact.[26]

The Conflict with Authority

At the very heart of the new philosophy was the importance of "sense experience" and necessary reasoning, particularly mathematical reasoning. This means that things are not to be taken on authority but must be checked by observation and calculation. Galileo emphasized that we really know very little and cannot, at least at present, answer all questions. He ridiculed those who appealed to authority alone, citing the experience of a certain expert in anatomy.

Turning to a gentleman whom he knew to be a Peripatetic philosopher, and on whose account he had been exhibiting and demonstrating everything with unusual care, he asked this man whether he was at last satisfied and convinced that the nerves originated in the brain and not in the heart. The philosopher, after considering for a while, answered: "You have made me see this matter so plainly and palpably that if Aristotle's text were not contrary to it, stating clearly that the nerves originate in the heart, I should be forced to admit it to be true."[27]

And again:

. . . what is more revolting in a public dispute, when someone is dealing with demonstrable conclusions, than to hear him interrupted by a text (often written to some quite different purpose) thrown into his teeth by an oppo-

[25]*Ibid.*, pp. 182–183.
[26]*Ibid.*, p. 210.
[27]Galileo Galilei, *Dialogue. . .* , *op. cit.*, p. 108.

nent? If, indeed, you wish to continue in this method of studying, then put aside the name of philosophers and call yourselves historians, or memory experts; for it is not proper that those who never philosophize should usurp the honorable title of philosopher.

. . . So put forward the arguments and demonstrations, Simplicio—either yours or Aristotle's—but not just texts and bare authorities, because our discourses must relate to the sensible world and not to one on paper.[28]

Although Galileo did not reject the civil and religious authority of civil and religious leaders, he did reject claims to authority where scientific truth is concerned. Indeed, he made it clear that truth is not subject to authority. Instead, he argued for a point of view usually regarded as inherent in the so-called scientific method, as in the following excerpts from his reply to "Sarsi" (the pseudonym of the noted Jesuit astronomer, Horatio Grassi).

In Sarsi I seem to discern the firm belief that in philosophizing one must support oneself upon the opinion of some celebrated author, as if our minds ought to remain completely sterile and barren unless wedded to the reasoning of some other person. Possibly he thinks that philosophy is a book of fiction by some writer, like the *Iliad*, or *Orlando Furioso*, productions in which the least important thing is whether what is written there is true. Well, Sarsi, that is not how matters stand. Philosophy is written in this grand book, the universe, which stands continually open to our gaze. But the book cannot be understood unless one first learns to comprehend the language and read the letters in which it is composed. It is written in the language of mathematics, and its characters are triangles, circles, and other geometric figures without which it is humanly impossible to understand a single word of it; without these, one wanders about in a dark labyrinth.

Sarsi seems to think that our intellect should be enslaved to that of some other man. . . .[29]

I cannot but be astonished that Sarsi should persist in trying to prove by means of witnesses something that I may see for myself at any time by means of experiment. . . . So when Sarsi brings in this multitude of authors it appears to me that instead of strengthening his conclusion he merely ennobles our case by showing that we have outreasoned many men of great reputation. . . .[30]

Galileo did not understand the nature of comets, the subject of the discussion, but he had no trouble with Sarsi.

If Sarsi wants me to believe with Suidas that the Babylonians cooked their eggs by whirling them in slings, I shall do so; but I must say that the cause of this effect was very different from what he suggests. To discover the true

[28]*Ibid.*, p. 113.

[29]Drake, *op. cit.*, pp. 237–238.

cause I reason as follows: "If we do not achieve an effect which others formerly achieved, then it must be that in our operations we lack something that produced their success. And if there is just one single thing we lack, then that alone can be the true cause. Now we do not lack eggs, nor slings, nor sturdy fellows to whirl them; yet our eggs do not cook, but merely cool down faster if they happen to be hot. And since nothing is lacking to us except being Babylonians, then being Babylonians is the cause of the hardening of eggs, and not friction of the air." And this is what I wished to discover. Is it possible that Sarsi has never observed the coolness produced on his face by the continual change of air when he is riding post? If he has, then how can he prefer to believe things related by other men as having happened two thousand years ago in Babylon rather than present events which he himself experiences? . . .

Sarsi says he does not wish to be numbered among those who affront the sages by disbelieving and contradicting them. I say I do not wish to be counted as an ignoramus and an ingrate toward Nature and toward God; for if they have given me my senses and my reason, why should I defer such great gifts to the errors of some man? Why should I believe blindly and stupidly what I wish to believe, and subject the freedom of my intellect to someone else who is just as liable to error as I am?[30]

Galileo is still a controversial figure. His biographer, Santillana, contends that Galileo was trying to keep the authorities from making fools of themselves and his Church, since he could see "the three forces of mathematics, physics, and astronomy converging rapidly toward a junction which would make them irresistible."[31] Arthur Koestler severely criticizes Galileo for his fight with the Jesuits and seems to blame Galileo's arrogance for facilitating the split between the scientific and spiritual sides of our culture.[32] Scientists generally consider Galileo as a champion of freedom of scientific investigation. We are not historians, but we find in the writings of Galileo the same scientific attitude that we are used to in this modern age. True, our attitudes toward the immutable mathematical truth of nature have altered somewhat since Galileo's day. Still, the combination of the practical, the workable, with the intellectual and the mathematical seems very much the scientific point of view of our own day. Consider this matter of the stone falling from the mast of the moving boat. To some of the ancient philosophers the application was unimportant. Such philosophers need have nothing to do with the operation of boats. If the sailor were as ignorant as the philosophers he would have drowned. In the "new philosophy," however, we are compelled to be as accurate in practical matters as in logic. It is indeed the possi-

[30]*Ibid.*, pp. 271–272.

[31]Santillana, *op. cit.*, p. 106.

[32]Koestler, *op. cit.*, pp. 425–426.

bility of practicable operation that determines truth, according to one point of view.

PROBLEM SOLVING

The concepts discussed in this chapter lend themselves to three main types of problems, involving (a) changing units; (b) uniform motion; and (c) relative motion. Some problems are a combination of these types.

Changing Units

The equations of physics are given in terms of the quantities of physics, such as distance, time duration, speed, and velocity. Each of the quantities in these equations may be expressed in a variety of units. For example, meters, yards, inches, miles, kilometers, fathoms, and cubits are all units of distance. A quantity given in terms of one unit (or set of units if it is a derived quantity) may be changed to another unit (or set of units) in a very simple way. In order to do this we need an expression or equation which gives one unit or set of units in terms of the other. We shall give three examples of this method.

EXAMPLE 1

Express a distance of 6.0 meters in terms of inches.

Solution. We need an equation that relates inches and meters. This equation is

$$\text{meter} = 39.37 \text{ inches}$$

Since there are 39.37 inches in one meter, there will be six times this many inches in the distance being considered. To obtain the answer we simply substitute 39.37 inches for *meter* in the expression *6.0 meters*, as follows:

$$\begin{aligned} 6.0 \text{ meter} = 6.0 \text{ (meter)} &= 6.0 \text{ (39.37 inch)} \\ &= (6.0 \times 39.37) \text{ inch} \\ &= 236.2 \text{ inches} \\ &= 240 \text{ inches} \end{aligned}$$

Since the least accurate number in our data (6.0) was valid only for two significant figures, the result is good to two significant figures.[33]

[33]See Appendix II-A for a discussion of significant figures. This appendix explains why we write 240 inches here instead of 236.2 inches.

Example 2

An object has a speed of 30.0 inches per second. What is its speed in meters per second?

Solution. We again begin by expressing the given unit in terms of the unit required:

$$\text{meter} = 39.37 \text{ inches}$$

Dividing both sides of the equation by 39.37 we obtain

$$\frac{\text{meter}}{39.37} = \text{inch}$$

We substitute this expression for *inch* in the given expression, as follows:

$$30.0 \frac{\text{inch}}{\text{second}} = 30.0 \frac{\text{meter}/39.37}{\text{second}} = \frac{30.0}{39.37} \frac{\text{meter}}{\text{second}}$$
$$= 0.762 \frac{\text{meter}}{\text{second}}$$

Example 3

An object has a speed of 5.0 inches per minute. What is its speed in meters per second?

Solution. We need one equation that relates inches and meters and another that relates second and minutes:

$$\text{meter} = 39.37 \text{ inches}$$
$$\text{minute} = 60 \text{ seconds}$$

These equations allow us to express the old units in terms of the new units that we desire to use. We proceed as follows:

$$\text{inch} = \frac{1}{39.37} \text{ meter}$$

Substituting the equivalent expressions for inch and minute we have

$$5 \frac{\text{inch}}{\text{minute}} = 5 \left(\frac{(1/39.37) \text{ meter}}{60 \text{ seconds}} \right)$$
$$= 5 \left(\frac{1}{39.37} \right) \left(\frac{1}{60} \right) \frac{\text{meter}}{\text{second}}$$
$$= 0.0021 \frac{\text{meter}}{\text{second}} = 2.1 \times 10^{-3} \frac{\text{meter}}{\text{second}}$$

Notice that we have kept only the first two figures in the answer.[34]

[34]See Appendix II-B for a description of the use of powers of 10.

mt = 39.37

Uniform Motion

In this section we deal only with uniform motion; that is, motion in which both the speed and the direction remain constant. Objects in this state of motion move along a straight line and travel equal distances in each equal interval of time.

An equation that describes uniform motion is

$$d = vt \tag{1}$$

where

d is the distance traveled
v is the velocity
t is the time of travel

This formula is useful for a single object in uniform motion.

It is convenient, especially for more complex problems, to establish a reference frame in which one of the axes is the line of motion of the object. In such a reference frame, uniform motion is described by the following equation rather than by equation 1:

$$d = d_0 + vt \tag{2}$$

In this equation d does not represent the distance traveled by the object, but, rather, the distance along the axis from the origin of the reference frame at the time t. The symbol d_0 represents the distance of the object from the origin when the time t is equal to zero. As before, v is the velocity of the object. When d_0 equals 0, equation 2 is like equation 1. One direction along the axis or line of motion will be considered positive and the opposite direction will be considered negative. A vector, such as velocity, along the axis of motion will be considered positive or negative depending on whether it points in the positive or negative direction. The distance, d or d_0, of an object from the origin will be positive or negative depending on whether the object is in a positive or negative direction from the origin. The use of a reference frame in treating uniform motion will be illustrated in the step-by-step procedure used to work the following examples.

EXAMPLE 4

An airplane starts in Chicago and travels due east for 20 minutes. How fast does it have to go at a constant speed in order to travel 5000 meters during that time? Express the answer in meters per second.

Solution.

STEP 1. Draw a diagram and set up a reference frame. In Figure 1-36b, the arrows indicate positive directions along each of the axes of the reference frame. We commonly refer to one axis as the x-axis and the other as the y-axis.

Because of our symmetry principles, it does not matter where we put the origin of the reference frame or which way we point the axes as long as they are perpendicular to each other. Therefore, we can set up the reference frame according to our convenience.

In this problem, it is convenient to put the origin at Chicago, where the motion starts, and to point one of the axes in the direction of the motion, which is eastward (see Fig. 1-36*b*). In this reference frame, all the motion takes place along the x axis.

STEP 2. List the given quantities and the unknown quantities. In this example we have:

$$d = 5000 \text{ meters}$$
$$t = 20 \text{ minutes}$$
$$d_0 = 0, \text{ since we start at the origin}$$
$$v = ?$$

STEP 3. Write down the equations which relate the quantities involved either as given information or as information to be found. Here, since we are dealing with uniform motion, the equation is

$$d = d_0 + vt$$

(These three steps may be done in a different order. For very simple problems it may be easier to decide what equations apply and then list the quantities which appear in these equations as either given or unknown.)

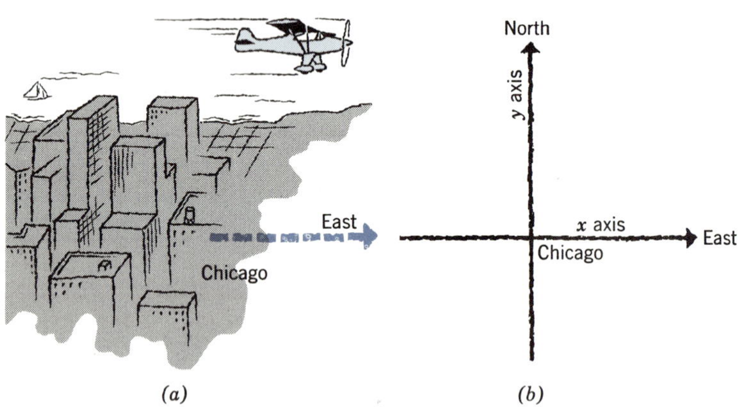

Fig. 1-36

STEP 4. Put the given quantities from Step 2 into the equation of Step 3, and solve the equation for the unknown quantities.

$$(5000 \text{ meter}) = 0 + v\,(20 \text{ minute})$$

$$\frac{5000 \text{ meter}}{20 \text{ minute}} = v$$

$$v = \frac{5000}{20}\,\frac{\text{meter}}{\text{minute}} = 250\,\frac{\text{meter}}{\text{minute}}$$

Now, 1 minute = 60 seconds, so

$$v = 250\,\frac{\text{meter}}{\text{minute}} = \frac{250}{60}\,\frac{\text{meter}}{\text{second}} = 4.2 \text{ meters per second}$$

EXAMPLE 5

A man is standing on a street corner. A car 1000 meters away is approaching at 20 meters per second. How long will it be before the car reaches the man?

Solution.

STEP 1. See Figure 1-37.

STEP 2. $d = 0$

$v = -20$ meter/second

$t = ?$

Notice that the reference frame has been set up so that it is centered on the person. That means that the car starts out at a distance $d_0 = 1000$ meters and ends up at the origin, where $d = 0$ meters. The car's motion is toward the origin, which means that v is negative.

Step 3. Again we have uniform motion.

$$d = d_0 + vt$$

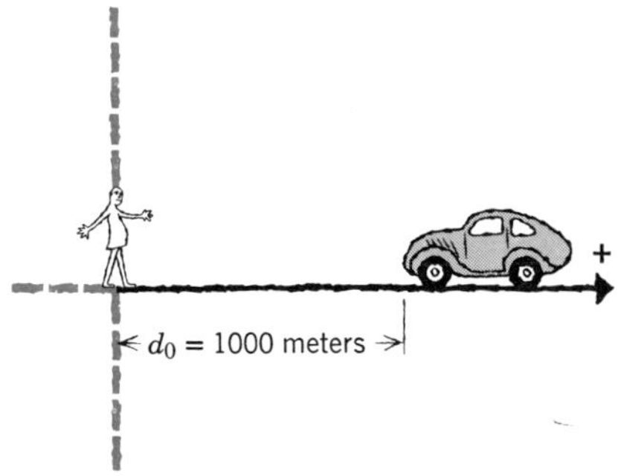

Fig. 1-37

STEP 4. $0 = (1000 \text{ meter}) + \left(\frac{-20 \text{ meter}}{\text{second}}\right) t$

Add $\left(\frac{20 \text{ meter}}{\text{second}}\right) t$ to both sides:

$$\left(\frac{20 \text{ meter}}{\text{second}}\right) t = 1000$$

$$t = \frac{1000 \text{ meter}}{20 \text{ meter/second}}$$
$$= 50 \text{ seconds}$$

Another way to solve this problem is to draw the reference frame so that the origin is located at the beginning point for the car (see Fig. 1-38). With this new reference frame, Step 2 is as follows:

$$d = -1000 \text{ meters}$$
$$d_0 = 0$$
$$v = -20 \text{ meter/second}$$
$$t = ?$$

Now, d is the end point at *minus* 1000 meters, since it is in the negative direction on the reference frame.

Therefore, for Steps 3 and 4,

$$d = d_0 + vt$$
$$-1000 = 0 + \left(-20 \frac{\text{meter}}{\text{second}}\right) t$$

Dividing both sides by −20 meters/second, we have

$$\frac{-1000 \text{ meter}}{-20 \text{ meter/second}} = t$$

and

$$t = 50 \text{ seconds.}$$

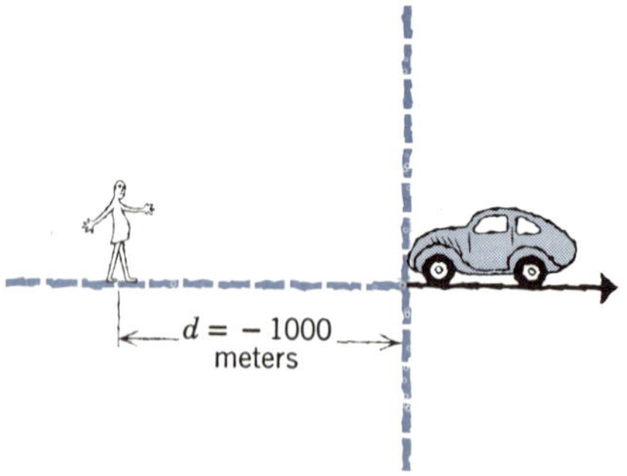

Fig. 1-38

Of course, the answer is the same no matter how the reference frame is placed.

EXAMPLE 6

Consider Example 5 again. This time, however, the observer, instead of standing on a street corner, is also in a car, traveling toward the other car at 30 meters per second. How long is it before they crash into each other?

STEP 1. See Figure 1-39.

STEPS 2 and 3. Two equations are needed, one for each car. This time we shall write the equations first (Step 3) and then do Step 2:

$$d_1 = d_{01} + v_1 t_1 \tag{3}$$

$$d_2 = d_{02} + v_2 t_2 \tag{4}$$

STEP 2 is as follows:

d_1 = position of car 1 at collision = ?
d_{01} = position of car 1 at start = 0
v_1 = velocity of car 1 = 30 meter/second
t_1 = time until collision for car 1 = ?
d_2 = position of car 2 at collision = ?
d_{02} = position of car 2 at start = 1000 meters
v_2 = velocity of car 2 = −20 meter/second
t_2 = time until collision for car 2 = ?

At the end, the two cars must be in the same place at the same time, otherwise, there will be no collision. Therefore

$$d_1 = d_2; \quad \text{we shall call it } d \tag{5}$$

$$t_1 = t_2; \quad \text{we shall call it } t \tag{6}$$

Equations 5 and 6 are additional conditions or constraints that apply to the given and unknown quantities. We need them to solve the

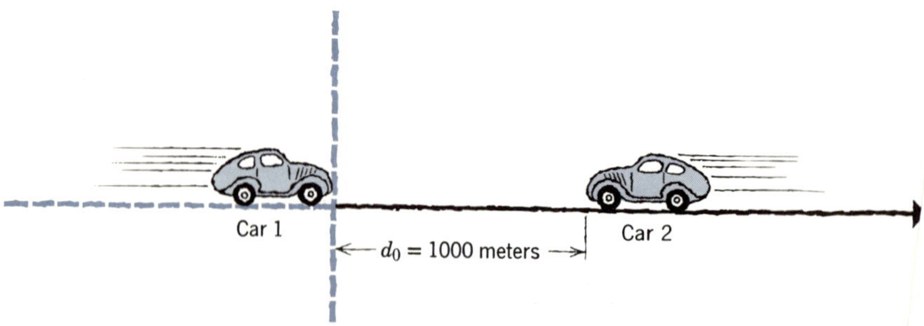

Fig. 1-39

problem. There are often such constraints applying to a particular problem in addition to the general equations that describe the type of motion.

STEP 4. Now we have

$$d = 0 + \left(30 \frac{\text{meter}}{\text{second}}\right) t$$

and

$$d = (1000 \text{ meter}) + \left(-20 \frac{\text{meter}}{\text{second}}\right) t$$

Since two things both equal to the same thing (in this case, d) must be equal to each other,

$$\left(30 \frac{\text{meter}}{\text{second}}\right) t = 1000 \text{ meter} - \left(20 \frac{\text{meter}}{\text{second}}\right) t$$

We then add (20 meter/second) t to both sides of the equation to obtain

$$\left(50 \frac{\text{meter}}{\text{second}}\right) t = 1000 \text{ meters}$$

and finally

$$t = \frac{1000 \text{ meter}}{50 \text{ meter/second}} = 20 \text{ seconds}$$

Relative Motion

EXAMPLE 7

An aircraft carrier is moving at 10 meters per second in a given direction. An airplane takes off from the deck of the carrier in the same direction at a speed of 50 meters per second, as measured from the carrier. How fast is the airplane going with respect to the water? (See Fig. 1-40.)

Solution. We add the velocity of the airplane relative to the ship to the ship's velocity relative to the water in order to get the velocity of the plane relative to the water. Thus

$$50 \frac{\text{meter}}{\text{second}} + 10 \frac{\text{meter}}{\text{second}} = 60 \frac{\text{meter}}{\text{second}}, \text{ or 60 meters per second}$$

If the plane had taken off in the opposite direction we would *subtract* the speed of the plane with respect to the ship from the ship's speed

with respect to the water. This seems to be a commonsense procedure, but actually, as we shall see in Chapter Ten, the rule is not quite correct. However, we shall adopt it for the present.

EXAMPLE 8

Rework Example 6 (p. 53), using the principle of relativity.

Solution. We redraw the situation (see Fig. 1-41) to indicate a moving reference frame.

Because of the special principle of relativity, we know that the laws of physics will be the same in this moving reference frame, provided that it is moving uniformly in a straight line. Therefore, we can use the same principles to solve the problem as when the reference frame is at rest. In this example it is convenient to locate our reference frame on one of the cars. We shall put our reference frame on car 1 and relate all motion to it. In this reference frame, the speed of car 1 is zero; in other words, it is not moving at all *in this frame.* The second car appears to be moving rapidly across the 1000-meter distance and seems to crash into car 1 while car 1 remains stationary. Moreover, in this frame car 2 appears to move at its own speed (20 meter/second) plus an additional speed (30 meter/second), which it acquires when we change to the moving reference frame. Thus car 2 appears to be approaching us at 50 meter/second. To solve the problem, therefore, we merely write the single equation for car 2:

$$
\begin{aligned}
d &= d_0 + vt \\
d &= 0 \\
d_0 &= 1000 \text{ meters} \\
v &= -50 \text{ meter/second} \\
t &= ?
\end{aligned}
$$

Note that we have again reversed the order of Steps 2 and 3.

Fig. 1-40

STEP 4 :

$$0 = 1000 \text{ meter} + (-50)\,\frac{\text{meter}}{\text{second}}\,t$$

$$50\,\frac{\text{meter}}{\text{second}}\,t = 1000 \text{ meters}$$

$$t = \frac{1000 \text{ meter}}{50 \text{ meter/second}} = 10 \text{ seconds}$$

This example illustrates the fact that the special principle of relativity sometimes allows us to simplify the problem by changing to a moving reference frame.

Summary

I. FUNDAMENTAL LAWS AND PRINCIPLES

A. THE FIRST LAW OF MOTION. An object continues in its state of rest, or of uniform motion in a straight line, unless it is compelled to change that state of motion by forces impressed upon it.

B. THE SPECIAL PRINCIPLE OF RELATIVITY. All laws of nature are the same in all reference frames moving uniformly.

C. OTHER SYMMETRY PRINCIPLES. The laws of nature are not changed if:

1. We observe nature from a different position in space.
2. We change our position in time.
3. We change our orientation in space.

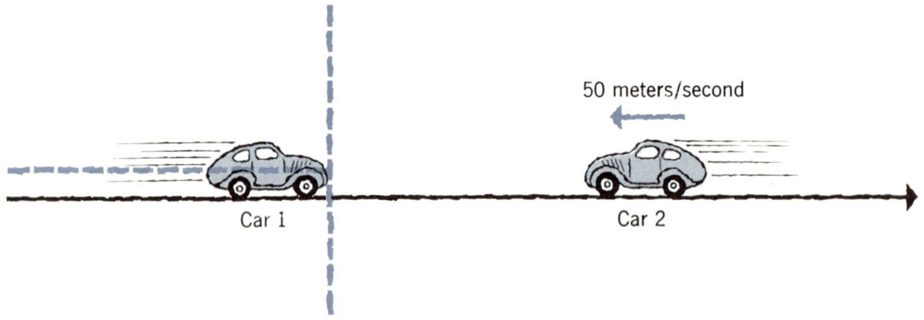

Fig. 1-41

4. We observe the world in a mirror.*
5. We reverse time.*

II. QUANTITIES. These are obtained by measurement or by combining other quantities that are obtained by measurement.

A. FUNDAMENTAL QUANTITIES

1. Distance, a scalar.
2. Time duration, a scalar.

B. DERIVED QUANTITIES

1. Area, a scalar.
2. Volume, a scalar.
3. Speed, a scalar.
4. Velocity, a vector.

III. APPLICATIONS

A. FRICTION

B. DETECTING THE MOTION OF THE EARTH

1. Foucault pendulum.
2. Stellar aberration.
3. Stellar parallax.

IV. METHODS OF SCIENCE. We discuss Galileo, Kepler, and others and the place in scientific thought of:

A. REASON: especially mathematics and thought experiments.

B. EXPERIMENT: necessary to verify laws and theories.

C. AUTHORITY: not acceptable as proof of scientific ideas.

V. PROBLEM SOLVING. The use of frames of reference.

A. CHANGES OF UNITS

B. USE OF THE FORMULAS FOR UNIFORM MOTION

C. CALCULATION OF RELATIVE VELOCITIES AND USE OF THE PRINCIPLE OF RELATIVITY

Questions

1. Is there any actual limit to the number of derived quantities that could be obtained? Is there any practical limit?

2. Are all fundamental quantities vector quantities?

*There are possible exceptions to principles 4 and 5.

3. What is the difference between a standard and a unit?
4. Which of the following are continuous and which are discrete?
   (a) Area.
   (b) Speed.
   (c) Spoons.
   (d) Rocks.
   (e) The stuff rocks and spoons are made of.
5. Can something really keep moving without being "pushed"?
6. If an elephant and an ant were both moving at different speeds on a level, frictionless surface, which would stop first? (See Fig. 1-42.)
7. If physics is an experimental science, why do we use general principles, such as the first law of motion, which describe experiments that can never be performed?
8. Why do airplanes land and take off into the wind?
9. Would you expect the laws of physics to turn out to be the same if you did your experiments on Mars? In a diving airplane? In a rocket being propelled upward at an increasing speed? In an airplane flying smoothly on the level? In a laboratory fixed upside down on the ceiling of your room? If you looked at the results of the experiments in the mirror? If the results were reflected in two mirrors? In three mirrors?
10. Why would a bomb dropped from an airplane in horizontal flight remain directly under the plane as it fell? What could prevent the bomb from remaining directly under the plane?
11. Galileo considered the thought experiment shown in Figure 1-43. If we put the peg in as shown, where will the pendulum bob swing?

Fig. 1-42

12. Find all the symmetries of an equilateral triangle. (An equilateral triangle has all sides equal in length.) That is, find all the ways in which one can change the rigid triangle by turning or flipping it so that it will look the same (see Fig. 1-44).

13. If you set up your laboratory on a merry-go-round, would all the laws of physics be the same? What would happen to the first law of motion?

14. What does a right-hand glove look like when it is reflected in a mirror?

15. List some activities that would be impossible or very difficult if there were no such thing as friction. List some that would be possible.

16. You are suddenly transported to a new planet that is shrouded in clouds. How would you determine its speed of rotation?

17. List some processes that are not time-reversible. List some that you think might be time-reversible.

18. Would the first law of motion still be true in a time-reversed world?

19. Make the best case you can for and against the motion of the earth.

20. Sketch the motion of a ball released a few meters above the ground as it appears from a reference frame that is (a) fixed to the earth, (b) fixed to a flatcar in uniform motion passing nearby, (c) fixed to the sun.

21. A ball is placed on a board that is resting on a frictionless surface. If we suddenly move the board, what happens to the ball? If the ball and board are moving across the surface as illustrated in Figure 1-45, and the board strikes a post, what happens to the ball?

22. A law suit developed as a result of the collision of a car and a pie truck. The truck driver claimed that he was stopped at an intersection when the car rammed into him from behind, causing the pies in the back of the truck to slide forward into the driver's seat. The car driver, on the

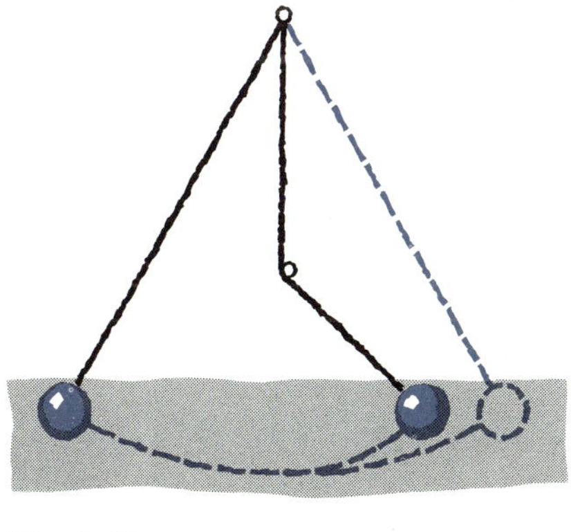

Fig. 1-43

other hand, claimed that both the truck and the car were stopped at the intersection when the truck driver suddenly backed up and rammed into the car, which happened to be an open convertible. The impact caused the pies to slide from the truck into the front seat of the car. Who do you think really got hit with the pies?

Problems

1. A marble is rolling at a speed of 20 inches per jiffy. What is this speed in meters per second if there are 10 seconds to the jiffy?
2. A building is 300 inches away from where I am standing. How many meters away is it?
3. If I can run 100 inches in one second, what is my speed in meters per second?
4. If I can run 100 inches in one minute, what is my speed in meters per second?
5. How many meters are there in a mile?
6. How many miles are there in a meter?
7. A boy grows two inches in a year. What is his rate of increase of height in meters per second?
8. Sound travels at about 1100 feet per second. What is its speed in meters per second?
9. To escape from the earth a rocket must have a so-called escape speed of about seven miles per second. What is this speed in meters per second?
10. The fastest rate at which information can be transferred is the speed of light in a vacuum, which is 186,000 miles per second. What is this speed in meters per second?

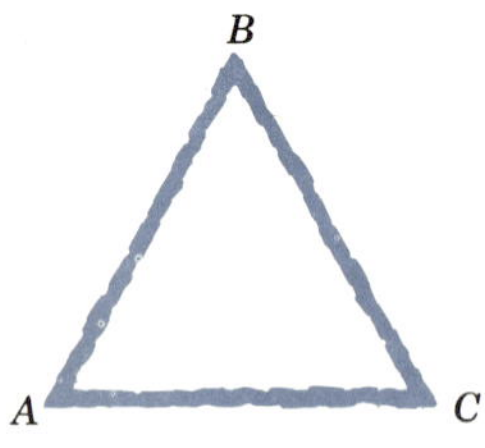

Fig. 1-44

11. In Example 6 (p. 53), find the point at which the two cars will meet.
12. An airplane travels 60,000 meters in 15 minutes. What is its speed?
13. A car travels 100 meters in three seconds. What is its speed?
14. How long would it take the car in Problem 13 to go 620 meters?
15. How long is a light year in meters? (A light year is the distance that light can travel in one year.)
16. A car travelling 30 meters per second passes a car travelling 40 meters per second going in the opposite direction. How soon after the cars pass each other will they be 350 meters apart?
17. A car covers half of a 10,000-meter trip at 30 meters per second. How fast will it have to go during the last half of the trip in order to make the whole trip in 22 seconds?
18. A furlong is 220 yards. A fortnight is 14 days. Express 10 furlongs per fortnight in meters per second.
19. A motor boat is traveling downstream at a rate of 15 miles an hour with respect to the water. The water is flowing at a rate of 25 miles an hour with respect to the land. What is the speed of the boat with respect to the land? What is it if the boat is traveling upstream?

More Difficult Questions and Problems

1. A man is walking at a rate of two meters per second due west at noon. What is his speed with respect to the sun? (The sun is 93,000,000 miles from the earth, and the sense of the earth's rotation is the same as its revolution around the sun.)
2. An airplane is landing on an aircraft carrier which is heading into the wind and traveling at 10 meters per second with respect to the water. The wind is blowing at 20 meter/second. The airplane requires a landing speed of 50 meter/second with respect to the air and a landing distance in still air of 800 meters.

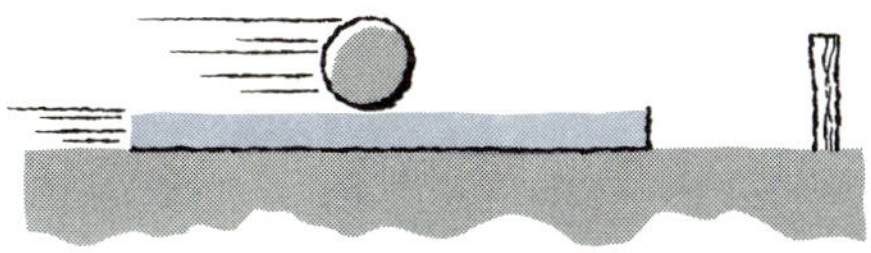

Fig. 1-45

(a) What is the plane's landing speed with respect to the carrier and the water, assuming that the plane lands into the wind?

(b) How long must the carrier be for a safe landing?

3. Why does a mirror reverse right and left and not top and bottom?

4. A blob of putty sliding at a velocity v on a frictionless surface strikes and sticks to an identical blob of putty which is originally at rest. What is the resulting velocity of the combined blob, and why?

5. How can the first law of motion be considered a rule for measuring time?

6. If motion can be validly described from any frame of reference (even an irregularly moving one), how can the first law of motion be valid?

2

THE SECOND LAW OF MOTION

The formulation of the laws of motion and gravitation by Isaac Newton is ranked among the most significant intellectual achievements of man. Newton admittedly "stood on the shoulders of giants," drawing on the work of his predecessors. Nevertheless, it was he who successfully stated the laws of motion. When he had finished, a new physics had emerged. The motion of matter, especially that of falling objects and wandering planets, could be accounted for by three laws of motion and a law describing the force of gravitation.

The first of these fundamental laws of motion, which evolved from the earlier investigations of Galileo, was discussed in Chapter One. In this chapter we shall introduce the second law of motion. We shall also discuss another fundamental quantity, mass, and two more derived quantities, force and acceleration; and we shall present an arithmetic for vector quantities. In Chapters Three and Four we shall discuss the third law of motion and the force law that describes gravitation, respectively.

Acceleration

In Chapter One we defined *velocity* as the vector quantity that specifies both the speed and the direction of the motion of an object. The unit of velocity that we use is the *meter per second* (meter/second). The velocity of an object remains constant when no external influences are exerted on the object. External influences may cause the velocity to change in speed or direction or both. The quantity which describes such changes is called *acceleration*. The acceleration is the rate at which a change in velocity takes place.

First let us consider motion along a straight line. Let us assume that a certain object—for example, a car—is moving along a straight road (see Fig. 2-1). It can speed up, stop, slow down, or back up. For the moment, however, we will not allow it to turn any corners. In this case the acceleration is familiar to us because of our experience with cars. We say that a car with great acceleration is one that reaches a high speed very rapidly. That is, it changes its speed at a rapid rate. Most people do not regard braking to a stop as an acceleration, but in fact it is. There is an acceleration whenever the speed of the car is changing. The acceleration expresses the rate at which the speed is changing.

Now suppose that the speed of the car is constant, but its direction of motion is changing. Is it accelerating? According to our definition, it is. We have defined acceleration as the rate of change of velocity, and velocity is a directional quantity. Therefore, when the direction of motion is changing, there is an acceleration. Often the motion of an object changes in both speed and direction simultaneously.

The unit of acceleration is *meter per second per second*. The occurrence of "per second" twice in this expression requires some explanation. Let us again consider the special case of motion in a straight line, as illustrated in Figure 2-1. Suppose that a car initially at rest acquires a speed of three meters per second after one second. If it continues to accelerate—that is, to change its speed—at the same rate, it will be going six meters per second after two seconds, nine meters per second after three seconds, and so on. In this example, the speed changes by three meters per second *each second.* We express this by saying that the acceleration is "three meters per second per second." As long as the acceleration has this value, we know that no matter what speed the object has at a given time, one second later its speed will have increased by three meters per second. The quantity acceleration describes how much the quantity velocity (which already involves time) changes in a certain amount of time. Therefore, the time must occur twice in the unit of acceleration. In our discussion of the unit of velocity or speed, we saw that the phrase "meter per second" is expressed mathematically as meter/second. The unit of acceleration, then, can be expressed mathematically as follows:

$$\text{Meter per second per second} = \frac{\text{meter/second}}{\text{second}}$$

According to the rule for performing mathematical operations with fractions, this is equal to

$$\frac{\text{meter}}{\text{second}} \cdot \frac{1}{\text{second}} = \frac{\text{meter}}{(\text{second})^2}$$

Mass

The fundamental quantity *mass* may be defined roughly as the amount of matter contained in an object. More precisely, this quantity is a measure of the response of an object to an interaction with another object or system of objects. Figure 2-2 shows what happens to a young man who wanders down the street kicking cans. Unknown to him, one of the cans is filled with lead. The interaction of his foot

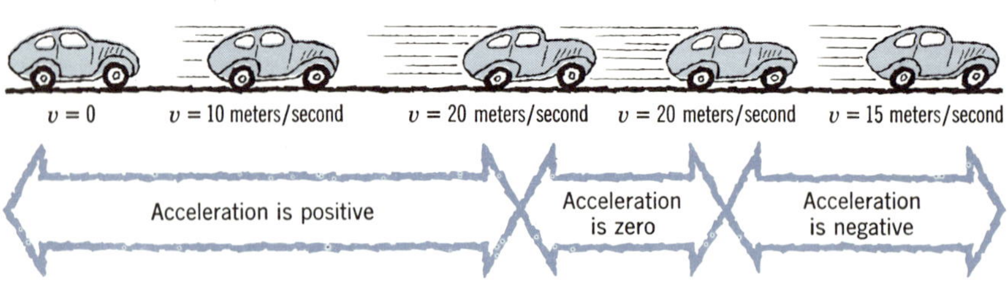

Fig. 2-1

with this lead-filled can is much different from the interactions with the cans that are "empty" (i.e., filled with air). There is, in a certain sense, more matter in the lead-filled can, as is shown by the fact that it has a lesser response to the same kick. Thus we say that the lead-filled can has more of the quantity mass than the other cans.

A Standard and Unit of Mass

In order to accurately define the quantity mass, we must first choose some object as a standard, establish a unit of measurement, and then devise some method of comparing the mass of other objects with that of the standard. We choose for a standard of mass the mass of a certain volume of pure water. The volume we choose is one thousandth of a cubic meter. This is the volume, for example, of a cube that has sides one tenth of a meter long (see Fig. 2-3). We call the mass of this standard one *kilogram*, and adopt the kilogram for our unit of mass. Mass is a scalar quantity. (Actually we must specify the temperature and pressure of the water, as well as its volume, to specify the standard exactly. Temperature and pressure will be discussed in later chapters.)

The Measurement of Mass

There are several procedures that might be used to compare the mass of an object with the mass of the standard. The method that we shall now describe is not so practical as others, but it helps to make the concept of mass clear.

As illustrated in Figure 2-4, we imagine a kilogram of water in a

(a) (b)

Fig. 2-2

very thin, rectangular container. We assume that the mass of the container is so small compared to the mass of the water that it can be disregarded. The container is resting on a perfectly smooth, frictionless surface. A string is attached to the container so that we can pull on it. We wish to pull on the container in such a way that we can continue to pull exactly as hard even while the container is moving. The exact device by which we achieve this constant pull is not important. However, to show that this is feasible we have illustrated a method by which this may be done. We have inserted a spring in the string as shown. We pull on the string in such a way that the spring remains stretched to the same length while the block is moving. Our basic operation is to pull on the block with an *unchanging* pull. The result of this pull is that the block accelerates.

As long as the pull on the block is the same, the acceleration is also the same. If, however, we change the mass of the object by pouring water into the container, the acceleration will no longer be the same. In fact, it will be less. The more mass (matter) there is in an object, the smaller the resulting acceleration if the pull remains the same. In other words, the mass is inversely related to the acceleration. A two-kilogram block of water (or the equivalent amount of matter in any shape or form) will accelerate at half the rate of our standard, the one-kilogram mass. An object with 1/3 kilogram of mass will have three times the acceleration as the standard, etc. Therefore, the mass of an object can be measured by comparing its acceleration to the acceleration of our standard mass. Any two objects of equal mass will receive the same amount of acceleration from the same amount of pull.

We wish to make it perfectly clear that the mass of an object is not dependent upon its volume alone but also depends on the type of

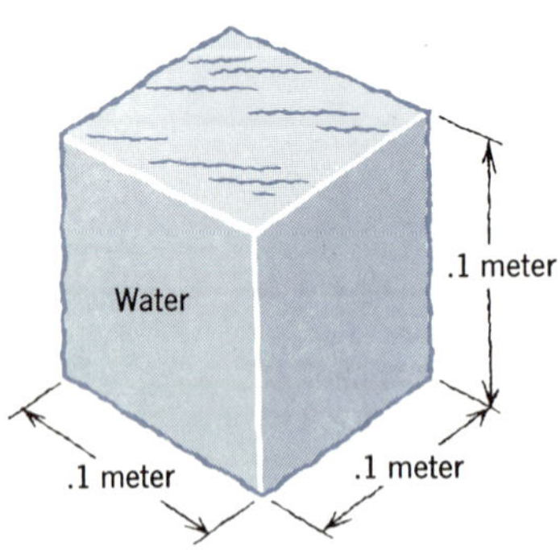

Fig. 2-3

material of which the object is composed. When we are dealing with only one kind of matter, such as the water in our previous example, we can increase the mass by increasing the volume of the material used. However, equal volumes of different kinds of material can have greatly different masses. For example, a cubic meter of lead has more than eleven times the mass of a cubic meter of water.

The Conservation of Mass

Mass is considered to be one of the basic realities of the universe. Mass can be neither created nor destroyed. Although matter can be changed from one form to another, the total amount of its mass does not change. According to all known experiments relating to this subject, there is just as much mass after any change or transformation as before. We express this fact by saying that *mass is conserved.*

If we boiled away a kettle-full of water, for example, it might appear that the mass of the water had disappeared. In fact, of course, if we were to collect all of the steam that had come from the kettle we would find that it had the same mass as the missing water. Although the form of the matter can be changed, the mass remains the same. Mass is one of several quantities in physics that are conserved.

FORCE

Whenever we push or pull on an object, we are exerting what is called a *force* on that object. As we shall see, there are other sources of force besides human muscle power. In this section, we shall

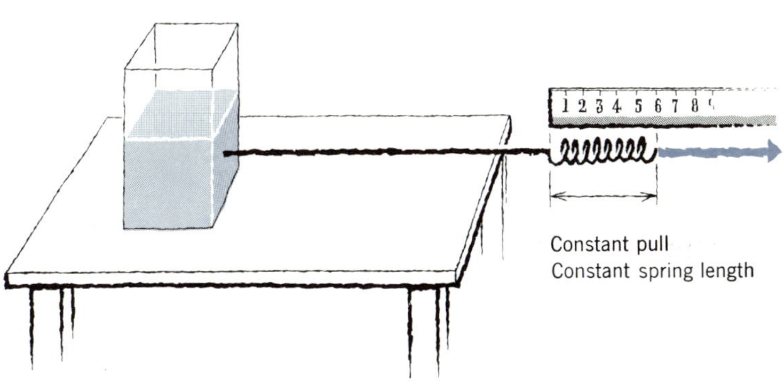

Fig. 2-4

develop the concept that the force on an object is the cause of its acceleration. Force is a derived quantity. Since it has a direction, it is a vector quantity also.

In the previous section we described a system that was capable of exerting a constant pull, or force, on a container of water. This force was responsible for the acceleration of the water. By keeping the force the same, applying it to various objects, and observing the acceleration produced, we established a way of measuring mass. Now we wish to compare the accelerations of the same object caused by different amounts of force.

Once again, our object is a container of water resting on a level frictionless surface (see Fig. 2-5). The mass of the water is the same in all three parts of the figure. Each of the strings shown pulls with the same force F. This means that the force in each string, acting alone, would produce the same acceleration of the container of water as the force in any other string, acting alone. We observe that doubling the force, as in Figure 2-5b, doubles the acceleration also. Tripling the force, as in Figure 2-5c, triples the acceleration. In fact, the acceleration is proportional to the force acting on the object. (For a discussion of proportionality, see Appendix II-C.) Therefore, one way of measuring a force is to measure the acceleration caused by that force on an object of standard mass.

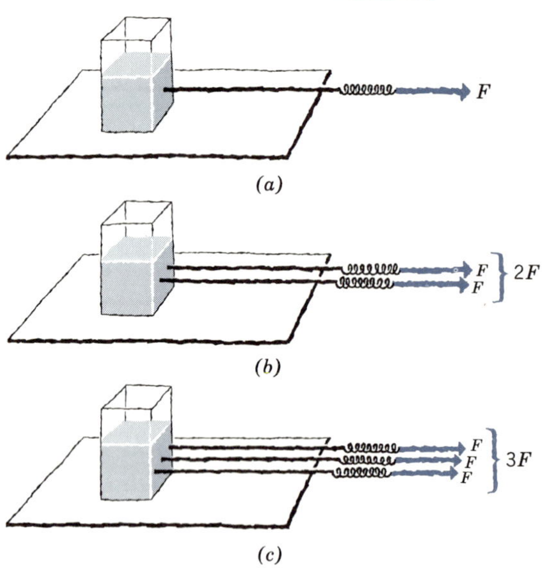

Fig. 2-5

The unit of force is called a *newton*. It is defined as the force needed to cause one kilogram of mass to accelerate at the rate of one meter per second per second. Two newtons of force give a one-kilogram mass an acceleration of two meters per second per second, and so on.

In the foregoing pages, we have found two relations between the force on an object, its mass, and its acceleration. We first investigated what happens when the force is kept the same in different experiments while the mass and acceleration are allowed to change. (See Fig. 2-4.) Then, when discussing force, we kept the mass the same (in the three experiments illustrated in Fig. 2-5) but allowed the force and acceleration to vary. There is a third relation that expresses what happens when the forces and masses are carefully selected so that the acceleration turns out to be the same in different experiments as, for example, in the three experiments illustrated in Figure 2-6. We notice in such experiments that a force of one newton acting on an object of one-kilogram mass gives the same acceleration as a force of two newtons acting on a mass of two kilograms, etc. Therefore, for the acceleration to remain the same, the force must be directly proportional to the mass of the object. This gives us three relations between force, acceleration, and mass which arise from experimental facts.

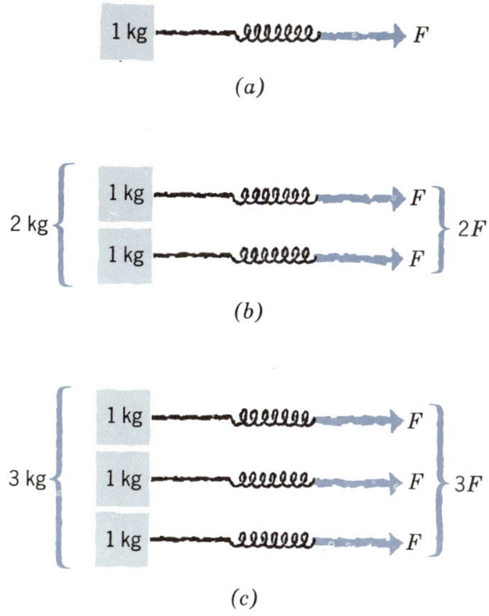

Fig. 2-6

THE SECOND LAW OF MOTION

The three relations among mass, force, and acceleration discussed above are all contained in one comprehensive law: Newton's second law of motion. For the quantities and units we have discussed, this law may be stated as follows.

The force on an object is equal to its mass multiplied by its acceleration.

(We are assuming that the mass of the object does not change while it moves.)

It should be noted that both force and acceleration are vector quantities, and that the acceleration of the object is in the direction of the force applied to it. We can write the second law of motion in equation form as follows:

$$F = ma \tag{1}$$

in which

F is the force on the object
m is the mass of the object
a is the acceleration of the object

The second law of motion tells us that a force causes an object to accelerate. It is a force, a push or a pull, from an external source that is responsible whenever an object undergoes a change in its motion. If there are no forces acting on the object, there is no acceleration.

The law also implies that the direction of the acceleration is the same as that of the applied force. If the force is in the direction of motion, the acceleration is also in that direction, and we observe the acceleration as a change in speed in that direction. This situation is illustrated in Figure 2-7. If the force is not along the direction of

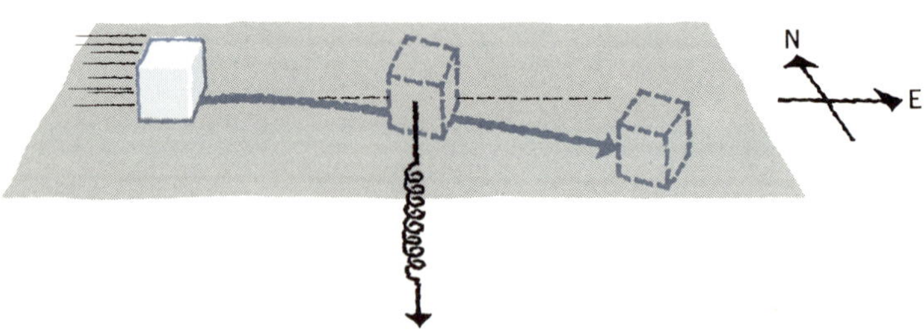

Fig. 2-7

motion, the object will follow a curved path, whether it changes its speed or not. In either case, the *change* in motion is in the direction of the force, although the new motion may or may not be. For example, the object shown in Figure 2-7 was traveling due east. After being pulled toward the south, however, it has some motion toward the south. Its new direction of motion is southeast. Thus, the *change* in its motion was in the direction of the force.

We can use the second law of motion to express the unit of force in terms of the units of mass, length, and time. We recall from the last section that one newton of force will give one kilogram of mass an acceleration of one meter per second per second. We simply insert these values for force, mass, and acceleration in the second law of motion, as follows:

$$\text{Force} = \text{mass} \times \text{acceleration}$$
$$\text{Newton} = (\text{kilogram}) \times (\text{meter/second}^2)$$

Therefore

$$\text{Newton} = \frac{\text{kilogram meter}}{\text{second}^2}$$

Since we are able to use the second law of motion to obtain the unit of force in this way, we may regard force as a derived quantity. We have derived it from the fundamental quantities mass, length, and time by using the second law of motion.

We can use the equation form of the second law of motion to show that this law does indeed include the three relations between mass, force, and acceleration that we have discussed in this chapter.

$$F = ma$$

Since the equation must be true in all circumstances, we cannot change one side without changing the other also. For example, if the acceleration is constant (as in Fig. 2-6), the force and mass must change proportionally to preserve the equality. If one is increased, the other must be increased also for this case. If, on the other hand, the mass is held constant the force and the acceleration must change proportionally. Finally, if the force is held constant (as in Fig. 2-4), the left side of the equation cannot change. Therefore, the right side of the equation must not change either, and the mass and the acceleration must change inversely; that is, if one is increased, the other must be decreased.

Vectors

Vectors and Scalars

We have already distinguished between the quantity speed (a scalar) and the quantity velocity (a vector). You will recall that the vector quantity, velocity, could not be specified without giving a direction, while the scalar quantity merely required a number. The number, with its units, is called the *magnitude* of a quantity. Thus velocity necessarily has a magnitude and a direction, while speed has a magnitude alone. In fact, the scalar quantity speed is just the magnitude of the vector quantity velocity. Many, but not all, scalar quantities are magnitudes of vector quantities. In general, then, we here define a vector quantity as a quantity that has both a *magnitude* and a *direction* and obeys the special rules of arithmetic discussed below. A *scalar* quantity is here defined as a quantity that has a magnitude only and obeys ordinary rules of arithmetic.

Distance and Displacement

We now define and discuss the vector quantity that we call *displacement*. This quantity is a measure of the difference in position between two points or places. The magnitude of this vector quantity is the straight-line distance between these two points. The direction of the displacement is the direction from one of these points (usually the first one mentioned) to the other. In Figure 2-8, for example, two points, P and Q, are shown that are some distance apart. We draw a straight line from P to Q. We then draw an arrowhead so that the line points to Q as shown. The length of the straight line from P to Q gives the magnitude of the displacement, which is, of course, just the ordinary (straight-line) distance between the two points P and Q. The direction of the arrow is the direction of the displacement from P to Q. If we had chosen the opposite direction—that is, the direction

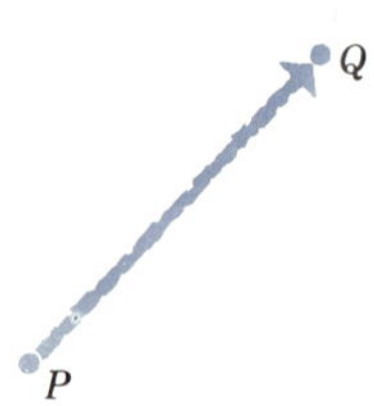

Fig. 2-8

from Q to P—we would not have the same vector. The magnitude would be the same, but not the direction. The displacement from P to Q is not equal to the displacement from Q to P even though the distance is the same, because the direction is different. We can generalize this principle as follows: two vectors are equal to each other if and only if they have the same magnitude and the same direction.

The displacements shown in Figure 2-9 are all equal to each other; they have the same length, and they are pointing in the same direction. In order for the direction of two vectors to be the same, it is not necessary that they lie on the same straight line, but merely that they lie on parallel straight lines, and that the arrows point in the same direction.

Figure 2-10, which contrasts the flight of a butterfly with that of a conventional bee, illustrates the displacement vector and permits us to compare distance and displacement. Both insects leave flower P and travel to flower Q. The butterfly, depending on the wind and his whim, might travel a very long path en route. Furthermore, he could make many flights of exactly the same path length and arrive at a different destination each time, possibly even right back where he started from. The "distance" traveled, the total length of the path, is a scalar; taken by itself it does not determine the end point of a particular journey.

The bee, on the other hand, lives up to his reputation for making "beelines" by traveling the shortest path between the two flowers. This directed quantity, a distance in a specific direction, is the displacement from P to Q. As before, we represent the displacement from P to Q by an arrow drawn in the direction of the line joining the two points with a length equal to the beeline distance. By traveling along this beeline, the bee travels a much shorter path than the butterfly. Nevertheless, the displacements shown in the diagram for

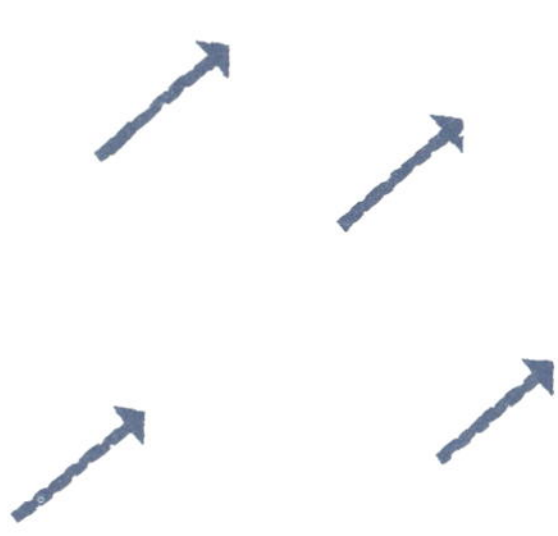

Fig. 2-9

the two insects are exactly the same, since they both start at the same point and end at the same point. Displacement is concerned only with the direction and the straight-line distance between two points—that is, with an overall change in position. The actual path length traveled in between is irrelevant as far as displacement is concerned. This is the reason we can represent the displacement vector as a straight arrow between the two points, pointing in the proper direction.

Addition of Vectors

The arithmetic of vectors is different from the arithmetic of scalar quantities. Scalars obey the familiar rules of addition and subtraction, but vectors do not. We illustrate the rules for adding together two vectors in Figure 2-11, which shows the three displacements **A**, **B**, and **C**. **A** is the displacement from *P* to *Q*, **B** is the displacement from *Q* to *R*, and **C** is the displacement from *P* to *R*. Of course, the displacement **C** may be obtained by going directly from *P* to *R*. However, we can also get from *P* to *R* by going first from *P* to *Q* and then from *Q* to *R*. The result obtained by joining displacement **A** to displacement **B** is exactly the same as if we had taken the displacement **C** alone. In each case we have started at *P* and ended at *R*. Since the displacement depends only on the starting point and the end point, we have accomplished the same displacement in two different ways. We say, then, that the displacement **C** is the sum of the two displacements **A** and **B**, or

$$\mathbf{C} = \mathbf{A} + \mathbf{B}$$

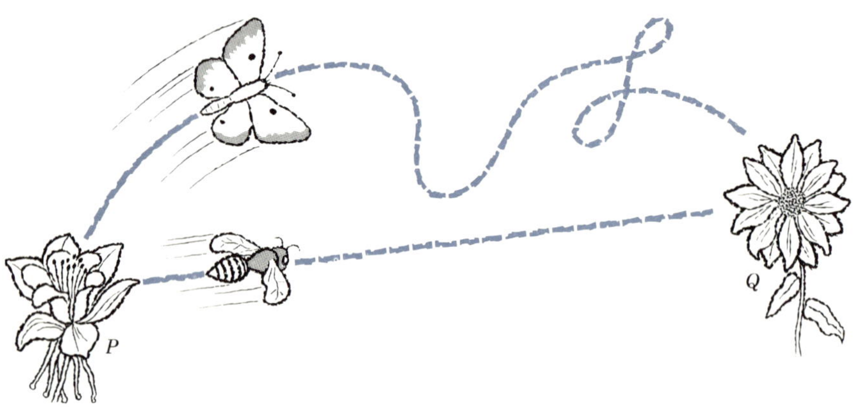

Fig. 2-10

We now present the rule for adding two vectors together. (We illustrate this addition for displacements, but the procedure can easily be adapted for vectors in general.) Consider the two vectors **A** and **B** in Figure 2-12. We wish to obtain the sum **A** + **B**. The first step is to construct a vector equal to **B**, beginning at the arrowhead of vector **A**, as shown in Figure 2-12*b*. This step is sometimes described as moving vector **B** to the end of vector **A**. Care must be taken to see that the direction of **B** is the same after we move it as before. Next we connect the beginning point of **A** with the ending point of **B** in it's new position by a straight line. We draw the arrowhead on this new line as shown. This procedure puts the head of **C** at the same point as the head of **B** the second vector, and puts the tail of **C** at the same point as the tail of **A**, the first vector. This new line with its new arrowhead gives us a new vector. This new vector **C** is called the *resultant* or *sum* of **A** and **B**.

In the above procedure, we brought the tail of **B** to the head of **A**. If we had instead brought the tail of **A** to the head of **B** as illustrated in Figure 12*c*, and then found our new vector, we would have obtained an equal resultant. It does not matter in what order the two vectors are added. In other words, we have the familiar-looking fact that

$$\mathbf{A} + \mathbf{B} = \mathbf{B} + \mathbf{A}$$

When we add vectors together, the magnitude of the resultant is not necessarily equal to the sum of the magnitudes of the individual vectors from which we obtained the resultant. Figure 2-13 gives some examples of this fact. In Figure 2-13*a* the resultant has a magnitude that is even less than either the magnitude of **A** or **B** alone. In Figure 2-13*b* the resultant is zero, although neither **A** nor **B** has a zero magnitude. In Figure 2-13*c* **A** and **B** are pointing in exactly the same direction; in this situation only, the resultant has a magnitude equal to the sum of the magnitude of **A** and the magnitude of **B**.

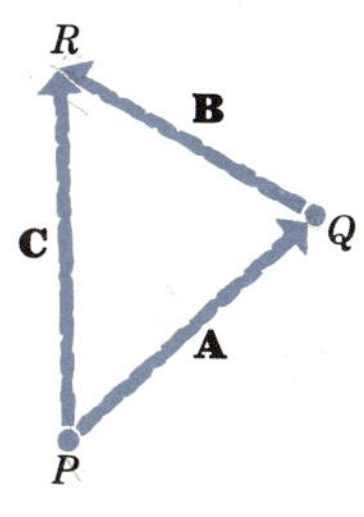

Fig. 2-11

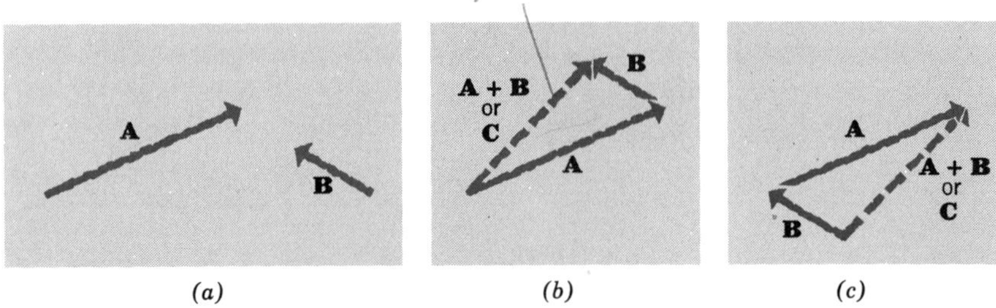

Fig. 2-12

Figure 2-14 shows how we can add more than two vectors together. To add three vectors together, for example, as is done in Figure 2-14*a*, we first add two of the three vectors together, and then add the third to the resultant of the first two. This simply amounts to putting all three of them in a head-to-tail chain as shown in Figure 2-14*b*. This same rule works for any number of vectors; Figure 2-15 shows the addition of five vectors.

All vector quantities can be added by this same technique. For example, velocity may also be represented on a diagram (see Fig. 2-16). We simply let one unit of velocity, one meter per second, be represented by some convenient length, shown by a scale included on the diagram. We then construct arrows representing each of the velocity vectors to be added. The length of each arrow is determined

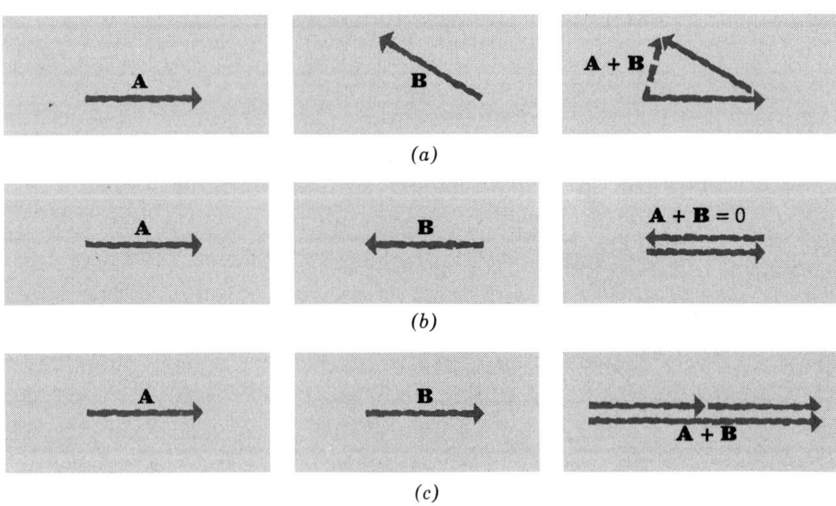

Fig. 2-13

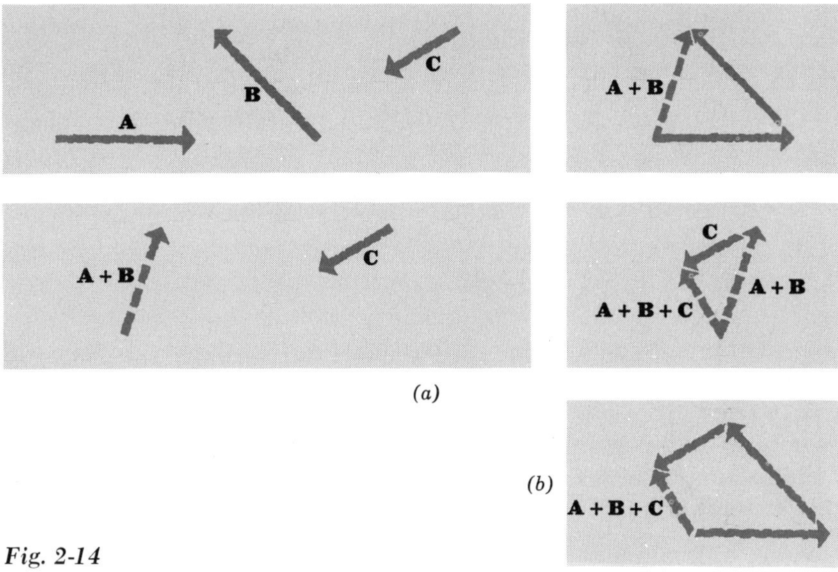

Fig. 2-14

from the scale on the diagram and is therefore proportional to the magnitude of the vector velocity represented by the arrow. Then we add the vectors in the usual way and obtain the resultant. The magnitude of the resultant velocity is obtained by using the scale to convert the length of the resultant arrow on the diagram to meters per second. The direction of the resultant velocity is the same as the direction of the resultant on the diagram. By using such scale drawings, we may also find the resultants for other vector quantities besides velocity and displacement.

Resultant Force and the Second Law of Motion.

In our earlier discussion of the second law of motion, we pointed out that it is force that causes an object to accelerate, and we noted

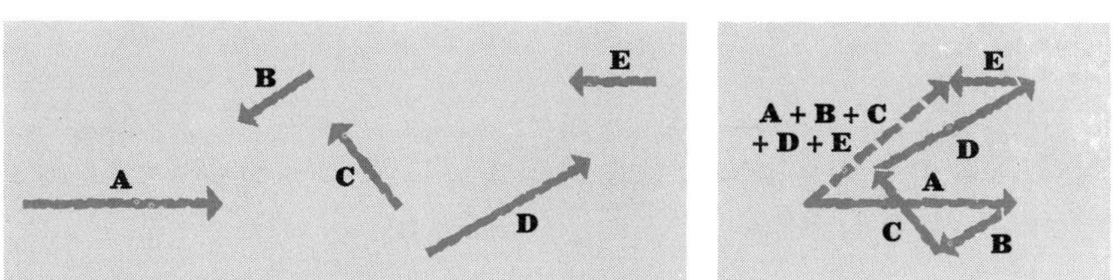

Fig. 2-15

that both force and acceleration are vector quantities. We described the second law of motion for the case in which a single force is acting on an object. Now let us consider the application of the second law of motion when an object is acted on by more than one force at the same time. In this case, the acceleration of the object is produced by the *resultant* of all the forces acting on the object, and this resultant force (**F**) is found by the method of vector addition outlined above.

For example, suppose an object in space is being pulled by two astronauts using ropes, as shown in Figure 2-17. To obtain the resultant force and the acceleration it causes, we proceed as follows.

1. Draw a diagram representing all of the forces on the object as arrows, using a convenient scale.
2. Add the vectors and obtain the resultant force on the object.
3. Apply the second law of motion, using the resultant force to obtain the acceleration. The object will accelerate in the direction of this resultant force.

Steps 1 and 2 have been carried out in Figure 2-17*b*. Step 3 can be accomplished if we know the mass of the object. The magnitude of the acceleration is equal to the magnitude of the resultant force

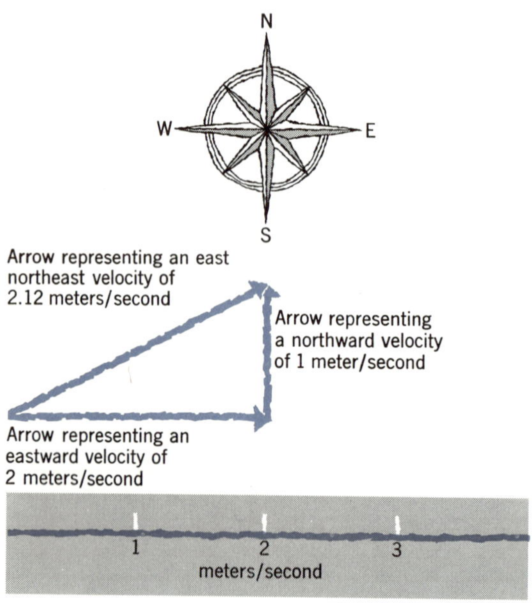

Fig. 2-16

divided by the mass. That is,

$$a = \frac{F}{m}$$

This equation is obtained from equation 1 by dividing each side by the mass m. Therefore, once we have found the resultant force, it is easy to find the acceleration. The direction of the acceleration is the same as the direction of the resultant force. If the object in Figure 2-17 were at rest to begin with, it would accelerate as shown in Figure 2-17*c*. If it were moving when the forces were applied, it would start to curve toward the direction of the resultant force, as shown in Figure 2-17*d*.

If the resultant of all the external forces acting on an object is zero, there is no acceleration and the object is said to be in *motion equilibrium*. This condition will be discussed further in Chapter Five.

Let us consider another concept. Figure 2-18 shows a bead which

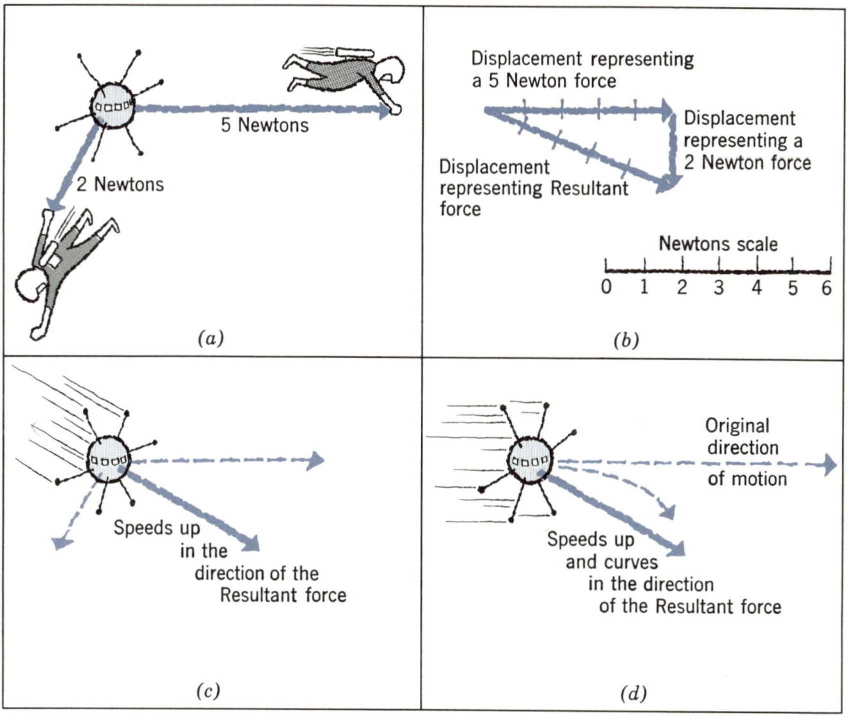

Fig. 2-17

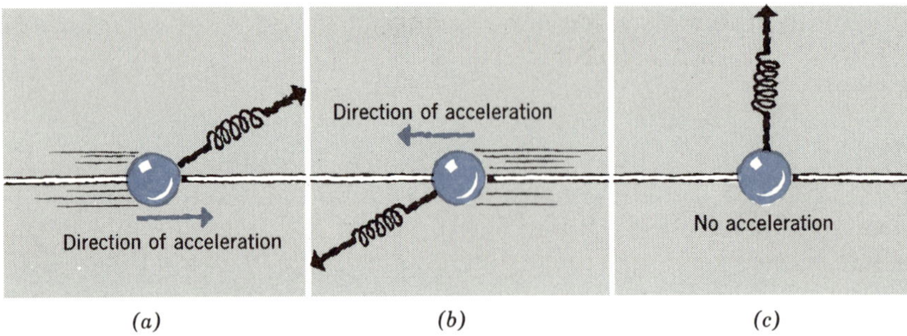

Fig. 2-18

is free to slide on a stiff frictionless rod. If we pull it by the string it will move along the rod in one direction or the other as shown. The rod exerts forces which prevent the bead from moving in any direction except along the rod. Suppose that we pull with the string in a direction exactly perpendicular to the rod. Now which way will the bead move? Intuitively we guess that the bead will not move at all, since there would be just as much reason for it to move to the right as to the left. The symmetry of the situation indicates that no motion will result. This turns out to be correct, and it illustrates an important principle: a force has no direct effect on motion in a direction perpendicular to the direction of the force.

Other Vector Operations

THE NEGATIVE OF A VECTOR. The negative of a vector is formed simply by reversing the arrow so that the head is on the other end (see Fig. 2-19). The negative of a vector, then, is a vector of the same magnitude but pointing in the opposite direction.

SUBTRACTION OF ONE VECTOR FROM ANOTHER. We have described a way to find the sum of two vectors:

$$\mathbf{A} + \mathbf{B}$$

Suppose we wish to find the difference between these two vectors:

$$\mathbf{A} - \mathbf{B}$$

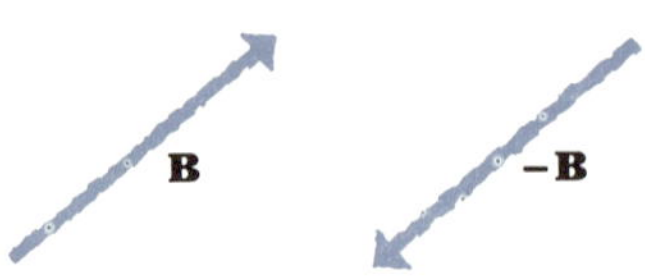

Fig. 2-19

To do this we simply reverse the second vector **B** to form its negative **−B** and add it to **A** (see Fig. 2-20):

$$\mathbf{A} - \mathbf{B} = \mathbf{A} + (-\mathbf{B})$$

THE MULTIPLICATION OF A VECTOR BY A NUMBER OR A SCALAR. We can also multiply a vector by a number or a scalar quantity, as illustrated in Figure 2-21. If we multiply the vector by a positive number, we merely multiply the length or magnitude of the vector by that number, keeping the direction the same. For example, if we multiply a vector by 3 we get a vector three times as long in the *same* direction. If, on the other hand, we multiply the vector by a negative number, we multiply the magnitude or length of the vector by that number but reverse the direction. For example, if we multiply a vector by *minus* 3, we get a vector three times as long, but in the opposite direction.

*Vector Components

We have already described how two given vectors can be added together to obtain a third. A vector can also be "broken up" into two other vectors at right angles to each other which, when added together, yield the original vector. That is, given the vector **C**, we can find two vectors **A** and **B** at right angles to each other such that **A** + **B** = **C**. Under these special circumstances, **A** and **B** are called the components of **C** (see Fig. 2-22).

Figure 2-23 shows how we obtain the components of a vector and how they can be useful. In this figure we have a reference frame made up of two axes, one running north and south, and the other running east and west. We also have a vector **C** drawn on the same

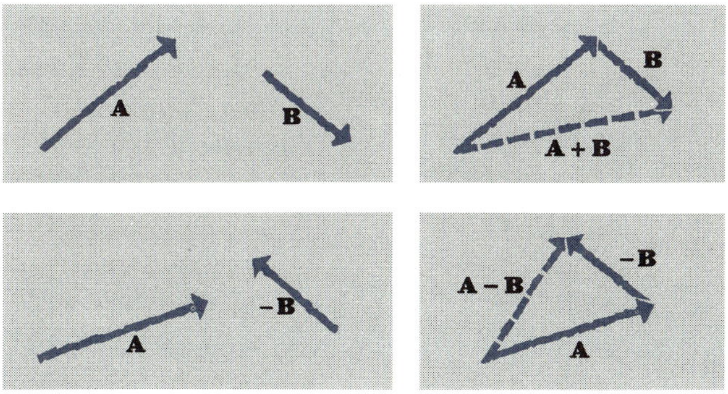

Fig. 2-20

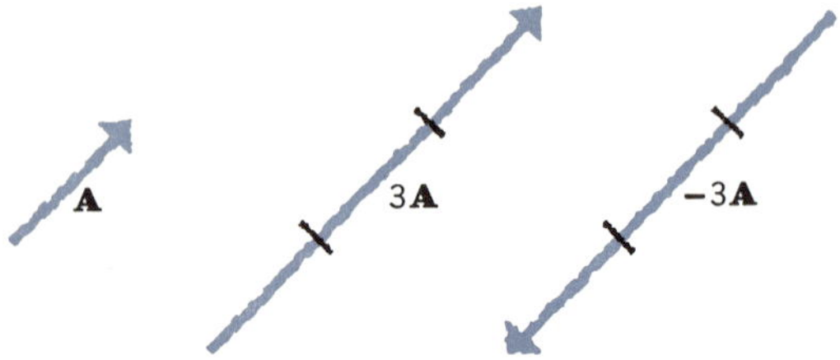

Fig. 2-21

diagram from the center or origin. We call the north-south axis the y-axis and the east-west axis the x-axis.

Now suppose that we are not interested in the total displacement but only in the displacement to the east. That is, we do not care about the whole distance we must travel to get from O to P, but are interested merely in how far east of the origin we must go in order to reach point P. The eastward vector **A** (from O to X) is just as good as **C** for any question that concerns only the displacement to the east or west. Similarly, the northward vector **B** (from O to Y) is just as good as **C** for questions involving only the north-south displacement. The vector **A** is called the component of the vector **C** along the x axis, or in the x direction. The vector **B** is called the component of the vector **C** along the y axis or in the y direction. The two vectors **A** and **B** are the components of the vector **C** in this reference frame. They are perpendicular to each other. If we add them together we obtain **C**:

$$\mathbf{A} + \mathbf{B} = \mathbf{C}$$

The vector component **A** can be found by drawing a line from the tip of the vector **C** so that this line is perpendicular to the x axis. The component **B** is found by drawing a similar line perpendicular to the y axis. This procedure can be used to find the component of any vector along any axis. In Figure 2-24, for example, we wish to find the component of a vector **A** in a given direction. To do this we draw a line in that direction through one endpoint of the vector. Then we draw a line from the other end-point of the vector perpendicular to

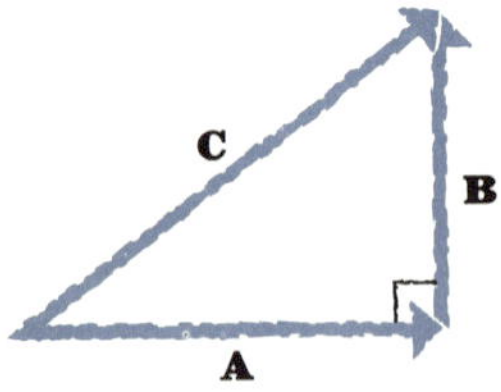

Fig. 2-22

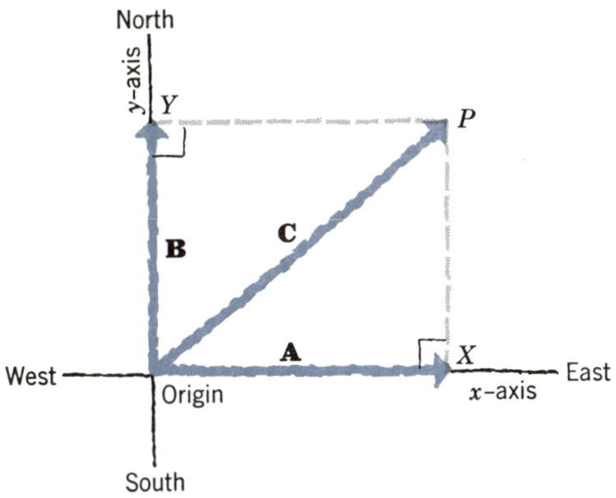

Fig. 2-23

the first line, to determine the desired component. This procedure allows us to find the component of a given vector in any direction.

In any reference frame, we can form the components of a given vector along each of the mutually perpendicular axes. If we then form the vector sum of all the components of the given vector along each of these axes, we obtain the given vector back again. A vector is equal to the sum of all its components along the perpendicular axes of a reference frame. This fact is often useful. If for some reason we prefer not to deal with a given vector, we can choose some convenient reference frame, find the components of the vector in that reference frame, and use the components instead of the original vector.

If we are concerned with one direction only, we can use the component of our vector in that direction and ignore the other components. For example, in Figure 2-25 we again have a bead which is

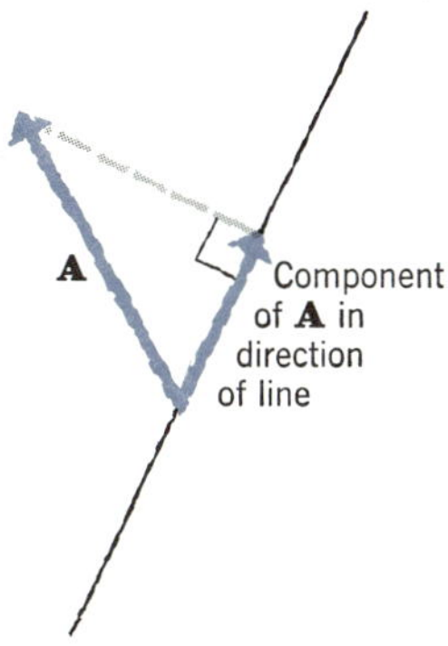

Fig. 2-24

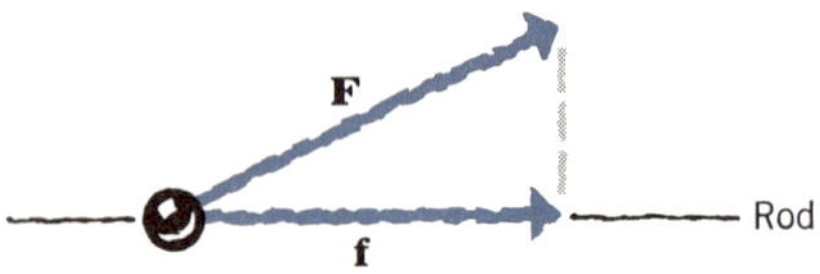

Fig. 2-25

free to move (without friction) along a rod. The bead is being pulled by a string which acts at an angle to the rod as shown. We wish to find the acceleration of the bead, knowing its mass. Since we know that the bead must move along the rod, we are interested only in the direction along the rod. Therefore, we are interested only in the component of the force in the string that acts in the direction of the rod. It is *not* the force F in the string which we use to find the acceleration of the bead. Rather, it is the force f, which is the component of F in the direction of the rod. Thus, $f = ma$, and the acceleration is easily obtained as $a = f/m$. Any equation of physics which holds between vectors must also hold between the components of those vectors along any axis or direction. Let us use the reference frame of Figure 2-26 and write the equation $F = ma$ for each axis:

$$F_x = m\mathbf{a}_x$$

$$F_y = m\mathbf{a}_y$$

The small x's and y's are used as subscripts on the letters F and a to indicate that we are referring to the components of the forces and accelerations along the x and y axes, respectively. These expressions tell us that the component of acceleration in the x direction is caused by the component of force in the x direction, and that the component of acceleration in the y direction is caused by the component of force in the y direction.

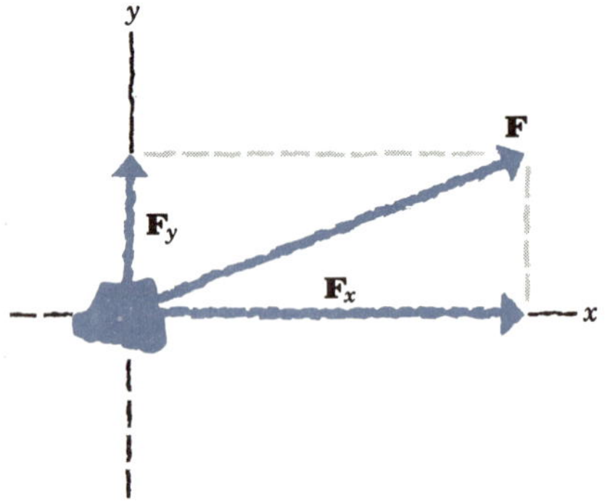

Fig. 2-26

If we were concerned with all of space, instead of just the flat surface, we would have another equation for another direction, the z direction, as illustrated in Figure 2-27. Thus, we can think of a vector equation as being three *independent* equations written as one. Since our reference frame may be oriented in any direction we please, it follows that the component of force in *any* given direction is equal to the mass times the component of acceleration in that same given direction.

PROBLEM SOLVING

The principles introduced in this chapter are useful in solving problems involving (1) force, mass, and acceleration and the second law of motion and (2) vector operations.

The Second Law of Motion

EXAMPLE 1

What is the force required to give a mass of 5 kilograms an acceleration of 30 meters per second per second?

Solution. We apply the steps introduced in Chapter One. For this simple problem we do not need a diagram or reference frame. Therefore Step 1 here corresponds to Step 2 in Chapter One.

STEP 1. List the given quantities and the unknown quantities:

m = mass = 5 kilograms
a = magnitude of the acceleration = 30 meter/second²
F = magnitude of the force = ?

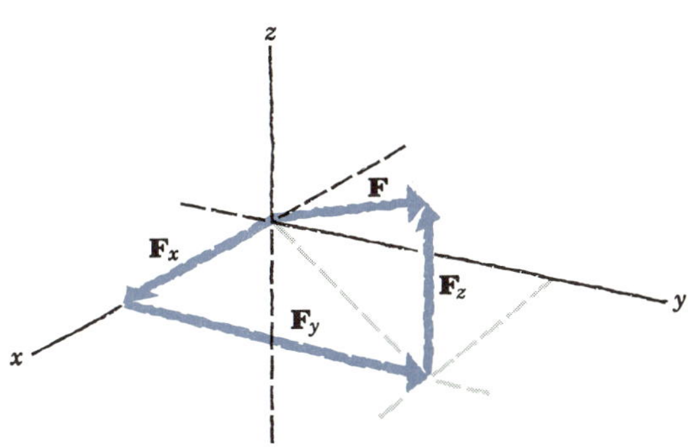

Fig. 2-27

STEP 2. Write, in equation form if possible, the physical laws that apply. Since this example deals with force, mass, and acceleration, we should expect the second law of motion to be useful:

$$F = ma$$

STEP 3. Enter the given information from Step 1 into the equation from Step 2 and solve for the required quantities:

$$F = (5 \text{ kilogram})\ (30 \text{ meter/second}^2)$$

$$F = 5 \text{ kilogram} \cdot \frac{30 \text{ meter}}{\text{second}^2}$$

$$= 150 \frac{\text{kilogram meter}}{\text{second}^2} = 150 \text{ newtons}$$

EXAMPLE 2

What acceleration would the force in Example 1 give to an object with only one kilogram of mass?

Solution:

STEP 1. F = magnitude of the force = 150 newtons (from the results of Example 1)

m = mass = 1 kilogram

a = acceleration = ?

STEP 2. $F = ma$

STEP 3. 150 newton = (1 kilogram) a

$$\frac{150 \text{ newton}}{1 \text{ kilogram}} = a$$

$$a = \frac{150 \text{ newton}}{1 \text{ kilogram}} = \frac{150 \text{ kilogram meter/second}^2}{\text{kilogram}}$$

$$a = 150 \text{ meters/second}^2$$

(Note that we have expressed the quantity *newton* as kilogram meter/second^2 in order to obtain the acceleration a in meter/second^2. It is often convenient to express all units in terms of the fundamental quantities mass, distance, and time duration in order to simplify answers.)

The problem illustrated in Example 2 may be solved more quickly and more simply by recalling the principle that when the force is constant, the acceleration must be inversely proportional to the mass. The force is the same in Examples 1 and 2. Therefore, since the mass in Example 2 is one fifth the mass in Example 1, the acceleration must be five times higher than in the first example. Five times 30 meter/second^2 gives us the required acceleration, 150 meter/second^2.

This method is obviously less cumbersome than the computation shown above. By using ingenuity or applying simpler principles, one can often obtain answers to problems more quickly and easily than by slavishly following steps. However, the steps listed are useful in solving a wide variety of more complex problems.

EXAMPLE 3

A car traveling on a straight highway at a speed of 20 meters per second begins to accelerate at a constant rate until it reaches a speed of 32 meters per second. If this acceleration takes 4 seconds, what is the magnitude of the acceleration of the car?

Solution. For motion along a straight line we can determine the acceleration by finding the rate of change of the speed. The total change of speed from 20 meters to 32 meters per second is 12 meters per second, obtained by subtracting the first speed from the final speed. This change occured in 4 seconds, so the magnitude of the acceleration is

$$\frac{12 \text{ meter/second}}{4 \text{ second}} = \frac{12}{4} \cdot \frac{\text{meter}}{\text{second} \cdot \text{second}}$$

$$= 3 \, \frac{\text{meters}}{\text{second}^2}$$

Problems Involving Vector Operations

ADDITION USING DIAGRAMS.

EXAMPLE 4

Using a diagram, find the resultant of a displacement of two inches east and a displacement of one inch northeast.

Solution. Figure 2-28 shows the two vectors drawn from the same origin in a reference frame. We have drawn each vector to the proper length and in the proper direction. We have labeled the two-inch vector **A** and the one-inch vector **B**. Now we construct the one-inch vector **B**′ equal to **B**, so that its tail is at the head of vector **A**. This gives us the proper head-to-tail diagram for adding the two vectors. We are careful that **B**′ is equal to **B** in both magnitude and direction. Now we draw a vector from the tail of **A** to the head of **B**′. This vector

C is the resultant of **A** and **B′** and also is the resultant of **A** and **B**, since **B** is equal to **B′**. Measuring with a ruler, we find that **C** has a magnitude of about 2.8 inches. The direction is approximately east northeast.

Example 5

Two forces are exerted on an object. One of them is a force of 3 newtons due east, and the other is a force of 2 newtons northwest. What is the total vector force on the object?

Solution. Since we want to draw the vectors, we must first choose a scale. (The vector is a force, but we can draw only displacements.) We choose the scale of one newton to the inch; that is, one inch on our paper will represent a force of one newton. Now we draw both vectors, as in Figure 2-29, labeling the 3-newton vector **A** and the 2-newton vector **B**. To add them we must put the tail of one to the head of the other. We choose to put the tail of **B** at the head of **A**.

We draw the vector from the tail of **A** to the head of **B** to get the resultant **C**. The magnitude of this vector can be determined by measuring it, using the same scale as before. The vector we have drawn is 2.2 inches long; therefore, the resultant force has a magnitude of 2.2 newtons. The direction is roughly northeast. (Since we were asked for the force, which is a vector, we must give the direction as well as the magnitude.)

Example 6

The two boys shown in Figure 2-30 are running along the banks of a frozen stream, pulling a sled as shown, using ropes of equal length.

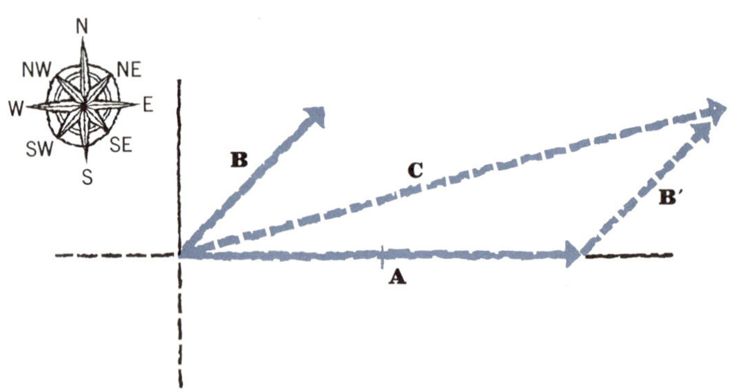

Fig. 2-28

The boys are running so that they are directly opposite each other across the stream, and so that the ropes make a right angle at the sled as shown. The sled has a mass of 100 kilograms. At a certain time, each boy is pulling on his rope with a force of 20 newtons. Disregarding the friction between the ice and the sled, what is the acceleration of the sled?

Solution. In problems of this kind, where there is more than one force on an object, we can avoid confusion by drawing a diagram of the object of interest separate from its surroundings. We have illustrated this in Figure 2-30*b*. Then we draw arrows on this diagram representing the magnitude and direction of each force on the object. We take the trouble to draw this separate diagram to be sure that we include all the forces that are actually acting on the object, and to be just as sure that we do not draw in any forces that do not act on this object but merely on its surroundings. In this example there are only two forces which we need to consider—the forces by which the two ropes pull on the sled. (There is a force of friction exerted on the sled by the ice, but we have decided that it is small enough to ignore in this problem. To be completely accurate, we must admit that there are other forces on the sled: gravitational force pulling it down onto the ice and the pressure between the ice and the sled—forces that

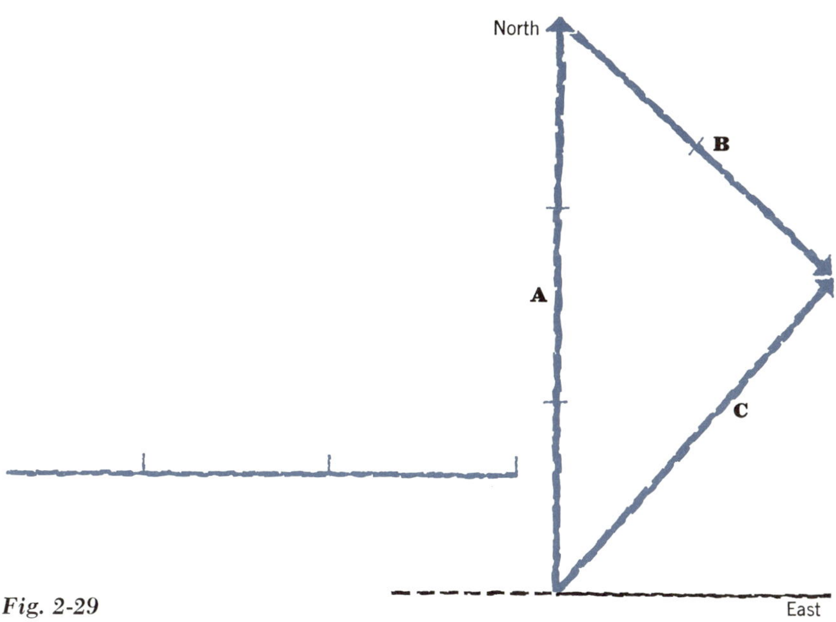

Fig. 2-29

we consider in more detail later. These forces, however, act in directions perpendicular to the direction of motion and therefore can be ignored in this example. It turns out, in fact, that the effects of these forces cancel each other out). Now we can consider the problem in the following steps:

STEP 1. Draw the object separately and indicate the forces on it by arrows drawn to scale (see Fig. 2-30*b*). In some cases the decision as to which object to isolate may be the most difficult part of the problem.

STEP 2. Add the vector forces on the object together to get the resultant (see Fig. 2-30*c*.) Then measure the resultant to obtain its magnitude, which here turns out to be 31 newtons.

STEP 3. Solve the equation $F = ma$, as we have done in previous examples, using in this case a force of 31 newtons and a mass of 100 kilograms:

$$
\begin{aligned}
F &= m \quad a \\
31 \text{ newton} &= 100 \text{ kilogram} \cdot a \\
\frac{31}{100}\,\frac{\text{newton}}{\text{kilogram}} &= a \\
a &= 0.31 \text{ newton/kilogram} = 0.31 \text{ meter/second}^2
\end{aligned}
$$

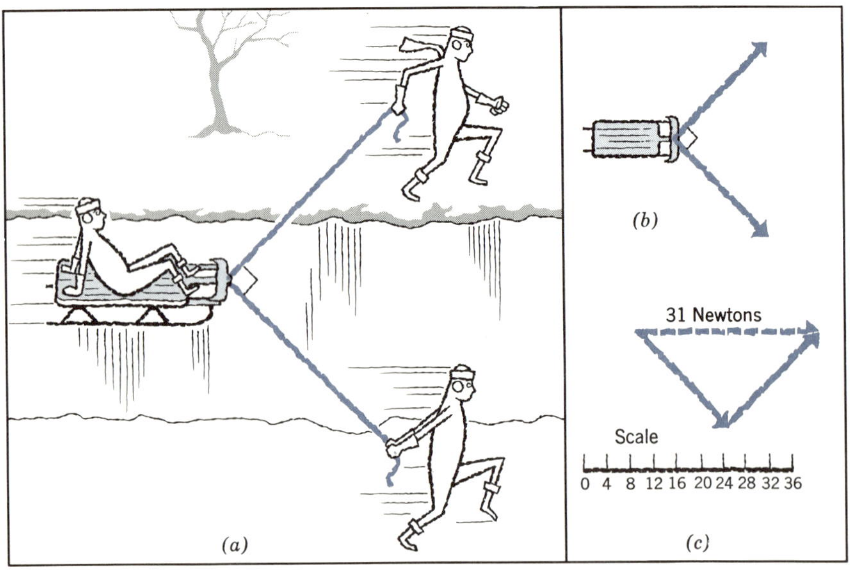

Fig. 2-30

*FINDING THE COMPONENTS OF A VECTOR USING DIAGRAMS.

EXAMPLE 7

A vector velocity is 3 meters per second southeast. Find its component along the north-south axis.

Solution. See Figure 2-31. First we draw the north-south axis. Then we draw in the vector so that its tail is on the north-south axis.

We have chosen a scale on which one inch equals two meters per second. Therefore, we draw the vector one and one-half inches long. Now we drop a perpendicular line to the axis from the head of the vector. It is essential that we make this line at right angles to the axis, but not necessarily at right angles to the vector. Then we draw in the component required, which is OP. We measure it and find that it is 1.05 inches long. Thus the component of velocity which we are looking for has a magnitude of 2.1 meters per second. We also give the direction of our component: south.

Summary

I. FUNDAMENTAL LAWS AND PRINCIPLES

A. THE SECOND LAW OF MOTION. This law includes the three proportionalities developed in the text while defining or discussing mass, force, and acceleration. $\mathbf{F} = ma$.

B. THE LAW OF CONSERVATION OF MASS

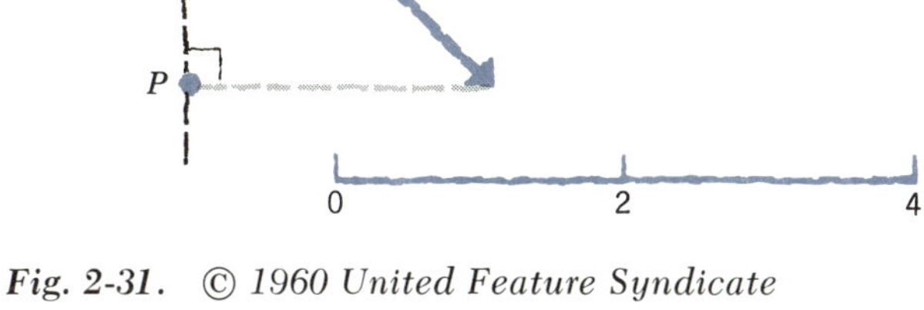

Fig. 2-31. *© 1960 United Feature Syndicate*

II. QUANTITIES

A. FUNDAMENTAL QUANTITIES

1. Mass, a scalar. The unit of mass is the kilogram.
2. Displacement, a vector, whose magnitude is distance.

B. DERIVED QUANTITIES

1. Acceleration, a vector. The unit of acceleration is the meter/second^2. Acceleration is the rate of change of velocity of an object.
2. Force, a vector. The unit of force is the newton, which is equal to a kilogram meter/second^2

C. VECTOR OPERATIONS

1. Vector addition.
2. * Other vector operations: obtaining the negative of a vector, subtraction, multiplication of a vector by a scalar, and obtaining vector components.

III. PROBLEM SOLVING

A. FORCE, MASS, ACCELERATION AND THE SECOND LAW OF MOTION

B. VECTOR OPERATIONS

Questions

1. A car is in an intersection facing east. Describe its motion under each of the following sets of conditions. In this question, vectors pointing east are considered positive and vectors pointing west are considered negative.

(a) Acceleration is zero; speed is positive.
(b) Acceleration is zero; speed is negative.
(c) Acceleration is zero; speed is zero.
(d) Acceleration is positive; speed is zero.
(e) Acceleration is negative; speed is positive.
(f) Acceleration is positive; speed is negative.

2. An object is accelerated by the only force acting on it until it reaches a speed of 20 meters per second. Then the force is removed. What would you expect the object to do after that?

3. Is mass related to size? Explain.

4. Describe a way to find the mass of an object. Show that your method is valid.

5. Suppose that you did not have a definition for mass in terms of a standard. By what procedure could you define force?
6. It is often said that "motion sickness" is responsible for making people carsick, airsick, etc. Is the term "motion sickness" precisely correct? Can you think of a more appropriate term? Does this have any bearing on the principle of relativity?
7. A certain automobile can reach a speed of 100 miles per hour from a stop in one minute if the automobile is pulling a trailer of equal mass. Estimate the time it will take the automobile to reach this speed if it is not pulling the trailer, assuming that the automobile exerts a constant force regardless of its speed or load. Justify your answer.
8. A rocket out in space becomes easier and easier to accelerate as it travels along. Why? (*Hint.* The same thing is true of an automobile traveling on a level road between stops for gasoline, although the effect is less marked.)
9. Why is acceleration a vector quantity?
10. How is it possible for two vectors to add up to zero if neither of them is equal to zero?
11. Suppose, for the moment, that all matter was made up of identical particles. How would you explain the fact that an object made of lead is more massive than an object of the same size made of wood?
12. List as many ways as you can think of to exert a force on an object.
13. Mass is conserved. List some processes in which mass appears to disappear as matter disappears or changes its form. Can you explain such situations?
14. How does the displacement from one position to another differ from the distance between the two positions?
15. Can we exert a single force on an object without changing its speed? Give an example where this occurs (approximately). Does this example disprove the second law of motion?
16. Why does the football player in Figure 2-32 tend to lose his balance and fall when the ball he intended to kick is unexpectedly removed?

*17. Prove the principle that a force has no direct effect on motion in a direction perpendicular to that force by using the information given about vector components.

*18. In Example 6, page 90, suppose that the boys are still pulling with equal lengths of rope, but that they are not opposite each other on the bank, so that the sled is not in the middle of the frozen stream. Which boy is pulling with more force, and why?

*19. A racing automobile is traveling around a circular track at a constant speed. What is its average velocity over one full lap?

Problems

1. An airplane lands on a runway at a speed of 6.0 meters per second. Half a minute later, it comes to rest. Assuming a constant rate of change, what is the acceleration of the airplane?
2. A car traveling at 60,000 meters per hour hears a friendly siren. Three minutes later it achieves the legal speed limit of 30,000 meters per hour. What was its acceleration?
3. If an object of mass m has an acceleration a when subject to a force F, what will be the acceleration of an object of $3m$ when it is subject to the same force?
4. An object has a mass of 0.25 kilograms and is subject to a force of 0.50 newtons. What is its acceleration?
5. An object has a mass of 7 kilograms and a constant acceleration of 3/7 meters per second. What is the force acting on it?

Fig. 2-32. Gospel According to Peanuts ©, *1960. Robert L. Short, United Features Syndicate.*

6. Find the sums of the following sets of vectors graphically.

(a) A vector southwest of magnitude 6 inches.
A vector northwest of magnitude 7 inches.

(b) A displacement north of magnitude 10 meters.
A displacement east of magnitude 5 meters.

(c) A force of 10 newtons north.
A force of 5 newtons north.

(d) A velocity of 9 meters per second east.
A velocity of 5 meters per second north.
A velocity of 7 meters per second north, northwest.

(e) A force of 10 newtons due north.
A force of 10 newtons 30 degrees south of east (one third the way from east to south).
A force of 10 newtons 30 degrees south of west (one third the way from west to south).

(f) A force of 5 newtons 30 degrees south of east.
A force of 7 newtons northwest.

7. An object is subject to three forces: a force f to the east, a force $2f$ to the north, and an unspecified force. The object does not change its motion as a result of the forces. Draw a diagram more or less to scale illustrating how you would find the third force, and estimate its magnitude and direction.

8. Figure 2-33 illustrates the parallelogram rule for addition of vectors. Show that it is equivalent to the triangle rule.

9. In Example 6, page 90, suppose that each boy is pulling on the rope with a force of 30 newtons, and that in addition there is a force of friction of 5 newtons between the sled and the ice, which tends to retard the sled. What is the acceleration of the sled under these conditions?

10. If we double the mass of a given object and triple the force on it, what happens to the acceleration?

*11. Find graphically the components of the following vectors on the north-south axis. North is positive and south is negative.

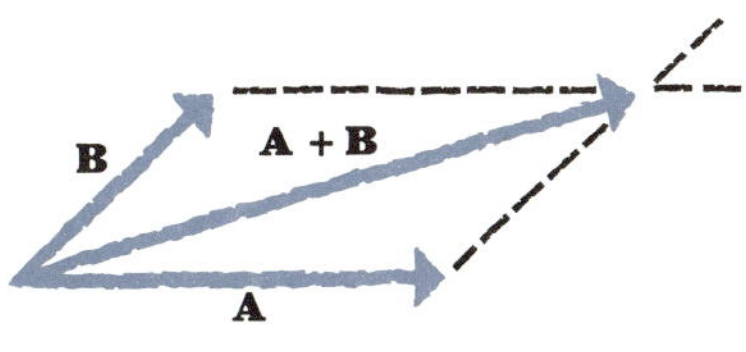

Fig. 2-33

(a) Ten newtons northwest.
(b) Seven newtons east.
(c) Six newtons south.
(d) Ten newtons southwest.

*12. Work Example 6, page 90, and Problem 9 above by finding the components of the forces in the appropriate direction.

*13. Find the acceleration of the bead on the rod in Figure 2-25, page 86, if the mass of the bead is 2 kilograms and the force on it is 4 newtons, acting at a 45-degree angle to the wire. You may solve the problem graphically. (See Appendix II-D for a discussion of the 45° right triangle.)

14. Show that the acceleration of an object is the same in all reference frames that are moving at a constant velocity with respect to each other. (Assume that the object is moving along a straight line in one, and therefore all, reference frames.)

3

THE THIRD LAW OF MOTION

INTERACTIONS

Whenever one object disturbs the motion of another object, it is itself disturbed. The *action* of one object on another is always accompanied by a *reaction* of the second object on the first; thus, we have an *interaction* between the two objects and both are disturbed. The action and reaction are simultaneous; we speak of the "first" and "second" object only for the sake of convenience. In this chapter we shall study the third law of motion, which governs such interactions. We will also introduce a new, derived, vector quantity, *momentum*, and show that it is conserved.

We all know better than to double up our fist and punch a brick wall. The action of our fist against the wall may or may not cause some damage to the wall, but the reaction of the wall on our fist will certainly give us some discomfort (see Fig. 3-1). When we push on a wall or any other object, the wall or object pushes back on us. For another example, consider our can-kicking friend illustrated in Figure 2-2 (p. 67). He cannot kick the can without receiving some reaction on his foot. If the can is massive and he kicks it hard, he may injure his foot or at least stub his toe. He cannot disturb the can with his foot without having the can disturb his foot in return.

As another example, suppose that there are two balls moving on a smooth (frictionless) surface, as shown in Figure 3-2. The balls would continue to move along straight lines if they did not collide. However, when the balls collide, the motion of both balls changes, not just the motion of one alone. The motion of each ball changes because of a force exerted on it by the other ball. In each interaction, then, there must be at least two forces.

THE THIRD LAW OF MOTION

Newton's third law of motion tells us exactly how the forces in an interaction are related. He stated this law as follows.

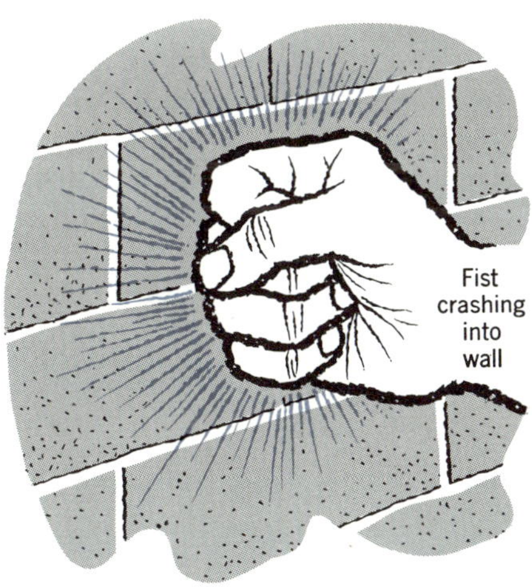

Fig. 3-1

To every action [force] there is always opposed an equal reaction [force]: or, the mutual actions of two bodies upon each other are always equal, and directed to contrary parts.[1]

This law tells us that if two objects interact, the forces they exert on each other are precisely the same in magnitude and exactly opposite in direction. The forces in an interaction never occur alone but always in pairs. We call such pairs *action-reaction force pairs*. The two forces in an action-reaction pair are equal in magnitude and opposite in direction. It is important to realize that when we are considering the forces of interaction between two objects, we never find that both forces of the action-reaction pair act on just one of the two objects. One of the forces acts on one object and the other acts on the other object. When we are concerned with the entire interaction it does not matter which of the two forces in the pair we call the action force and which the reaction force.

In summary, for two forces to make up an action-reaction pair they must fulfill the following requirements.

(a) The two forces must arise in the same interaction.
(b) The two forces must be of the same magnitude.
(c) The two forces must act in opposite directions.
(d) The two forces must act on different objects in the interaction.

[1]Isaac Newton, *The Mathematical Principles of Natural Philosophy and His System of the World*, translated by Andrew Motte in 1729, rev. by Florian Cajori, Berkeley: University of California Press, 1934, p. 13.

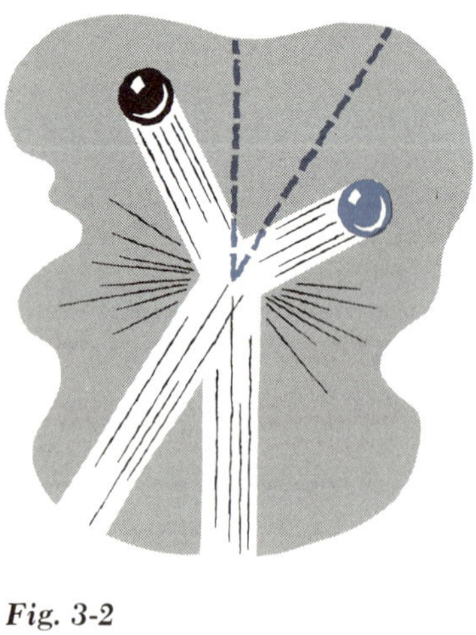

Fig. 3-2

Figure 3-3 illustrates these concepts in an interaction between two canoes, each bearing a passenger. In this case, each of the two objects is a system that includes both a canoe and its occupant. The two canoes in our illustration are of equal mass. A man is in one, and a girl of equal mass is in the other. Therefore, both objects or systems have the same mass. Suppose that the man reaches out his arm and pushes the other canoe away from him. This sets up an interaction between the two canoes. As a result of this interaction, not only the girl's canoe but the man's canoe also begins to move. The canoes move in opposite directions but at the same speed. In applying a force to the girl's canoe to set it in motion, the man automatically sets up an additional force which sets his canoe in motion at the same time. We can tell that the two forces are in opposite directions because the canoes accelerate in opposite directions. Therefore, these two forces fulfill the conditions of an action-reaction pair. They arise in the same interaction, they have the same magnitude, they are in opposite directions, and they act on different objects.

Suppose that a young man is throwing packages from a boat, as illustrated in Figure 3-4. This, of course, sets up an interaction between the man and the package. Assuming that the young man does not lose his balance and fall overboard, we can consider the man and boat as one object (system) and the package as the other. In order to throw the package to the dock, he must exert a force on it and thus set up an interaction. In this interaction his force on the package must be one of the two forces in an action-reaction pair. Therefore, there must be another force by which the package pushes back on him as he throws it. He feels this force in his hand as he throws the package, and it causes him (and the boat) to start moving away from the dock. As is required in any action-reaction force pair, the two forces act in opposite directions and on different objects. It is also true, as must be the case, that the magnitudes of the two forces are

(a) (b)

Fig. 3-3

equal. The reason that the motion of the package is changed more than that of the man and boat is that the package has much less mass.

This same principle can be used to explain the operation of a rocket. The rocket, of course, does not throw packages out behind it. It does, however, throw out large quantities of gas through the nozzle at the rear (see Figure 3-5). To expel the gas, the rocket must exert a force on it. Since this force must be part of a force pair, the gas exerts a similar force on the rocket, which pushes the rocket in the opposite direction from the gas. Again we have a typical situation. One force is exerted on the gas by the rocket to expel it. Another force of equal magnitude is exerted by the gas on the rocket in the opposite direction.

Momentum

Now we define a new, derived, vector quantity, called *momentum*. The momentum of any object is its mass multiplied by its velocity. The magnitude of the momentum of an object is its mass multiplied by its speed. The direction of the momentum of any object is the same as the direction of its velocity.

The letter **p** is commonly used to represent the momentum of an object. Using m to represent the mass of an object and **v** to represent its velocity, we may write

$$\text{momentum} = \mathbf{p} = m\mathbf{v}$$

(a) (b)

Fig. 3-4

Momentum is closely related to velocity, but it is more useful in some situations. Figure 3-6 illustrates a circumstance in which the momentum of an object is more important than its velocity alone. A small car is facing immediate interaction with a runaway truck. (A collision is a form of interaction.) In both of the cases illustrated, the velocity of the truck is the same. In the second case, however, a trailer

Fig. 3.5. Courtesy of NASA.

carrying 10 tons of steel is still attached to the cab. We recognize this case as far more frightening than the first. This illustrates that momentum, mass times velocity, is an important quantity in a collision.

Conservation of momentum

If we add up the momentum of all the objects involved in an interaction, we find that this sum is the same before, during, and after the interaction. The *total momentum* of the system of objects is *conserved* in an interaction. To illustrate this principle, let us refer again to Figure 3-3. Before the man exerts a force to push the girl's canoe away, both canoes are motionless in the water. In this condition each canoe has a momentum equal to zero because the velocity of each is zero. The total momentum for the overall system, which includes both canoes (and both people), is also zero, since it is obtained by adding together the momenta of the two objects. After the interaction, each canoe is moving, and therefore each has a nonzero momentum. Since the two systems have the same mass and speed, their momenta have the same magnitude. Since the canoes move in opposite directions, the resultant total momentum is zero. Thus momentum is conserved in this simple example.

We could, of course, do the same experiment with unequal masses, and momentum would still be conserved. Suppose, for example, that the girl in the second canoe has a mass several times the mass of the man. If the man then pushes the canoes apart, using the same force for the same length of time as he did in the first example, his canoe will move with the same speed as before, but the other canoe, since it is carrying more mass, will be accelerated to a lower speed. Never-

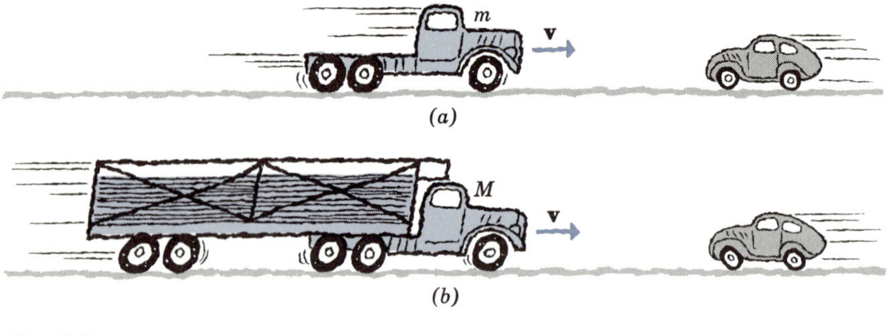

Fig. 3-6

theless, its momentum will still be equal in magnitude to the momentum of the man's canoe. The mass is greater, but the final speed is proportionally less, so that the momentum, the mass times the velocity, still has the same magnitude. Thus the total momentum of the whole system still adds up to zero. In this special case, the two objects in interaction have zero speed and therefore zero momentum to begin with. Had the canoes been moving before they were pushed apart they would probably have had some total momentum to begin with. In this case the total momentum after the interaction would not have added up to zero but, rather, to the same total momentum the two canoes had to begin with. The total momentum remains the same in an interaction.

The principle of conservation of momentum allows us to compare masses by a more direct means than we have used so far. Figure 3-7 illustrates one way in which this can be done. Two carts, carrying different masses, are held together by a string. When the string is cut, the spring exerts a force between the two carts, pushing them in opposite directions. We measure the speed of each cart after they separate. Since the total momentum is zero to begin with, it must be zero after the carts are set in motion. This means that the momenta of the two carts must be the same in magnitude and opposite in direction. Ignoring the directions, we write the equality that holds between the magnitude of the two momenta.

mass × speed of the cart on the left = mass × speed of the cart on the right.

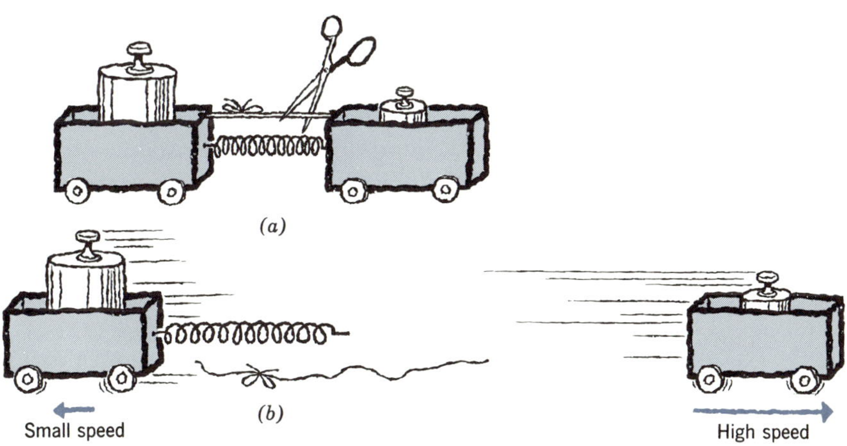

Fig. 3-7

If we use the letters M and S to represent the mass and speed of the cart on the left, and the letters m and s to represent the mass and speed of the cart on the right, we may write the above equation as follows:

$$MS = ms$$

or, dividing by S,

$$M = m(s/S)$$

Let us assume that we know m. Perhaps it is our standard mass. Then, by measuring the speeds s and S, we can find M, an unknown mass. A more accurate means of measuring mass will be explained later.

*Conservation of Momentum and the Laws of Motion

Newton originally stated the second law of motion in terms of momentum rather than acceleration. The law read as follows.

> The change of momentum is proportional to the motive force impressed; and is made in the direction of the straight line in which that force is impressed.[2]

This means that the force on an object is proportional (in our system of units, actually *equal*) to the rate of change of momentum of the object. Using **f** for force, m for mass, and **v** for velocity, we may express the law as follows:

$$\mathbf{f} = \text{rate of change of } (m\mathbf{v})$$

For most of the circumstances we have been discussing, the mass itself changes very little, if at all. Therefore, we have been concerned with changes of velocity. Under these circumstances, since the acceleration **a** is just the rate of change of the velocity **v**, we have:

$$\begin{aligned} \text{rate of change of } (m\mathbf{v}) &= m \times \text{rate of change of } \mathbf{v} \\ &= m\mathbf{a} \end{aligned}$$

Therefore, once again we obtain the law in the form more usually seen: $\mathbf{f} = m\mathbf{a}$.

[1] *Ibid.* Motte's translation uses the word *motion* rather than *momentum*. However, the quantity referred to is defined as we define momentum. We also use the term *straight line* rather than Motte's *right line*.

It is easy to prove that momentum must be conserved in interactions if Newton's laws hold. The third law of motion tells us that forces of interaction always occur in pairs. The two forces in such a pair have the same magnitude but opposite directions. According to the second law, each of these forces is equal to the rate at which it is changing the momentum of some object. Since the forces in a pair are equal in magnitude, they must produce a change of momentum of equal magnitude in the two objects they act on. However, the forces in a pair are in opposite directions. Therefore, the change of momentum caused by one of the forces is in the opposite direction to the change of momentum caused by the other force. The total momentum is not changed at all. Thus, Newton's laws of motion imply that momentum is conserved in interactions.

Problem solving

In this section we will illustrate the solution of problems involving calculation of the momentum of objects and systems and also problems dealing with the conservation of momentum.

Calculation of Momentum

Example 1

A car with a mass of 1500 kilograms is moving north at a speed of 32 meters per second. Calculate its momentum.

Solution. The momentum of an object is equal to its mass times its velocity, $m\mathbf{v}$. To find the magnitude of the momentum of the car, we multiply its mass, 1500 kilograms, by its speed, 32 meters/second:

$$\begin{aligned}\text{momentum} &= (1500 \text{ kilogram}) \left(32 \frac{\text{meter}}{\text{second}}\right) \\ &= 1500 \cdot 32 \frac{\text{kilogram meter}}{\text{second}} \\ &= 48{,}000 \frac{\text{kilogram meters}}{\text{second.}}\end{aligned}$$

The momentum of the car is 48,000 kilogram meter/second directed to the north.

Example 2

Three speedboats are engaged in a race, as shown in Figure 3-8. One boat has already passed the buoy and is on the return trip. The masses and speeds of the boats are as follows.

	Mass in Kilograms	*Speed in Meters/Second*
Boat 1	9.00×10^2	43.0
Boat 2	7.50×10^2	41.0
Boat 3	8.00×10^2	40.0

Calculate the total momentum of the system made up of the three speedboats.

Solution. The first step is to find the magnitude of the momentum of each boat. We do this by multiplying the mass of each boat by its speed.

Boat 1	900 kilogram · 43.0 meter/second = 38,700 kilogram meter/second
Boat 2	750 kilogram · 41.0 meter/second = 30,800 kilogram meter/second
Boat 3	800 kilogram · 40.0 meter/second = 32,000 kilogram meter/second

(Since the data were given with three significant figures, we have retained three significant figures in the results.)

Fig. 3-8

Now that we have found the magnitude of the momentum of each boat, we must specify the direction. We see from Figure 3-8 that the race is in an east-west direction. We designate the eastward direction as positive, which makes the westward direction negative. Therefore, boats 2 and 3 have a positive momentum and boat 1 has a negative momentum. Finally, we obtain the total momentum of the system by adding together the momentum of the three boats. The momentum of boats 2 and 3 added together is

$$(30{,}800 + 32{,}000)\ \frac{\text{kilogram meter}}{\text{second}} = 62{,}800\ \frac{\text{kilogram meter}}{\text{second}}$$

We now add to this sum the momentum of boat 1. This *decreases* the total, since the momentum of boat 1 is negative:

$$(62{,}800 - 38{,}700)\ \frac{\text{kilogram meter}}{\text{second}} = 24{,}100\ \frac{\text{kilogram meters}}{\text{second}}$$

Therefore, the total momentum is 24,100 kilogram meters/second to the east.

Example 3

Figure 3-9*a* shows two airplanes, one flying north and the other flying west. The northbound airplane has a momentum of 50,000 kilogram meters/second and the westbound airplane has a momentum of 87,000 kilogram meters/second. Find graphically the total momentum of the system that includes both airplanes.

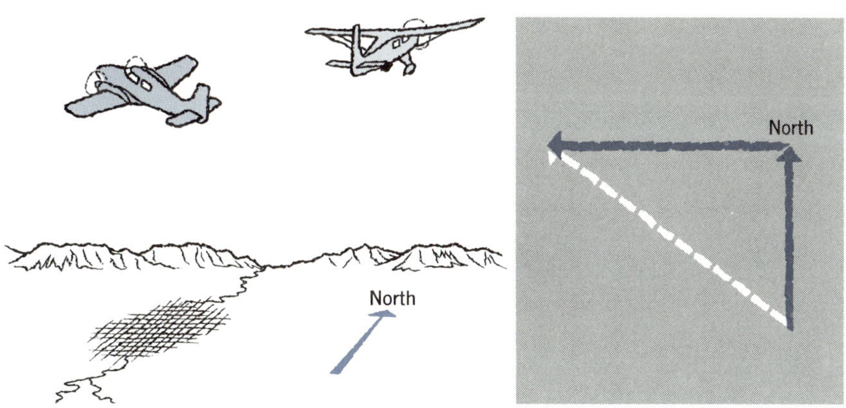

(*a*) (*b*)

Fig. 3-9

Solution. This problem requires the addition of two vectors. In Figure 3-9*b* we have represented the momentum of each of the airplanes by an arrow drawn to scale. We obtain the resultant of these two vectors by using the "head to tail" method described in Chapter 2. Measuring the length of the resultant arrow, we find that the magnitude of the total momentum is equal to 100,000 kilogram meters/second. The direction of the momentum is approximately west northwest. If we were to measure it more accurately with a protractor, we would find this direction to be very close to 60 degrees west of north.

Conservation of Momentum

The basic procedure in solving problems that involve the principle of conservation of momentum is to set the total momentum of a system of objects before an interaction equal to the total momentum of the same system after the interaction. This procedure can be broken down into the following steps.

STEP 1 Draw a diagram including a reference frame that shows the condition of the objects both before and after the interaction.

We must choose two situations (1) and (2) which represent the "before" and "after" conditions. The second situation will usually involve what we want to know. The first situation will usually, but not always, involve what we already know.

STEP 2 Write the law of conservation of momentum in equation form so that it will include all the objects in the interaction. Ordinarily, one side of the equation will give the total momentum of the system before the interaction and the other side will give the total momentum of the same system after the interaction. That is,

$$\text{total momentum before} = \text{total momentum after}$$

STEP 3 List the known and unknown quantities.

STEP 4 Substitute the known quantities of Step 3 into the equation obtained in Step 2 and solve the equation for the unknown.

These steps are common to the solution of most problems dealing with the conservation of momentum. However, some problems will require greater ingenuity for their solution.

EXAMPLE 4

An object with a mass of 5 kilograms is traveling to the right on a level surface at 10 meters/second. This object strikes a second object of mass 20 kilograms at rest. The first mass rebounds at a speed of 2 meters/second. What is the resulting velocity of the second mass? (Disregard the effect of friction between the bodies and the surface.)

Solution.

STEP 1 See Figure 3-10. We have chosen to designate the direction to the right as positive. The directions of motion of the objects are indicated by the arrows. The two diagrams represent conditions before and after the collision.

STEP 2 We are concerned only with motion in the right and left direction. Since there is no friction, the only forces (or components of forces) in this direction are the forces between the two objects during the collision. Therefore, as far as the left-right direction is concerned, there is only one interaction and it involves only the two objects. We may then write the principle of conservation of momentum as follows:

$$m_1v_1 + M_1V_1 = m_2v_2 + M_2V_2$$

The small letters represent the first object and the capital letters represent the second. The subscript 1 refers to conditions before the collision, and the subscript 2 refers to conditions after the collision.

In this problem, and in many others, the masses of the two objects are the same before and after the collision; the equation may therefore be written more simply as follows:

$$mv_1 + MV_1 = mv_2 + MV_2$$

In writing this equation it is important that none of the objects we are considering has a force of interaction (or component of a force of

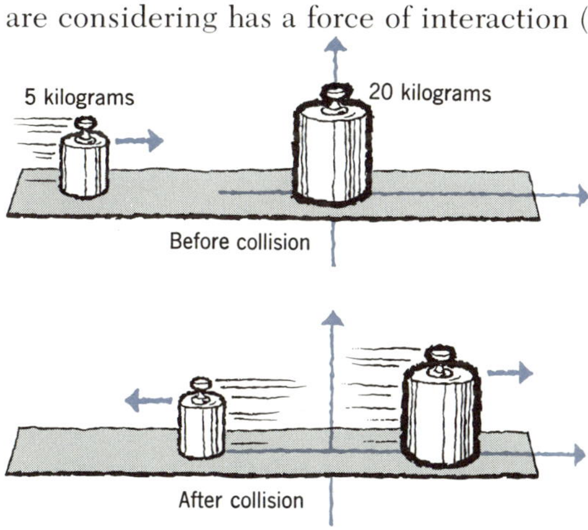

Fig. 3-10

interaction) in the direction in which we are interested with any object that is not included in the equation.

STEP 3 We list the known and unknown quantities in the equation of Step 2.

$$\begin{aligned} m &= 5 \text{ kilograms} \\ M &= 20 \text{ kilograms} \\ v_1 &= 10 \text{ meter/second} \\ v_2 &= -2 \text{ meter/second} \\ V_1 &= 0 \\ V_2 &= ? \end{aligned}$$

Notice that v_2 is negative, since the object rebounded to the left.

STEP 4 We enter these values in the equation and solve for V_2:

$$(5 \text{ kilogram}) \left(10 \frac{\text{meter}}{\text{second}}\right) + (20 \text{ kilogram})\ (0) = (5 \text{ kilogram}) \left(\frac{-2\text{meter}}{\text{second}}\right) + (20 \text{ kilogram})\ V_2$$

$$-10 \text{ kilogram} \frac{\text{meter}}{\text{second}} + 20V_2 \text{ kilogram} = 50 \text{ kilogram} \frac{\text{meter}}{\text{second}} + 0$$

$$20V_2 = 50 \frac{\text{meter}}{\text{second}} + 10 \frac{\text{meter}}{\text{second}} = 60 \frac{\text{meter}}{\text{second}}$$

$$V_2 = \frac{60}{20} \frac{\text{meter}}{\text{second}} = 3 \text{ meters/second}$$

EXAMPLE 5

A mass $3m$ with speed 100 meters per second strikes a mass m at rest and adheres to it; both move off together. There is no friction from the level surface they are sliding on. What is the velocity of the combined mass after the collision?

Solution.

STEP 1 See Figure 3-11.

STEP 2 The only interaction involving these objects in the right-left direction is the collision. Therefore, momentum is conserved for the system composed of these two objects as far as the right and left motion is concerned. We write the equation as follows:

$$m_1 v_1 + M_1 V_1 = m^* v^*$$

Here, the combined mass acts as a new particle. Thus, there is only one object of concern in the second condition (Fig. 3-11*b*). We let m^* and v^* represent the mass and velocity, respectively, of this new object.

STEP 3 $m_1 = 3m$
$M_1 = m$
$m^* = 3m + m = 4m$
$v_1 = 100$ meter/second
$V_1 = 0$
$v^* = ?$

STEP 4 We put these values in the equation and solve for v^*:

$$(3m)\ 100\ \frac{\text{meter}}{\text{second}} + m\ (0) = 4mv^*$$

$$4mv^* = 300m\ \frac{\text{meter}}{\text{second}}$$

$$v^* = \frac{300}{4}\ \frac{\text{meter}}{\text{second}} = 75 \text{ meters/second}$$

EXAMPLE 6

Two flatcars collide. The car on the left, initially traveling at a speed v, strikes the second car, which is at rest. As a result of the collision, the first car shifts its load of mass m onto the second car. The second car plus load then moves at a speed V. Find the velocity of the first flatcar after the collision. The two flatcars are of equal mass m when empty.

Solution.

STEP 1 See Figure 3-12.

STEP 2 If we ignore the friction with the rails, there are no outside forces on the system comprising the load and the two flatcars as far as the direction of motion is concerned. Therefore, for this system in this direction, momentum is conserved.

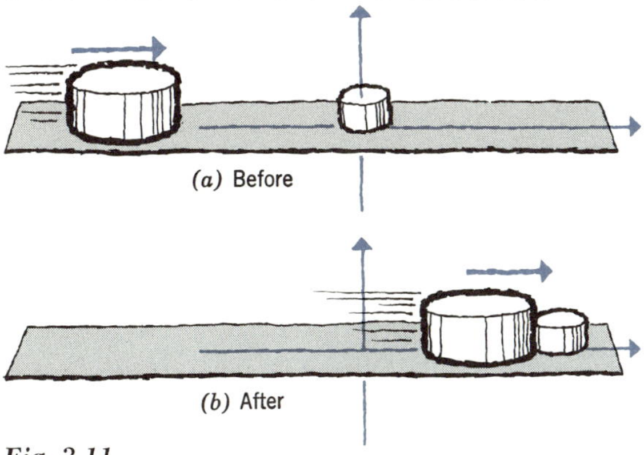

Fig. 3-11

We can choose to treat the total system as three objects that do not change their mass (the two cars and the load) or as two objects that do change in mass. We choose to use the latter technique. Thus we have

$$m_1v_1 + M_1V_1 = m_2v_2 + M_2V_2$$

STEP 3
$m_1 = m + M =$ mass of first car plus load
$m_2 = m =$ mass of first car without load
$M_1 = m =$ mass of second car without load
$M_2 = m + M =$ mass of second car with load
$v_1 = v =$ velocity of first car before collision
$V_1 = 0 =$ velocity of second car before collision
$V_2 = V =$ velocity of second car after collision
$v_2 = ? =$ velocity of first car after collision

STEP 4 Putting in the known values and solving for v_2, we have

$$(m + M)\,v + m\,(0) = mv_2 + (m + M)\,V$$
$$mv_2 + (m + M)\,V = (m + M)v$$
$$mv_2 = (m + M)\,v - (m + M)\,V$$
$$mv_2 = (m + M)\,(v - V)$$
$$v_2 = \frac{(m + M)\,(v - V)}{m}$$

Summary

I. FUNDAMENTAL LAWS AND PRINCIPLES

A. THE THIRD LAW OF MOTION. Forces in an interaction occur in pairs. The forces in a pair are equal in magnitude, opposite in direction, and are exerted on different objects.

B. CONSERVATION OF MOMENTUM

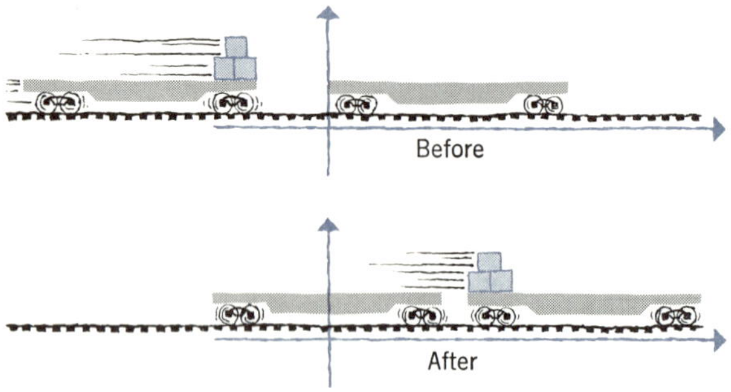

Fig. 3-12

II. QUANTITIES

A. MOMENTUM; a derived vector quantity, equal to mass times velocity.

III. PROBLEM SOLVING

A. CALCULATING MOMENTUM

B. CONSERVATION OF MOMENTUM

Questions

1. Why can I hurt my foot when I kick a rock?
2. Why does a gun recoil?
3. Explain the operation of a rocket.
4. In the example illustrated in Figure 3-4, in which a man is throwing a package from a boat, explain what forces are operating in that act and how they are related.
5. Show that in an interaction between two objects the total momentum would not necessarily remain the same if the third law of motion did not hold.
6. If the man in Figure 3-3 were able to push very quickly on the other canoe, could he exert a greater force on it than it exerts on him? Explain.
7. A man, fully dressed, is standing in the middle of a frictionless ice sheet. How can he get off the ice?
8. Would you expect a rocket to work better in an atmosphere or outside an atmosphere? Why?
9. Why would it not be possible for a rocket to have some device to catch the escaping gases after they had slowed down and use them over again, so that it would not have to take so much fuel along?
10. An astronaut drifting in space has a lead weight fastened to a rope. Can he move himself through space by throwing out the weight and then hauling it back in again by the rope? Explain.
11. A man claims to have perfected a device which, when operating on internal batteries, will rise off the ground, hover, and move around. The box is sealed, nothing enters or leaves it, and it does not exert any forces on its surroundings. Would it be wise to invest money in this invention? Explain.
12. A bullet strikes a block of wood resting on a frictionless surface. Under which of the following conditions will the block receive the greatest

velocity, and under which the least velocity, and why? (Assume that the bullet was initially moving horizontally.)

(a) The bullet passes through the block.

(b) The bullet gets stuck in the block.

(c) The bullet bounces backward off the block.

*13. Under what circumstances is the momentum of a system conserved?

*14. When a snowball strikes a wall, it appears to stop dead, losing its velocity and, therefore, its momentum also. Does this mean that the momentum of the snowball is not conserved? Explain.

Problems

1. Calculate the momentum of a rifle bullet of mass 0.010 kilograms traveling 400 meters per second.
2. Calculate the momentum of an automobile of mass 1200 kilograms traveling 30 meters per second.
3. An automobile of mass 1300 kilograms traveling at 20 meters per second strikes head on another automobile of mass 1500 kilograms traveling at 15 meters per second. What is the total momentum of the system comprising the two cars?
4. In Problem 3, suppose that the two cars lock bumpers in the collision. Which one continues to move in its original direction?
5. If, in the example shown in Figure 3-3, the man pushes on the girl's canoe with a force of 10 newtons northeast, what other force must be involved in the interaction and what does it act on?
6. A bomb explodes into four pieces of equal mass. They thereupon travel at the same speed, each in one of the directions of the compass, north, south, east, and west. What is the total momentum of the system if the mass and speed of each piece are m and s, respectively?

More Difficult Questions and Problems

1. A mass of velocity v traveling on a frictionless surface strikes an equal mass and sticks to it. What is the resulting velocity of the two masses?
2. What is the velocity of the combined masses in Problem 1 if the second mass is twice the magnitude of the first mass?
3. Solve Problem 1 for an arbitrary ratio of masses m/M.

4. A stationary bomb explodes into three equal masses. One travels north at a speed v; one goes east at a speed v. In what direction and at with what speed does the third one go?

5. A man of mass m is standing on a frictionless ice sheet. He catches a girl of mass M sliding by with a velocity v. He heaves her back in the direction she came from at the same speed she arrived with. What is his final speed?

6. A moving block of speed V and mass m strikes a stationary block of mass M. As a result of the collision, the first block receives a speed $V/2$ in the opposite direction. What is the final velocity of the block of mass M? The two blocks are sliding without friction on a level surface.

7. A body of mass 10 kilograms traveling to the right at 20 meters per second strikes a second body of mass 40 kilograms at rest. The first mass rebounds at a speed of 4 meters per second. What is the resulting velocity of the second mass? Disregard friction.

8. A loaded rifle is lying on a frictionless ice sheet next to a block of wood of mass 3 kilograms. The gun spontaneously discharges, shooting a bullet which lodges in the block. The bullet has a mass of 0.03 kilograms and comes out of the rifle at a speed of 600 meters per second. The rifle has a mass of 4 kilograms. What is the final speed of each item? (Disregard the momentum of the hot gases from the burnt powder.)

9. An astronaut separated from his space capsule sees it drifting away from him at the speed of 0.1 meter/second. How fast must he throw away a 1-kilogram wrench to keep up with the capsule, and in what direction must he throw it? The astronaut, dressed for space, has a mass of 110 kilograms.

10. The cannon shown in Figure 3-13 fires a ball at a speed S. The cannon has mass M and the ball mass m. What is the recoil speed of the gun?

11. Two beads are free to slide without friction on a wire, as shown in Figure 3-14. Mass m is 10 kilograms, and M is 20 kilograms. The lighter mass has a velocity of 14 meters per second to the right, and the other has a velocity of 7 meters per second to the left. The two beads adhere when they collide. What is the resulting speed of the combined beads?

Fig. 3-13

Fig. 3-14

12. Two objects travelling at the same speed collide. One has three times the mass of the other. As a result of the collision, the object with the smaller mass doubles its speed. Find the speed of the larger mass after the collision.

*13. A cannon of mass M, shown in Figure 3-15, is fired at a 45-degree angle. The cannonball, of mass m, comes out at a speed v. What is the recoil speed of the cannon assembly?

*14. Suppose that a force f acts on a free object for a time t. How much momentum does the object gain?

15. Show that the laws of motion, as expressed, obey mirror symmetry.

16. Show that the laws of motion, as expressed, obey position and orientation symmetry.

17. Show that the laws of motion, as expressed, obey the special principle of relativity.

Fig. 3-15

4

GRAVITATIONAL INTERACTION

Newton was the greatest genius who ever lived, and the most fortunate, for there cannot be more than once a system of the world to establish. *(Joseph LaGrange)*

SEVERAL FUNDAMENTAL LAWS have been discovered that describe how objects or particles of matter interact with one another. One of these is Newton's law of gravitation, which describes the forces that arise because of the masses of objects. Through these gravitational forces, each particle of matter attracts every other particle of matter. The laws of motion which we have presented in the last three chapters allow us to predict the motion of an object when the forces acting on it are known. The law of gravitation describes some of these forces. Gravitational forces hold celestial objects such as the planets in their orbits and also influence objects thrown or shot by people on the surface of the earth. The behavior of these objects can be predicted with great accuracy by using the law of gravitation together with the laws of motion.

In this chapter we shall discuss the law of gravitational interaction. We shall also discuss circular motion at constant speed, which is approximately the motion of the planets in their orbits around the sun and the moon in its orbit around the earth. This leads us to Newton's system of the world."

THE LAW OF GRAVITATION

According to Newton's law for the gravitational interaction, every material object attracts every other material object. This means that every particle of matter pulls and is pulled by every other particle of matter. In Figure 4-1 we have diagramed two objects, each pulling on the other with gravitational force. We emphasize that the magnitudes of these forces are exactly the same for both objects. The sun pulls on the earth, and the earth pulls back on the sun with a force of equal magnitude. The earth pulls down on rocks and men, and rocks and men pull up on the earth. You might not think that a pebble pulls as hard on the earth as the earth pulls on the pebble, but it does. Of course, the force on the pebble has a greater influence on its motion than the force on the earth has on *its* motion because the mass of the pebble is so much smaller than the mass of the earth. The force on each object in Figure 4-1 points directly toward the other object. In fact, the two forces, one on one object and one on the other, are an action-reaction force pair and together make up the gravitational interaction between the two objects.

Several factors affect the strength of the gravitational forces. Among these factors are the masses of the two objects and the distance between them. If we increase the mass of either object, we *increase* the attractive force on both objects. For example, if we triple one of the masses we triple the force on both objects. Therefore, the force is proportional to either one of the masses. If we increase the distance between the two objects, we *decrease* the attractive force on both objects. In fact, the force between the two objects decreases very rapidly as we increase the distance between them. If, for example, we triple the distance between the two objects we find that the force is *not* 1/3 of what it was before; rather, it is $(1/3)^2$, or 1/9th, of its former value.

These relationships between force, mass, and distance can be expressed in a single formula:

$$F_g = \frac{Gm_1 m_2}{d^2} \tag{1}$$

(The quantity G, equal to 6.67×10^{-11} newton meters^2/kilograms^2, will be discussed in the next section.) This equation gives the magnitude of the gravitational force on either object. Using this formula, we can calculate the gravitational force between any two objects when we know the mass of both and the distance between them.

In order to apply this formula to large celestial bodies such as the earth and the sun, Newton had to show that the distance d should be taken as the distance between their centers. This may be done when dealing with spherical objects whose mass is distributed in a sufficiently uniform manner around the center. It cannot be done in the unusual circumstances when one body is inside the other.

The Gravitational Constant G

Unlike the other quantities in Equation 1, which may have different values in different situations, the quantity G always has the same value. Quantities that always have the same value are called constants, or constant quantities; hence, G is called the gravitational constant. This quantity is the first of several *fundamental constants* which we will introduce.

The importance of the fundamental constant G can be understood by examining Equation 1, which gives the magnitude of the forces in the gravitational interaction. Suppose, however, the constant G were twice as big as it actually is. Then, according to the equation, the

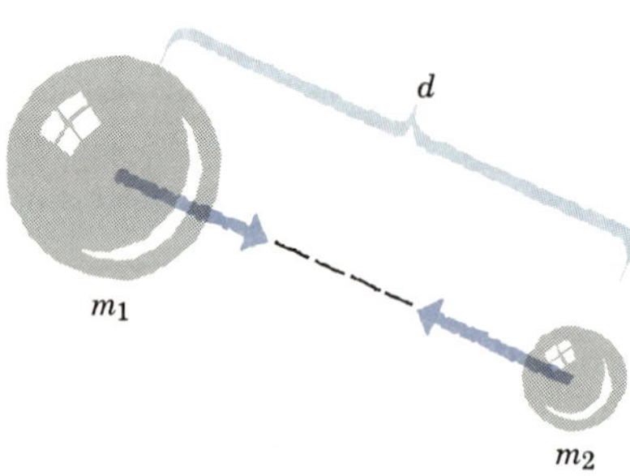

Both forces are of magnitude F_g

Fig. 4-1

gravitational forces would have twice the magnitude that they actually have. From this we see that, in a given situation, the value of the constant G fixes the magnitude of the forces in the gravitational interaction. We can imagine another universe in which this constant, and therefore all the gravitational forces, might be different. In our universe, however, this quantity is always the same, as far as our experiments tell us. Later, we shall be concerned with other interactions and force laws whose strength is governed by other fundamental constants. We find that the gravitational interaction is the weakest interaction that has been yet observed.

The gravitational force is so weak that for most practical purposes we ignore it, unless at least one of the objects in the interaction has a mass as great as that of the earth, the moon, or some other celestial object. However, by using a very sensitive device, we can measure gravitational forces between objects small enough to hold in your hand. In so doing, we can also measure the value of the constant G.

Such a device, called the Cavendish balance, is shown in Figure 4.2. The masses A and B are fixed to the ends of a thin, stiff rod, suspended by a thin wire, as shown. If two additional masses C and D are placed near the masses A and B, they will pull on these masses because of gravitational forces. This will cause masses A and B to move slightly, twisting the wire. In practice, the masses C and D are first put in one position, which twists the wire one way, and then moved to the other position, shown by the dotted lines in the figure. This twists the wire in the opposite direction. In order to detect the amount of twist, which is very small, a mirror is mounted on the wire. As the wire is twisted, the mirror rotates. A beam of light

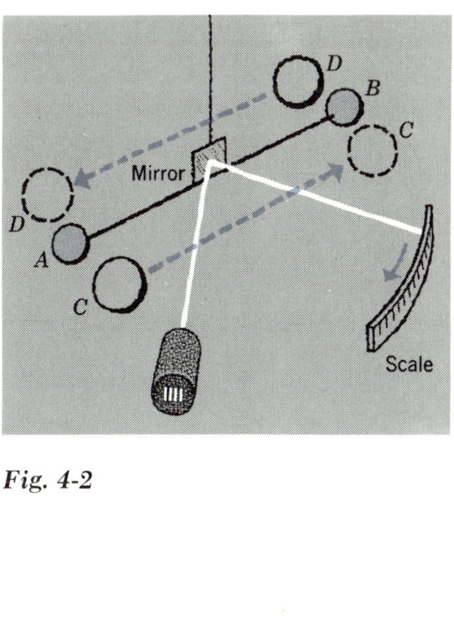

Fig. 4-2

reflected from the mirror moves along a scale as the mirror turns. This enables us to detect and measure the amount of "twist" in the wire caused by the gravitational force. From this twist we can actually evaluate the gravitational force, since we know from various prior experiments how much force it takes to twist the wire. Knowing the gravitational force, the masses of the balls, and the distance between them, we can calculate the gravitational constant G, using Equation 1.

Weight

We define the weight of an object to be the gravitational force acting on it. For an object on the surface of the earth, the weight is essentially the gravitational pull of the earth on the object. On the moon, however, the weight of the object is essentially the gravitational pull of the moon acting on it. Since these gravitational pulls are not equal, the same object has a different (smaller) weight on the moon than on the earth. Note that the weight of an object is *not* an invariant property of that object. The weight of an object is different in different places in the universe and depends upon its position with respect to other massive objects. In fact, the weight of an object is even slightly different at different places on the surface of the earth (see Fig. 4-3).

The mass of an object is different from its weight. We shall consider the mass of an object to be the same no matter where in the universe the object is located. Therefore, we emphasize that weight and mass are distinctly different quantities. The weight of an object is a force —in fact, the gravitational force on that object. Mass, on the other hand, is a fundamental quantity. The mass of an object determines

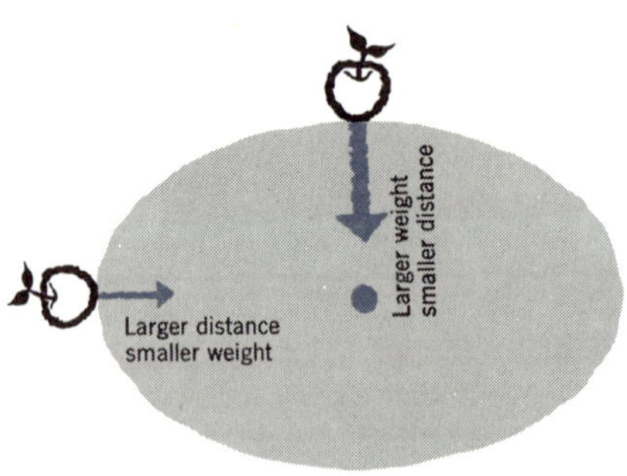

Fig. 4-3

the acceleration produced by a force, whether it be the force of gravitation or any other force (see Chap. 2, pp. 66–69).

Even though mass and weight are different quantities, they are closely related in practical applications. The most practical (and most accurate) method used to determine the mass of an object is the process known as "weighing." In this process we normally obtain the mass of an object by actually measuring its weight. This process works because, at a particular place, the mass of a given object is proportional to its weight. At the same location, an object with twice that mass has twice the weight, and so on.

We shall illustrate the process of weighing by discussing the balance shown in Figure 4-4. Similar balances are used in scientific work to measure masses to an accuracy of one ten-millionth of a kilogram.

The beam with its two pans is balanced on a knife edge. The pans are suspended from the beam at equal distances from the knife edge. In the process of weighing, an unknown mass is placed in one pan and known masses are placed in the other pan until the system is again balanced. This will occur when the gravitational force on the unknown mass is the same as the gravitational force on the known masses. In this process of weighing, therefore, we are comparing the weight (gravitational force) of the known mass with the weight (gravitational force) of the unknown mass. Since, for a given location, the mass is proportional to the weight, the masses will be equal when their weights are equal. (There are other weighing devices and

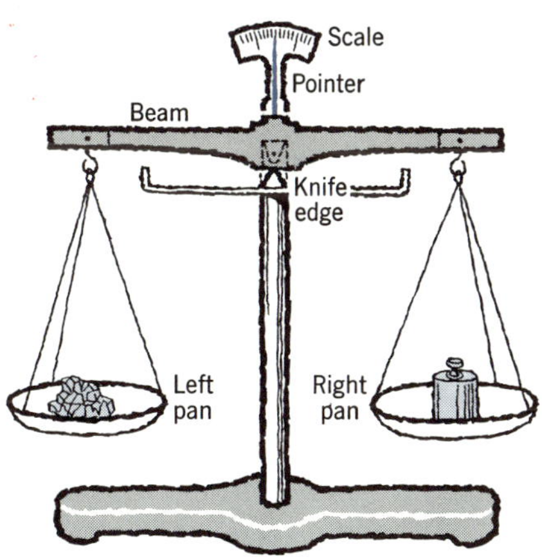

Fig. 4-4

also other ways to use this device that do not require such symmetry in construction.)

We can easily show that the weight of an object is proportional to its mass by looking again at Equation 1, which describes the forces in a gravitational interaction. We write the equation so that the first mass is the mass of the earth, m_E, and the second mass is the mass of the object, m_{ob}. We place our weighing apparatus on the surface of the earth; therefore, we take the distance d to be the radius R of the earth—the distance from the center of the earth to its surface. Thus, we obtain for the force of gravity on the object, the expression

$$F_g = \frac{G \cdot m_E \cdot m_{ob}}{R^2} = \left(\frac{G \cdot m_E}{R^2}\right) m_{ob} \qquad (2)$$

In this last expression, the quantities within the parentheses are all constant as long as we do not move from our position on the surface of the earth. As long as we stay in this one place, therefore, the weight of the object, F_g, is equal to a constant $\left(\frac{Gm_E}{R^2}\right)$ multiplied by the mass of the object, m_{ob}. This means that the weight F_g is proportional to the mass m_{ob}.

Two Kinds of Mass or One?

In the preceding section, we discussed another and more practical method of measuring mass, weighing. This method is based on the law of gravitational interaction and apparently has nothing to do with the laws of motion. On the other hand, the other methods of measuring mass, described in Chapters 2 and 3, are deeply involved in the development and results of the laws of motion and appear to be completely independent of the law of gravitation. Thus, it seems that we have two wholly different and independent ways of measuring mass experimentally. It is very important to find out whether these two methods of measuring mass give the same results. If they do not, then we must have two kinds of mass—gravitational mass, obtained by weighing, and ordinary mass, obtained by the other methods. In fact, it turns out that, according to very accurate experiments, the gravitational mass of an object is equal to its ordinary mass. Therefore, to the limits of accuracy of our experiments, we claim that the two possible kinds of mass are really the same. (Actually, we assumed this to be the case when we stated the law of gravitation, and we shall continue to assume it throughout this book.)

The fact that ordinary mass and gravitational mass are the same has this very interesting consequence: all objects (in the same position with respect to the earth) receive the same acceleration from the gravitational pull of the earth, *regardless of their mass*. This means that a feather and an elephant dropped at the same time from the same height will hit the ground at the same time, in the absence of air friction and other interfering forces (see Fig. 4-5). Of course, if we actually tried this experiment the elephant would hit the ground long before the feather because the retarding effect of the air is more important for the feather than for the elephant. The retarding effect of the air is also important in the time of descent for the two men shown falling in Figure 4-6. When, in the absence of air, gravitation acts alone, skydivers, parachutists, feathers, and elephants all fall together with the same acceleration. This acceleration is called the gravitational acceleration. Near the surface of the earth the gravitational acceleration is about 9.8 meters/second². We shall use the symbol a_g for this quantity:

$$a_g = 9.8 \frac{\text{meters}}{\text{second}^2}$$

Air resistance is often negligible for such objects as marbles, billiard balls, elephants, and other things much more compact than feathers, falling for short distances at low speeds. These objects can therefore be used to demonstrate that objects of different mass fall with the same acceleration. We can do better than this, however, by using the apparatus shown in Figure 4-7. Nearly all the air can be removed from the large transparent tube by means of an ordinary air

Fig. 4-5

pump. Once we have removed the air, we can actually observe the coin and the feather inside the tube falling together side by side. Galileo and others, from their study of physics, realized that this would occur in a vacuum although air pumps had not yet been invented to create one.

The common acceleration of gravity, a_g, can be obtained from the laws of motion and the law of gravitational force. As before, we write the law of gravitational force for an object falling near the surface of the earth (see Equation 2). The gravitational force on such an object is

$$F_g = \left(\frac{G \cdot m_E}{R^2}\right) m_{ob} \tag{3}$$

This force is the cause of the observed acceleration of the falling object. A falling object, of course, falls with the acceleration a_g. According to the second law of motion ($F = ma$), the force and the acceleration must obey the equation

$$F_g = m_{ob} \cdot a_g \tag{4}$$

Since it is the same force that appears in the two equations, and

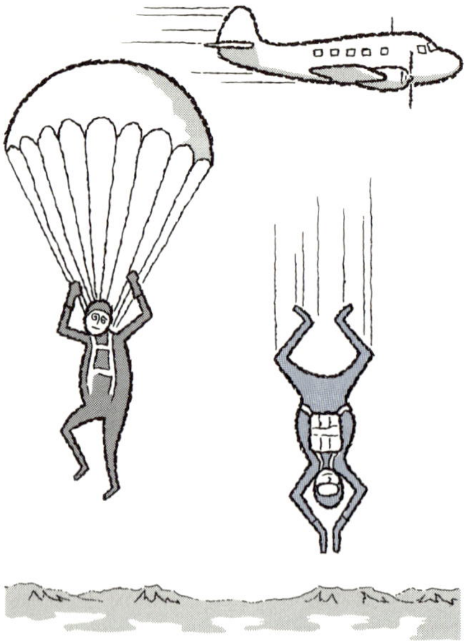

Fig. 4-6

since quantities that are equal to the same thing are equal to each other, we can write

$$m_{ob} \cdot a_g = \left(\frac{Gm_E}{R^2}\right) m_{ob} \tag{5}$$

If we divide both sides of the equation by m_{ob}, the result is

$$\frac{m_{ob} \cdot a_g}{m_{ob}} = \left(\frac{Gm_E}{R^2}\right) \frac{m_{ob}}{m_{ob}} \tag{6}$$

$$a_g = \left(\frac{Gm_E}{R^2}\right) \tag{7}$$

Thus the mass of the falling object disappears from the equation entirely; the acceleration depends only on the quantities on the right-hand side of the equation, not on the mass of the falling object. The acceleration is therefore the same for all objects regardless of mass.

If there were a difference between gravitational and ordinary masses, we could not cancel the mass m_{ob} from the equation as we did above, since the symbol would refer to two different quantities

Fig. 4-7

on the two sides of Equation 5. This would mean that we could not predict that different objects would fall with the same acceleration. Since our experiments indicate that they do, we are strengthened in our belief that the two kinds of mass are really the same.

THE LAW OF GRAVITATION AND "THE SYSTEM OF THE WORLD"

After Newton had developed his laws of motion and his law of gravitation, he proceeded to apply them by demonstrating "the frame of the system of the world."[1] Before discussing the motion of the planets in this system, let us consider circular motion.

Circular Motion at Constant Speed

In Chapter 2 we pointed out that we must exert a force on an object if we wish to change its direction of motion. Force is necessary to change the direction even if the speed of the object remains the same. In this section we will discuss a very special case of accelerated motion: motion in a circle with a constant speed.

In Figure 4-8, which illustrates this kind of motion, a man is hold-

Fig. 4-8

[1]Isaac Newton, *The Mathematical Principles of Natural Philosophy and His System of the World*, translated by Andrew Motte 1729, rev. by Florian Cajori, Berkeley: University of California Press, 1934, p. 397.

ing onto one end of the rope and his son, sitting in a pan, is holding the other end. We assume that the ice is so smooth that friction does not slow down the boy's motion. As the boy moves in a circle his speed remains constant. The direction of his velocity, of course, is changing all the time—that is, as long as both man and boy hang onto the rope. The force causing the change in direction is the force exerted through the rope on the boy. This force is along the direction of the rope and pointed toward the center of the circle. The boy would not move in a circle if this force were not maintained. If, for example, either the boy or his father lets go of the rope, the boy will cease moving in a circle and start moving in a straight line, in the direction in which he was moving at the instant that he or his father let go of the rope.

An object moving in a circle with constant speed is changing its direction of motion at all times, and a force acting on it at all times is necessary to maintain this kind of accelerated motion. This force must act toward the center of the circle. Whenever a force causes this kind of motion, it is called a *centripetal* force. The centripetal force is the force actually exerted on an object that is performing circular motion. In the previous example, the centripetal force is the force exerted on the boy. It is *not* the force exerted on his father. As we have stated, the direction of the centripetal force is always toward the center of the circle. This means that it is perpendicular to the velocity. Therefore, the direction of the centripetal force is also always changing as the object moves in a circle (see Fig. 4-9). The magnitude of the centripetal force, however, does not change as long as we are dealing with an object with the same mass traveling in

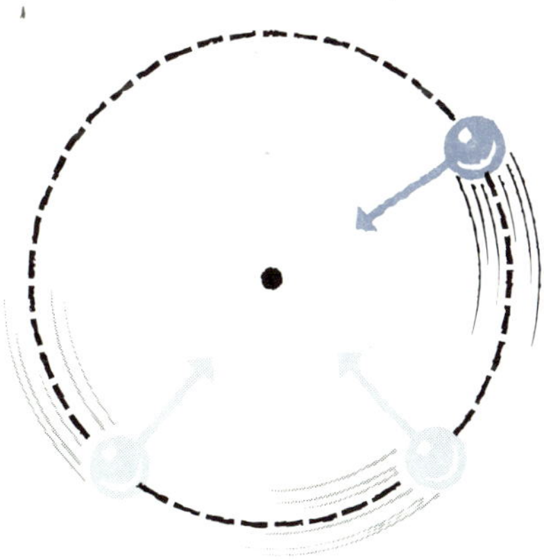

Fig. 4-9

the same circle at the same speed. It *does* depend on the mass m of the object, the speed s of the object, and the radius r of the circle. The following equation gives the magnitude of the centripetal force F:

$$F = \frac{ms^2}{r} \tag{8}$$

A force of this magnitude is required for circular motion regardless of how the force arises, whether from gravitation, a man pulling on a rope, or any other source.

The Planetary Model

In Newton's day, people had begun to believe that the earth and the other planets moved in orbits around the sun. The orbits of these planets, as we have noted, are not perfect circles but ellipses. However, they are so close to perfect circles that we can consider them as such in order to explain the main features of planetary motion.

Figure 4-10 shows the planets moving in their orbits around the sun in approximately circular motion. As we have explained, a centripetal force is required to maintain this motion. What interaction, acting across millions of miles of space, can supply this force on the planets, pulling them toward the sun? Newton realized that it was the gravitational force that was responsible. The same force that controls the motion of falling objects also controls the motion of the planets and even the stars. There is a gravitational interaction be-

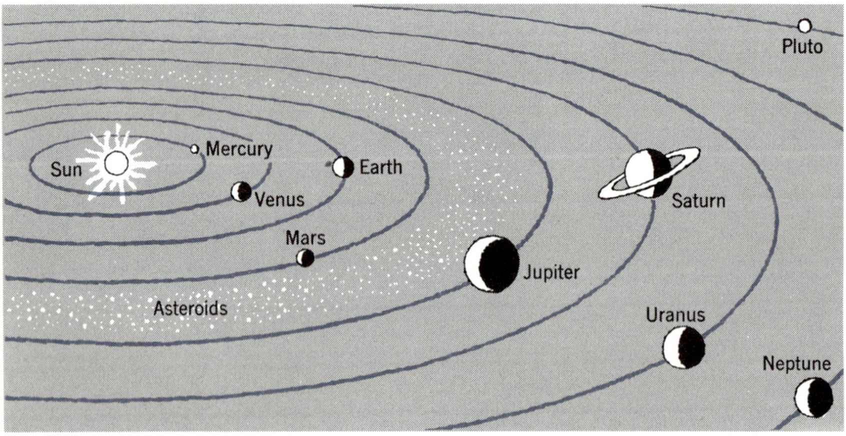

Fig. 4-10

tween each planet and the sun. This causes a gravitational force on each planet that acts in the direction of the sun. The gravitational force is the centripetal force that maintains the nearly circular motion of the planets.

The planets, of course, also exert a gravitational force on the sun, which has the same magnitude as the force exerted on them by the sun. The sun's mass is so large, however, that the effect of this force on the motion of the sun is small.

The moon, pulled by the gravitational attraction of the earth, moves in a nearly circular orbit around the earth, just as the planets do around the sun. Other planets also have moons, or natural satellites, orbiting about them. On a still greater scale, the sun and billions of other stars form a system, called a galaxy. The stars move around the center of the galaxy. The sun is about two thirds of the way out from this center, a distance of about 35,000 light years. (A light year is the distance that light travels in one year, at a speed of 300 million meters per second – 186,000 miles per second.) Galaxies themselves occur in groups. The nearest galaxy in our group is supposed to be about two million light years away. The modern picture of the universe is thus one of systems of systems of systems, held together by gravitational forces, moving through the "immensity of space".

Newton, in his law of gravitation, made no attempt to explain why it is that the earth or any other mass can exert a gravitational force. Nevertheless, using this law we can predict, with beautiful precision, the accelerations of the moon and the planets. For example, the planet Uranus showed variations in its path that could not be accounted for by the gravitational forces of the other known planets. Therefore, the gravitational law predicted the existence of an unknown mass, a planet even more distant from the sun than Uranus, that was responsible for the disturbance. Faith in the prediction led Leverrier to compute the required orbit of this new planet and to direct the astronomer Galle to look at the precise spot in the heavens where the new planet should be located. The planet was found as predicted and named Neptune. Much later, the planet Pluto was discovered in a similar manner. Again, the bright star Sirius was observed to have variations in its motion with respect to the other stars in the heavens. Bessel's faith in the law of gravitation led him to predict an invisible star, which he called "Little Dog." Later the star, though very faint, was found at the predicted position.

The list of successful predictions resulting from Newton's laws of motion and his law of gravitation is impressive. The time finally came, however, when a planet was predicted and not found. A planet named Vulcan was predicted to explain very small variations in the orbit of the planet Mercury, but Vulcan was never observed. Today we use a new theory of gravitation devised by Albert Einstein (see Chap. 10) to account for these small variations.

Artificial Satellites and Weightlessness

In the course of his discussion of planetary orbits, Newton showed the possibility of artificial satellites.[2] In doing this he used a diagram similar to Figure 4-11. We imagine a mountain so high that the air near its top is too thin to cause much resistance to the motion of objects. An object thrown horizontally from this tall mountain will fall to the earth in a curved path. The faster it is thrown, the further it will travel before hitting the ground. In fact, given sufficient speed, it will travel clear around the earth in a nearly circular path. (It *falls* clear around the earth.) In order for this to happen, the object must attain a speed of about 7,700 meters/second (about 5 miles/second). Still greater speeds will cause the object to travel in an elliptical path around the earth. If the object is given a speed greater than about

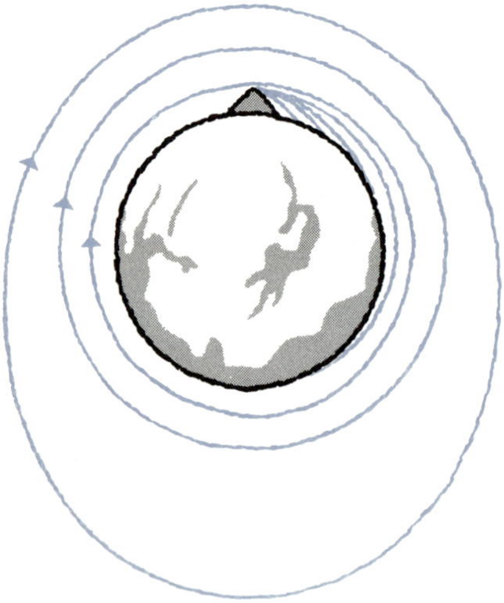

Fig. 4-11

[2]*Ibid.*, p. 551.

11,000 meters per second (about 7 miles per second), it will leave the earth and not return.

The technique used in putting a satellite in orbit, then, is to first send it high enough for the air resistance to be very small, then get it pointed in a nearly horizontal direction, and finally increase its speed to at least five miles per second. For artificial satellites, just as for the moon, it is the gravitational force of the earth that acts as a centripetal force holding the satellite in nearly circular motion around the earth, once it has attained the proper speed in a nearly horizontal direction.

This orbital speed of five miles/second is the same regardless of the mass of the object that is in orbit. (We can prove this from the equation expressing the law of gravitation, Equation 1, and the equation for centripetal force, Equation 8. It is, again, basically the principle that the acceleration of an object acted upon by a gravitational force does not depend upon its mass. See Example 3, page 165.)

Since the orbiting speed is the same for all objects orbiting at the same distance from the earth, it will seem to an astronaut as though he and the other objects in his capsule are "weightless." The astronaut will observe no tendency for any of these objects to "fall" or move away from him, since both he and they are moving at essentially the same speed in essentially the same orbit. Actually, of course, these objects are not weightless, since it is their weight (the force of gravitation) that keeps them in orbit. Their weight supplies the centripetal force that holds them in circular motion. A person or object in orbit is already in free fall, whether the orbit is circular or elliptical, because gravitation is the only force acting on him (see again Fig. 4-11). Thus he cannot drop a pencil or a cup, since these objects are already falling with him. The effect is essentially the same as if a man falling from a building tried to drop a rock. If he let go of the rock, it would still fall along with him.

The Newtonian Model

We have now completed our discussion of the first comprehensive mathematical theory of the universe. It is made up of the following principles and their consequences.

I. The three laws of motion.

II. The law of gravitational interaction:

$$F_g = GmM/d^2$$

If we know the positions and velocities of the various objects of the universe at a given time, and if the gravitational interaction supplies the only force acting on them, we can, in principle, accurately predict the future (and past) behavior of the objects. We do this by solving equations that express the above laws. (The solutions may be extremely difficult to obtain.) Since the objects in the heavens—the earth, the stars, the sun, and the planets—interact mainly through gravitation, we can predict with amazing accuracy how they behave. Knowing them as they are now, we can reconstruct the appearance of the heavens many thousands of years in the past and in the future, except for the effects of unknown objects such as previously unobserved comets. Thus we can replace observation, in many instances, by mathematical computation and prediction. We can obtain information about the universe by studying our equations instead of looking at the universe itself.

Today we might refer to the Newtonian mathematical theory of the universe as the Newtonian *model* of the universe. Why, we might ask, should we refer to scientific theories as models?

If we have a scale model of an airplane, we can obtain information about the plane by studying the model instead of the plane itself. The model will probably not be exact or complete—there will be some differences between the model and the real airplane. For some purposes the model will be completely satisfactory, and for others it will be only approximately valid. Similarly, our scientific theories enable us to learn something about the universe by looking at the theory rather than the actual universe. From the mathematical laws of motion and gravitation that comprise the Newtonian model of the universe, we can learn many things about the universe. Like most models, however, this is not perfect. There are some inadequacies in it, as we shall see later. Nevertheless, it is so accurate that for a long time most scientists thought that it was precisely true in every detail. This mathematical model has been one of the most successful intellectual constructions of mankind. It is still satisfactory for most celestial motions and for normal motions on earth which do not approach the speed of light.

Among other remarkable things about the Newtonian model is the fact that it has time-reversal symmetry (see Chap. 1 p. 33). This means that, according to this model, the same laws of motion and the same law of gravitation would hold if time ran backward. If Newton had lived in a world where time was reversed, he should have come up with the same laws.

*Rotation

Angular Momentum

In discussing the planets and other objects in nature, we often describe the motion of rotating or revolving bodies. The earth spins (rotates) on its axis as it revolves around the sun. On earth, we often deal with the spinning wheels of automobiles and other machines.

There is a conserved quantity which is closely related to the rotations and revolutions of objects. This quantity is called *angular momentum.* We may consider the conservation of angular momentum very roughly as a conservation of the spin or rotation of objects and systems of objects. A top, once set spinning, tends to continue rotating forever unless it is slowed down by friction. The earth continues to spin, the planets continue to revolve around the sun, etc.

Let us consider the behavior of a rotating object, such as the spinning bicycle wheel shown in Figure 4-12. A rod forms the fixed axis around which the bicycle wheel rotates. Most of the mass of this particular wheel is located on the rim, since the spokes have relatively little mass. For this discussion, therefore, we consider all the mass to be on the rim, at a distance r from the axis. There are three quantities which are essential in considering the rotation of this wheel: its mass m, the speed of the rotating rim, s, and the

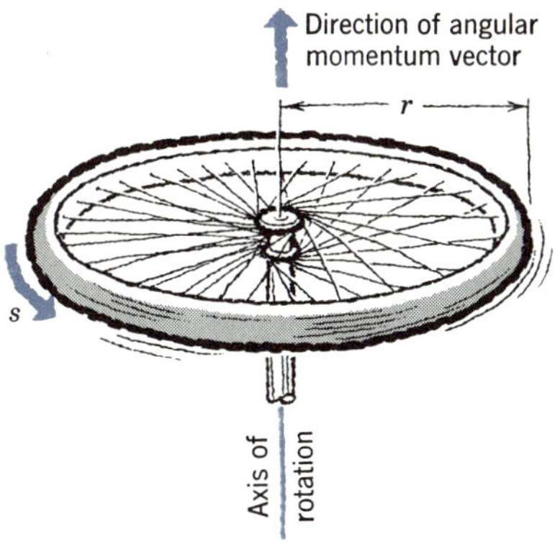

Fig. 4-12

radius r. The derived quantity, angular momentum, has a magnitude which is defined as follows:

$$\text{Magnitude of angular momentum} = msr. \tag{9}$$

Angular momentum is a vector quantity; we must therefore define its direction as well as its magnitude. The axis gives us the line of direction; we must now decide which direction along the axis will be used. The custom is to specify that if the object appears to turn counter-clockwise about its axis as we look at it along the axis of rotation, the direction of the angular momentum is toward us. If the object appears to turn clockwise, the direction is away from us. Notice that the direction of the angular-momentum vector is perpendicular to the plane in which the wheel is spinning (see Fig. 4-12).

The concept of angular momentum can be extended to bodies that are not actually rotating about any particular axis. We can find the angular momentum of any object around any straight line chosen as an axis. For simplicity, however, we shall restrict ourselves to a situation similar to that illustrated in Figure 4-13, in which the axis is perpendicular to a plane in which the object is moving. In Figure 4-13 the object is moving in the plane of the page, and the axis intersects the page perpendicularly at the point marked with a heavy dot. The angular momentum of this particular object with respect to this particular axis is simply msr. The quantity m is again the mass of the object, and the quantity s is again its speed. The quantity r, however, is not the distance from the axis to the object but, rather, the perpendicular distance from the axis to the line of motion of the object. If the object were moving in a circle about this axis, the two distances would be the same. The angular momentum of this object will be different about different axes because the distance r will be different.

Although we can take the angular momentum about any axis, there will often be some particular axis that is most convenient in a given situation. For the bicycle wheel considered above, and for similar

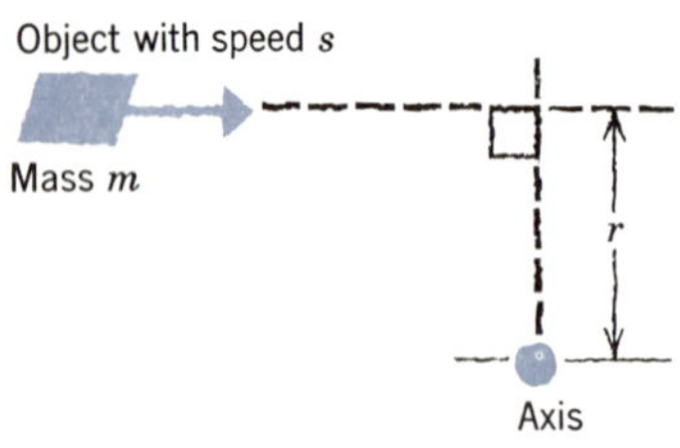

Fig. 4-13

objects, it is most convenient to choose the axis through the center.

To find the *total angular momentum* of a group or system of objects around a certain axis, we simply find the angular momentum of each object around that same axis and add these values together as vectors. A rotating object such as the disk in Figure 4-14 does not have all its mass at the same distance from the axis. Also, different parts of the disk will be moving at different speeds. To find the angular momentum of the disk we must add up the contributions, *msr*, of each small part of the disk. This, of course, explains why we chose the bicycle wheel at the beginning of our discussion—nearly all the mass is at the same distance from the axis and is moving at nearly the same speed. The total angular momentum will be *conserved* for any system that has no interaction with objects outside the system. (Actually, we can make a stronger statement than this, since forces can be applied to a system in such a way that they do not change its angular momentum.)

*Spin and Orbital Angular Momentum

If the object in Figure 4-13 happens to be spinning while it moves, there will be an additional angular momentum associated with its spin. This *spin angular momentum* will be the same around any stationary axes parallel to each other. In addition to the spin angular momentum, there is, of course, still the angular momentum discussed in the previous section. This angular momentum *msr*, is often called the *orbital angular momentum* of the object. The total angular momentum of an object around an axis can be separated into these two parts—spin angular momentum and orbital angular momentum.

We can illustrate the difference between these two parts of an object's angular momentum by referring once again to the motion of the earth. The orbital angular momentum of the earth is associated with its motion in following its yearly orbit around the sun. Its spin

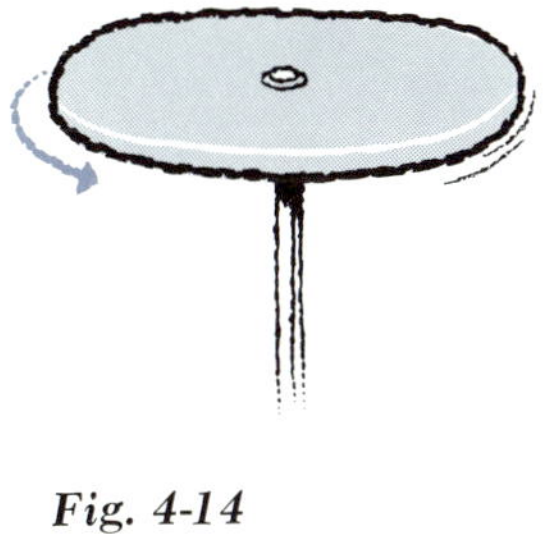

Fig. 4-14

angular momentum arises from its rotation around its own axis once a day. The angular momentum of the earth around the sun includes both orbital and spin angular momentum added together as vectors. The angular momentum of the earth about its own axis of rotation includes only spin angular momentum. The bicycle wheel we discussed earlier had only spin angular momentum. If the bicycle wheel were actually on a bicycle, moving from place to place, there would probably be orbital angular momentum as well, depending on the axis chosen.

Torque

We now need to discuss the kind of interaction that will change the angular momentum of an object about an axis. Of course, it is necessary to exert a force, but the force alone does not determine the change. In Figure 4-15 we illustrate attempts to change the angular momentum of a wheel by applying force at different distances from the axis of rotation. In Figure 4-15*a* and *b*, we apply the force along a straight line that runs right through the axis. This fails to rotate the wheel. (This is a special case, however, because the wheel is held by its center axle, which is not allowed to move.) The general rule for any object is that a force on that object directed exactly toward or away from an axis cannot change the angular momentum of the object around that axis.

In the next parts of Figure 4-15 we apply the same force at different distances from the axis. Now we do indeed cause the object to begin to rotate. Furthermore, the greater the perpendicular distance r from the axis to the line along which the force acts, the

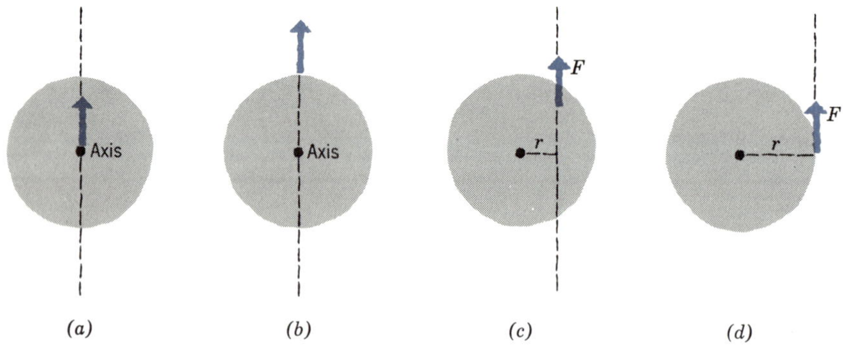

Fig. 4-15

greater the effectiveness of the force in causing the wheel to start to rotate. Thus, the force in Figure 4-15*d* is more effective than the force in Figure 4-15*c*. We can define a new vector quantity, ***torque***, in terms of the force F and the perpendicular distance r. Its magnitude is given as follows:

$$\text{Magnitude of torque} = Fr \tag{10}$$

The distance r is often called the lever arm of the force F. (Torque is important in theories of the lever and other simple machines.) Torque may be thought of, roughly, as the amount of twist exerted around an axis. The greater the distance r, the more effective the twist obtained from a given force. This explains why the longer wrench, in Figure 4-16, makes it easier to loosen the nut. The same force is more effective in exerting a twist when applied at a greater distance from the axis of rotation.

The definition of torque is similar in some ways to the definition of angular momentum. For one thing, the perpendicular distance r is determined in a similar way. For another, the direction of the vector torque is determined in a similar way. The direction of the torque is along the axis. A clockwise twist means a direction of torque away from the observer; a counter-clockwise twist means a direction toward the observer. The torque in Figure 4-17*a* is the negative of that in Figure 4-17*b*, even though equal forces are used in each case.

Torque exerted on an object causes a change in angular momentum. The angular momentum of an object or system of objects about a certain axis cannot change unless a torque about that same axis is applied to the object or system. The situation is very similar to the relationship between force and momentum. Just as a force is necessary to change the momentum of a system, so a torque about an axis is necessary to change the angular momentum about that axis. Thus,

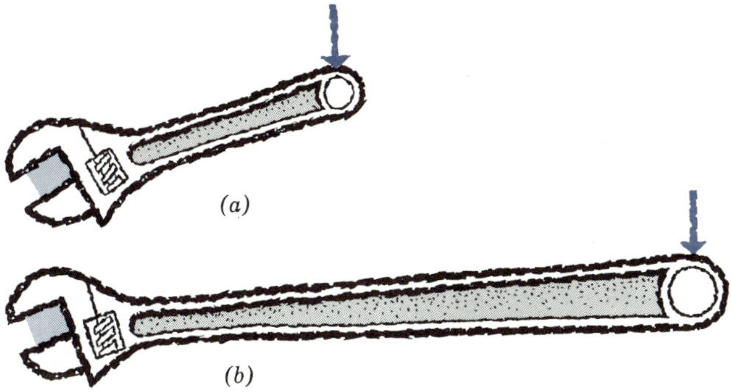

Fig. 4-16

Force (F) changes momentum (magnitude ms)
Torque (Fr) changes angular momentum (magnitude msr)

Examples of Conservation of Angular Momentum

The law of conservation of angular momentum is easily illustrated. In Figure 4-18 we see a man sitting on a stool. The stool is so designed that the seat can rotate freely around a vertical axis. When the seat of the stool is rotating, there will, of course, be some friction, causing a torque, which tends to slow it down. We will assume that this friction is so small that it can be ignored. Therefore, we assume that there are no torques around the vertical axis.

In our first experiment a man is sitting on the seat, rotating with it. He holds a large mass in each of his outstretched hands. What happens when he pulls in his hands? We find that his rotation speeds up. The law of conservation of angular momentum explains why.

The angular momentum msr must be conserved. When the man pulls in his hands, the r (or distance out) for the mass in his hands is reduced. In order for the quantity msr to remain the same, something else in the formula must increase. Since the mass m is constant, the speed s must increase. Thus the speed of rotation of the whole system must increase. (Actually, it is the angular momentum of the whole rotating body that must be conserved, not the angular momentum of the masses alone. The net effect is that the number of rotations per second increases.)

Now consider a second demonstration. The man is standing stationary on the same stool, holding the bicycle wheel above his head like an umbrella (see Fig. 4-19). At first neither the wheel nor the man on the stool is rotating. Then the man reaches up and begins to spin the wheel. What happens? Since the angular momentum of the system must remain the same, the increased rotation of the wheel must be balanced by an opposite rotation somewhere else in the system. As the man spins the wheel, he and the seat of the stool begin to rotate in the opposite direction.

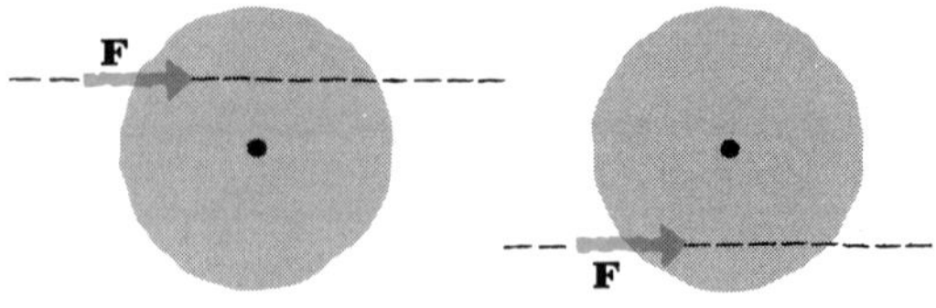

(a) (b)

Fig. 4-17

A third demonstration (Fig. 4-20) begins the same way as the second except that the bicycle wheel is already spinning. (The man and the stool are stationary.) Then the man turns the bicycle wheel upside down as shown, so that it is now spinning in the opposite direction with respect to the stool. What happens in this case? The angular momentum of the wheel has been reversed by turning the wheel upside down, and therefore this change in its angular momentum must be balanced by a change in the angular momentum of the man on the stool. The man and the stool seat with him begin to rotate in the original direction of the wheel. The man and stool seat must finally have twice the angular momentum the wheel had to begin with, since they must make up not only the original angular momentum of the wheel but also the reversed angular momentum the wheel has after it is turned over.

Pseudovectors and Mirror Reflection

We mentioned in Chapter 1 that the laws of physics, with certain exceptions, should be the same if our experiments were viewed in a mirror. The process of mirror reflection demonstrates a curious difference between the kind of vector represented by velocity and the angular momentum vector just discussed.

We illustrate this difference in Figure 4-21, which shows a particle in motion with its vector velocity heading to the right toward the mirror. It also shows a ball spinning so that its angular momentum vector is pointing into the mirror, to the right. When we look into the mirror to see the reflected image of these two objects, the particle heading into the mirror gives an image that appears to be going in the opposite direction—that is, to the left. The image of the spinning ball, however, appears to be spinning in the same direction as the ball itself. Thus the angular momentum vector of the ball as seen in

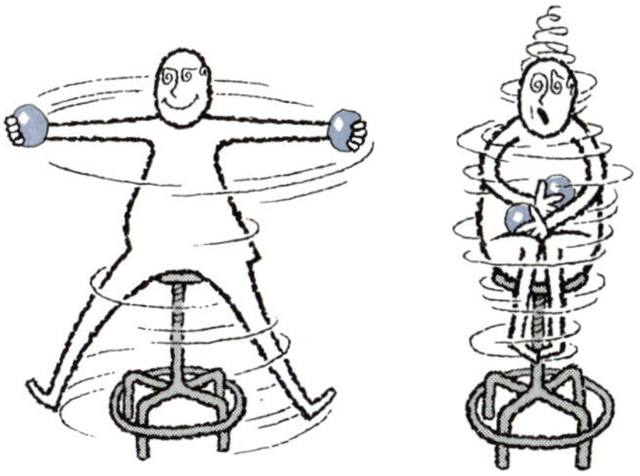

Fig. 4-18

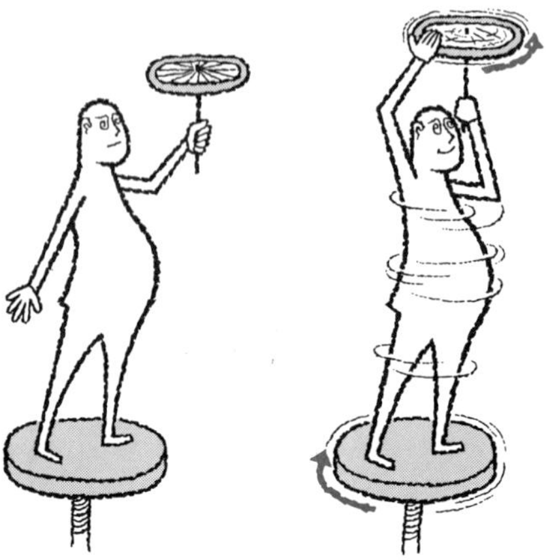

Fig. 4-19

the mirror must also be taken in the same direction, to the right. Of course, the image of the ball itself will be reversed as usual by the mirror, but not its direction of rotation. In this way we obtain a curious result: the velocity vector is reversed by the mirror, but the angular momentum vector is not.

Actually this result depends upon the orientation of the mirror: there is a way of locating the mirror so that the angular momentum vector is reversed by the mirror while the velocity vector is not. The important point is that the two vectors are reflected differently. There are, thus, two kinds of vector. Those that are reflected like the velocity vector are called *true vectors*. Those that are reflected like

Fig. 4-20

the angular momentum vector are called *pseudovectors*. (This terminology does not, of course, imply moral inferiority on the part of pseudovectors, which are perfectly respectable physical quantities.) Sometimes pseudovectors are called ***axial vectors***, because they are associated with an axis.

**The Experimental Philosophy

Since the ancients (as we are told by Pappus) esteemed the science of mechanics of greatest importance in the investigation of natural things, and the moderns, rejecting substantial forms and occult qualities, have endeavored to subject the phenomena of nature to the laws of mathematics, I have in this treatise cultivated mathematics as far as it relates to philosophy. . . . . therefore I offer this work as the mathematical principles of philosophy, for the whole burden of philosophy seems to consist in this—from the phenomena of motions to investigate the forces of nature, and then from these forces to demonstrate the other phenomena. . . . I give an example of this in the explication of the System of the World. . . . I derive from the celestial phenomena the forces of gravity with which bodies tend to the sun and the several planets. Then from these forces, by other propositions which are also mathematical, I deduce the motions of the planets, the comets, the moon, and the sea. I wish we could derive the rest of the phenomena of Nature by the same kind of reasoning from mechanical principles, for I am induced by many reasons to suspect that they may all depend upon certain forces by which the particles of bodies, by some causes hitherto unknown, are either mutually impelled towards one another, and cohere in regular figures, or are repelled and recede from one another. These forces being unknown, philosophers have hitherto attempted the search of nature in vain; but I hope the principles here laid down will afford some light either to this or some truer method of philosophy. (Isaac Newton)[3]

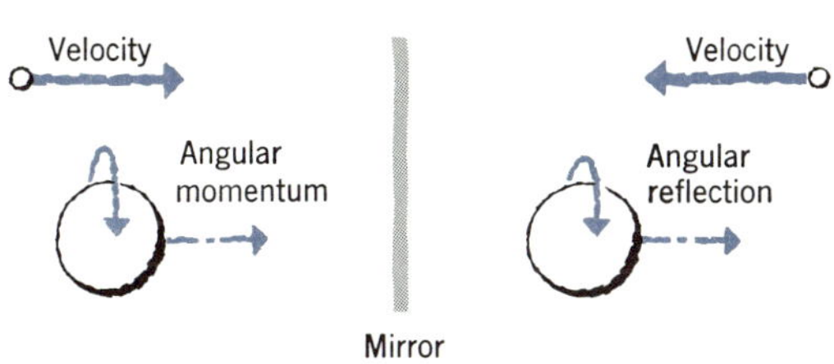

Fig. 4-21

[3]*Ibid.*, p. xvii.

The Impact of the Newtonian Synthesis

Newton knew that the laws of Kepler for planetary motion were experimentally correct. He also knew that the laws of Galileo for falling bodies near the earth were experimentally correct. He adopted his three laws of motion as axioms, or assumptions, and then proceeded to obtain his law of gravitation by reasoning from these laws and experimental results. He showed that the observed behavior of falling objects, and also the behavior of comets, planets, and other celestial objects, could be precisely predicted using his three laws of motion and the law of gravitational force. In this way he combined his results with those of his predecessors into a simple, comprehensive system—the famous "Newtonian synthesis." It was and is considered one of the most important achievements of the human mind. It complemented the geometrical system of Euclid, who had derived the properties of space from simple assumptions, often thought to be "self-evident." From simple assumptions Newton derived the properties and behavior of matter in motion. The result was a successful mathematical treatment of mechanics—the behavior of matter in motion under the influence of forces.

The men of the eighteenth-century Enlightenment had reason to be impressed with the new physics. When had the world known such an achievement? The universe, both the heavens and the earth, operated according to the same precise mathematical laws. All one had to know was the position and velocity of the particles of matter and the forces between them. Then one could, in principle at least, predict exactly what would happen to the particles at any time in the future—or, for that matter, determine what had happened to them in the past. Thus it is that our modern planetariums can reproduce the face of the sky, the positions of the planets and the stars, for any day in history, past or future. Of course, there is always some experimental error in determining the starting position and velocity of the particles, and there are too many particles in the universe to keep track of them all. In principle, however, there should be an exact position and velocity for every particle, and the laws of motion and force should exactly determine the behavior of the particles.

This type of physics almost inevitably suggested a mechanistic philosophy, in which everything was reduced to particles of matter in predictable motion according to exact and irrevocable mathematical law. It also suggested a deterministic type of philosophy, in which the future is in principle predictable and all "behavior is caused." In this philosophy, because of the invariable and predictable operations of natural law, conditions in the future were irrevocably determined by conditions in the present.

Thus the universe appeared to be a "world machine," in which each piece behaved in a perfectly predictable and determined way. Newton had thought that the gravitational effects of the planets on each other might make the whole solar system unstable, requiring God to make adjustments. Pierre Simon LaPlace (1749–1827) worked out the consequences of the Newtonian physics and decided that this would not be necessary; the machine could run by itself. (This question is still discussed.)[4] Newton had also noted that the planets and all the then-known satellites had the same direction or sense of revolution. (Satellites with contrary behavior have since been discovered.) He felt that this order must have been impressed by the Creator. LaPlace originated the nebular hypothesis to explain this. (This theory has been superseded.) When Napoleon noted that there was no place for the Creator in LaPlace's system of the world, LaPlace is reported to have said: "Je n'avait pas besoin de cette hypothèse-là." (I didn't need that hypothesis.)[4] Natural law was supreme.

Such a philosophy inevitably had ramifications and effects outside of physics: it clearly contained the foundations of a challenge to free will and other spiritual values. In a world in which machines were becoming more and more important in manufacturing, the world itself was beginning to be considered a machine. Such a world view would influence the other sciences, the humanities, religion, and the general culture. Of such influence, C. C. Gillispie has written:

> "Ye shall know the Truth, and the Truth will make you free"—it is one of the great themes of the Enlightenment. But for most people emancipation was not enough. It never has been. They are not content to take science for what it is intellectually, a great creation, a description of how the physical world works, beautiful and admirable in itself, but empty of morals and lessons. They want more. They want reassurance about the existence of God from the design of nature, and about His loving care from His continuing to repair its imperfections. They want a license to rugged individualism in the theory of evolution. They want a crumb of comfort about free will in the unpredictability of the electron. They want studies of the mind considered as a digital computer. In short, they want science to give us a world we can fit, as Greek science did, and not just a world like any external object that we can first measure, and then destroy.
>
> In moments of discouragement, therefore, one is tempted to think that the history of the influence of science in culture is bound to be the history of a

[4]*Ibid.*, Appendix by F. Cajori, pp. 677–678. See also Werner Heisenberg, "Nonlinear Problems in Physics," *Physics Today*, **20**, 5 (May 1967), pp. 27–33, esp. p. 29.

misunderstanding, in which what changes is the way in which the import of science is misunderstood. But this would be too despairing a view. . . .[5]

Although we do not agree with all the suggestions in these remarks, the way in which the Newtonian synthesis influenced the general culture might indeed cause one to be appalled or amused, depending on one's temperament.

There is a temptation to blame the French Revolution on the philosophers of the Enlightenment: "It is the fault of Rousseau," "It is the fault of Voltaire." Could Newton have had anything to do with it? Voltaire was an avid fan of Newton; he even considered it worthwhile to write a book about Newton's philosophy. In the Age of Reason, Newton was evidentally considered an apostle of Reason, and therefore the influence of his ideas was important among philosophers.

Maximilien de Robespierre, a powerful figure during the Terror of the French Revolution, said, "What does this mysterious science of government and legislation amount to? To putting into the laws the moral truths culled from the works of the philosophers."[6] The exact laws of nature as found in physics evidentally suggested that exact laws could also be found for ethics, psychology, politics, etc. If these laws were precisely obeyed (and precisely enforced) then the ideal society might be created. There was, fortunately or unfortunately, some difficulty in carrying this out in practice.

What about the American Revolution? Some have suggested a history involving the following elements. The psychological and political doctrines of the English philosopher John Locke (1632–1704) were evidently constructed with the successful Newtonian philosophy as a model. Newton spoke of "experimental philosophy"; Locke proposed a philosophy in which the source of knowledge was experience. These theories influenced the British Whigs, whose practice was analyzed by the French jurist and philosopher, Montesquieu, who in turn influenced the leaders of the American Revolution. Woodrow Wilson remarked:

> The government of the United States was constructed upon the Whig theory of political dynamics, which was a sort of unconscious copy of the Newtonian theory of the universe. In our own day, whenever we discuss the structure or development of anything, whether in nature or in society, we consciously or unconsciously follow Mr. Darwin; but before Mr. Darwin they followed Newton. . . .

[5]C. C. Gillispie, *The Edge of Objectivity*, Princeton, N. J., Princeton University Press, 1960, p. 155.

[6]"The French Revolution," *Encyclopedia Britannica*, Fourteenth Ed., 1929, Vol. 9, p. 805.

> The makers of our federal Constitution followed the scheme as they found it expounded in Montesquieu, followed it with genuine scientific enthusiasm. . . . Politics is turned into mechanics under his touch.
>
> The trouble with the theory is that government is not a machine, but a living thing. . . . It is accountable to Darwin, not to Newton. . . . Living political constitutions must be Darwinian in structure and in practice.[7]

So the game goes on.

We can perhaps recognize some familiar notions about natural law, natural rights, causes and necessities in these lines, written mainly by Thomas Jefferson, from the Declaration of Independence.

> When in the Course of human events it becomes necessary for one people to dissolve the political bands which have connected them with another, and to assume among the powers of the earth, the separate and equal station to which the Laws of Nature and of Nature's God entitle them, a decent respect to the opinions of mankind requires that they should declare the causes which impel them to the separation.—We hold these truths to be self-evident, that all men are created equal, that they are endowed by their Creator with certain unalienable Rights, that among these are Life, Liberty, and the pursuit of Happiness. . . . That whenever any Form of Government becomes destructive of these ends, it is the Right of the People to alter or to abolish it, and to institute new Government, laying its foundation on such principles and organizing its powers in such form, as to them shall seem most likely to effect their Safety and Happiness.

We certainly do not contend that the ethical and political principles that arose in the Enlightenment are dependent upon Newton's mechanics. It is interesting, however, to note the influence of "natural law" upon the expression and defense of these principles.

In the "Romantic Reaction" which followed the "Age of Reason," and at other times, Newton and the mechanistic world view came in for some criticism. An early example comes from the writings of the visionary poet William Blake (1757–1827).

> Mock on, Mock on, Voltaire, Rousseau:
> Mock on, Mock on; 'tis all in vain!
> You throw the sand against the wind,
> And the wind blows it back again.
>
> * * *
>
> The Atoms of Democritus
> And Newton's Particles of light

[7]Woodrow Wilson, *Constitutional Government in the United States*, New York: Columbia University Press, 1921, pp. 54–55. On this general subject we must acknowledge our debt to the following authors without holding them responsible for our opinions: J. Bronowski and Bruce Mazlish, *The Western Intellectual Tradition from Leonardo to Hegel*, New York: Harper Torchbooks, 1962; and Gillispie, *op. cit.*

Are sand upon the Red Sea shore,
Where Israel's tents do shine so bright.[8]

From a scientific point of view, however, the nineteenth century saw an increase in mechanistic views and what appeared to be a near-triumph for exact mathematical law. Lord Kelvin, one of the ranking scientists of the age, wrote in 1884:

> I am never content until I have constructed a mechanical model of the object I am studying. If I succeed in making one I understand; otherwise I do not. Hence, I cannot grasp the electromagnetic theory of light. I wish to understand light as fully as possible, without introducing things I understand still less.[9]

As we shall see, however, the "electromagnetic theory of light," as developed by James Clerk Maxwell, was a fitting addition to Newton's laws from the standpoint of precise mathematical prediction.

Another example of mechanistic ideas was the theory of the evolution of living things proposed by Charles Darwin (1809–1882) and Alfred Wallace (1823–1913). They suggested a mechanism by which, in succeeding generations, animals and plants could change to produce new species. Within the same species there exist many varieties of individuals, differing from one another. Since not all individual plants and animals survive to reproduce, there is a selection automatically made by nature between those varieties that do reproduce and those that do not. As the conditions change, or as populations move into different surroundings, there is a change in the type of individual that will best survive and reproduce. This "natural selection" was a mechanism that might help to account for changes in species. This theory, of course, was not mathematical in the Newtonian sense; the precise calculation or prediction of the future was not a characteristic part of it.[10] However, it did allow one to think about changes of species mechanistically, as manifestations of matter in motion according to law.

The combination of Newtonian and Darwinian theories strongly

[8]William Blake, from *Verses and Fragments from the Rossetti and Pickering Manuscripts,* second series, 1800–1810, quoted in Alfred Kazin, ed., *The Portable Blake,* New York: Viking Press, 1946, p. 142.

[9]William Thomson, Lord Kelvin, as quoted by Julien Ripley, *The Elements and Structure of the Physical Sciences,* New York: John Wiley & Sons, 1964, p. 370.

[10]See George Gaylord Simpson, "Biology and the Nature of Science," *Science,* **139**, 3550 (January 1963), pp. 81–88; and John Kemeny, *A Philosopher Looks at Science,* Princeton, N.J.: Van Nostrand, 1959, Chap. 12. See also George Gaylord Simpson, *The Meaning of Evolution,* New York: New American Library of World Literature, Special Rev. and Abr. Ed., 1952, Chaps. 2–8.

influenced philosophy and informed public opinion in the succeeding decades of the nineteenth and twentieth centuries. Evolution could be thought of as a form of progress; however, its mechanistic characteristics could also have other implications. Mark Twain, for example, satirized the idea of holding a machine such as man morally responsible.[11] Another example can be found in these words of the lawyer Clarence Darrow regarding the punishment of criminals, written in 1922.

> That man is the product of heredity and environment and that he acts as his machine responds to outside stimuli and nothing else, seem amply proven by the evolution and history of man. But, quite aside from this, logic and philosophy must lead to the same conclusions. This is not a universe where acts result from chance. Law is everywhere supreme. Every process of nature and life is a continuous sequence of cause and effect. . . . Chance, accident and whim have been banished from the physical world.
>
> . . . Man's every action is caused by motive. . . . If two or more motives pulled in opposite directions, he could not have acted from the weakest but must have obeyed the strongest.
>
> . . . Assuming that the scientific attitude toward crime should be adopted. . . .[12]

We cannot discuss all the mechanistic, materialistic, or deterministic philosophies of the nineteenth and twentieth centuries, nor would we contend that they all depended upon Newtonian physics. The unprecedented success of mechanistic and deterministic physics through the nineteenth century, however, must have helped to spread and strengthen these philosophies.

Mechanistic ideas are probably still among the most important and influential effects of physics on the general culture, even though they are no longer dominant in physics itself. There are, for example, less mechanistic theories of evolution than Darwin's, but they are not generally regarded as quite scientific. It is doubtful that extrasensory perception will be considered wholly respectable until some mechanism is suggested to account for its alleged effects. Freudian psychology might also be reproached for its lack of a material mechanism—behavioristic psychology seems more "scientific" to some. It is hard for some scientists to look at an animal, for example, which appears to be made up of material particles, and not believe that its properties and behavior can be explained in terms of those particles. The science of molecular biology attempts to do this as far as possi-

[11]Mark Twain, *Letters from the Earth*, Greenwich, Conn.: Fawcett Publications, Inc., 1966, pp. 32, 170–172.

[12]Clarence Darrow, *Crime, Its Cause and Treatment*, New York: Thomas Y. Crowell, 1922, pp. 34, 35, 124.

ble, and has been amazingly successful in uncovering the chemical mechanisms by which molecular particles determine the nature and heredity of the living cell.[13] This tends to reinforce the attempt to understand all things, man included, in terms of matter and motion. One scientist, Kendon Smith, has written,

> Living objects are not composed of a magical "lifestuff." Protoplasm is made up of molecules; molecules are made up of atoms; and atoms are made up of the basic, sub-atomic particles. What a living object is, then, is exactly what Democritus said it was: a swarm of particles in space. This is the message borne by modern biochemistry, and the message is amply confirmed by a great body of factual information about the accidental beginnings, the toilsome history, and the present, awesomely peripheral, status of life in the cosmos.
>
> Although there is in fact no good reason to doubt the validity of the notion of complete determinism, distinguished men have felt impelled to contest it. . . .[14] [Smith then mentions, among those who have contested determinism, the distinguished physicists Neils Bohr, Max Planck, and Erwin Schrodinger.]

So determinism is still with us. According to Jacques Barzun, the "articulate" have been crying out for the past thirty years "against the tyranny of scientific thought, the oppression of machinery, the hegemony of things, the dehumanization brought about by the sway of number and quantity."[15]

The strict determinism of Newton's laws is gone from physics today. Strict prediction is also gone, although prediction is even more important in physics than ever. Today, however, physicists make *approximate* predictions, forecasting a probability rather than a certainty. Knowledge of the future is of great practical importance, since it is only the future that we can change, not the past or the present. The models of the universe incorporated in our scientific theories can help us to do this. We can use a creation of the psyche to explain the world, or perhaps even to help explain the psyche itself.

To close this section, we quote from a modern poet, Robert Frost. (A scientist might prefer the word "model" to "metaphor.")

> Let me ask you to watch a metaphor breaking down here before you. Somebody said to me a little while ago, "It is easy enough for me to think of the universe as a machine, as a mechanism."

[13]See, for example, J. D. Watson, *The Molecular Biology of the Gene*, New York: Benjamin, 1965.

[14]Kendon Smith, "The Naturalistic Conception of Life," *The American Scientist*, **46** (December 1958), pp. 413–423.

[15]Jacques Barzun, *Science, the Glorious Entertainment*, New York: Harper and Row, 1964, p. 1.

I said, "You mean the universe is like a machine?"

He said, "No, I think it is one. . . Well, it is like . . ."

"I think you mean the universe is like a machine."

"All right. Let it go at that."

I asked him, "Did you ever see a machine without a pedal for the foot, or a lever for the hand, or a button for the finger?"

He said, "No—no."

I said, "All right. Is the universe like that?"

And he said, "No. I mean it is like a machine, only . . ."

". . . It is different from a machine," I said.

He wanted to go just that far with that metaphor and no further. And so do we all. All metaphor breaks down somewhere. That is the beauty of it. It is touch and go with metaphor, and until you have lived with it long enough you don't know where it is going. You don't know how much you can get out of it and when it will cease to yield. It is a very living thing. It is as life itself.[16]

Newton's "Rules of Reasoning in Philosophy"

RULE I. *We are to admit no more causes of natural things than such as are both true and sufficient to explain their appearances.*

To this purpose the philosophers say that Nature does nothing in vain, and more is in vain when less will serve; for Nature is pleased with simplicity and affects not the pomp of superfluous causes.

* * *

RULE IV. *In experimental philosophy we are to look upon propositions inferred by general induction from phenomena as accurately or very nearly true, notwithstanding any contrary hypothesis that may be imagined, till such time as other phenomena occur by which they may either be made more accurate or liable to exceptions.*

This rule we must follow, that the arguments of induction may not by evaded by hypotheses.[17]

Several characteristics of Newton's experimental philosophy are embodied in these passages.

1. *Simplicity.* The first rule tells us to get rid of excess baggage in scientific explanations. Only those causes that are necessary to explain the experimentally observed appearances will be allowed, and

[16]From "Education by Poetry" from *Selected Prose of Robert Frost* edited by Hyde Cox and Edward Connery Lathem. Copyright ©, 1966 by Holt, Rinehart and Winston, Inc. Reprinted by permission of Holt, Rinehart and Winston, Inc.

[17]Newton, *op. cit.*, pp. 398–400.

no more of those than are true and necessary. This principle is, of course, well known in philosophy.

The commentary on the first rule also claims that nature itself is simple. In a commentary on a second rule, Newton added that nature wants to be "always consonant with itself."[18] This expresses the idea of the uniformity of nature, which makes the discovery of laws possible. If every event in nature were different in every way, we would have a very difficult time learning by experience. Indeed, it would be doubtful that any useful law or principle could be discovered or known. The fact that similar processes happen in the same way over and over again enables us to discover laws.

2. *No certain truth.* Rule IV points out that the principles obtained in this philosophy are regarded as only approximately true and are subject to revision. One does not obtain absolutely exact or certain truth from the experimental philosophy.

3. *Knowledge comes from experience.* Rule IV makes it clear that the statements we make in this philosophy must come from experience, or experiments, with the phenomena, and not from "vain fictions of our own devising."[19]

4. *Hypotheses are rejected.* A hypothesis is defined by Newton in a way that does not correspond exactly to the way in which the term is normally used today.

> And the word "hypothesis" is here used by me to signify only such a proposition as is not a phenomenon nor deduced from any phenomena, but assumed or supposed without any experimental proof.[20]

By this definition, a hypothesis is a statement which does not come from the experimental data and is therefore rejected.

5. *Induction.* Induction is the process of using inductive logic. We usually distinguish between two types of logic, deductive and inductive. We first present an example of deductive logic.

A. All swans are white.
B. John is a swan.
Therefore: John is white.

We are not here concerned with whether or not these statements

[18] *Ibid.*
[19] *Ibid.*
[20] From Newton's letters to Cotes, originally published in J. Edleston, ed., *Correspondence of Sir Isaac Newton and Prof. Cotes*, London, 1713, pp. 154–157, and quoted in H. S. Thayer, ed., *Newton's Philosophy of Nature, Selections from His Writings*, New York: Hafner, 1953, p. 6.

are true (statement A is not). We assume the truth of the first two statements, and then inquire as to whether or not the truth of the last statement follows. In this case it does, and therefore the deduction is valid. If the first two statements are true, the third one must be true.

The example above involves the application of a general statement (A) to a particular case (John). Another type of valid deduction allows us to obtain limited statements from experimental results.

A. I have seen twenty-five swans.
B. The first swan was white.
The second swan was white.
The third swan was white.
The fourth swan was white, etc.

. . .

The twenty-fifth swan was white.
Therefore: all the swans I have seen are white.

This is also a perfectly valid deduction since the conclusion follows necessarily from the two assumptions, A and B. The conclusion, however, does not tell us anything that was not contained in the assumptions. This is a general property of deductions. Our example helps us to understand what Newton meant by deducing a principle from the phenomena. Statement A says that I have observed some swans. Statement B gives the data I have collected about these swans. Thus statement A and B together give the information I have about swans. The last statement, the conclusion, has been deduced from this information.

Let us now consider an example of inductive logic.

A. All the swans which I have seen are white.
Therefore: all swans are white. (Or, more conservatively, the next swan I see will be white.)

This argument is clearly false, deductively speaking. It is not necessarily true that all swans are white just because all the swans I have seen are white. From a deductive point of view, in fact, this is a logical fallacy. From an inductive point of view, however, the example is a correct and valid induction. In inductive logic, we seek a conclusion that does not necessarily follow from the assumptions — we seek something new. In inductive logic we want the conclusion to contain more information than is contained in the assumptions. If we are to obtain useful knowledge from experience, our knowledge must go beyond that experience. For example, we may wish knowledge about the future in order to make sensible decisions. The price we pay for these useful predictions is uncertainty. We expect to be wrong in some cases. We expect, however, that the odds are with us; if something has happened in the same way a large number of times, we have a reasonable expectation that it will probably happen in the same way again. We make an educated guess. Thus we obtain probable knowledge, which becomes more and more certain as our experience widens but never becomes absolutely certain. Newton tells us that "these principles are deduced from phenomena and made general by induction, which is the highest evidence that a proposition can have in this philosophy."[21] Thus, we devise principles which hold for all we have experienced and assume that they hold in general. This assumption is a process of induction. Newton further explains the process:

> . . . For hypotheses are not to be regarded in experimental philosophy. And although the arguing from experiments and observations by induction be no demonstration of general conclusions, yet it is the best way of arguing which the nature of things admits of, and may be looked upon as so much the stronger by how much the induction is more general. And if no exception occur from phenomena, the conclusion may be pronounced generally. But if at any time afterward any exception shall occur from experiments, it may then begin to be pronounced with such exceptions as occur. By this way of analysis we may proceed from compounds to ingredients and from motions to the forces producing them, and in general from effects to their causes and from particular causes to more general ones, till the argument end in the most general. This is the method of analysis; and the synthesis consists in assuming the causes discovered and established as principles, and by them explaining the phenomena proceeding from them and proving the explanations.[22]

Thus, once we have discovered our principles by induction, we use these same principles as assumptions and deduce the phenomena from them. After using induction to get the general principles, Newton built upon these principles a deductive system similar to ordinary Euclidian geometry. It is interesting to notice, in the quotation above, that Newton obtained the law for forces acting between objects by observing the behavior of matter in motion. Then the force law was assumed and used to show what motions should occur.

We have explained that learning from experience, by the process

[21] *Ibid.*

[22] Isaac Newton, *Opticks, or a Treatise of the Reflections, Refractions, Inflections and Colours of Light*, 4th Ed. as quoted by Thayer, *op. cit.*, p. 178.

of induction, gives probable, not certain, knowledge. The Scottish philosopher David Hume (1711–1776) asked why we should have any expectation, probable or otherwise, that our knowledge of the past has anything whatsoever to do with future events. After all, the past and the future are different. To say that this expectation has been justified in the past is really begging the question, since it is the relation of the past to the future that is at issue. To give an experimental reason for the validity of the experimental philosophy *would* seem to be begging the question, albeit consistently.

We have here raised what many regard as an unsolved question in philosophy. In science we use inductive logic, if only because we have little choice. We assume (and look for) some kind of uniformity in nature.

It is interesting to see how Newton's "rules of reasoning" and philosophy led him to handle two difficult problems. The first is the problem of action at a distance. The gravitational forces of the planets and the sun act on each other over very large distances. How can the influence of gravitation cross these great distances if there is no matter in between? How can it act across these distances instantaneously? (Today we think that gravitational influences travel at the speed of light.) Newton admitted that this was a problem.

> That gravity should be innate, inherent, and essential to matter, so that one body may act on another at a distance through a *vacuum*, without the mediation of anything else, by and through which their action and force may be conveyed from one to another, is to me so great an absurdity that I believe no man who has in philosophical matters a competent faculty of thinking can ever fall into it. Gravity must be caused by an agent acting constantly according to certain laws, but whether this agent be material or immaterial I have left to the consideration of my readers.[23]

But he also said:

> . . . Hitherto I have not been able to discover the cause of those properties of gravity from phenomena, and I feign no hypothesis; for whatever is not deduced from the phenomena is to be called a hypothesis, and hypotheses, whether metaphysical or physical, whether of occult qualities or mechanical, have no place in experimental philosophy. In this philosophy particular propositions are inferred from the phenomena, and afterwards rendered general by induction. Thus it was that the impenetrability, the mobility, and the impulsive force of bodies, and the laws of motion and gravitation were discovered. And to us it is enough that gravity does really exist, and act according to the laws which we have explained, and abun-

[23]Isaac Newton, in a letter to Richard Bentley, Feb. 25, 1692/3, as quoted in Thayer, *op. cit.*, p. 54.

dantly serves to account for all the motions of the celestial bodies, and of our sea.[24]

The mathematical laws predict the correct behavior. We do not make guesses about hypothetical causes and give them the same status as the laws.

The second problem concerns the special principle of relativity. Newton did not accept this principle, although, as he knew, his laws obeyed it. He knew that ordinary time and space, as measured experimentally, involved relationships between one object and another, or one event and another. He apparently believed, however, that underlying and supporting relative time and space was absolute time and space. Motion in absolute time through absolute space would also be absolute. He knew that this concept challenged his philosophy, because absolute time, absolute space, and absolute uniform motion in a straight line were experimentally unobservable. He wrote, however, that "the thing is not altogether desperate"[25] because accelerated motion, being caused by a force, is absolute. A person spinning, for example, feels the dizzying effects of centripetal force. Presumably he would not feel these effects if he were at rest and the universe were spinning around him.

The Irish philosopher George Berkeley (1685–1753) believed that " to be is to be perceived." In a religiously motivated attack on the new mathematics, he questioned "whether the notions of absolute time, absolute place, and absolute motion be not most abstractly metaphysical? Whether it be possible for us to measure, compute, or know them?"[26] Newton, however, seems to have had theological motivation for his doctrine of space and time.

> . . . He is not eternity and infinity, but eternal and infinite; he is not duration or space, but he endures and is present. He endures forever and is everywhere present; and by existing always and everywhere, he constitutes duration and space. Since every particle of space is *always*, and every indivisible moment of duration is *everywhere*, certainly the Maker and Lord of all things cannot be *never* and *nowhere*.[27]

[24]Newton, *Mathematical Principles . . .*, *op. cit.*, p. 547. We have used the word "feigned" instead of "framed," following Koyre, see Gillispie, *op. cit.*, p. 147.

[25]*Ibid.*, p. 12.

[26]George Berkeley, *The Analyst*, included in James R. Newman, ed., *The World of Mathematics*, Vol. I, New York: Simon and Schuster, 1956, p. 293.

[27]Newton, *Mathematical Principles . . .*, *op. cit.*, p. 545. For a discussion of this question, see E. A. Burtt, *The Metaphysical Foundations of Modern Physical Science*, rev. ed., Garden City: Doubleday Anchor Books, 1932, Chap. V, Section F, and Chap. VII, Section 4.

The ancient materialists—Democritus, Epicurus, and Lucretius—who evidently believed that everything that exists is matter and that all that happens is motion, left certain unresolved problems. How can the void, which is nothing, be said to exist? Newton speculated that the void and bodies also might be filled with a "subtle spirit," which transmitted electrical and other impulses.[28]

Newton the Person

About 1684 the English astronomer Edmund Halley asked Newton if Kepler's first law would follow from the now familiar variation of the gravitational force with distance. Newton said that indeed it would, but that he had lost the calculation. As C. C. Gillispie has so aptly put it, "While others were looking for the law of gravity, Newton had lost it."[29] Evidently he had mentioned the discovery to no one. Newton appeared to be more interested in his private researches than in publishing their results, which could lead to controversy. In the course of an unpleasant conflict with another very able English scientist, Robert Hooke (1635–1703), he stated:

> Philosophy is such an impertinently litigious Lady, that a man had as good be engaged in lawsuits, as have to do with her. I found it so formerly, and now I am no sooner come near her again, but she gives me warning.[30]

He threatened to suppress part of the *Principia* because of this dispute.

Newton also did important research in the nature and behavior of light, which Halley convinced him to publish in the *Opticks*. In addition he is responsible for one of the great advances in mathematics—the differential and integral calculus. This is the kind of mathematics needed to deal with quantities which change in a nonuniform way. (The calculus was also discovered independently by the German philosopher-mathematician G. W. von Leibnitz (1646–1716), and a dispute over priority resulted.) Newton's powers of concentration were evidently very great; when asked how he made his discoveries, he replied, "By always thinking on them."[31]

In 1696 Newton left Cambridge University to accept the post of Warden of the Mint (later Master of the Mint). According to Derek Price, this post was intended as a sinecure, but Newton took it "so

[28]Newton, *ibid.*, p. 547.
[29]Gillispie, *op. cit.*, p. 137.
[30]As quoted in E. C. Da C. Andrade, "*Isaac Newton,*" in Newman, *op. cit.*, p. 265.
[31]See Gillispie, *op. cit.*, p. 117.

conscientiously and seriously that one must credit quite a lot of the later economic strength and security of England to [his] efficient reforms and administration."[32] Even at Cambridge, Newton was absorbed by nonscientific studies; indeed, his work in physics and mathematics seems to be the result of discoveries which he made very early, in the years 1665–1666. He was fascinated by the esoteric lore of alchemy. He was evidently considered a very able theologian, and wrote a commentary on the Apocalypse of John and the prophecies of Daniel. Apparently he disagreed with the standard doctrine of the Trinity but had to keep this secret.

About Newton's nonscientific interests the economist John Maynard Keynes commented:

> In the eighteenth century and since, Newton came to be thought of as the first and greatest of the modern age of scientists, a rationalist, one who taught us to think on the lines of cold and untinctured reason.
>
> I do not see him in this light. I do not think that anyone who has poured over the contents of that box which he packed up when he finally left Cambridge in 1696 and which, though partly dispersed, have come down to us, can see him like that. Newton was not the first of the age of reason. He was the last of the magicians, the last of the Babylonians and Sumerians, the last great mind which looked out on the visible and intellectual world with the same eyes as those who began to build our intellectual inheritance rather less than 10,000 years ago. Isaac Newton, a posthumous child born with no father on Christmas Day, 1642, was the last wonder-child to whom the Magi could do sincere and appropriate homage.[33]

We may see Newton as one of the greatest mathematicians of the world, or perhaps the greatest physicist of all time; as a magician, "Copernicus and Faustus in one," as Keynes saw him; or as the rationalist of the eighteenth century, the mechanist of the nineteenth century, or even as the almost positivist and empiricist of our own century. He himself said:

> I do not know what I may appear to the world; but to myself I seem to have been only like a boy, playing on the seashore, and diverting myself, in now and then finding a smoother pebble or a prettier shell than ordinary, while the great ocean of truth lay all undiscovered before me.[34]

[32]Derek J. de Solla Price, "Newton as an Elder Statesman," *Science*, **158** (December 8, 1967), p. 1298, a book review of J. F. Scott, ed., *The Correspondence of Isaac Newton, Vol. 4, 1694–1709*, published for the Royal Society by Cambridge University Press, 1967.

[33]John Maynard Keynes, "Newton the Man," in Newman, *op. cit.*, pp. 277, 285.

[34]Andrade, *op. cit.*, p. 271.

Some of that "ocean of truth" left undiscovered by Newton will concern us in later chapters.

PROBLEM SOLVING

The problems in this chapter develop understanding of the formulas for gravitational force and centripetal force. The main method of solution for the easier problems is simply to choose the proper formula or formulas, identify the known and unknown quantities, and solve for the unknown.

EXAMPLE 1

The distance between a boy and a girl seated on a sofa is one meter. Their masses are 50 and 100 kilograms, respectively.

(a) What is the gravitational force exerted on the boy by the girl?
(b) What is the gravitational force exerted on the girl by the boy?
(c) If the sofa is frictionless, what is the acceleration caused by this gravitational force on the person of smaller mass?

Solution, part a. The formula for gravitational attraction is

$$F_g = \frac{Gm_1 \cdot m_2}{d^2}$$

The known and unknown quantities are as follows:

m_1 = mass of the boy = 50 kilograms
m_2 = mass of the girl = 100 kilograms
d = distance between girl and boy = 1 meter
G = gravitational constant = 6.67×10^{-11} newton meter²/kilogram²
F_g = gravitational force = unknown

Putting the known values into the equation we have

$$F_g = \frac{6.67 \times 10^{-11} \times 50 \times 100}{(1)^2} \quad \frac{\text{newton meter}^2\text{kilogram kilogram}}{(\text{kilogram})^2\ (\text{meter})^2}$$

$$F_g = 3.3 \times 10^{-7} \text{ newton}$$

This force on the boy is directed toward the girl.

Solution, part b. The force on the girl is of the same magnitude as the force on the boy but in the opposite direction; that is, toward the girl.

Solution, part c. In order to find the acceleration caused by a force, we use the formula

$$F = ma$$

The known and unknown quantities are as follows:

F = gravitational force (found in solving a) = 3.3×10^{-7} newton
m = the smaller mass (that of the boy) = 50 kilograms
a = the acceleration of the boy = unknown

Putting these values in the equation we have

$$3.3 \times 10^{-7} \text{ newton} = (50 \text{ kilogram}) \cdot a$$

Solving for a:

$$\frac{3.3 \times 10^{-7} \text{ newton}}{50 \text{ kilogram}} = a$$

$$a = 6.6 \times 10^{-9} \, \frac{\text{meter}}{\text{second}^2}$$

Example 2

The boy shown in Figure 4-8 has a mass of 25 kilograms (including the mass of the pan), the rope is 3 meters long, and the boy has a speed of 3 meters per second. Find the magnitude of the force in the rope.

Solution. The boy is held in his circular path by the force in the rope, which thus acts as a centripetal force. The formula for centripetal force is

$$F = \frac{ms^2}{r} \tag{11}$$

The known and unknown quantities are as follows:

F = the force in the rope = unknown
m = the mass of the boy = 25 kilograms
s = the speed of the boy = 3 meters/second
r = the length of the rope = 3 meters

Inserting these values into Equation 11 above, we have

$$F = \frac{25 \cdot (3)^2}{3} \cdot \frac{\text{kilogram}}{\text{meter}} \left(\frac{\text{meter}}{\text{second}}\right)^2$$

$$= 75 \, \frac{\text{kilogram meter}}{\text{second}^2}$$

$$= 75 \text{ newtons}$$

EXAMPLE 3

Suppose that the moon were moving in a perfectly circular orbit around the earth at a distance of 400 million meters. The mass of the earth is 6×10^{24} kilograms.

(a) At what speed would the moon be moving?

(b) What would be its acceleration?

Solution, part a. The moon is held in its orbit by the gravitational attraction of the earth. This gravitational attraction, therefore, acts as a centripetal force. The formulas for gravitational attraction and centripetal force, respectively, are as follows:

$$F = \frac{Gm_1m_2}{d^2} \tag{12}$$

$$F = \frac{m_1s^2}{d} \tag{13}$$

In these formulas, m_1 and m_2 are the masses of the moon and the earth, respectively, and d is the distance from the center of the earth to the center of the moon.

It is convenient to do some of the mathematics before putting in the numbers. Since the two forces are the same in this case, we may set them equal:

$$\frac{Gm_1m_2}{d^2} = \frac{m_1s^2}{d}$$

We divide both sides by m_1 and multiply by d:

$$\frac{Gm^2}{d} = s^2$$

or

$$s = \sqrt{\frac{Gm_2}{d}} \tag{14}$$

This gives the speed of the moon or any other satellite traveling in a circular orbit around a planet of mass m_2 at a distance d from the center of the planet. If, for example, we were interested in the speed of an artificial satellite near the earth, then d would be approximately the radius of the earth. The equation does not involve the mass of the satellite but only the mass of the earth. Therefore the speed of the satellite does not depend on its mass.

To find the speed of the moon, we list the known and unknown quantities as follows:

G = gravitational constant
 = 6.67×10^{-11} newton meter²/kilogram²
d = distance from the center of the earth to the center of the moon.
 = 4×10^{8} meters
m_2 = mass of the earth 6×10^{24} kilograms
s = ?

Inserting the known quantities into Equation 14 we have

$$s = \sqrt{\frac{6.67 \times 10^{-11} \text{ newton meter}^2/\text{kilogram}^2 \times 6 \times 10^{24} \text{ kilograms}}{4 \times 10^{8} \text{ meters}}}$$

$$= 10^{3} \frac{\text{meter}}{\text{second}}$$

Solution, part b. The acceleration caused by a force is given by the formula

$$F = m_1 a \tag{15}$$

We have used the mass m_1 since we are interested in the acceleration of the moon. The force is the force on the moon and is the same as F in Equations 12 and 13. We can use either of these equations together with Equation 15 to find the acceleration of the moon. We choose to use Equation 13, and we shall delay putting in the numbers until we are nearly finished. The equations we shall use, then, are

$$F = m_1 a$$
$$F = \frac{m_1 s^2}{d}$$

Since the F's are the same, we equate them:

$$m_1 a = \frac{m_1 s^2}{d}$$

The mass of the moon again cancels out, leaving us with a formula for acceleration under centripetal force that is valid for any object moving in circular motion at a constant speed:

$$a = s^2/d \tag{16}$$

Now we insert the values of s and d to find a:

$$a = \frac{10^6}{4 \times 10^8 \text{ meter}} \frac{\text{meter}^2}{\text{second}^2}$$

$$= 2.5 \times 10^{-3} \frac{\text{meter}}{\text{second}^2}$$

Example 4

Kepler's third law gives a relation between the distance of a planet from the sun and the time it takes for the planet to go once around the sun. Assuming that the planet travels in circular motion around the sun, derive this relation from the laws of motion and gravitation.

Solution. Once again we have gravitational force acting on a planet as a centripetal force to keep it in a circular orbit. The equations for gravitational force and centripetal force are again

$$F = \frac{Gm_s m_p}{d^2} \tag{17}$$

$$F = \frac{m_p s^2}{d} \tag{18}$$

Here the quantities are:
F = gravitational force of the sun on the planet which acts as the centripetal force.
m_s = mass of the sun
m_p = mass of the planet
G = gravitational constant
d = distance of the planet from the sun
s = speed of the planet
From Equations 17 and 18 we have

$$\frac{m_p s^2}{d} = \frac{Gm_s m_p}{d^2} \tag{19}$$

Canceling out the mass of the planet, and multiplying both sides by d, we have

$$s^2 = Gm_s/d \tag{20}$$

This gives us a relation between the speed s of the planet, and the distance d of the planet from the sun. However, we would like the time of revolution, which we shall call T, instead of the speed s. We can easily find the time T it takes the planet to go around the circle of its orbit since we have the radius d of the orbit and the planet's speed s. The total circumference of the orbit of radius d is $2\pi d$. This distance is traversed at a speed s. Therefore

$$\text{distance} = \text{speed} \cdot \text{time}$$
$$2\pi d = sT$$
$$2\pi d/T = s$$

Inserting this value for s in Equation 20, we obtain

$$(2\pi d/T)^2 = Gm_s/d$$

Therefore
$$\frac{4\pi^2 d^2}{T^2} = \frac{Gm_s}{d}$$

$$\frac{d^3}{T^2} = \frac{Gm_s}{4\pi^2}$$
$$\frac{T^2}{d^3} = \frac{4\pi^2}{Gm_s} \qquad (21)$$

The right side of Equation 21 is constant. Therefore we may write:

$$T^2/d^3 = \text{constant} \qquad (22)$$
$$T^2 = (\text{constant}) \cdot d^3 \qquad (23)$$

That is, the square of the time of revolution is directly proportional to the cube of the radius of the planet's orbit.

***EXAMPLE 5**

Calculate the angular momentum of the boy in Example 2 about an axis through his father perpendicular to the ice.

Solution. The formula for angular momentum is

$$\text{Magnitude of angular momentum} = msr$$

Here, m = mass of the boy = 25 kilograms
s = speed of the boy = 3 meters/second
r = distance from the axis = 3 meters

Inserting these values into the formula, we have

Magnitude of angular momentum

$$= 25 \times 3 \times 3 \text{ kilogram } \frac{\text{meter}}{\text{second}} \text{ meter}$$
$$= 225 \text{ kilogram } \frac{\text{meter}^2}{\text{second}}$$

***EXAMPLE 6**

Suppose that the man in Example 5 pulls in on the rope, thus shortening the distance between him and the boy to two meters.

What is the resulting speed of the boy?

Solution. This is an example of conservation of angular momentum. When the man hauls in the rope he exerts an additional force on the boy, but this force is directed toward the axis and therefore does not change the angular momentum about this axis. Therefore, the angular momentum of the boy after the rope is pulled in, msr, is equal to the angular momentum before the rope is pulled in, 225 kilogram meter²/second;

$$msr = 225 \text{ kilogram meter}^2/\text{second}$$

In this formula,

m = the mass of the boy = 25 kilograms
s = the new speed of the boy = unknown
r = the new distance from the axis = 2 meters

Inserting these values in the equation,

$$25 \text{ kilogram} \times s \times 2 \text{ meter} = 225 \text{ kilogram meter}^2/\text{second}$$

Solving this for s we have

$$s \times 50 \text{ kilogram meter} = \frac{225 \text{ kilogram (meter)}^2}{\text{second}}$$

$$s = \frac{225}{50} \frac{\text{kilogram meter}^2}{\text{kilogram meter second}}$$

$$= 4.5 \text{ meters/second}$$

EXAMPLE 7

Show that the gravitational force, F_g, on an object is given by the equation $F_g = ma_g$.

Solution:

The force, F, on any object is given in terms of its mass, m, and its acceleration, a, by the equation $F = ma$. If the gravitational force, F_g, is the only force acting on the object, then the acceleration, a, is the gravitational acceleration, a_g. Hence the gravitational force acting on the object (its weight) is given by:

$$\text{Weight} = F_g = ma_g$$

This equation is correct for the weight even when there are other forces on the object. In such a case, however, since F_g is not the total force, a_g is not the actual acceleration.

Summary

- I. FUNDAMENTAL LAWS OR PRINCIPLES
  - A. THE LAW OF GRAVITATIONAL INTERACTION. Every two objects attract each other with a force of magnitude $F_g = Gm_1m_2/d^2$. The gravitational masses m_1 and m_2 in this equation represent, within experimental error, the same quantity as the ordinary mass m in the equation $F = ma$.
  - B. CONSERVATION OF ANGULAR MOMENTUM
- II. QUANTITIES
  - A. DERIVED QUANTITIES
    - *1. Angular momentum, a vector (actually a pseudovector).
    - *2. Torque, a vector (actually a pseudovector).
  - B. FUNDAMENTAL CONSTANTS
    - 1. The gravitational constant $G = 6.67 \times 10^{-11} \dfrac{\text{newton meter}^2}{\text{kilogram}^2}$
- III. APPLICATIONS
  - A. WEIGHT
  - B. ACCELERATION OF GRAVITY. a_g, the same for all objects at the same location.
  - C. CIRCULAR MOTION AND CENTRIPETAL FORCE
  - D. "SYSTEM OF THE WORLD"
  - E. SATELLITES AND "WEIGHTLESSNESS"
- IV. METHODS OF SCIENCE
  - A. MODELS—THE NEWTONIAN MODEL
  - B. THE EXPERIMENTAL PHILOSOPHY

Questions

1. What else is constant when an object moves in a circle at constant speed? What is not constant?
2. Suppose that each of the following objects is traveling in circular motion: (a) a rock in a sling, (b) water in a spinning bucket, (c) an airplane in a horizontal circle, and (d) an automobile on a circular race track. How is the centripetal force applied in each case?
3. What are the units of weight?
4. What is the difference between weight and mass?
5. Did Newton know how to orbit a space capsule?
6. Describe the motion of an astronaut who leaves his capsule for a space

"walk." Is he massless? Is he weightless? Why does he not fall to the ground?

7. Does an apple exert a gravitational force on the earth? How does this force compare with the force exerted on the apple by the earth?

8. A cannonball and a postage stamp are allowed to fall in a vacuum. They are released at the same time from the same height. Describe the motion of each. What is the acceleration of each object? Are the forces on the two objects the same as they are falling? Why is there no contradiction between the answers to the last two questions?

9. If you were suddenly transported to another planet, how would you determine its mass?

10. We usually think of the sun as stationary and the earth as moving around it. Actually, if the earth were the only planet, the earth and the sun would both revolve about a common center which is inside the sun, close to its center. How do we know that both the sun and the earth revolve to some extent? (Assume that Newton's laws of motion and his law of gravitation are correct.)

11. Will the system in Figure 4-22 move if there is no friction but M is a much greater mass than m?

12. Why will a flywheel fly apart if it is spun too rapidly?

13. Why is it difficult to turn an automobile which is moving along an icy road?

14. In Figure 4-23 the monkey and the object have the same mass. What happens if the monkey starts to climb the rope? (Ignore friction in the pulley.)

15. In Figure 4-23, we now allow the monkey to be more massive than the object. Is there any way in which the monkey can get the object to rise by climbing his end of the rope?

16. A ball is whirled in a horizontal circle on the end of a string. There is a force in the string constantly pulling the ball toward the center of the

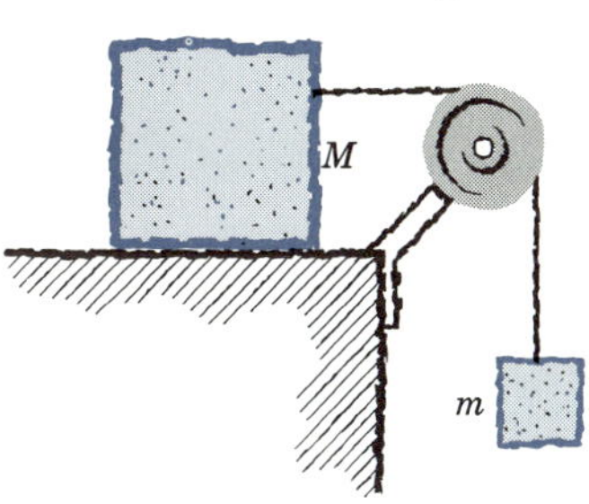

Fig. 4-22

Fig. 4-23

circle; yet the ball keeps going around at the same speed. How can this fact be reconciled with Newton's second law of motion, which states that force causes an acceleration?

17. In what way is gravitational mass different from ordinary mass?

18. In what way is gravitational mass the same as ordinary mass?

°**19.** Figure 4-24 shows two ways of hanging a swing. If we use the same kind of rope for each, which swing will support more weight, and why?

*__20.__ The string attached to a lighter-than-air ballon is just strong enough to hold it down when the air is calm. What happens when the wind starts to blow horizontally, and why?

*__21.__ Can we ever pull a flexible clothesline so tight that it becomes perfectly straight?

*__22.__ Figure 4-25 shows the same picture hung in two different ways. In which way is there less tension in the rope, and why?

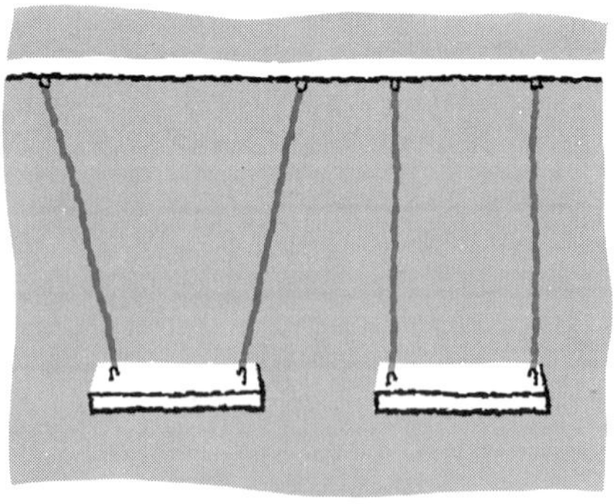

Fig. 4-24

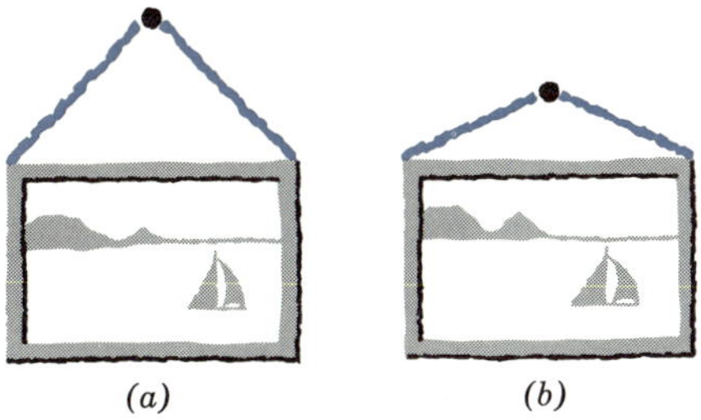

Fig. 4-25

23. Explain why mass can be accurately measured by weighing.
24. Would the process of measuring mass by weighing give the same result on the moon as it does on the earth?
25. Is the acceleration of gravitation always and everywhere the same? Explain.
26. The air resistance to a moving object increases with speed. How could this account for the fact that falling objects generally reach a maximum velocity if they fall far enough?
*27. Why does a comet travel fastest in its orbit when it is nearest the sun?
*28. Suppose that a rotating ball of hot gas cools down and contracts. What happens to its rate of rotation? Its angular momentum?
*29. Does a car traveling down a straight street have any angular momentum around the axis defined by a nearby telephone pole? Explain.
*30. In Figure 4-20, suppose that the man merely turned the wheel halfway, until its axis was horizontal. What would be the resulting motion of the system, compared with the experiment in which the wheel is turned completely upside down?
*31. What forces are exerted on the man's hands by the spinning bicycle wheel in Figure 4-20 as he turns it upside down?
*32. Which of the following vectors are true vectors and which pseudovectors: force, displacement, momentum, acceleration, torque.
*33. Find a position of the mirror in Figure 4-21 that will reverse the angular-momentum vector but not the velocity vector.
*34. Could there be an equation expressing a physical law in which one side of the equation was a vector quantity and the other side was a pseudovector quantity? Explain.
35. What sort of effect would you expect if the gravitational constant were ten times its actual value?
36. What is the velocity of a ball thrown straight up when it is at its highest point? What is its acceleration at that point?

37. What is the force on the ball in Question 36 when it is at its highest point?

38. In what sense can Newton's system of the world be regarded as a model?

*39. If the man in Figure 4-8 pulls in or lets out on the rope, thus changing the distance at which his son swings around him, is angular momentum conserved in the process? If he shortens the distance at which the boy swings around him by simply standing still and letting the rope wind around his own body, is the angular momentum of the boy still conserved? Explain.

40. In Figure 4-19, after the man has reached up and started the bicycle wheel spinning, he turns the wheel upside down, without stopping it from spinning. What happens to his own rotation?

41. The law of gravitational force is expressed in Equation 1 without any reference to a reference frame. Is this a mistake?

Problems

1. What is the weight of a 50-kilogram boy?

2. What is the weight of a 50-kilogram boy on the moon? (The moon has a mass of 6.7×10^{22} kilograms and a radius of 1.74×10^{6} meters.)

3. What is the gravitational acceleration on the moon? (See Problem 2, above.)

4. An object of six kilograms mass is traveling in a circle of radius five meters at a speed of ten meters per second. Find the centripetal force and the acceleration of the object.

5. If I wish to keep a 500-kilogram object moving in a circular orbit of radius 10,000 kilometers at a speed of five kilometers per second, how much force must I exert on it? (A kilometer is equal to 1000 meters.)

6. How large is the gravitational force of attraction between two astronauts in space if they both have a mass of 80 kilograms and they are 40 meters apart? What is the acceleration resulting from this attraction?

7. What is the gravitational force of attraction between two atoms in space if each has a mass of 2×10^{-27} kilogram and they are 10^{-7} meter apart? What is the acceleration resulting from this force?

8. What is the mass of a boy who weighs 600 newtons?

9. A 20-kilogram rocket is shot upward. The thrust developed by its engine is 600 newtons. What is its acceleration?

10. In Example 2, page 164, suppose that the force in the rope is 100 newtons, and the mass of the boy and his distance from his father are as before. What would the speed of the boy have to be? What would be his acceleration?

12. The planet Mars travels in a nearly circular orbit around the sun at a distance of about 2.3×10^{11} meters. One complete orbit takes 1.88 years. What is the mass of the sun?

13. The acceleration due to gravity on the moon is about one sixth that on the earth. The radius of the moon is 0.27 times that of the earth. How does the mass of the moon compare with that of the earth?

14. The acceleration due to gravity on Jupiter is 25.9 meters/second^2, and the radius of Jupiter is 7×10^7 meters. What is the mass of Jupiter?

***15.** In Example 6, page 168, suppose that the father lets out on the rope an additional meter. What is the resulting speed of the boy?

16. Calculate the approximate speed of a satellite orbiting near the surface of the earth.

***17.** Calculate the angular momentum per kilogram of the satellite in Problem 16 with respect to an axis through the center of the earth perpendicular to the plane of the satellite's motion.

***18.** Find the angular momentum of a hoop of mass m and radius r rolling on the ground at a speed s. Take the axis in the ground but perpendicular to the hoop.

***19.** In Figure 4-26 four men are turning a mill. Each man exerts a force of 400 newtons at a distance of three meters from the axis. How much torque does each man exert? What is the total torque on the mill exerted by all the men together? What is the total force on the mill exerted by all the men together?

***20.** The wheel in Figure 4-27 is originally at rest. A man sets it in motion by pulling on the string for 20 seconds with a force of 10 newtons. How

Fig. 4-26

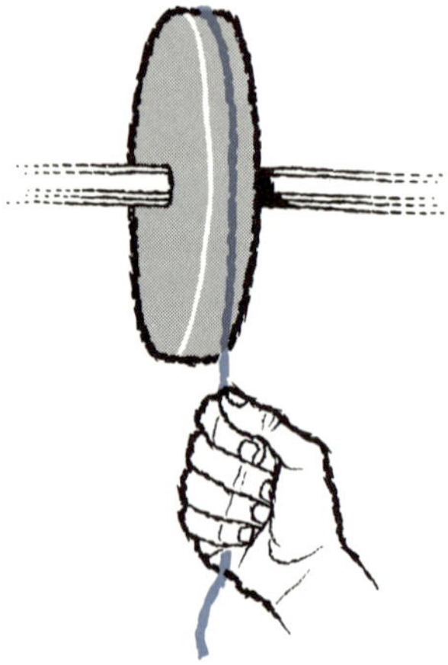

Fig. 4-27

much angular momentum does this supply to the wheel? The radius of the wheel is 0.4 meters. (Disregard friction between the wheel and the axle.)

*21. Show that Kepler's second law (p. 39) is a law of conservation of angular momentum for planets.

22. Show that the law of gravitational interaction as expressed obeys position, mirror, and orientation symmetry.

5

*FURTHER APPLICATIONS OF THE LAWS OF FORCE AND MOTION

FORCES ON OBJECTS

The Resultant Force Responsible for Motion

Frequently we are interested in the motion of an object that is subject to more than one force. In order to apply the laws of motion to such an object and to find its acceleration, it is necessary to add up all of these forces as vectors. It is this vector sum, or resultant force, that determines the acceleration of the object. To obtain the resultant we must be sure that we take into account all of the forces acting on the object.

In order to find all the forces on an object, we first determine all the interactions in which the object participates. All of the interactions we have discussed up to now have been the result of the direct contact of an object with other objects (through rods, strings, touching, etc.) or of the gravitational interaction. Later we will discuss other kinds of interaction. Each of the interactions the object has with another object will produce *one force* on the object. (Each interaction will also produce one force on the other object, but since this force does not affect the motion of the object with which we are concerned, we can ignore it here.) Different parts of the same object can have interactions with one another, but these interactions may be ignored when we are studying the motion of the object as a whole.

To make sure that all of the forces are considered, it is helpful to draw a diagram of the object alone, separate from its surroundings. The next step in the procedure is to represent each of the forces acting on the object by an arrow drawn on the diagram. The vector resultant of these forces is then obtained by the method of vector addition described in Chapter 2. This resultant force is responsible for the acceleration of the object.

Before considering an example, let us outline the procedure for finding the resultant force on an object, the force responsible for its acceleration. The steps are as follows.

1. Select the object in whose motion you are interested.
2. List each interaction that this object has with other objects or bodies.
3. Draw a diagram that isolates the object, and draw an arrow on the diagram for each force acting on the object.
4. Obtain the resultant force by vector addition.
5. If desired, use the equation $F = ma$ to obtain the acceleration of the object from its mass and the resultant force.

As an example of this procedure, we shall find the resultant force on a boat that is being pulled upstream by two boys (Fig. 5-1). In accordance with the first step above, we have selected the boat as the object in whose motion we are interested. The second step is to list all of the interactions the boat has with other objects or bodies. Since we know that the acceleration of the boat is in a horizontal direction, we know that the resultant force must also be in a horizontal direction. Therefore, we ignore all vertical forces caused by the interaction of the boat with its surroundings. For example, we ignore the gravitational force of the earth pulling down on the boat. We will show later that it is all right to ignore these vertical forces when we

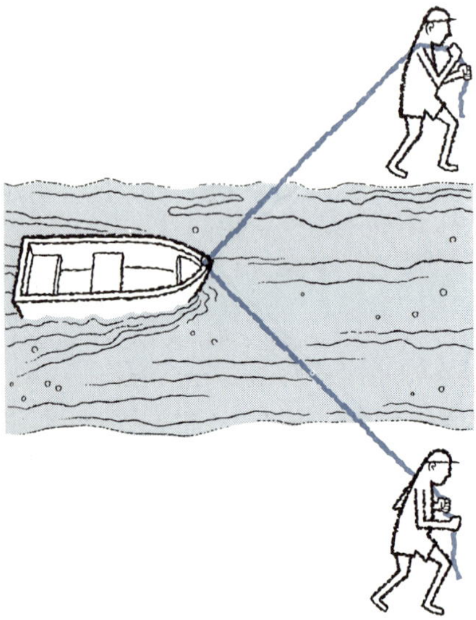

Fig. 5-1

are dealing with motion in a horizontal plane. We find that there are just three interactions that produce horizontal forces on the boat. We list these interactions as follows.

(a) The interaction with the rope being pulled by the boy on the right bank.
(b) The interaction with the rope being pulled by the boy on the left bank.
(c) The interaction with the water due to the retarding friction as the water passes by the boat.

In Step 3 we draw a diagram of the isolated object together with

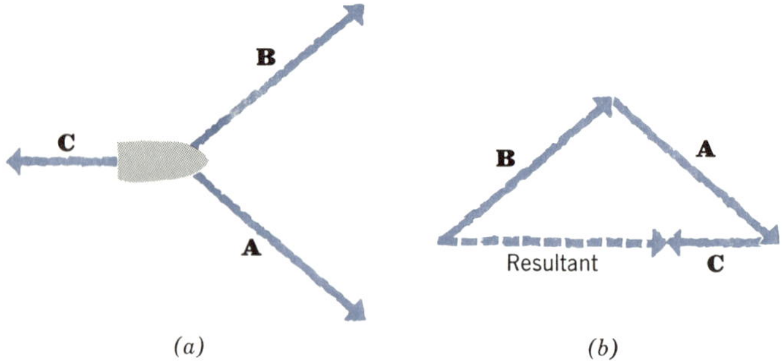

Fig. 5-2

arrows representing the forces acting on it. Since there are three interactions, there will be three forces acting on the boat. This step is illustrated in Figure 5-2*a*. We are assuming that the magnitudes and directions of the forces are known so that the vectors can be drawn to scale, as indicated in the figure. Step 4, obtaining the resultant force, is illustrated in Figure 5-2*b*. (We are using the "head to tail" rule for adding vectors, discussed on p. 77.) If we performed Step 5 of the procedure, we would find, of course, that the acceleration is in the same direction as this resultant force.

In the example we have chosen, the resultant force turned out to be upstream. If the three forces exerted by the boys and the water were different, however, the resultant force could turn out to be downstream, to the side, or even zero. These possibilities are illustrated in Figure 5-3.

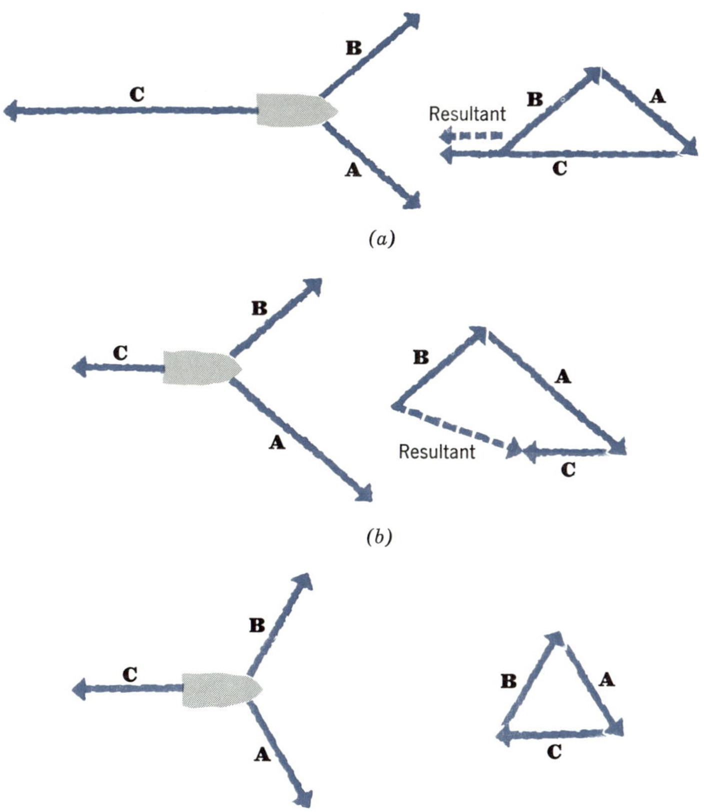

(*a*)

(*b*)

(*c*)

Fig. 5-3

Equilibrium

An object is in *equilibrium* whenever all of the forces acting on it add up to zero. When this resultant force is zero, the acceleration is also zero. Conversely, any object which is not accelerating is in equilibrium, and the forces acting on it must add up to zero. Since an object can be moving without accelerating an object in equilibrium does not have to be at rest. It may either be at rest or in motion with a constant velocity. When the vector forces on an object in equilibrium are added together by the "head to tail" procedure, the head of the last vector comes right up to the tail of the first vector, forming a closed figure (see Fig. 5-3*c*).

A simple example of an object at rest is the motionless moose shown in Figure 5-4. Following the steps outlined above, we first identify the object we are interested in—namely, the moose. Then we list the interactions between the moose and other objects.

(a) The gravitational interaction between the earth and the moose.
(b) The interaction of direct contact between the moose's body and the ground.

This second interaction consists of a pair of forces. One of these forces is exerted by the moose pressing down on the ground, and the other is exerted by the ground pressing up on the moose. The force of the ground pushing up on the moose is responsible for the difference in the motion of the moose in Figure 5-4 and its motion in Figure 5-5, where the ground has been removed. In Figure 5-5 the moose is involved in only one interaction, that of gravitation, and he is therefore accelerating. The moose in Figure 5-4 is at rest; the force of the ground pushing up on him is required to keep him from falling.

We next draw a diagram in which we isolate the moose and represent the two forces acting on him by means of arrows (see Fig 5-6).

Fig. 5-4

Since the moose remains at rest, he is in equilibrium, and the resultant of the two forces acting on him must be zero. The force of the ground pushing up must have the same magnitude as the gravitational force of the earth pulling him down. This same analysis can be applied to any object on the earth that has no vertical acceleration. This explains why we were able to eliminate the discussion of all vertical forces in the example shown in Figure 5-1. The force of gravitation pulling down on the boat is balanced by the force of the water pushing up on it.

The Moose-Log Paradox

We now hook the moose to a log and ask him to pull it for us (Fig. 5-7). The learned moose objects, however, quoting from Isaac Newton.

> If a horse draws a stone tied to a rope, the horse (if I may so say) will be equally drawn back towards the stone; for the distended rope, by the same endeavor to relax or unbend itself, will draw the horse as much towards the stone as it does the stone towards the horse, and will obstruct the progress of the one as much as it advances that of the other.[1]

The moose correctly interprets this Newtonian wisdom to mean that however hard he pulls on the log, it will pull back on him just as

Fig. 5-5

[1]Isaac Newton, *The Mathematical Principles of Natural Philosophy and His System of the World*, translated by Andrew Motte, 1729; rev. by Florian Cajori, Berkeley: University of California Press, 1934, p. 14.

hard. "Therefore," says he, "it is useless for me to attempt to move this log (or anything else, for that matter), since any force I exert on the log to move it will be exactly balanced by the force with which the log pulls back."

In order to expose the fallacy propounded by this lazy beast, let us analyze the forces on the log as the moose begins to pull it. We have selected the log as the object of our interest. Since the motion of the log is horizontal, we shall again ignore all vertical forces. Thus there are only two interactions that we need to deal with: the interaction between the log and the moose, and the interaction between the log and the ground. The second interaction is made up of the frictional forces produced as the log is dragged across the ground. Thus, when we isolate the log in Figure 5-8, we have only two forces to consider. They are the force exerted on the log by the moose to the right, and the frictional retarding force acting on the log to the left. To start the log moving, all that is required is that the moose exert a force on the log greater than the retarding force of friction. In this condition there will be a resultant force to the right, and the log will accelerate in the direction of the moose. If, after the log is in motion, the moose wants to pull it along with a constant velocity, he must exert a force on the log equal in magnitude to the force of friction on the log. In this condition, the resultant force on the log will be zero, and therefore the log will be in equilibrium. It will continue to move, in equilibrium, with a constant velocity as long as this resultant is zero.

Notice that in this analysis we have not mentioned the force of the log on the moose. This force is not acting on the log; although it does influence the motion of the moose, it does not affect the motion of the log. The fallacy of the moose's statement arises from the fact that one of the forces he was talking about does not even act on the object he was supposed to move. The interaction between the moose and

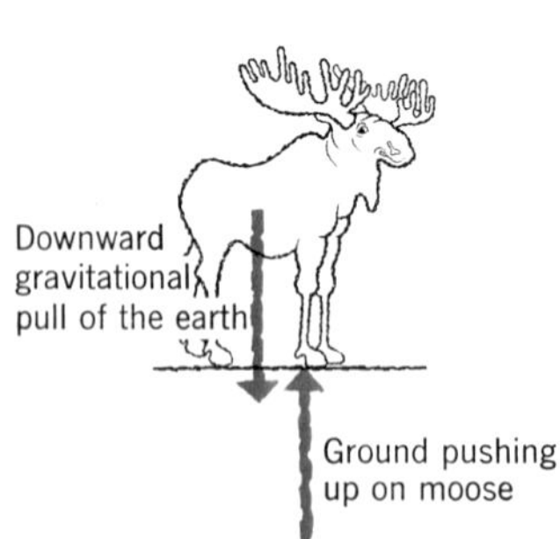

Fig. 5-6

the log does involve the two forces he described, but only one of them acts on the log.

*FORCES IN FLUIDS

Fluids at Rest

A fluid is a substance that can flow. Examples of fluids are liquids (such as water) that can be poured and gases (such as air) that tend to fill their containers completely. Two derived quantities, density and pressure, are especially useful in describing the properties of fluids.

DENSITY. The *density* of a fluid (or, indeed, of any body of matter) is equal to the amount of mass it has per unit of volume. Water, for example, has a greater density than air because there is more mass in a given volume of water than there is in the same volume of air—a bucket of water has more mass in it than the same bucket filled with air. Since the density is the mass per unit volume of a substance, we can express it in a formula as follows:

$$D = \frac{m}{V} \tag{1}$$

V is the volume of a certain amount of matter, m is the mass of this matter, and D is its density. For example, suppose that we wish to find the density of water. We happen to know that three cubic meters

Fig. 5-7

of water has a mass of 3000 kilograms. Since there are 3000 kilograms in three cubic meters there must be 1000 kilograms in one cubic meter. Therefore, the density of water is 1000 kilograms per cubic meter.

PRESSURE. When a force exerted on a surface is distributed over the surface (as is often the case in fluids) and not just applied at a single point, it is convenient to use a new derived quantity, *pressure.* The pressure on any region of a surface is equal to the total force on the region divided by the area of the region. That is to say, the pressure P is equal to the force F divided by the area A:

$$P = \frac{F}{A} \tag{2}$$

To show that pressure, rather than force, is the important factor in some physical effects, consider the dancer in Figure 5-9. If she weighs 506 newtons (115 pounds weight), she exerts a total force of 506 newtons on the floor, whether she is standing on the floor or sitting on it. The pressure, on the other hand, is much different in these two situations. When she is sitting down, her force is spread over a much more extensive area than it is when she rocks back on a single heel while dancing. If the bottom of her heel has a very small area, she might even dent the floor. The 506 newtons of force exert a much larger pressure when they are distributed over a smaller area.

PRESSURE IN FLUIDS. The philosopher and mathematician Blaise Pascal (1623–1662) studied fluids at rest, and as a result he obtained several principles that describe pressure in fluids. He found, for example, that the pressure in a fluid is perpendicular to any surface in contact with the fluid. In fact, one may consider that the pressure is perpendicular to any chosen surface in the fluid itself, regardless of its position or orientation. Furthermore, at any given point in the fluid, the magnitude of the pressure is the same in all directions. Figure 5-10 illustrates these facts. The piston in these illustrations is subjected to the same pressure in each of its orientations. In each case the pressure is perpendicular to the surface of the piston.

Pascal also found that the pressure in a fluid increases with depth

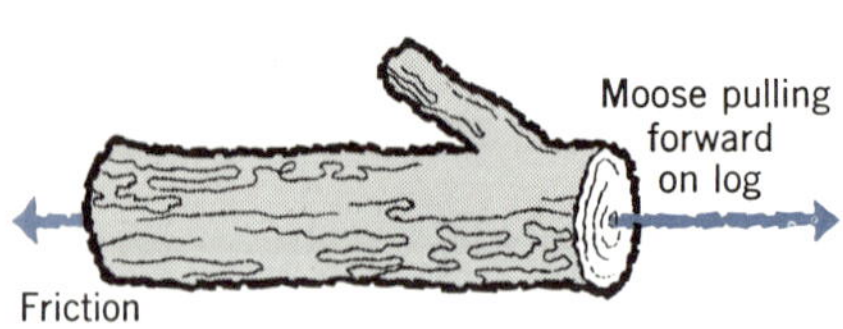

Fig. 5-8

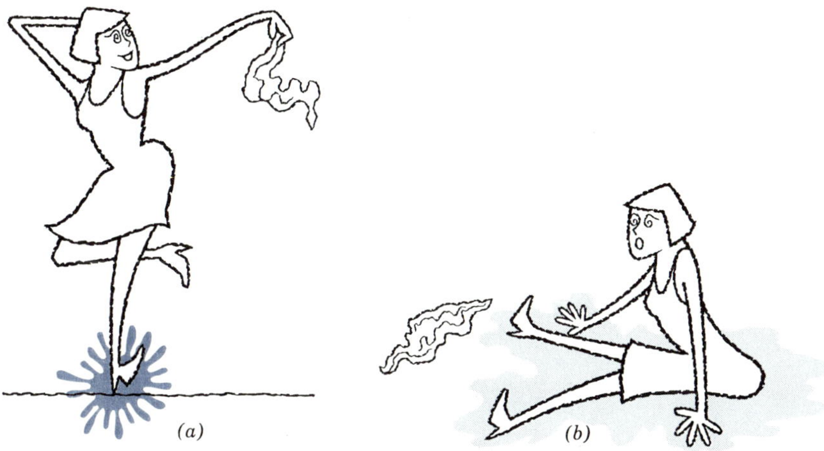

Fig. 5-9

and does not depend on the total volume of the fluid. For example, in the connected vessels shown in Figure 5-11, the pressure is the same at any point on the same level, as, for example, at points P_1, P_2, P_3, P_4, and P_5. This is true even though the vessels themselves differ greatly in size and shape. Because the pressure is the same at any point on the same level, the level of the surface is the same in all three vessels. (For small tubes, etc, there are exceptions to this rule.) If the pressure were not the same at the same level in all three vessels, water would flow from one vessel to another until the pressure was equalized. The pressure at any given level in the connected vessels is, of course, greater than the pressure at higher levels, since each level must help to support the weight of the water in all the levels above it.

The pressure in a fluid also depends on the density of the fluid.

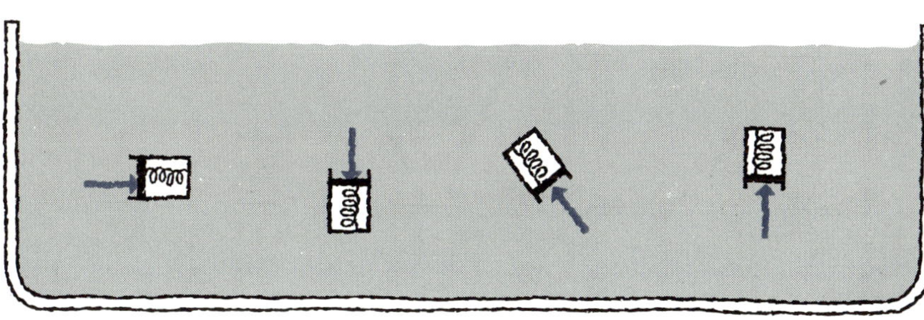

Fig. 5-10

Since mercury, for example, is 13.6 times more dense than water, the pressure at a given depth in a vessel containing mercury will be 13.6 times greater than the pressure at the same depth in water.

The entire atmosphere is a fluid sea of gas that covers us all, exerting a pressure of 100,000 newtons per square meter on objects on the surface of the earth. This atmospheric pressure arises from the weight of the air, from the surface to the "top" of the atmosphere. Since the earth's atmosphere merges with the atmosphere of the sun, its "top" is not easy to define. However, the density in the upper regions is so small that the weight of the air above 5×10^4 meters contributes less than one percent to the total weight of the atmosphere.

We measure atmospheric pressure by using an instrument known as a barometer. Figure 5-12 shows how to make one kind of barometer—a mercury barometer. A glass tube, about a meter long, is filled with mercury. Then the open end of the tube is covered (to prevent loss of mercury) while it is immersed in a vessel containing mercury. In this process no air is allowed to enter the tube. When the open end is again uncovered, some of the mercury flows out of the tube into the dish, leaving a column of mercury in the tube with a near vacuum above it at the top (Fig. 5-12*d*). The pressure is essentially zero in the vacuum. Therefore, the pressure at the point P is due only to the weight of the mercury column above it. The pressure at the point Q at the surface of the mercury in the dish outside the tube is due only to the atmospheric pressure. Since these two points are connected by fluid mercury and are at the same level, the pressure at the two points must be equal; that is, the pressure represented by the height of the mercury column is the same as the pressure of the atmosphere outside. If the atmospheric pressure increases, the mercury will rise higher in the tube. If the atmos-

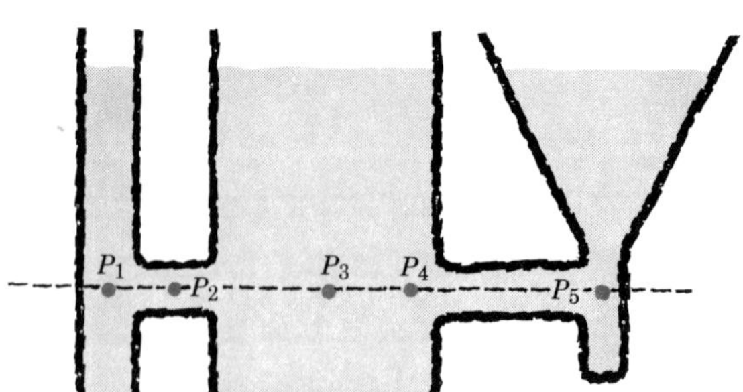

Fig. 5-11

pheric pressure decreases, the mercury will fall lower in the tube. Thus the mercury column can be used to measure air pressure. Changes in atmospheric pressure are used to help predict the weather. Since the atmospheric pressure also changes with altitude, the barometer can be used to measure altitude.

Air pressure also helps us to breathe, as illustrated in the simple model of the human breathing apparatus shown in Figure 5-13. To begin with, everything is in equilibrium and all pressures are equal to atmospheric pressure. The breathing muscles pull on the diaphragm in order to expand the volume inside the chest cavity. This reduces the pressure on the outside of the lungs, represented as two balloons in the figure. The pressure inside the lungs is still the atmospheric pressure, which therefore expands the lungs, bringing air into them. When the diaphragm reduces the volume enclosing the lungs, the pressure in the chest cavity exceeds atmospheric pressure, causing the lungs to collapse and thus forcing air out.

BUOYANCY. If you have ever tried to remove a large rock from a pool of water, you may have discovered that although the rock can be lifted easily while submerged, it is not so easily lifted out of the water and onto the bank. Somehow a fluid is able to exert a force which helps us to lift rocks or other submerged objects. This force is called a *buoyant* force.

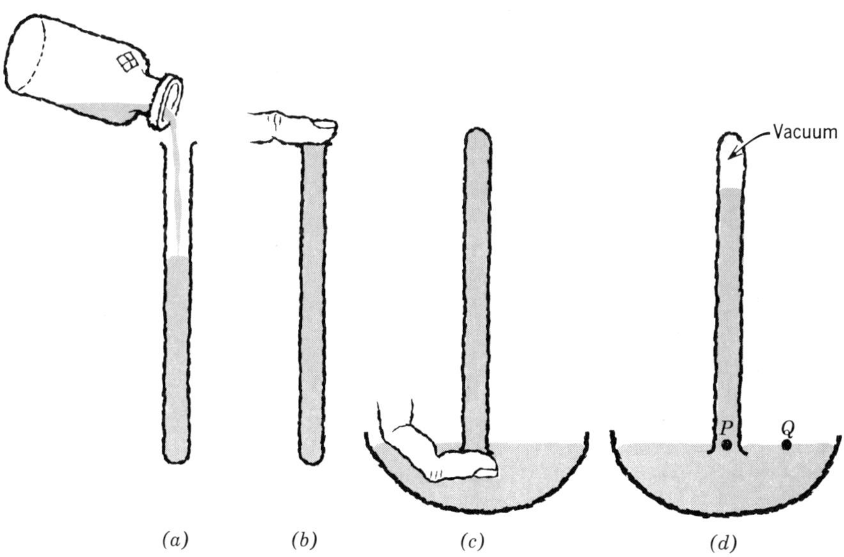

Fig. 5-12

Since the most important interaction between the water and the rock arises in the direct contact between the fluid and the rock, the buoyant force must come from the pressure exerted on the rock by the water. To see how the pressure in a fluid can produce a buoyant force, we first consider the forces on a small part of the fluid itself.

Figure 5-14*a* indicates the forces acting on a small volume of fluid which is submerged in a vessel of the same fluid. The pressure acts on the surface of this volume, perpendicular to it. The only other force on the fluid inside the volume is the gravitational force, which pulls it downward with a force equal to its weight. In order for this volume of fluid to remain at rest—that is, suspended in the fluid—the total resultant of all the forces arising from the pressure on its surface must balance this gravitational pull. Therefore, the pressure must cause a resultant upward force on the volume of fluid with a magnitude equal to the magnitude of its weight. This is the buoyant force. We can see that such a force should exist; since the pressure increases with depth in a fluid and, therefore, the pressure pushing up on the bottom of this small volume of fluid will be greater than the pressure pushing down from the top. The result of the pressure from all directions is a net upward force—the buoyant force.

We have found the direction and magnitude of the buoyant force that pushes up on a certain volume submerged in a fluid. The direc-

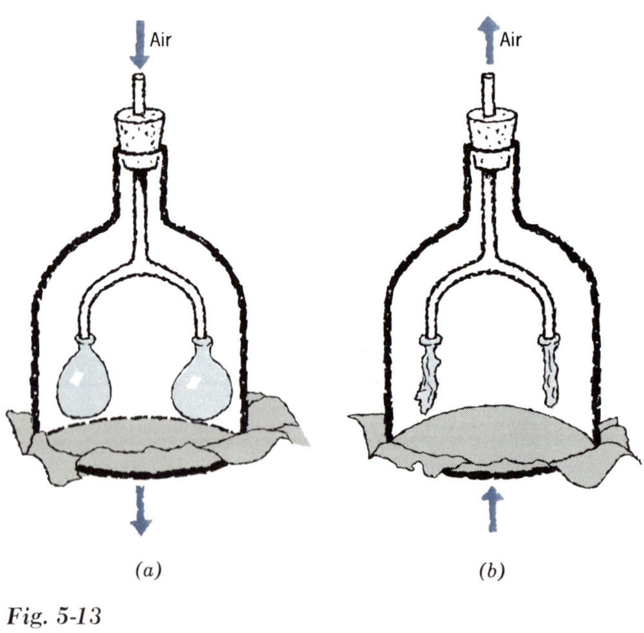

Fig. 5-13

tion is upward and the magnitude is equal to the magnitude of the weight of the same volume of fluid. Suppose now that this same volume shown in Figure 5-14*a* is filled not with water but with wood, as in Figure 5-14*b*. The pressure of the fluid is not changed by the fact that the volume is filled with wood instead of water, since the depth to each point of the volume's surface is still the same. Therefore, the buoyant force, caused by the pressure of the surrounding fluid, will still be the same. The buoyant force depends on the volume of the object and not on its nature.

We shall now explain, using the above principles, why some objects sink and some objects float. Figure 5-15 represents three objects submerged in water. One of them is wooden, one is a rock, and one is a small volume of water itself. All of these objects have the same volume and, therefore, they will all have the same buoyant force exerted on them by the surrounding water. They do not all have the same weight, however. The wood has less weight than the same volume of water, and the rock has more weight than the same volume of water. The water, needless to say, has the same weight as the same volume of water. The buoyant force on our small volume of water exactly balances the weight of the water inside; therefore, this small volume of water is in equilibrium. However, the weight of the wood is less than the buoyant force, and the resultant of these two forces pushes the wood up. The weight of the rock is greater than the buoyant force, and therefore the resultant pulls the rock down and it sinks.

Instead of comparing the weights of the wood and the rock with the weight of water, however, we could have compared their masses, since their masses are proportional to their weights. If we do this, it is the mass for a given volume that determines whether an object will rise or sink. Since it is the mass for a given volume and not just the mass alone that is involved, it is really the density that determines whether an object will rise or sink. Objects that are more

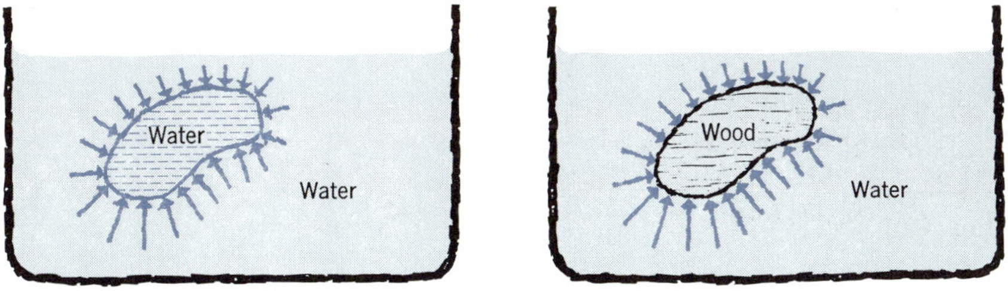

(*a*) (*b*)

Fig. 5-14

dense than water will sink in water. Objects that are less dense than water will rise in water. The forces of weight and buoyancy also account for the equilibrium of floating bodies. A floating body sinks partly into the water and therefore occupies a certain volume below the surface level of the fluid (see Fig. 5-16). The weight of this same volume of fluid is equal to the weight of the floating object. This is true because the buoyant forces around this volume must be sufficient to support an equal volume of fluid.

The theory of buoyant forces was first developed by Archimedes of Syracuse (287–212 B.C.), one of the greatest mathematicians of all time, whose studies of equilibrium inspired Galileo. The story is that Archimedes was soaking in a public bath, while contemplating a problem set him by King Hiero of Syracuse. The king suspected that the gold in a crown had been diluted with a less dense metal. The mass of the crown could be obtained very easily by weighing it, but to check the density would require a very accurate measurement of the volume as well. One could find this by melting the crown, of course, but this was hardly a satisfactory solution. Archimedes realized that his own loss of weight in the water of his bath depended on his volume, and in this he saw the solution to the case of the king's crown. Elated with this discovery (so it is told), Archimedes jumped from his bath and ran home, shouting "Eureka!" (I have found it!).[2]

Pressure in Moving Fluids

In this section we shall discuss an effect first discovered by Daniel Bernoulli (1700–1782), a member of a famous Swiss family of scien-

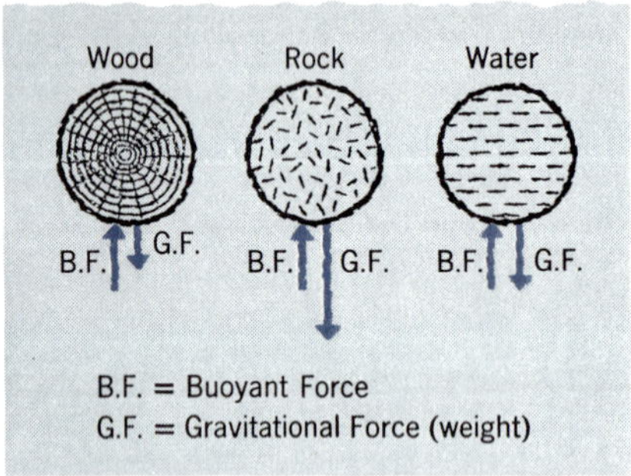

Fig. 5-15

[2]See Vitruvius, *On Architecture*, included in James R. Newman, ed., *The World of Mathematics*, Vol. I, New York: Simon and Schuster, 1956, p. 185; and Galileo Galilei, *The Little Balance*, included in Laura Fermi and Gilberto Bernardini, *Galileo and the Scientific Revolution*, Greenwich, Conn.: Fawcett Publications, 1965, p. 114.

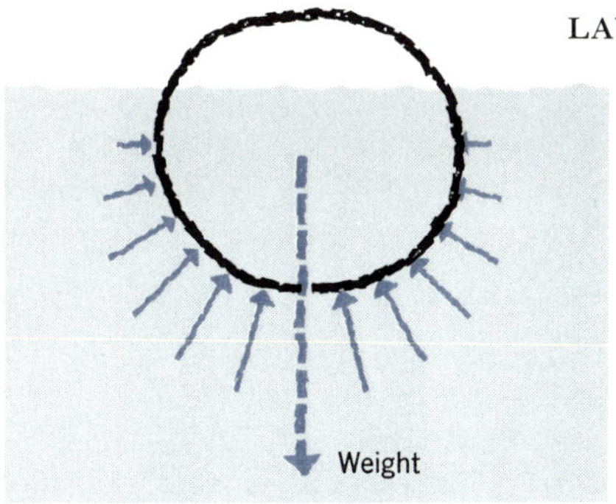

Fig. 5-16

tists. According to the Bernoulli effect, in a moving fluid the pressure in the low-speed regions is higher than the pressure in the high-speed regions. Within the stream, therefore, the higher the speed, the lower the pressure. This may seem rather paradoxical, but it is correct. (Pressure effects resulting from increased depth and other matters may obscure this effect.)

Figure 5-17 shows water flowing smoothly through a tube. The tube is narrower in the center section than it is at either end; and therefore, the fluid must flow faster through this region. (If the same amount of fluid must flow through a smaller channel without being compressed, it must move faster.) Bernoulli's effect states that the pressure in this section must be less than it is on either side. We can actually show that this must be true by using the second law of motion. Suppose that we have a small volume of fluid just entering the narrow region of the tube. Since it is entering a high-speed region, it must accelerate. The only forces on this volume of fluid arise from the pressure of the surrounding fluid. The pressure must be greater behind it than in front of it, or there will not be a resultant force on it to accelerate it. Where the tube becomes wider, the situation is reversed.

Another example of this effect is illustrated in Figure 5-18, which shows a pingpong ball suspended in an air jet. We are not concerned

Fig. 5-17

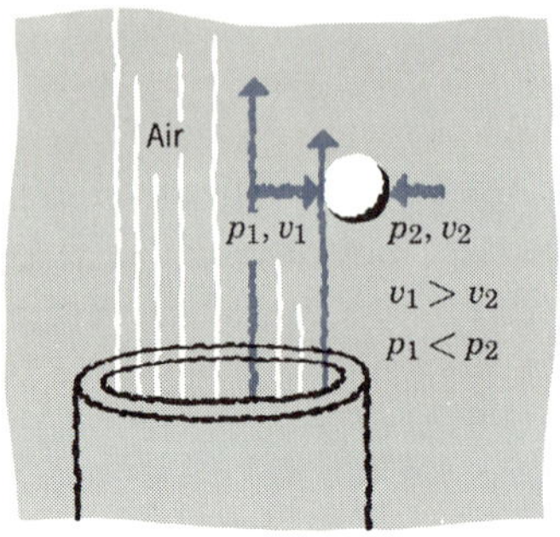

Fig. 5-18

with what keeps the ball up (the force of the air pushing against it from below balances its weight). The real question is why it does not fall out of the air jet to the side even when the air jet is moved around or tilted. Bernoulli's effect helps us to explain why the ball stays in the jet. Suppose the ball were at the edge of the jet, in danger of moving out of the stream of air. The air of the jet, on one side of the ball, is moving faster than the air outside the jet, on the other side of the ball. The faster air has the lower pressure; therefore, the ball is pushed back into the jet.

Still another example is illustrated in Figure 5-19. One cannot make the cardboard fall away from the bottom of the spool by blowing smoothly through the tube. According to Bernoulli's effect, air flowing between the bottom of the spool and the cardboard must have a lower pressure than the stationary air on the other side of the cardboard. This difference in pressure balances the weight of the cardboard.

Bernoulli's effect also accounts for the upward force of lift on the

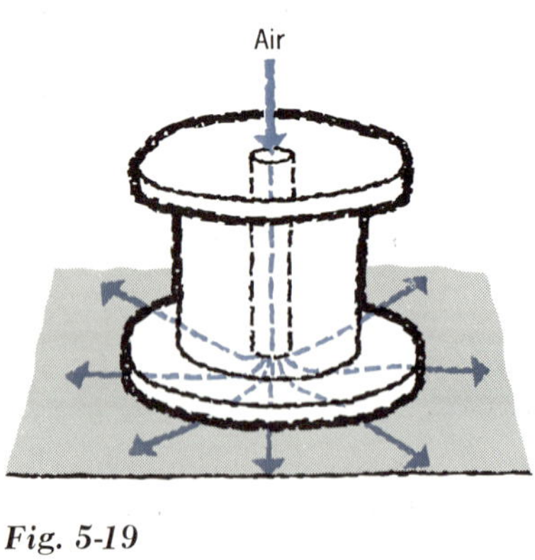

Fig. 5-19

wing of an airplane. Figure 5-20 shows an airplane wing in cross section, with air streaming above and below it. The wing is designed so that the airstreams above and below the wing reunite smoothly at the trailing edge of the wing. Under these conditions, the air above the wing travels a greater distance in the same time than the air below the wing; therefore it travels at a higher speed. This higher speed above the wing means that there will be a lower pressure above the wing, and the result is an upward force.

THE RELATION BETWEEN DISTANCE, VELOCITY, AND TIME FOR MOVING OBJECTS

Motion Under Zero Force

If an object is subject to no interactions, or if the resultant of all the forces acting on the object is zero, then that object is in equilibrium and its velocity is constant. According to the first law of motion, it will continue to move in a straight line at a constant speed. Use of the second law of motion, $F = ma$, also gives this same result. If the force on an object of mass m is equal to zero, then the acceleration will also be zero. This means that the object cannot change its velocity, and must either remain at rest or in uniform motion (motion in a straight line at a constant speed).

The distance traveled by an object in uniform motion is proportional to the time of travel. This relationship is easily expressed in the following equation:

$$d = vt \tag{3}$$

Distance, velocity and time are represented by d, v and t, respectively.

It may be convenient to use a reference frame in which the moving

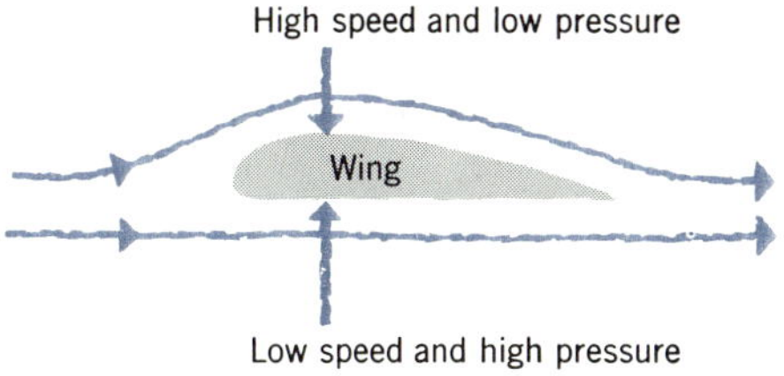

Fig. 5-20

object is already a distance d_0 from the origin of our reference frame when the time t equals zero. In this situation the proper equation for uniform motion in a straight line becomes

$$d = d_0 + vt \tag{4}$$

In the special case where $d_0 = 0$, Equation 4 is the same as Equation 3. These equations have already been introduced in Chapter 1, page 49. Here we wish to emphasize that in these equations the velocity v is a constant.

We can use these equations to draw a graph that relates distance and time (Fig. 5-21). Since Equation 3 expresses the fact that distance is proportional to time, the graph is the simple graph we expect for quantities that are proportional to each other (see App. II-C). The straight line goes through the origin of the graph. Equation 4 also has a simple graph (Fig. 5-22), but the straight line no longer goes through the origin. We may obtain the straight line in Figure 5-22 by raising the straight line in Figure 5-21 upward a distance d_0.

Motion in a Straight Line due to Constant Force

In this section we shall allow a nonzero resultant force to act on the object, but we shall make the restriction that neither the mass of the object nor the resultant force on it changes with time. Under these conditions, the acceleration is not zero, but it does not change with time either; in fact, all of the quantities in the equation $F = ma$ are constant. We make the further restriction that the motion must be along a straight line. This means that the force, the velocity, and the acceleration must be along this straight line also.

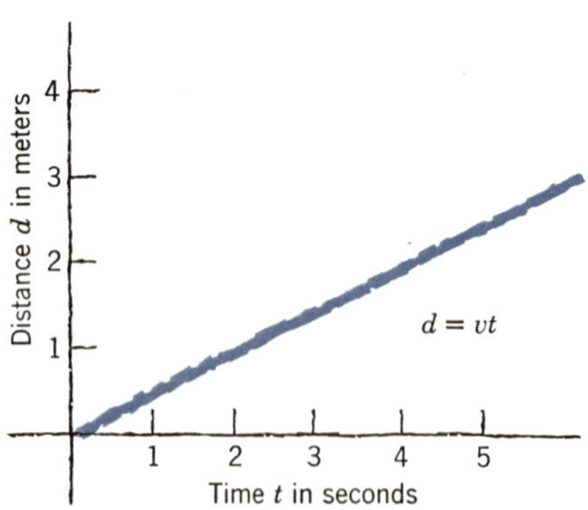

Fig. 5-21

THE MEANING OF CONSTANT ACCELERATION. Although acceleration is defined as the rate of change of velocity, it is possible for the acceleration itself to be unchanging, or constant. Indeed, as we have seen, if the mass of an object and the force on it are constant, the acceleration *must* be constant.

Suppose that we are in an automobile, stopped at a stop sign. We start out from rest and gain a velocity of five meters per second in the first second. We gain another five meters per second in the second second, giving us a velocity of 10 meters per second. Suppose that in the third second we gain another five meters per second, making our total velocity 15 meters per second. Although our velocity is continuously and smoothly changing, it is changing by the same amount in each second. Under these conditions, we say that the rate at which the velocity is changing is the same during the whole three seconds; that is, the rate of change of velocity is constant. The rate of change of velocity is the acceleration. Therefore, when we refer to a constant acceleration, we mean that the change of velocity is the same in each equal interval of time. This means that if we start from rest, as in our example, the velocity we attain will be proportional to the time that has elapsed since we started. After twice the time we will have twice the velocity, etc. Because of this proportionality, we can write the relation between velocity v and time t as follows for a constant acceleration:

$$v = at \tag{5}$$

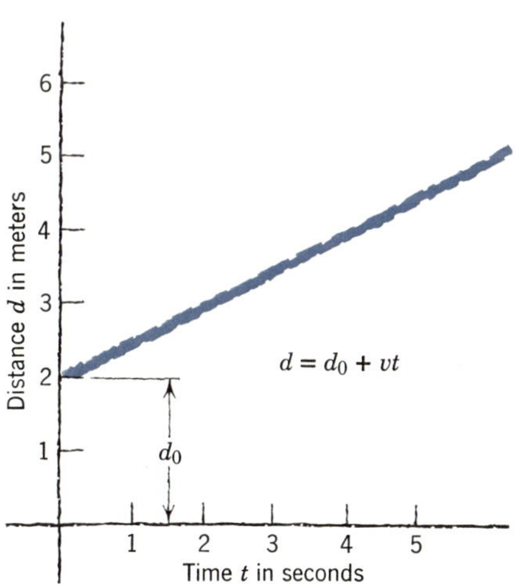

Fig. 5-22

It may happen that our automobile did not start from rest but already had a velocity, v_0. In this case, we simply add the starting velocity v_0 to the right side of the equation:

$$v = v_0 + at \tag{6}$$

Equations 5 and 6 have certain similarities to Equations 3 and 4. However, we should remember that these two sets of equations were devised for different situations. Equation 3 or 4 holds when the velocity is *constant*. Equation 5 or 6 holds when the velocity is *changing* with a constant acceleration. Of course, Equations 5 and 6 still hold if the acceleration is zero, although they become rather trivial.

We can draw a graph (Fig. 5-23) which expresses the simple proportionality between velocity and time given by Equation 5. The straight line goes through the origin, as required for proportionality. Equation 6 can also be graphed, as shown in Figure 5-24. The straight line is raised by an amount v_0 in this graph to allow for the initial velocity v_0. These straight-line graphs of velocity and time are characteristic of constant acceleration. If the acceleration were not constant, the graphs would not be single straight lines but might have many different shapes.

CONSTANT ACCELERATION IN NATURE. Where can motion with constant acceleration be found in nature? Falling objects near the surface of the earth closely approximate this kind of motion. The student may wonder at this, since he knows that the earth's gravitational pull varies with distance, according to the formula:

$$F = \frac{Gm_1m_2}{d^2} \tag{7}$$

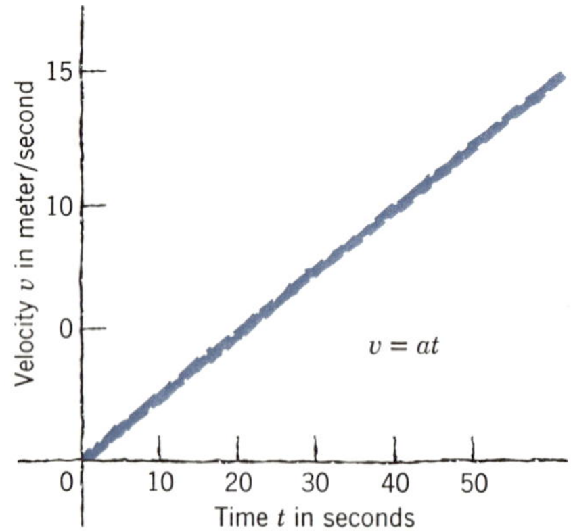

Fig. 5-23

How can the acceleration be constant when the force itself varies as the object falls? We usually deal with objects that fall very small distances compared with the radius of the earth. Therefore, their distance from the center of the earth remains very nearly the same and the gravitational force remains very nearly the same also as they fall. This almost constant force causes these objects to have an almost constant acceleration. The friction of the air will also influence these falling objects. If the falling objects are sufficiently compact, dense, and do not fall too fast, however, the air friction is small. (We could not ignore air friction if we were dealing with such things as falling feathers, or space satellites reentering the atmosphere.) We are also ignoring small variations in gravitational force as we go from place to place on the earth's surface, as well as "Foucault pendulum" type effects that arise because of the earth's rotation and revolution. If we make all these approximations and set all these conditions, we can consider that falling bodies fall with a constant acceleration. It was this type of motion that Galileo studied so effectively.

DISTANCE—TIME RELATIONS IN CONSTANTLY ACCELERATED MOTION. Equations 5 and 6 give us the relations between velocity and time for motion with constant acceleration. We now wish to find the relation which holds between distance and time for this same kind of motion. To do this we use a method used by Galileo and known to medieval scientists.

We shall start the motion of the object from rest, so that the equation $v = at$ will be true. If we knew that there was no acceleration,

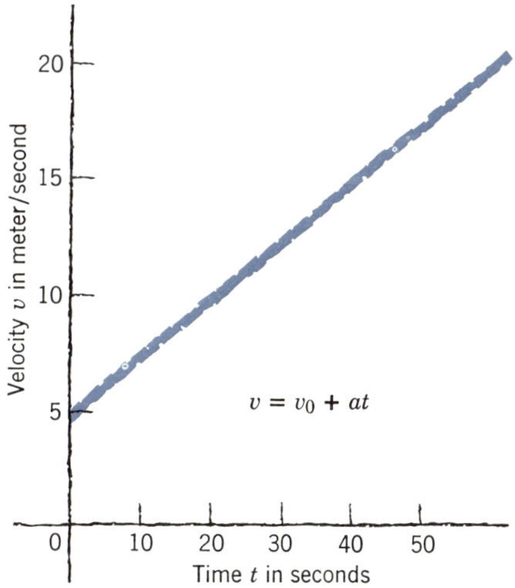

Fig. 5-24

we could use Equation 3, $d = vt$. Unfortunately, however, there is an acceleration—a constant acceleration, in fact. Therefore, the equation $d = vt$ does *not* hold. If, however, we were able to average the velocity over the whole time period t, we might expect that the following equation, using the average velocity, would be correct:

$$d = (\text{average } v) \cdot t \tag{8}$$

We can find the average velocity from the velocity-time graph shown in Figure 5-25. We are interested in the average velocity over a time interval T, as shown in the figure. From the graph it appears that the average of v over the whole time interval T must be given by the level of the horizontal straight line we have drawn through the halfway point of the graph. For each point where the velocity is a certain amount higher than this average in the second half of the graph, there will be a symmetrically placed point in the first half of the graph where the velocity is the same amount lower. This average velocity, being at the halfway point, is just half the velocity attained at the time T. According to the equation $v=at$, the velocity at time T is aT. Therefore, the average velocity must be

$$(\text{average } v) = 1/2\ aT \tag{9}$$

At the time T, therefore, the distance, as given by Equation 8 should be:

$$d = (\text{average } v)T \tag{10}$$

$$d = 1/2\ aTT \tag{11}$$

$$d = 1/2\ aT^2 \tag{12}$$

It is now convenient to return to using the symbol t instead of T. We may say, then, that the object starting from rest and moving with

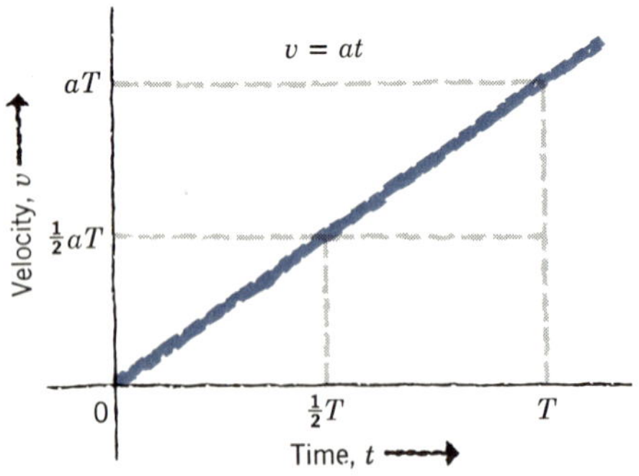

Fig. 5-25

a constant acceleration *a* will have traveled through a certain distance *d*, after a time *t*. This distance *d* is given by the equation

$$d = 1/2\, at^2 \tag{13}$$

The fact that the time is squared in this equation means that the distance traveled is proportional, not to the time, but to the *square* of the time. After two seconds the object will have traveled (two)^2 (or four) times as far as after the first second. After three seconds the object will have traveled nine times as far, and so on.

DEMONSTRATION OF THE DISTANCE-TIME RELATIONS FOR FALLING BODIES. Theoretical results should be checked in practice to see if they really hold. It is rather difficult to check the distance-time relations for falling bodies using rulers and stopwatches. Fortunately, however, there is a device that makes it relatively easy to demonstrate these relations. The device is called a stroboscope.

The stroboscope produces a flashing light. The flashes are very short, so that a falling object moves very little during a single flash. So short is the flash that it appears to the onlooker to stop the object in its flight (see Fig. 5-26). The flashes of the stroboscope are repeated at equal, known time intervals, and thus the flashes appear to stop the object at successive positions in its path. By obtaining a photograph such as Figure 5-26, and then measuring the distance from one image to the next, it is possible to check the distance-time relations for falling bodies.

GALILEO'S TEST OF HIS LAWS OF FALLING BODIES. Galileo, of course, did not have a stroboscopic light, but because of his ingenuity he was still able to obtain and test the laws of falling bodies. Since falling objects fell too fast for him to measure, he used balls rolling down inclined planes, as illustrated in Figure 5-27. (He had to use hard balls and smooth tracks so that friction would not significantly affect the results.) By making the slope of the inclined plane very gradual, he was able to get the balls rolling slowly enough so that he could measure the time necessary for them to roll various distances. This meant that he could check the distance-time relations. He actually wanted to test the velocity-time relations, but he could not measure the changing speed accurately enough. Therefore, he used the graph technique we have discussed to obtain distance-time relations he *could* check from velocity-time relations which he could *not* check.

Since stopwatches were not yet available, he used a water clock. The water clock is a tank so designed that water flows from it at a constant rate (see Fig. 5-28). Galileo measured time intervals by

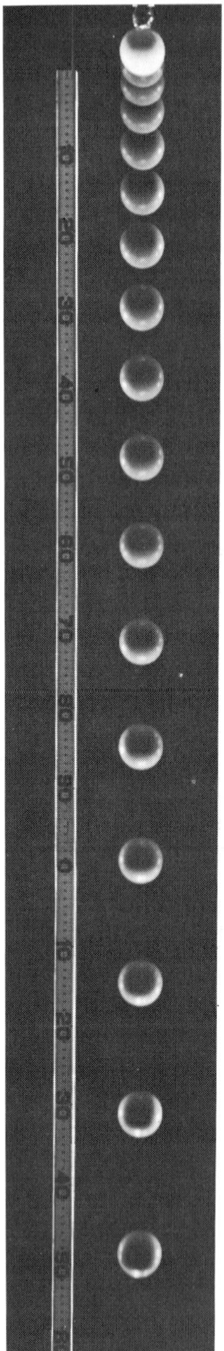

Fig. 5-26 *From PSSC Physics, D. C. Heath and Company, Boston, 1965. Used with permission.*

weighing the amount of water that escaped from his water clock.

Another problem had to be solved before Galileo could be sure of his results. He needed to be sure that the ball rolling on the inclined plane described the same kind of motion as the more rapid motion of a freely falling object. He reasoned that the speed of a ball rolling down an inclined plane depended only on the height through which it had fallen, not on the slope of the inclined plane. This conclusion can be justified by using Figure 1-9 (p. 17) in reverse (see Fig. 5-29). We let the balls roll down the various inclined planes on the left and up the inclined plane on the right. If they start from the same level on the planes to the left, they must reach the same level on the plane to the right. This means that they must all have the same speed at the bottom; otherwise they would climb to different heights on the plane to the right. Thus, the speed depends only on the vertical height through which the balls have fallen, and not on the steepness of the inclined plane.

In the limiting case, when the plane becomes vertical, the motion is essentially free fall. (The rolling ball is turning; the falling ball is not. Ideally, we should consider an object sliding without friction, not rolling, on the inclined plane.) From such reasoning Galileo could see that an object falling freely would have the same speed as the object on the plane after descending from a given height. Therefore he could justify the use of the inclined plane in his study of freely falling objects. The moving objects in each case undergo constant acceleration.

Galileo's researches on falling bodies required creative ingenuity, clear reasoning and experiment, and not just one of these items alone. The results were not achieved just by dreaming, just by rigorous reasoning from self-evident principles, or just by going out and measuring everything in sight—although all these activities may play some part in science. Galileo's philosophy insisted that both experiment and reasoning are important.

***EQUATIONS FOR MORE GENERAL SITUATIONS OF CONSTANT ACCELERATION.** We obtained the following equations for motion

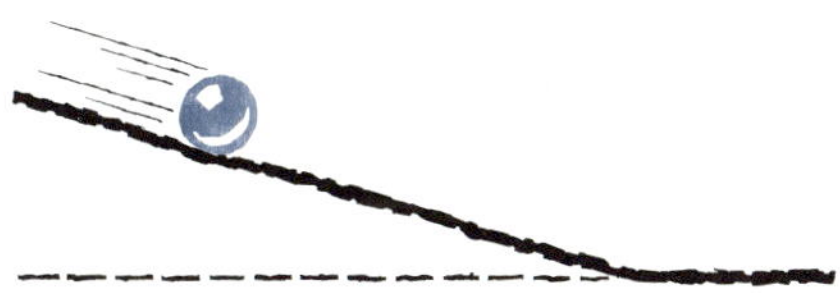

Fig. 5-27

under constant acceleration when the starting velocity and the starting distance are zero:

$$v = at \tag{14}$$

$$d = 1/2\ at^2 \tag{15}$$

From these equations we may easily find, using algebra, that

$$v^2 = 2ad \tag{16}$$

The general equations that hold for motion with constant acceleration when there is both a starting distance, d_o, and a starting velocity, v_o, are:

$$v = v_0 + at \tag{17}$$

$$d = d_0 + v_0 t + 1/2\ at^2 \tag{18}$$

The first of these equations has already been given as Equation 6. From these two equations, by using algebra, we may also obtain another equation:

$$v^2 = v_0^2 + 2a(d - d_0) \tag{19}$$

Equation 19 is very useful for situations in which one neither knows nor desires any information about the time t. In Equations 14 to 19, the letters symbolize the following properties of the moving object:

d = distance of the object from the origin of the reference frame at the time t

d_0 = distance of the object from the origin of the reference frame when the time t equals zero. This is the so-called starting distance.

v = the velocity of the object at the time t

Fig. 5-28

v_0 = the velocity of the object when the time t equals zero. This is the so-called starting velocity

a = the acceleration, which must be constant if these equations are to be valid

t = the time that has elapsed

Although Equations 14 to 19, properly defined, can be regarded as vector equations, we shall not do so here. Since motion has been restricted to a single straight line (or axis), we can regard the quantities v, v_0, a, d, and d_0 as positive or negative scalar quantities instead of vectors. These quantities will be positive if directed in the positive direction along the line, and negative if directed in the opposite direction. The quantity v^2 is, then, simply the square of the speed.

Projectile Motion under Constant Force

In this section we shall study the type of motion that can be illustrated by such things as baseballs and cannonballs. More precisely, we shall be concerned with objects that are dropped or thrown near the surface of the earth and are acted on only by the essentially constant vertical force of gravitation. In the last section we were concerned with motion along a straight line. The principles and equations developed were sufficient to describe what happens when an object is dropped or thrown straight up or straight down. In these cases the object describes a straight-line path. Now, we will not restrict the motion to a line but will allow the objects to move in any direction near the earth.

THE MOTION OF OBJECTS THROWN HORIZONTALLY. Let us consider the motion of a bomb that has been dropped from an airplane. We assume that the airplane is flying along horizontally and uniformly, and not engaging in aerobatics (see Fig. 5-30). The bomb is released while it is moving horizontally with the airplane; therefore, it behaves like an object with a starting velocity in a horizontal direction.

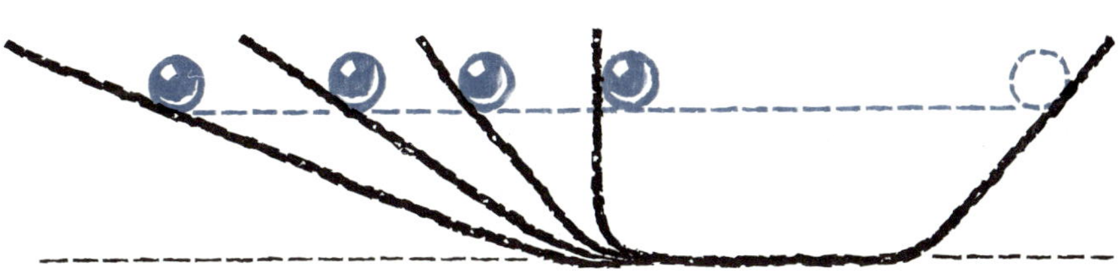

Fig. 5-29

One way of analyzing the motion of the bomb is to use the principle of relativity. (See again Chapter 1 pp. 21-25). From the uniformly moving reference frame attached to the airplane, the bomb falls straight down and is directly under the plane when it strikes the ground. In the reference frame attached to the earth, however, the bomb continues to move horizontally (to the right in our figure), with the same speed at which it started. At the same time it is gaining a downward velocity because of its acceleration by the gravitational force. Its vector velocity, therefore, is the resultant of two vector velocities—a vector velocity to the right (which remains the same all the time) and a vector velocity straight down (which is continually increased by the pull of the earth and, in fact, obeys Equations 14 to 19). One of the consequences of this fact is that the bomb should take the same time to reach the ground as it would if it were dropped from a stationary balloon at the same height and had no forward motion.

The situation can also be analyzed by using some of the principles discussed in this chapter. The gravitational force on the bomb is vertical and therefore has no effect on the horizontal motion. Since there are no other forces on the bomb, its horizontal velocity cannot change while the bomb is falling. Figure 5-31 shows what happens to the vector velocity for two bombs, one dropped from an airplane and one dropped from a stationary balloon. Notice that the velocity to the right never changes for either bomb during the time of fall. It

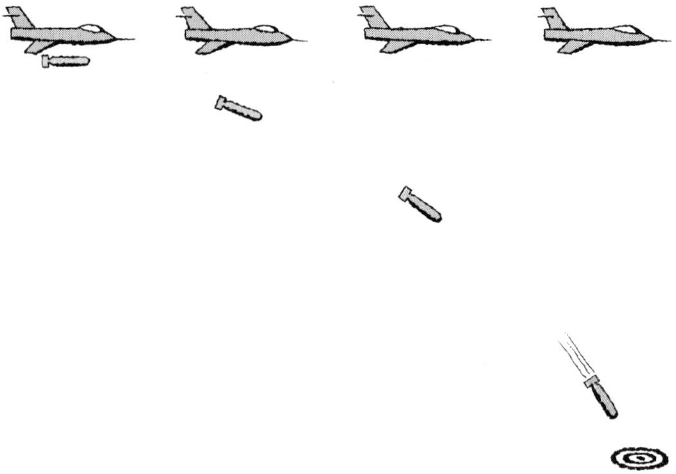

Fig. 5-30

is zero for the bomb dropped from the stationary balloon, and equal to the velocity of the airplane for the bomb dropped from the airplane.

We can describe this situation in terms of the vector components introduced in Chapter 2 (pp. 83-87). There is no component of force acting in a horizontal direction. Therefore, there is no acceleration in that direction, and the horizontal component of velocity remains the same. There *is* a force acting downward, and so the component of velocity in that direction does change. The equation for uniform motion (constant velocity) holds for the component of velocity in the horizontal direction. The equations for constantly accelerated motion hold for the component of velocity in the downward direction. To find the total resultant velocity of the bomb at any point, we must add these two components together as vectors.

OBJECTS THROWN AT AN ANGLE. A more complex situation arises when an object is thrown neither horizontally nor straight up and down but at an angle. In practice, most projectiles are thrown or fired in this manner. A shot-putter, for example, throws the shot "up and out," so that it is rising at the same time it travels away from him.

To understand the motion of such objects, consider the well-known puzzle illustrated in Figure 5-32. A hunter spots a monkey hanging from the branch of a tree. He aims his gun at the monkey in such a way that the barrel points straight at the monkey—if he looked through the barrel he would see the monkey. (We have here a very

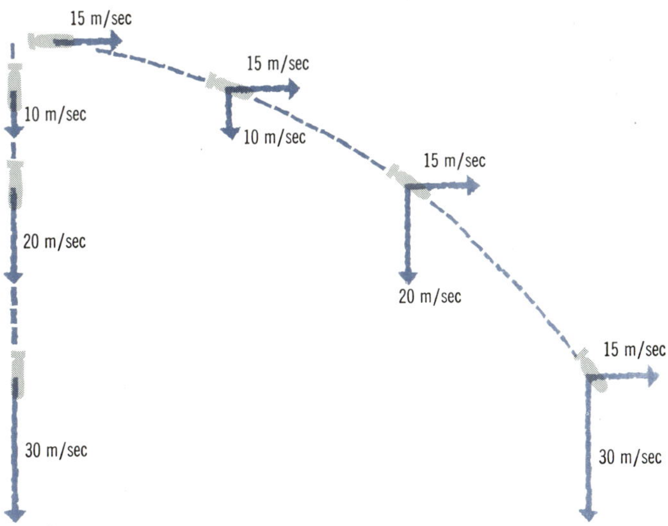

Fig. 5-31

naive hunter who does not realize that gravity will pull the bullet down below this line of sight.) Just as the hunter pulls the trigger, the monkey lets go of the branch and begins to fall to the ground. Thus the bullet travels toward the monkey while the monkey is falling to the ground. Both the bullet and the monkey are influenced by the gravitational pull of the earth. *Question:* Will the bullet hit the monkey?

We can answer this question by using the principle of relativity. Let us suppose that we were riding with the bullet when it came out of the gun, but that we, unlike the bullet, were not influenced by the pull of the earth. In other words, our reference frame follows the straight-line path the bullet would have taken if there were no gravitation. The origin of our reference frame travels on the dotted straight line shown in Figure 5-32. (Remember, we must use a reference frame that travels in a straight line at a uniform speed. This is one reason for not choosing a reference frame fixed to the bullet, since the bullet is influenced by gravitation and does not follow a straight line.)

The bullet appears to us to be at rest in the center of our reference frame when it comes out of the barrel, but it feels the gravitational pull and begins to fall straight down. When our reference frame reaches the branch, therefore, the bullet will have fallen some distance directly below the center of our frame. How far will this be? It will be the same distance any object would fall from rest in the time elapsed since the bullet left the gun. Since this is exactly the same time that the monkey has been falling from rest, he will be this same distance below the branch. This puts both the monkey and the bullet

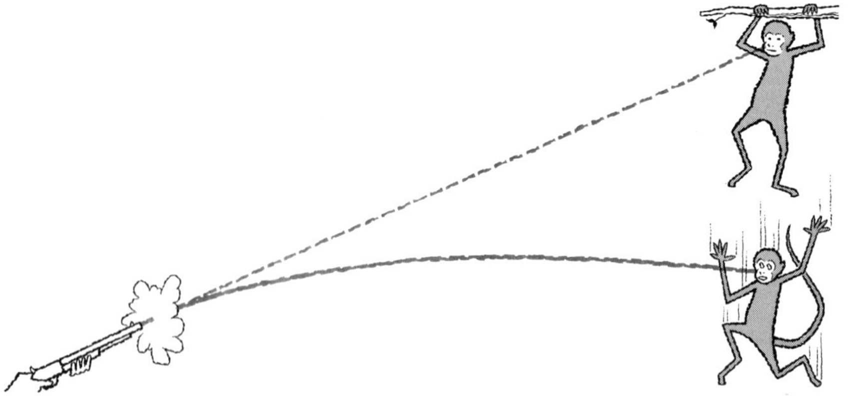

Fig. 5-32

the same distance below the branch at the some time. Thus the bullet does indeed hit the monkey.

This same result can be obtained by using the equations for accelerated motion as applied to vector components, but we shall not do this here. The general result is that the bullet, or any projectile, falls away downward from the path it would have taken if there were no gravitation, at the same rate it would have in falling from rest.

*Motion under Variable Forces

COMET ORBITS. We have already considered one kind of motion under variable force, circular motion at constant speed (Chap, 4, pp. 132-134). True, the *magnitude* of the centripetal force remains the same in such motion, but the force must always be directed toward the center. This means that the direction of the force changes as the object moves around.

As a further example of an object moving in response to a variable force, consider the motion of a comet traveling in its orbit around the sun. We choose a comet because its motion is not even approximately circular, and therefore it is different from the motions we have already discussed.

Figure 5-33 shows a comet at various points in its orbit. The comet's distance from the sun changes a great deal as it moves in its orbit. The gravitational force exerted on the comet by the sun

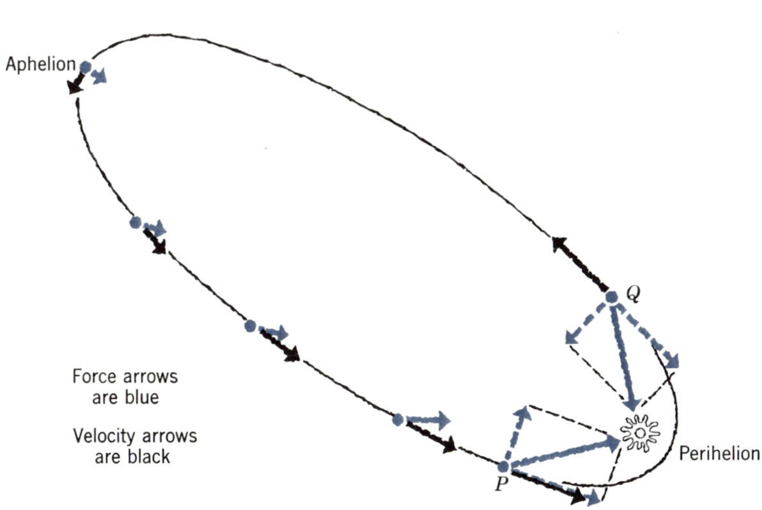

Fig. 5-33

changes in both magnitude and direction as the comet goes around the sun. Because of this nonconstant force, the comet has a nonconstant acceleration. The force on the comet is greatest when it is nearest the sun, and smallest when it is furthest from the sun.

The arrows in Figure 5-33 represent the vector forces on the comet and also the vector velocities of the comet at various points in its orbit. The vector velocities are all tangent to the orbit; this is always true for an object moving along a smooth path. The force vectors all point to the sun, since the gravitational pull of the sun on the comet is in the direction of the sun. At the point of closest approach to the sun (the perihelion) and the point of farthest distance from the sun (the aphelion), the force on the comet is exactly perpendicular to its velocity. At these points the force serves only to change the direction of the comet, and not to increase its speed. At no point in the comet's orbit does the force vector point in the same direction as the velocity vector. Yet the force does cause the speed of the comet to change, as well as its direction, since except for the two points noted it is not exactly perpendicular to the velocity.

CENTRIPETAL FORCE. If we find the component of the force along the direction of the velocity, we can see why the gravitational force causes the speed of the comet to change. At point P, for example, the force has a component in the same direction as the velocity. This force acts to increase the velocity as the comet comes closer to the sun. At point Q, on the other hand, the component of force is in the opposite direction from the velocity. This component of force slows down the comet as it heads away from the sun. The component of force in the direction of the velocity is the force that changes the speed of the comet. In fact, we may state:

$$\text{(Component of force along the direction of the velocity)} = \text{(mass)} \times \text{(rate of change of speed)} \quad (20)$$

This equation is similar to the second law of motion. Instead of the force, this equation deals with a certain component of the force; instead of the rate of change of velocity (that is, acceleration), it deals with the rate of change of speed.

There is another interesting relation which involves the component of force perpendicular to the velocity (see Fig. 5-34). For this component of force we may say:

$$\text{(Force component that is perpendicular to velocity)} = \frac{\text{(mass)} \times \text{(speed)}^2}{\text{(radius of curvature)}} \quad (21)$$

The only unfamiliar item in this equation is the "radius of curvature." If the object is traveling in a circle, the radius of curvature is simply the radius of the circle. The comet, however, is not traveling in a circle, and it requires more mathematics than we wish to introduce to define the radius of curvature for paths that are not circles. We may say, however, that the more tightly curved parts of an object's path have a smaller radius of curvature than the less tightly curved parts. Equation 21 informs us that the component of force perpendicular to the velocity is responsible for the change in direction of the comet.

Equations 20 and 21 apply to any object traveling in a smooth path, and not just to comets.

Problem Solving

In this section we shall discuss the solution of problems that (a) require us to find the resultant force on an object under conditions of both equilibrium and nonequilibrium; (b) treat pressures in fluids; and (c) use the distance, velocity, and time relations for objects moving with constant acceleration, including (d) trajectory problems.

Finding the Resultant Force

The procedure for solving problems of this nature has already been outlined (pp. 178-183). Simple examples of the steps to be followed in vector addition and the use of formula $F = ma$ are given at the end of Chapter 2. Here we shall consider a few more difficult examples.

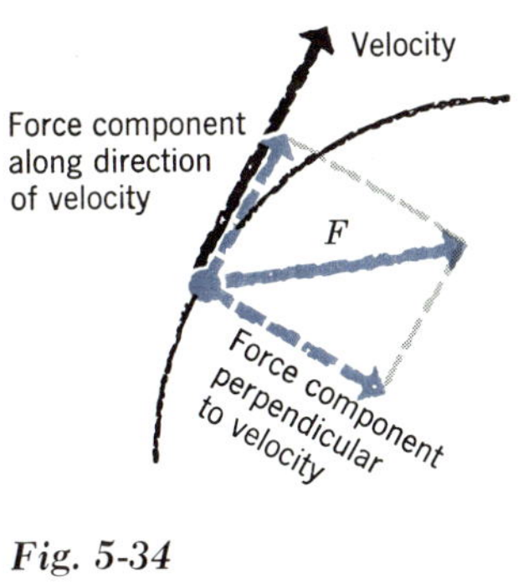

Fig. 5-34

EXAMPLE 1

In Example 6 of Chapter 2 (p. 90), suppose that the sled, instead of accelerating, is moving at a constant velocity as a result of the force exerted by the two boys. What is the force exerted by the ice on the sled under these circumstances?

Solution. The procedure for isolating the object (the sled) and identifying the interactions and forces, shown in Figure 5–35, is similar to that illustrated in Figure 5-2 (p. 180). The two forces exerted by the boys still add up to 31 newtons in the direction in which the sled is traveling, but now the resultant of all the forces, including the friction of the ice, must be zero. Therefore, the ice must exert a force of 31 newtons exactly backwards.

*EXAMPLE 2

In Figure 5-36, what is the magnitude of the total force exerted by the ground on the peg?

Solution. There are three interactions acting horizontally on the peg.

1. The interaction with the rope attached to the larger elephant.
2. The interaction with the rope attached to the smaller elephant.
3. The interaction with the ground.

In Figure 5-37 we have isolated the peg and indicated the three forces on it. Since the peg is in equilibrium, the interaction with the ground must supply a force that balances the resultant of the other two forces. It remains, then, only to find the magnitude of the resultant of the other two forces. For an elementary problem, this can be done graphically, by drawing the two forces to scale and measuring the resultant. Here we calculate the magnitude of the resultant.

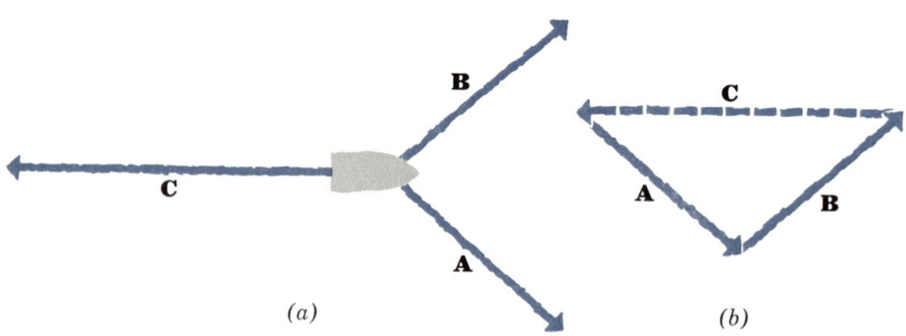

Fig. 5-35

Since the two forces are at right angles, their resultant is the hypotenuse of a right triangle which has the two given forces as legs. We can use the Pythagorean theorem (see App. II-D) in the following way. Letting h be the magnitude of the unknown force:

$$(5{,}000)^2 + (12{,}000)^2 = h^2$$
$$25{,}000{,}000 + 144{,}000{,}000 = h^2$$
$$169{,}000{,}000 = h^2$$
$$h = 13{,}000 \text{ newtons}$$

***Example 3**

Consider the inclined plane in Figure 5-38*a*. We place a mass m on the plane and allow it to slide without friction down the plane, under the influence of gravity. What is the acceleration of the mass?

Solution. In Figure 5-38*b* we have isolated the mass and put it at the center of a frame of reference. Because of the symmetry principle relating to direction in space, we know that we can take our axes in any direction. Since we know that the mass is going to slide down the plane, it is convenient to tilt our reference frame so that one of the axes points down the plane. In fact, it is usually convenient to take one of the axes along the direction in which we know the object will move.

Next we have drawn vectors representing all the forces that are actually acting on the body. The force ma_g exerts a downward pull because of gravitation. (See Example 7, Chap. 4, p. 169.) There is also

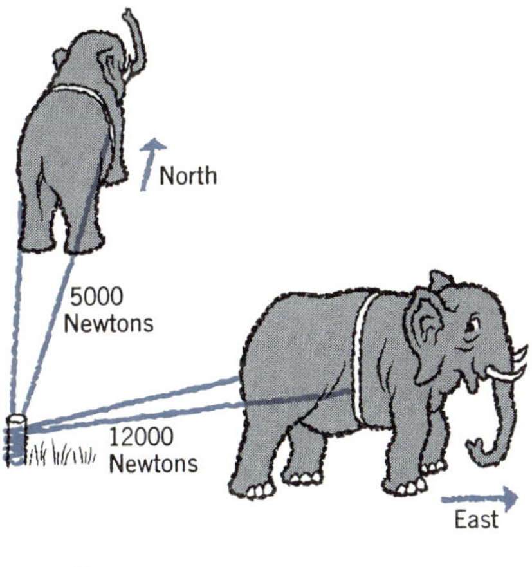

Fig. 5-36

the force caused by the contact pressure between the object and the plane. Now we see one reason our choice of reference frame was convenient. Since the unknown contact force exerted by the plane is perpendicular to the motion of the mass, it has no component in the direction of motion. Therefore we can ignore it. Only the force of gravitation has a component that is effective. We need, then, to find the component of the gravitational force directed down the plane.

Referring to Figure 5-38*b* again, we see that we have drawn a triangle including the force ma_g. The component we want is simply the short leg of this 30°–60° right triangle, and the force ma_g is the hypotenuse. Because of this triangle, the component we want is half of the force ma_g (see App. II-D). That is, the force down the plane is

$$F = ma_g/2$$

But since $F = ma$

we have $ma = ma_g/2$

and $a = a_g/2 = (9.8/2)\text{ meter/second}^2 = 4.9\text{ meter/second}^2$

Pressures and Forces in Fluids

Example 4

What is the total force arising from the air pressure on the outside face of a television tube which is a half meter square? (Air pressure is approximately 1.0×10^5 newtons per meter^2 under normal atmospheric conditions.)

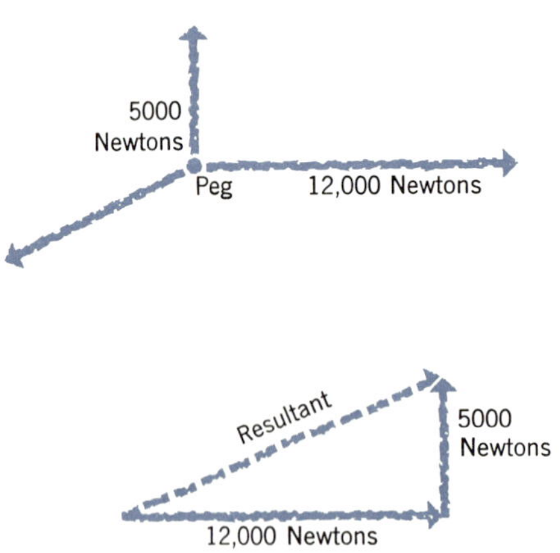

Fig. 5-37

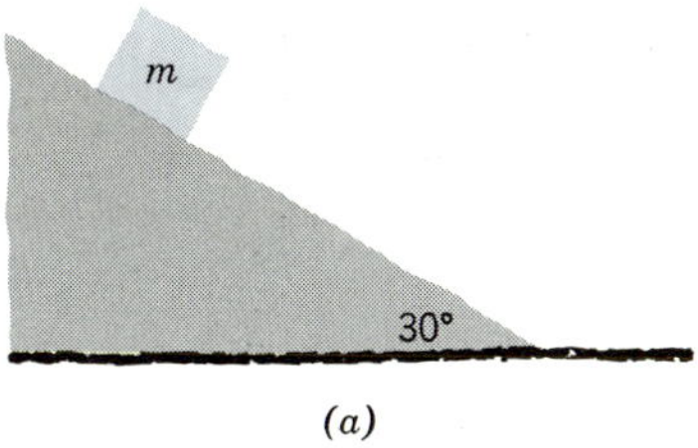

*Fig. 5-38*a

Solution. The force is defined by the equation: $F = PA$, in which

P = pressure of the air = 1.0×10^5 newton/meter²
A = area = ?
F = force on the face = ?

We know the length of the side of the square face of the tube, which is 0.5 meter. The area of the tube, then, is (0.5 meter)² = 0.25 meter². We put the values for P and A into the equation $F = PA$:

$$F = (1.0 \times 10^5)\,\frac{\text{newton}}{\text{meter}^2}\,(0.25)\ \text{meter}^2$$
$$F = 2.5 \times 10^4\ \text{newtons}$$

Example 5

The piston in Figure 5-39*a* is free to move without friction in the cylinder. The pressure inside the cylinder is 2.5×10^5 newtons/meter². The cylinder is exposed to the air on the other side of the piston. The piston has a mass of 2.0 kilograms and an area of 0.12 square meters. Find the acceleration of the piston.

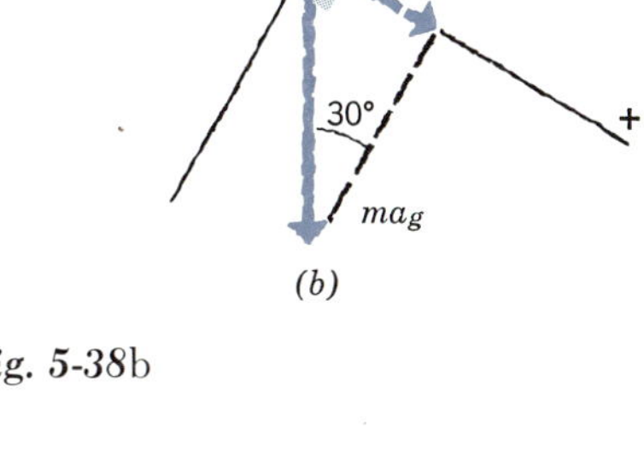

*Fig. 5-38*b

Solution. First we draw a diagram isolating the piston, as shown in Figure 5-39*b*. Then we list the interactions in which the piston is involved. We can ignore the gravitational forces and the pressure between the piston and its cylinder, since they are perpendicular to the horizontal motion of the piston. This means that there are two interactions to be considered.

1. The interaction between the piston and the gas inside.
2. The interaction between the piston and the air outside.

Thus, there are two forces to be considered. We find the force exerted by the gas inside the cylinder from the pressure inside the cylinder, as follows:

$$F_1 = P_1A = (2.5) \cdot (10^5)\,\frac{\text{newton}}{\text{meter}^2}\,(0.12\ \text{meter}^2) = 3.0 \times 10^4\ \text{newtons}$$

The force caused by the air outside the cylinder is found from the outside air pressure:

$$F_2 = P_2A = (1.0) \cdot (10^5)\,\frac{\text{newton}}{\text{meter}^2}\,(0.12)\ \text{meter}^2 = 1.2 \times 10^4\ \text{newtons.}$$

Since the force pushing to the left is greater than the force pushing to the right, the piston will accelerate to the left. The resultant force on the piston is found as follows:

$$F = F_1 - F_2 = (3.0 \times 10^4\ \text{newtons}) - (1.2 \times 10^4\ \text{newtons})$$
$$= 1.8 \times 10^4\ \text{newtons to the left}$$

To find the acceleration from this force we must use the equation

$$F = ma$$

in which:

$$F = 1.8 \times 10^4\ \text{newtons}$$
$$m = 2.0\ \text{kilograms}$$
$$a = ?$$

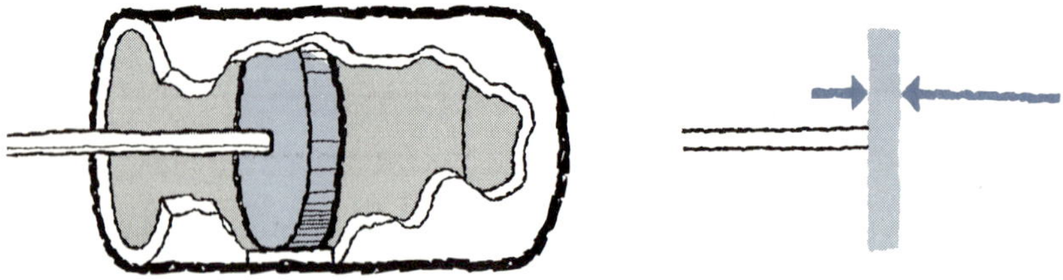

(*a*) (*b*)

Fig. 5-39

Putting these values in the equation, we obtain:

$$1.8 \times 10^4 \text{ newton} = (2.0 \text{ kilogram})a$$

$$\frac{1.8 \times 10^4 \text{ newton}}{2.0 \text{ kilogram}} = a$$

$$a = \frac{0.90 \times 10^4}{\text{kilogram}} \text{ kilogram } \frac{\text{meter}}{\text{second}^2}$$

$$= 9.0 \times 10^3 \frac{\text{meters}}{\text{second}^2} \text{ to the left.}$$

Example 6

A piece of wood of volume 1.0×10^{-3} meter^3 has a weight of 7.0 newtons. Find (a) the buoyant force on the wood when it is totally submerged in water; (b) the total force on the wood in this case; and (c) the buoyant force on the wood when it is floating.

Solution, part *a*. We have found that the buoyant force on a submerged object is equal to the weight of the fluid that would occupy the volume of the submerged object. The submerged wood in this example has a volume of 1.0×10^{-3} meter^3. This volume of water has a mass of 1 kilogram (see pp. 185-186). A mass of one kilogram weighs 9.8 newtons. Therefore the buoyant force is 9.8 newtons.

Solution, part *b*. In Figure 5-40 we have isolated the wood and drawn arrows to represent the two forces acting on it—its weight and the buoyant force. The buoyant force, found in *a*, is 9.8 newtons. This is greater than the weight, 7.0 newtons. Therefore, the resultant is upward and has a magnitude of

$$F = 9.8 \text{ newtons} - 7.0 \text{ newtons} = 2.8 \text{ newtons.}$$

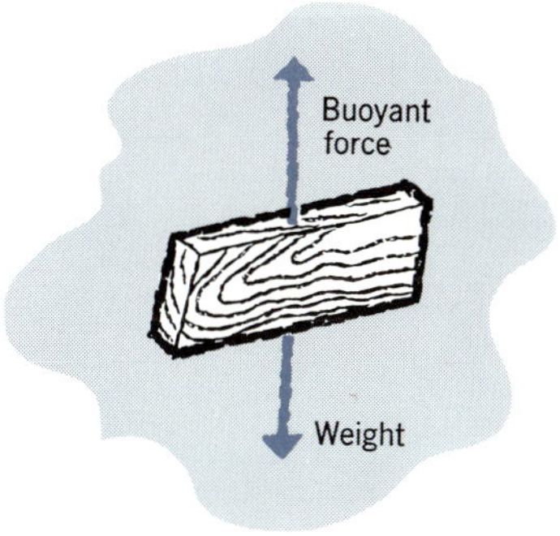

Fig. 5-40

Solution, part *c*. When the object is floating, the buoyant force must be equal in magnitude to the weight of the object, since the object is in equilibrium and the two forces must balance. Thus the buoyant force is equal to 7.0 newtons, directed upward.

EXAMPLE 7

What is the density of the wood in Example 6?

Solution. Density is defined by the equation

$$D = m/V \tag{22}$$

in which

D = density = ?
m = mass of the wood = ?
V = volume of the wood = 1.0×10^{-3} meter^3

In order to find the density, we first need to find the mass of the wood from its weight. The mass and the weight of an object are related by the equation:

$$f = ma_g$$

in which

f = weight of the wood = 7.0 newtons
a_g = gravitational acceleration = 9.8 meter

Inserting these values in our equation, we have

$$7.0 \text{ newton} = m\,(9.8 \text{ meter second}^2)$$

$$7.0\,\frac{\text{kilogram meter/(second)}^2}{9.8 \text{ meter/(second)}^2} = m$$

$$m = \frac{7.0}{9.8} \text{ kilogram} = 0.71 \text{ kilogram}$$

Now we can insert the values for m and V into Equation 22 to find the density:

$$D = \frac{0.71 \text{ kilogram}}{1.0 \times 10^{-3} \text{ meter}^3}$$

$$= 0.71 \times 10^3 \frac{\text{kilogram}}{\text{meter}^3}$$

$$= 7.1 \times 10^2 \frac{\text{kilograms}}{\text{meter}^3}$$

Constant Acceleration

In Chapter 1, we discussed the solution of problems involving uniform motion. Here we deal with somewhat more complex problems, but our treatment will be similar, involving the following steps.

1. Draw a diagram with a convenient reference frame. (The symmetry principles permit different ways of setting up a reference frame.) The solution of the problem may depend on how ingenious one is in choosing a reference frame, since this will determine the distances, and also the components of velocities, distances, accelerations, and forces.
2. List the given information and the information required by the problem.
3. Write down all the equations that involve the quantities in Step 2 and any additional quantities introduced by these equations. Note any other relations established by the nature of the problem that may not be in equation form.
4. Put the given information from Step 2 into the equations from Step 3. Then solve the problem, using the rules of algebra and logic.

This procedure will very often be valid and useful. The equations give us relations between what we know and what we wish to find. If we already knew the values of all the quantities, we would not need any of the equations that express the laws of physics.

We can look upon a problem as an input-output situation. Our input is the given information. Our desired output is the information required—the answer. The equations and other relations act as a machine to help us get the output information from the input information.

If things go wrong there are several possible mistakes to look for.

1. We forgot to use some of the given information—too small an input.
2. We neglected an equation or other relation that applies to the situation—too weak a machine.
3. We made a logical or mathematical error—that is, we pushed the wrong button on the machine. It is necessary to do exactly the right thing to get exactly the right answer, except by chance.
4. The problem was impossible or contradictory—not enough information was given to solve it, or so much was given that two inconsistent answers were possible. This does happen. In more interesting questions you may have to make a few reasonable assumptions. This usually is neither necessary nor permitted in routine numerical problems.

Generally we need two equations to solve for two unknowns, three to solve for three unknowns, etc. For problems involving constant acceleration, for example, we have Equations 17 and 18 (or sometimes 14 and 15). Therefore, if we are given four of the six quantities d, d_0, v, v_0, a, t, we can use these equations to find the other two. Each independent equation gives us a relation that enables us to solve for one more unknown. But note the word *independent*. Equation 19, for example, cannot be used along with Equations 17 and 18 to solve for three unknowns since it was deduced from the latter two Equations by using algebra; hence, it is not independent of them and gives us no information that the other two together do not give us. Nevertheless, it is a convenient formula, and it can make solving problems easier in some cases. Even if we have fewer equations than unknown quantities, however, we can often obtain some of the unknown quantities or relations between them.

EXAMPLE 8

A ball is dropped down a well. Five seconds later it strikes the bottom. How deep was the well, and how fast was the ball going when it hit the bottom?

Solution.

STEP 1 We draw a diagram as in Figure 5-41, setting up our reference frame as shown. Notice that the starting velocity is zero since the ball was simply dropped, not thrown, from rest. Therefore we can use the simple Equations 14 and 15 if we put the origin of our reference frame at the point where the motion starts. We have taken

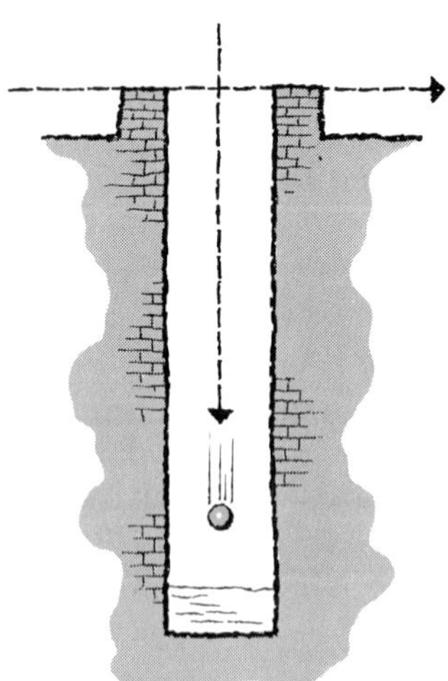

Fig. 5-41

one axis along the direction of motion, pointed downward, for convenience.

STEP 2 $d = ?$
$t = 5$ seconds
$a = 9.8$ meters/second^2
$v = ?$

STEP 3 $d = 1/2\ at^2$
$v = at$

Since we have two equations and two unknowns, we have enough equations to find the unknowns. We substitute the known values of the quantities into the equations as follows.

STEP 4

$$d = \frac{9.8(5)^2}{2} \text{ meter} = 4.9 \times 25 \text{ meter} = 120 \text{ meters}$$

$$v = 9.8(5)\,\frac{\text{meters}}{\text{second}} = 49\,\frac{\text{meters}}{\text{second}}$$

EXAMPLE 9

A ball is thrown straight up into the air at a speed of 100 meters per second. How high will it go, and how long will it take to come back?

Solution.

STEP 1 We draw the diagram and locate the reference frame as in Figure 5-42. Once again we place one of the axes along the direction of motion. Since we have a beginning velocity, we must use Equations 17 and 18. We take the origin where the particle started, to make $d_0 = 0$, which may be convenient.

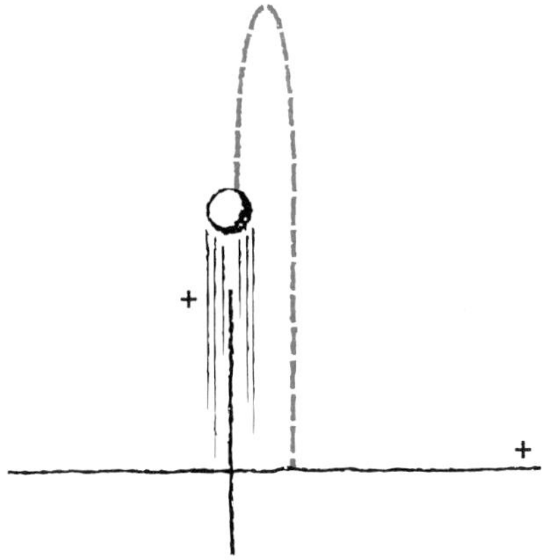

Fig. 5-42

STEP 2 Since we are interested in the time of the entire trip, the initial and final points are the same.

$$\begin{aligned} v_0 &= 100 \text{ meter/second} \\ a &= -a_g = -9.8 \text{ meter/second}^2 \\ h &= \text{maximum height} = ? \\ d_0 &= 0 \\ d &= 0 \\ v &= ? \\ t &= ? \end{aligned}$$

Note that d and d_0 are zero, since both refer to the origin of our reference frame. The fact that the ball has traveled in between does not matter, since d and d_0 do not refer to path lengths. Notice also that a is negative, since the gravitational force pulls down and our reference frame points up.

STEP 3 We use Equation 17 and 18. It will be convenient to use Equation 19 also:

$$\begin{aligned} v &= v_0 + at \\ d &= d_0 + v_0t + 1/2at^2 \\ v^2 &= v_0^2 + 2a(d - d_0) \end{aligned}$$

STEP 4 Substituting known quantities into our equations, we have:

$$\begin{aligned} v &= 100 \frac{\text{meters}}{\text{second}} - 9.8 \frac{\text{meters}}{\text{second}^2} (t) \\ 0 &= 0 + 100 \frac{\text{meter}}{\text{second}} (t) - 1/2 \left(9.8 \frac{\text{meter}}{\text{second}^2}\right) (t)^2 \\ v^2 &= \left(100 \frac{\text{meter}}{\text{second}}\right)^2 + 0 \end{aligned}$$

We notice immediately that the last equation tells us that the final speed coming down is 100 meters/second. This means that v must equal -100 meters/second, since it is in the negative direction. Knowing v, we can easily solve the first equation above for t:

$$-100 \frac{\text{meter}}{\text{second}} = v = 100 \frac{\text{meter}}{\text{second}} - 9.8 \frac{\text{meter}}{\text{second}^2} (t)$$

$$0 = 200 \frac{\text{meter}}{\text{second}} - 9.8t \frac{\text{meter}}{\text{second} \times \text{second}}$$

$$9.8t \frac{\text{meter}}{\text{second} \times \text{second}} = 200 \frac{\text{meter}}{\text{second}}$$

$$t \frac{9.8}{\text{second}} = 200$$

$$t = \frac{200}{9.8} \text{ second} = 20 \text{ seconds}$$

To find the height, we must start again with a different final position: the top of the trajectory. Step 1 is the same as before.

STEP 2

$$\begin{aligned} d_0 &= 0 \\ d &= h = ? \\ a &= -a_g = -9.8 \text{ meter/second}^2 \\ t &= ? \\ v &= 0 \\ v_0 &= 100 \text{ meter/second} \end{aligned}$$

The only quantities changed here are v, which is zero at the top of the path, and d, which now is the height.

STEP 3 We need only this equation:

$$v^2 = v_0^2 + 2a\,(d - d_0)$$

STEP 4

$$0 = (100)^2 \frac{(\text{meter})^2}{(\text{second})^2} - 2\left(\frac{9.8 \text{ meter}}{\text{second}^2}\right)(h)$$

$$\frac{(2 \times 9.8 \times \text{meter})\, h}{\text{second} \times \text{second}} = (100)^2 \frac{\text{meter} \times \text{meter}}{\text{second} \times \text{second}}$$

$$h = \frac{(100)^2 \text{ meter}}{2 \times 9.8} \approx \frac{(100)^2 \text{ meter}}{2 \times 10} = \frac{100}{20} \times \text{meter}$$

$$h = 500 \text{ meters}$$

If we had so desired, we could have taken a different approach after finding out that the final speed with which the ball hit the ground was 100 meters/second. The fact that this was the same speed with which the ball was thrown up into the air strongly suggests that it took the same time going up as coming back. This is true—the path back simply repeats the path up, but in opposite time order because the laws of physics are the same for reversed time. Therefore the time duration going up and the time duration coming down must be the same. Similar reasoning shows that all features of the path must be the same in magnitude (not necessarily in direction) for the up and down trips. This can be shown by calculation using the formulas. Henceforth we can take it for granted that the trajectory of a projectile is symmetric about the point at the top of the path, in both space and time.

Problems Involving Trajectories

EXAMPLE 10

An airplane is flying horizontally at a speed of 100 meters/second at an altitude of 490 meters. When it is directly over a crossroads a bomb is released. (a) How long will it take for the bomb to hit the ground, and (b) how far from the crossroads will it strike?

Solution, part *a*. Since the bomb had no vertical motion when it was released, it will fall to the ground in the same time it would take to fall if it had been released from rest.

STEP 1 (see Fig. 5-43). We have attached our reference frame to the moving airplane. In this reference frame the bomb simply drops vertically to the ground. For convenience, we have taken the positive direction downward.

STEP 2

$$\begin{aligned} v_0 &= 0 \\ d_0 &= 0 \\ a &= a_g = 9.8 \text{ meter/second}^2 \\ v &= \text{? (not required)} \\ t &= \text{?} \\ d &= 490 \text{ meters} \end{aligned}$$

STEP 3 The equation which relates these quantities is

$$d = d_0 + v_0 t + 1/2\, at^2$$

STEP 4 Substituting the information into the equation we have:

$$490 \text{ meter} = 0 + 0 + 1/2 \left(9.8 \frac{\text{meter}}{\text{second}^2}\right) t^2$$

$$490 \text{ meter} = 4.9 \frac{\text{meter}}{\text{second}^2}\, t^2$$

$$100 = \frac{t^2}{\text{second}^2}$$

$$t = 10 \text{ seconds}$$

Solution, part *b*. The bomb is moving horizontally at a constant velocity of 100 meters/second during the 10 seconds of its fall.

$$\begin{aligned} d &= vt \\ &= 100 \frac{\text{meter}}{\text{second}} \times 10 \text{ second} \\ &= 1000 \text{ meters} \end{aligned}$$

***EXAMPLE 11**

A cannonball has been shot out of a cannon at a 45-degree angle at a speed *s*. How far from the cannon will the cannonball land?

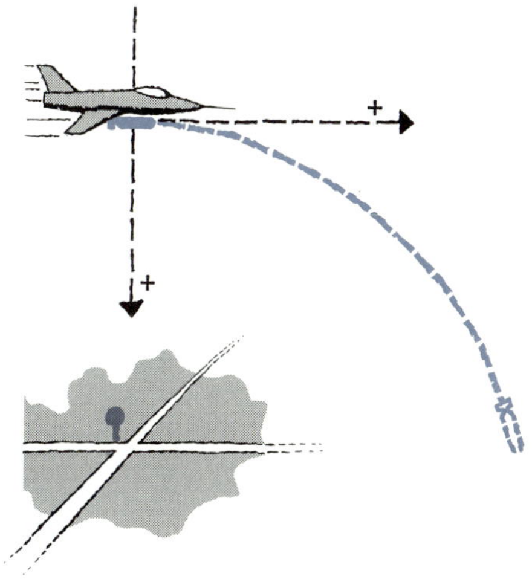

Fig. 5-43

Solution. This problem is complicated by the fact that two different things are happening in two different directions after the ball leaves the cannon. We have accelerated motion in the vertical direction. We have no horizontal force, so that the motion is uniform in the horizontal direction.

STEP 1 (see Fig. 5-44). The vector velocity with which the ball is shot from the cannon forms the hypotenuse of a 45-degree right triangle. We need to know the components of this vector in the direction of the axes of our reference frame. We have chosen the reference frame in this way because the force of gravity is downward. Thus one axis is along the direction of the force, and only the components of velocity along this axis will be affected by that force. In the horizontal direction there will be no change in velocity,

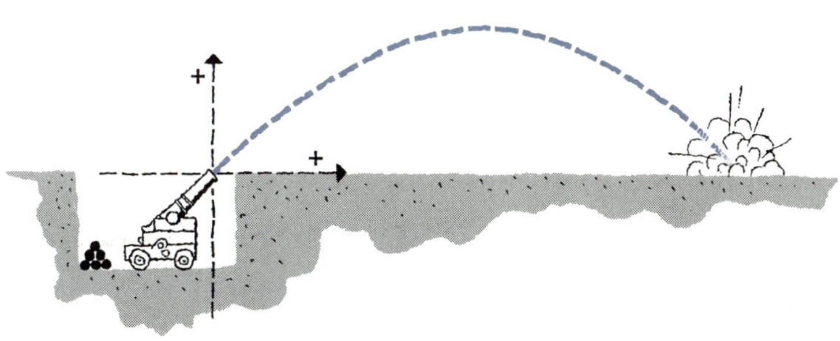

Fig. 5-44

since the gravitational force has no components in that direction. To find the components of the starting velocity, we use the information about 45-degree right triangles found in Appendix II-D. Both the upward component and the horizontal component of the starting velocity are equal to $s/\sqrt{2}$. We use the letter v_0 to refer to these components.

STEP 2 Consider first the vertical motion. This is accelerated motion under gravity. Thus, for vertical components:

$d_0 = 0$
$d = 0$ if we take the second point as the landing point
$v_0 = s/\sqrt{2}$
$v = -s/\sqrt{2}$
$t = ?$
$a = -a_g = -9.8$ meter/second^2

Now a word about strategy. We know that the velocity to the right will not change, and we know that it must be $s/\sqrt{2}$. If we know the total time of flight of the cannonball, we can easily find the horizontal distance, D, by using the formula $D = v_0 t$. Therefore, we want to find the time of flight. Now d_0 is zero because the starting point is the origin. But d is also zero, because it is only the vertical components of displacements that matter for up-and-down motion. The quantities in our constant-acceleration equations may be regarded as the components in the direction of the force. We proceed to find t.

STEP 3

$$d = d_0 + v_0 t + 1/2at^2$$
$$v = v_0 + at$$
$$v^2 = v_0^2 + 2a(d-d_0)$$

STEP 4

$$0 = 0 + \frac{s}{\sqrt{2}}t - \frac{a_g t^2}{2}$$
$$\frac{-s}{\sqrt{2}} = \frac{s}{\sqrt{2}} - a_g t$$
$$\frac{s^2}{2} = \frac{s^2}{2} - 2a_g\ (0-0)$$

We have from the first equation:

$$0 = \frac{st}{\sqrt{2}} - \frac{a_g t^2}{2} \tag{23}$$
$$\frac{a_g t^2}{2} = \frac{st}{\sqrt{2}}$$
$$t = \frac{2}{a_g}\frac{s}{\sqrt{2}} = \frac{s\sqrt{2}}{a_g}$$

The ball is in the air for a time $\frac{s\sqrt{2}}{a_g}$. All this time it is traveling to the right at a uniform speed $\frac{s}{\sqrt{2}}$. Since for uniform motion, $D = v_0 t$, we have

$$D = \frac{s}{\sqrt{2}}t = \frac{s}{\sqrt{2}}\left(\frac{s\sqrt{2}}{a_g}\right) = \frac{s^2}{a_g}.$$

Thus the cannonball lands a distance $\frac{s^2}{a_g}$ away from the cannon, disregarding air friction.

Notice one strange thing, however. Equation 23, which is the equation for distance from the starting point in the vertical direction, is also true if t equals zero. What does this mean? The ball is at zero distance twice—once when it starts out and once when it lands. Equation 23 gives us both answers. The equation does not "know" which time we want; it knows only that we set that distance equal to zero. It often happens that the mathematics of a problem will give two results in this way. The student must decide which to use by analyzing the problem situation.

Summary

I. DERIVED QUANTITIES

- **A.** DENSITY: a scalar, mass divided by volume.
- **B.** PRESSURE: a vector, force divided by area.

II. APPLICATIONS

- **A.** FORCES ON OBJECTS, FINDING THE RESULTANT FORCE: for equilibrium and nonequilibrium.
- **B.** FORCES IN FLUIDS
  1. Pressure in Fluids: pressure is perpendicular to any surface, is the same in all directions, and increases with depth.
  2. Barometer.
  3. Buoyant Forces.
  4. Bernoulli's effect in moving fluids.
- **C.** RELATIONS BETWEEN DISTANCE, VELOCITY, AND TIME FOR MOVING OBJECTS. We obtained the following equations for motion with constant acceleration.
  1. $v = v_0 + at$
  2. $d = d_0 + v_0 t + \frac{1}{2} at^2$
  3. $v^2 + v_0^2 + 2a(d - d_0)$
- **D.** PROJECTILE MOTION

III. METHODS OF SCIENCE. Galileo's use of inclined planes to study motion.

IV. PROBLEM SOLVING

A. RESULTANT FORCE ON AN OBJECT

B. PRESSURE, DENSITY, AND BUOYANCY IN FLUIDS

C. DISTANCE, VELOCITY, AND TIME RELATIONS FOR CONSTANT ACCELERATION

Questions

1. Can the number of forces acting on an object exceed the number of interactions in which the object is involved?
2. Can an object be in equilibrium if it is involved in only one interaction? In three interactions? Explain.
3. Why is it necessary to find the resultant force in order to find the acceleration of an object? Are there any exceptions or apparent exceptions to this rule?
4. List the interactions in which the following objects participate. Draw the object isolated from its surroundings, representing the forces on it by arrows, estimating the magnitude and direction of each force. Estimate the resultant force and its direction.
   (a) An airplane in level flight.
   (b) An artificial satellite in orbit.
   (c) An anchor being pulled up through the water.
   (d) A book resting on a table.
   (e) A man walking across the room.
   (f) A tree in a windstorm.
   (g) An automobile slowing down for a stoplight.
   (h) A skier sliding down a hill.
   (i) A shirt hanging on a clothesline.
   (j) A swinging pendulum.
5. In Question 4, estimate the relative magnitude and direction of the forces that the objects listed exert on their surroundings.
6. Figure 5-45 shows a tug-of-war. Identify all the interactions, and then the forces on the object and its surroundings, for the rope and each team in the following situations:
   (a) No motion.
   (b) One team is pulling the other at a constant speed.
   (c) One team is pulling the other faster and faster.
   Which team exerts the greater force on the other in each case?

Fig. 5-45

7. In Question 6 consider the rope and both teams together as a single object. Identify all the interactions and forces on this object.
8. What is meant by the statement that heavy objects sink? Is the statement true, false, or incomplete? Explain.
9. Why do we use mercury in a barometer instead of water?
10. Is it possible for one small boy to hold back the entire Atlantic Ocean by putting his finger in a hole in the dike? Explain. (See Fig. 5-46.)
11. In Figure 5-47, in which case is the force against the dam greater?
12. Does the buoyant force on an object depend on its density? Explain.
13. Does the buoyant force on an object depend on its depth in the fluid? (Consider that the object and the fluid may be compressed by pressure to varying degrees.)
14. Why is the atmosphere denser near the ground than at higher altitudes?
15. Why does the pressure in a fluid at a given depth depend on the density of the fluid?
16. What keeps the mercury in a barometer?

Fig. 5-46

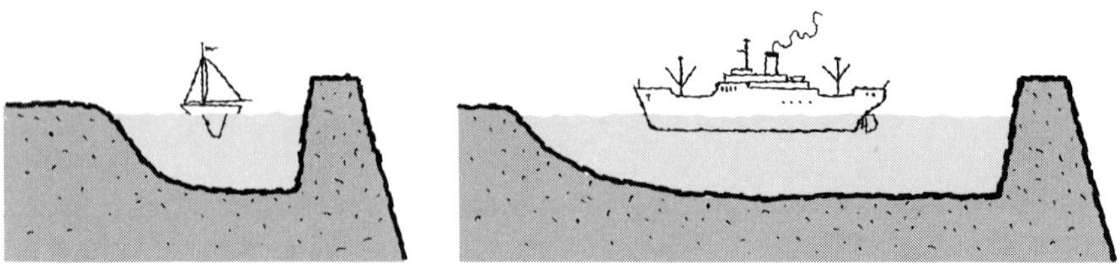

Fig. 5-47

17. What determines whether an object will float or sink?
18. Explain why the ping-pong ball in Figure 5-18 stays in the air jet.
19. Figure 5-48 shows how the air flows around a spinning baseball as it moves toward the batter. Will the ball rise or fall because of this flow? Why?
20. Explain how an airplane wing obtains lift.
21. Why is it that a bomb dropped from a moving airplane does not fall straight down?
22. How do you know that ice is less dense than water?
23. Is it possible for an object to float on the surface between two fluids in a container, as shown in Figure 5-49? Explain.
24. For which orientation of the pencil shown in Figure 5-50 is the buoyant force greater?
25. The barometer in Figure 5-51 is tilted as indicated by the dotted lines. Which of the two possibilities correctly describes the behavior of the mercury level, and why?
26. How would you measure the height of a building using a barometer?
*27. How does the force on an object increase its speed? How does it change the direction of motion?

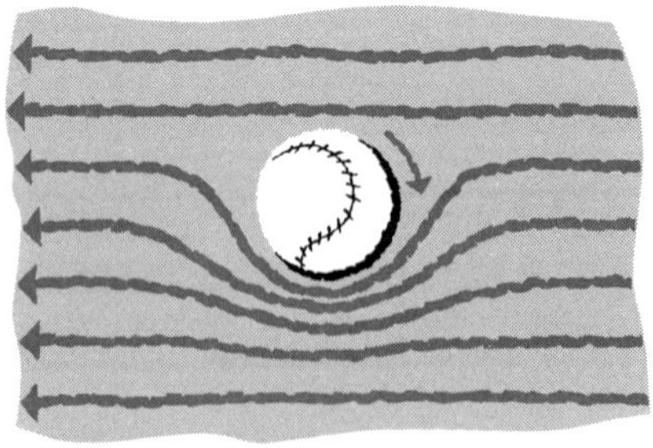

Fig. 5-48

Fig. 5-49

*28. The mass on the pulley on the clothesline in Figure 5-52 will come to equilibrium at the center of the clothesline. Explain why.

Problems

1. If we made a barometer using water instead of mercury, how high would the water stand in the tube, assuming that there was a perfect vacuum in the tube above the water column? Actually, some of the water would evaporate, forming a vapor in the tube above the water level; how would this affect the height of the water?
2. In Example 1, p. 212, suppose that the sled is caught on a twig so that it cannot move, but that the boys are pulling on it just as before. What would be the force supplied to the sled by the twig?

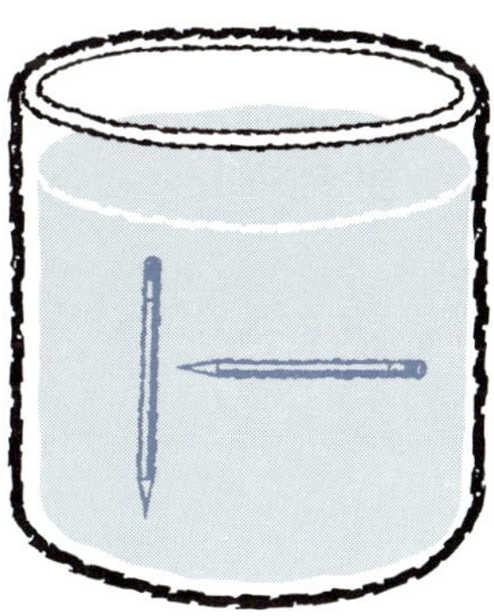

Fig. 5-50

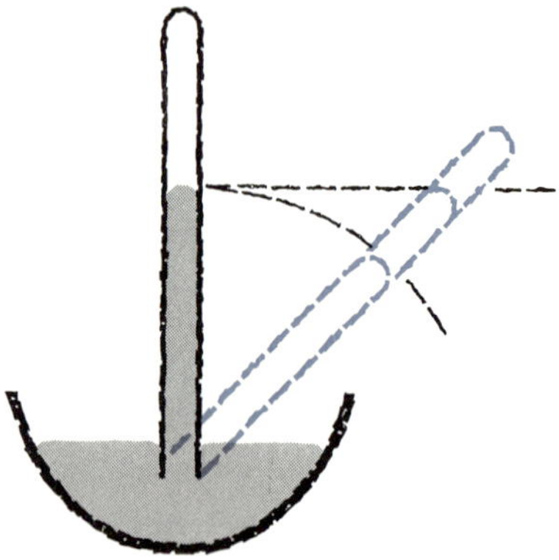

Fig. 5-51

3. The piston in Figure 5-53 is free to slide without friction in the cylinder. The cylinder outside the piston is open to the air. The piston has a mass of 15 kilograms and an area of 1.3×10^{-3} meter^2. What is the pressure inside the cylinder?
4. What is the buoyant force on a submarine whose volume is 1.2×10^5 meters^3? Assume that the submarine is in fresh water. Salt water has a higher density than fresh water; how would this affect the buoyancy of the submarine?
5. A block of wood has a mass of 50 kilograms and a density of 600 newtons per cubic meter. What is its volume?
6. How much water would the block of wood in Problem 5 displace if it were floating in water?
7. The block of wood in Problem 5 displaces 1.02 times as much fluid when it is floating in fresh water than when it is floating in a certain sample of salt water. What is the density of this sample of salt water?
8. Calculate the pressure exerted on the floor by the young lady described on page 186 when she rocks back on one heel. Assume that the area of

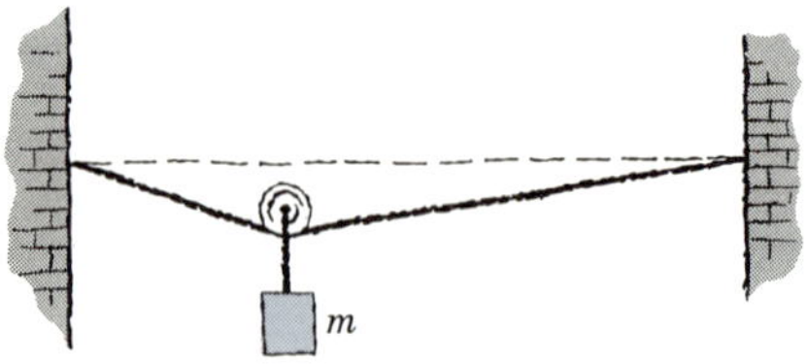

Fig. 5-52

the heel is 1.5×10^{-4} meter^2. Would it be possible for her to exert more pressure than that with the same heel area in contact with the floor?

9. From the equations $v = at$ and $d = \frac{1}{2}at^2$, obtain the equation $v^2 = 2ad$.

More Difficult Questions and Problems

*1. A boat that can travel at a speed v in still water travels a distance d downstream and back when the water is flowing at a speed V. Then it travels directly across stream a distance d and back. How long does it take to make each round trip? Which is the faster trip?

2. A mass m is pushed by a force f through a distance s. How long does it take the mass to travel this distance, starting from rest?

3. In Figure 4-22 (p. 171) find the acceleration of the masses if M is 30 kilograms and m is 10 kilograms.

4. How does putting a spin on a ping-pong ball affect the game?

5. Find the equation which gives the pressure in terms of the depth of the fluid, the density of the fluid, and a_g. Assume that the fluid is incompressible. (Hint: What is the pressure at the bottom of a vertical cylinder filled with water?)

6. The two vessels in Figure 5-54 have the same area on the bottom and are filled with water to the same depth. Therefore, the total force on the bottom of each vessel must be the same. Yet their weights are not the

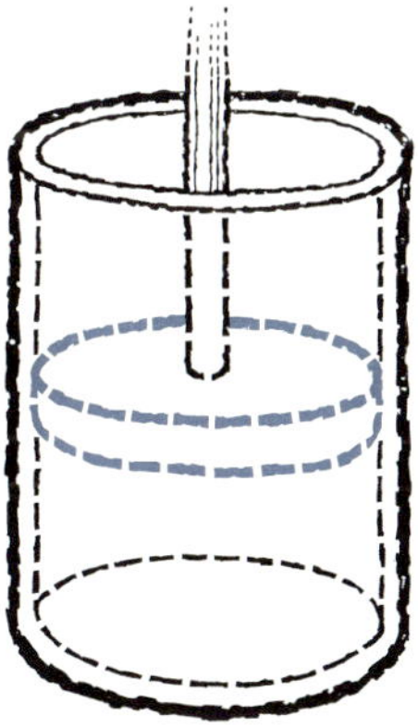

Fig. 5-53

same, since one contains less fluid than the other. How can the force on the bottom be the same if the weights are different?

7. Why does a comet move faster as it nears the sun?
8. Why are round wheels better than square ones?
9. Why is an angle iron stronger than a solid straight metal bar of circular cross section of the same length containing the same amount of iron?
10. Since the only force that causes a ball to roll down an inclined plane is the downward force of gravitation, why does the ball move to the side as well as down as it rolls down the plane?
11. Is a spinning top in equilibrium?
12. Why may we disregard the interactions between different parts of the same object when we are trying to determine the resultant force on the object?
13. What is the velocity of a ball thrown directly upward from the ground when the ball is at the top of its path? Is it in equilibrium?
14. We have stated that when the resultant of the forces on an object is zero the object is in equilibrium as far as acceleration of the whole object is concerned. Does this mean that the object cannot move at all? Explain. Does it mean that it cannot accelerate at all?
15. Obtain the equation $v^2 = v_0^2 + 2a(d-d_0)$ from the equations $v = v_0 + at$ and $d = d_0 + v_0t + 1/2\ at^2$.
16. What is the speed of the object whose distance is graphed in Figure 5-21?
17. What is the acceleration of the object whose velocity is graphed in Figure 5-23?
18. What is d_0 for the object whose distance is graphed in Figure 5-22?

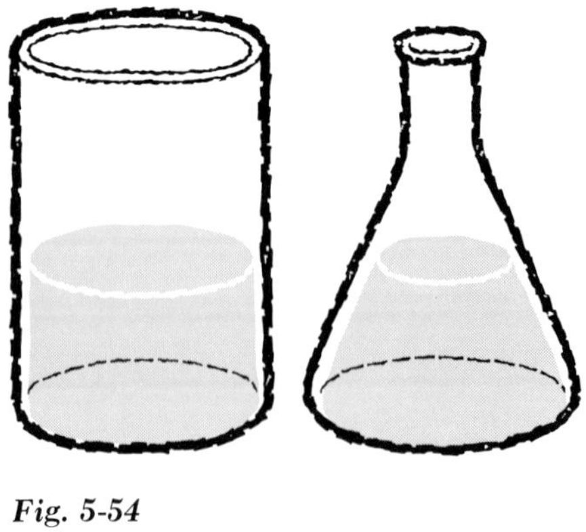

Fig. 5-54

19. What is v_0 for the objects whose velocity is graphed in Figures 5-23 and 5-24?

20. If the scales in Figure 5-26 are in hundredths of meters, what is the time between flashes of the stroboscope?

21. How do we know that the experiment shown in Figure 1-9, p. 17, can be reversed to get the experiment shown in Figure 5-29, p. 205?

22. Using symmetry principles, show that the resultant of the vectors in Figure 5-55 must be zero. Show that this is true for any number of vectors of the same magnitude arranged symmetrically in this way, except of course for one single vector. Why does the argument break down for one vector alone?

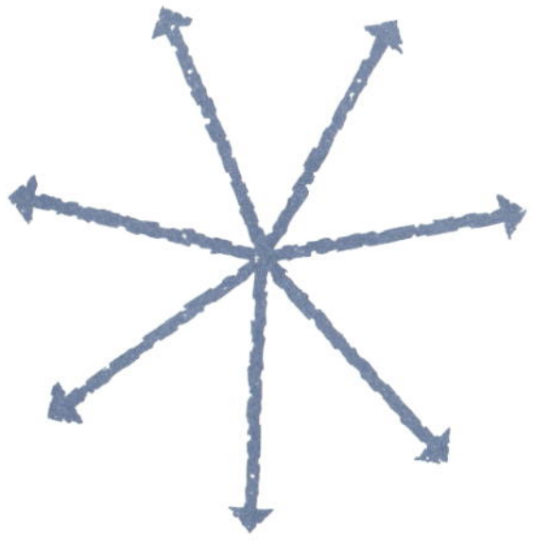

Fig. 5-55

6

CONSERVATION OF ENERGY

In Chapter 1 we described a thought experiment of Galileo in which he considered balls rolling on inclined planes. We pointed out that a ball placed at a certain level on one plane and allowed to run down that plane and up another would reach a level no higher than the level it started from (see Fig. 6-1).

Galileo knew that he could also show this same effect with a pendulum (see Fig. 6-2). A pendulum bob released from rest at a certain level rises no higher than its starting level, but swings back and forth beneath that level. The experimenter in Figure 6-2 is therefore perfectly safe; if he releases the bob from rest at the position shown, he will not be hit in the face by the returning pendulum bob, provided that he does not move. This is still true even if the motion of the string is obstructed by a peg, as shown in Figure 6-3.

Of course, friction prevents the rolling ball or the swinging pendulum bob from quite reaching its original level. The less friction there is, however, the closer it comes to this level. In the ideal situation, with no friction at all, the pendulum bob would swing forever, returning on every swing to its original level.

Fig. 6-1

Not only is the maximum height of the ball or pendulum bob the same for each swing, but the maximum speed is also the same. The speed of the ball at the bottom of the inclined plane depends on the height through which it has fallen, not on the slope of the plane. The same is true for the pendulum bob. Since the bob falls through the same height in each swing, the speed at the bottom must be the same for each swing. For both the ball on an inclined plane and the pendulum bob, the maximum speed and the maximum height are related. This maximum height and maximum speed keep

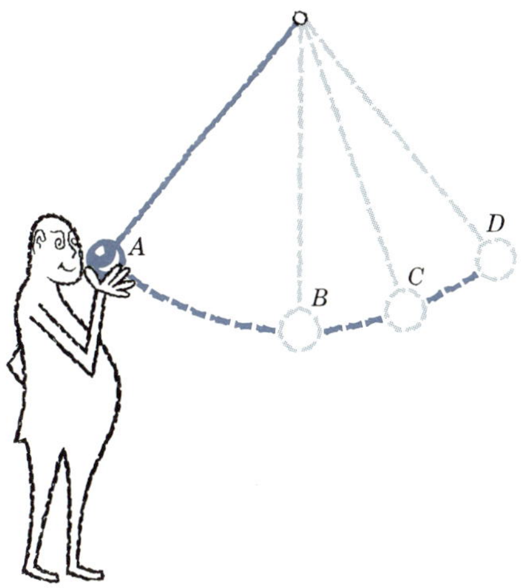

Fig. 6-2

recurring as time goes on. When the ball or the pendulum bob reaches its maximum height, the speed is zero. When the speed is maximum, the bob or ball is at its lowest point. As the speed increases, the height decreases, and vice versa.

It is very curious that these same heights and speeds keep recurring. One would suspect that there must be some quantity associated with the system which remains the same and is therefore conserved. There is indeed such a quantity. We call it *energy*, and it is a scalar quantity. This conserved quantity, energy, cannot be associated with either the height or the speed alone, since both of these quantities change during each swing. In fact, energy is found in more than one form. One form, called kinetic energy, depends on the speed of an object. Another form, called gravitational potential energy, depends on the height of an object above the surface of the earth. Neither of these two forms of energy alone is conserved in the motion of the ball and bob described above, but the *sum* of these two forms *is* conserved. This is one reason that energy is important and useful.

In this chapter we shall describe kinetic, gravitational potential, and internal energy. We shall discuss the way in which energy is transformed from one form to another and the way in which energy is transferred from one object or system to another.

Conservation of energy

The foregoing discussion strongly suggests that the quantities speed s and height h should each be associated with a form of a conserved quantity, energy. We now need to know precisely how the forms of energy depend on these two quantities. This information

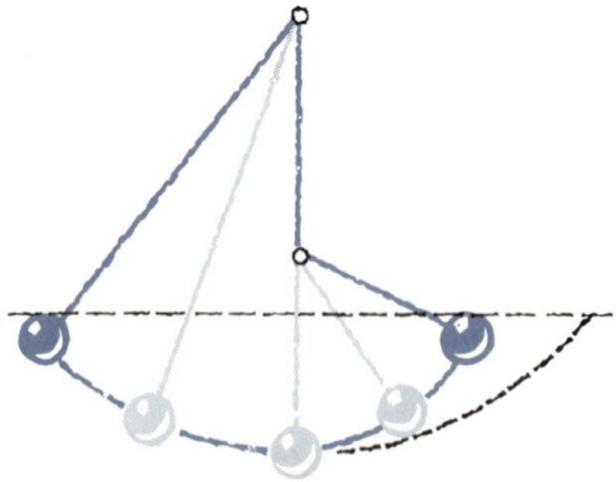

Fig. 6-3

has been obtained by physicists both theoretically and experimentally. Some of the results of their investigations, which established energy as a conserved, scalar quantity, will be discussed in this section.

Gravitational Potential Energy

As we have stated, the form of energy associated with the height of an object is called *gravitational potential energy*. It has been found that an object possesses an amount of gravitational potential energy equal to its weight w multiplied by its height h:

$$\text{Gravitational potential energy} = wh \tag{1}$$

Since the weight of an object is simply its mass m multiplied by the gravitational acceleration a_g we find that

$$\text{Gravitational potential energy} = wh = ma_gh \tag{2}$$

The formula ma_gh is more commonly used. From both formulas, however, we can see that the higher the object, the more of this kind of energy it has.

The *total gravitational potential energy* of a collection of objects is found by simply adding together all the gravitational potential energies of the individual objects. For example, a collection of three objects will have a total gravitational potential energy of

$$m_1a_gh_1 + m_2a_gh_2 + m_3a_gh_3 \tag{3}$$

In this discussion we have talked about the height h without specifying any reference frame; therefore, we have as yet no means of determining the value of h. We need a reference level from which to measure our heights. Actually, it does not matter where we put this reference level, since all of our applications will be dealing with *changes* in the gravitational potential energy, and the changes in height do not depend on the reference level. If, for example, I fall off a 300-meter cliff, I fall 300 meters; it does not matter whether I measure this change in height from a reference level at the top of the cliff, at the bottom, or anywhere in between. The physical effects are determined by the total *change* of height. In each application we are free to choose the most convenient reference level. In our definition, then, the height is simply the distance above or below this reference level. Distances above the reference level are positive; distances below it are negative. The discussion of gravitational potential energy in this section refers to ordinary operations with ordinary objects near the surface of the earth.

Kinetic Energy

The form of energy associated with the speed of an object is called *kinetic energy*. It has been found that an object of mass m and speed s has a kinetic energy of

$$\text{Kinetic energy} = 1/2\ ms^2 \tag{4}$$

From this equation we see that as the speed increases the kinetic energy increases. In fact it increases very rapidly because the square of the speed appears in the formula, and not just the speed itself.

The total kinetic energy of a collection of objects is found by simply adding together all the kinetic energies of the individual objects. For example, a collection of three objects will have a total kinetic energy as follows:

$$1/2\ m_1 s_1^{\,2} + 1/2\ m_2 s_2^{\,2} + 1/2\ m_3 s_3^{\,2} \tag{5}$$

The term kinetic energy was chosen to designate the form of energy associated with speed because the word *kinetic* comes from a Greek word meaning motion. The term potential energy refers to a kind of energy which an object possesses because of its position. The object with potential energy only is not moving, but it has the potential, or possibility, of moving. (If an object is already moving, its potential energy represents the potential it has for increased motion.) Gravitational potential energy refers to the potential energy an object has because of the gravitational forces that are acting on it. This is not the only form of potential energy, as we shall see.

The Unit of Energy

The unit in which energy is measured may be obtained from any of the formulas that express one of the forms of energy. We can use the formula ma_gh (for potential energy) for this purpose, simply replacing the symbol for each quantity by the unit in which it is measured. Thus

$$ma_gh = (\text{kilogram})(\text{meter/second}^2)(\text{meter}) = \frac{\text{kilogram meter}^2}{\text{second}^2} \tag{6}$$

This derived unit is called a *joule*, after the English physicist James Prescott Joule (1818–1889):

$$\text{joule} = \frac{\text{kilogram meter}^2}{\text{second}^2} \tag{7}$$

Total Energy

The total amount of energy possessed by an object may be found by adding up the amounts of each of the different forms of energy the object possesses. Thus far we have discussed two forms of energy. In situations where these are the only two important forms, we may obtain the total energy by adding the gravitational potential energy and the kinetic energy:

$$E = \text{total energy} = \begin{pmatrix}\text{gravitational}\\ \text{potential energy}\end{pmatrix} + \begin{pmatrix}\text{kinetic}\\ \text{energy}\end{pmatrix} \quad (8)$$

Where these two forms of energy are the only important ones, it is the total energy E, as given in Equation 8, which is conserved. For a single object of mass m, this will be

$$E = ma_g h + 1/2\ ms^2 \quad (9)$$

This equation applies to the objects illustrated in Figures 6-1, 6-2, and 6-3 if friction can be disregarded. For example, the total energy of the swinging bob of the pendulum does not change with time. Both the kinetic and the potential energy of the bob *do* change with time; in fact, the whole swinging process may be thought of as a continuous changing of one form of energy into the other. As the pendulum bob descends, its gravitational potential energy is converted into kinetic energy. As it rises, kinetic energy is converted back into potential energy. However, the total energy remains the same throughout the process.

Our understanding of many processes in nature is deepened by discovering how energy is changed from one form to another. This approach supplements our understanding of these processes in terms of force and acceleration. Many, if not most, modern physicists consider energy as more "real" than force. In fact, energy, although it is technically a derived quantity, appears to be one of the most fundamental and important "realities" that we study in physics. We have already seen that mass is one of the most fundamental and important "realities" in physics. The picture of mass as the measure of matter and energy as a measure of its motion, real or potential, is very appealing.

Another reason for thinking of mass and energy as important "realities" is that both of these quantities are conserved. Both of them can appear in many forms. With some limitations, they can be transformed from one form to another, but the total amount of each of these quantities remains the same. In a later chapter we shall discuss some surprising relationships between these two quantities.

The concepts of conservation and conserved quantities are considered very important in modern physics, for practical reasons in addition to the aesthetic and philosophical reasons mentioned above. Frequently we are interested in an interaction or a process that either is so complex or happens so quickly that we cannot observe details of the event. We may not even know the exact law that governs the interaction. All we can do is study the conditions that exist before and after the event. Fortunately, we find in nature that there are measurable quantities, such as mass, energy, and momentum, that are unchanged by an interaction. There is as much of each of these quantities after as there was before. This fact gives us information about the conditions after the interaction and allows us to make predictions about them.

*Derivation of Energy Forms and Conservation

We can actually obtain the formulas given above for kinetic and gravitational potential energy and also show that the total energy is conserved for the frictionless motion of such things as projectiles, pendulums, and objects moving on inclined planes. To do this we use Equations 16 and 19 from Chapter 5.

Consider first the special case of a body starting from rest and falling through a height *h*. The acceleration is constant, and we can use Equation 16 of Chapter 5:

$$v^2 = 2ad \tag{10}$$

Here the distance d is equal to the total height h; the acceleration a is equal to the acceleration of gravitation, a_g; and v^2 is equal to s^2, the square of the speed at the bottom. Making these substitutions we have

$$s^2 = 2a_g h \tag{11}$$

This equation actually holds for the frictionless motion from rest of falling objects, pendulum bobs, and objects sliding on inclined planes. In these circumstances, as we have already noted in this chapter, the speed s at the bottom depends only on the vertical height h through which the object has descended.

If we multiply both sides of Equation 11 by $(1/2)m$ we obtain

$$1/2\ m s_2^2 = m a_g h_1 \tag{12}$$

We have added the numbers 2 and 1 as subscripts to the letters s and h in this equation to emphasize that we are referring to two different points in the object's path; h_1 is the height of the object at the start of its motion, but s_2 is the speed of the object later, when it has fallen through the height h_1. Equation 12 informs us that the kinetic energy at the bottom (left side of equation) is equal to the potential energy at the top (right side of equation). For this special case, the beginning kinetic energy and the final potential energy are both zero.

We can find the law of conservation of energy for a more general case in which the object begins with a certain speed s_1 at a height h_1 and later has a speed s_2 at a height h_2 (see Fig. 6-4). For this case we use Equation 19 of Chapter 5:

$$v^2 = v_0^2 + 2a(d - d_0) \tag{13}$$

Using the reference frame shown in Figure 6-4, the acceleration a becomes *minus* a_g, since gravitation pulls downward on the object. The distances d and d_0 become the heights h_2 and h_1, and v^2 and v_0^2 becomes s_2^2 and s_1^2, respectively. Making these substitutions we have

$$s_2^2 = s_1^2 - 2a_g(h_2 - h_1) \tag{14}$$

$$s_2^2 = s_1^2 - 2a_g h_2 + 2a_g h_1 \tag{15}$$

If we add $2a_g h$ to both sides of the equation we have

$$s_2^2 + 2a_g h_2 = s_1^2 + 2a_g h_1 \tag{16}$$

Next we multiply both sides by 1/2 m:

$$1/2\ ms_2^2 + ma_g h_2 = 1/2\ ms_1^2 + ma_g h_1 \tag{17}$$

This last equation is just the law of conservation of energy for the

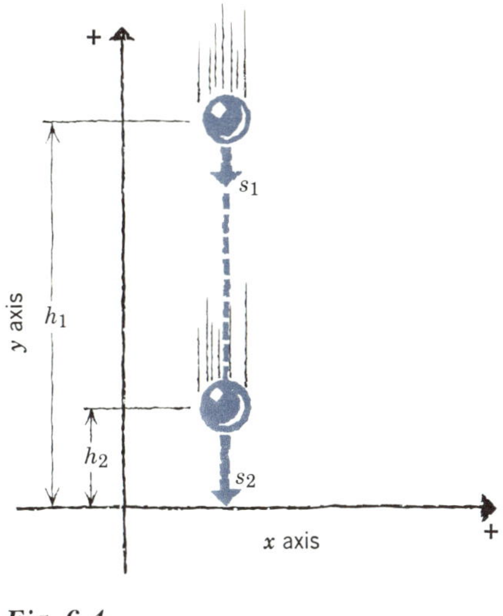

Fig. 6-4

object in question. We have obtained it from the material discussed in earlier chapters without assuming any new laws or principles. Equation 17 holds for the frictionless motion of falling objects, pendulum bobs, and objects sliding on inclined planes or other stationary surfaces.

Internal Energy

The existence of friction should convince us that there are more forms of energy than we have yet discussed. Let us consider, for example, a real process in which friction is present (see Fig. 6-5). A block sliding along a horizontal slab finally comes to a stop. Since the slab is horizontal, there has been no change in the potential energy of the block. Nevertheless, the block loses kinetic energy as it comes to rest. Does the energy of its motion really disappear entirely, as it appears to do? Where could the energy go if it is not lost? The only obvious possibility is that it somehow disappears into the slab or into the block or into both. Is there any observable change in these objects that would justify the notion that there is an increase of energy inside these objects? Is there reason to believe that there is an increase in what we shall call their *internal energy*? Yes, in fact there is an observable change: as a result of the friction, the two objects get hotter. The kinetic energy, which is lost because of friction, becomes part of the internal energy of the block and the slab. This change in internal energy causes the temperature of the block and the slab to rise. In general, when two objects rub together, most of the energy apparently lost because of friction is transformed into internal energy and usually causes the objects to become hotter. This additional internal energy leaves the objects if they are cooled.

Changes in internal energy may be responsible for other changes besides changes of temperature—for example, melting and boiling. Accurate experiments justify this concept of internal energy. There are other ways in which objects can acquire internal energy. For

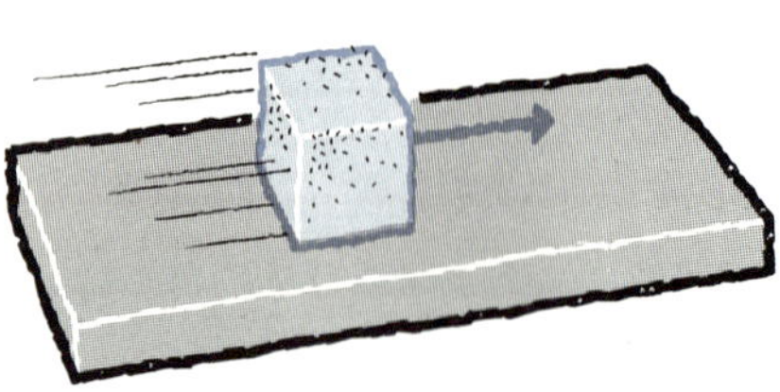

Fig. 6-5

example, the internal energy of objects is ordinarily increased in collisions.

If we take into account this new form of energy, internal energy, we can more accurately understand the behavior of objects subject to friction. We may now allow friction to exist in the pendulum and inclined plane in the examples with which we began this chapter. The total energy of these systems must include the internal energy.

$$\text{total energy} = \text{gravitational potential energy} + \text{kinetic energy} + \text{internal energy} \tag{18}$$

It is *total* energy that is conserved, although the amount of each form will generally change. Actually there are still other forms of energy that we have not yet discussed. For example, energy is associated with sound, with light, and with electricity.

Energy transformations

In this section we shall discuss two ways in which energy can be *transferred* from one object to another or *transformed* from one form to another (It is possible for energy to be transferred and transformed in the same process.)

Heat Flow

If we put a warm object in contact with a cold one, we find that the cold object gets warmer and the warm object gets colder (see Fig. 6-6). This means that some of the internal energy of the warmer object is flowing into the colder object. An energy transfer of this kind is called a *heat flow*. We reserve the term "heat" to refer to this kind of energy transfer. We do not say that an object "contains heat"; we prefer to say that it "contains internal energy."

Work

Energy may be transferred or changed in form if a force is exerted on (or by) an object as it moves through a distance. This kind of energy transfer or transformation is called "doing *work*." Work requires both a force and a distance. In this section we shall assume that the force is either in the direction of motion or in the opposite direction.

Figure 6-7*a* illustrates how energy is supplied to an object by doing work on it. The block is first at rest on a frictionless surface; then it is pushed by a force of magnitude F. This force keeps acting on the block until it has gone a distance d. We define the amount of *work* done by a force to be

$$\text{Work} = Fd \tag{19}$$

This is a general equation for work when the force is in the direction of motion. Thus, the work done is equal to the magnitude of the force multiplied by the distance through which it acts. In this example the work is equal to the energy that has been transferred to the block. The energy supplied to the block takes the form of kinetic energy: the change of kinetic energy is equal to the work done.

In Figure 6-7*b* we see a similar example, except that the block is already moving in one direction and the force acts in the opposite direction. This force slows the block down, and therefore decreases its kinetic energy. In a case like this, when the force on the object is opposite the direction of motion, energy is taken out of the object. The energy taken out is just equal to the work. That is, in such a case, the work is equal to the energy transferred *out*. In this case, again, the force multiplied by the distance gives us the work, and this work is equal to the energy transferred.

Another example of putting energy into an object by doing work is shown in Figure 6-8. In order to lift an object we must exert a force on it equal to its weight. (It will, of course, require a slightly larger

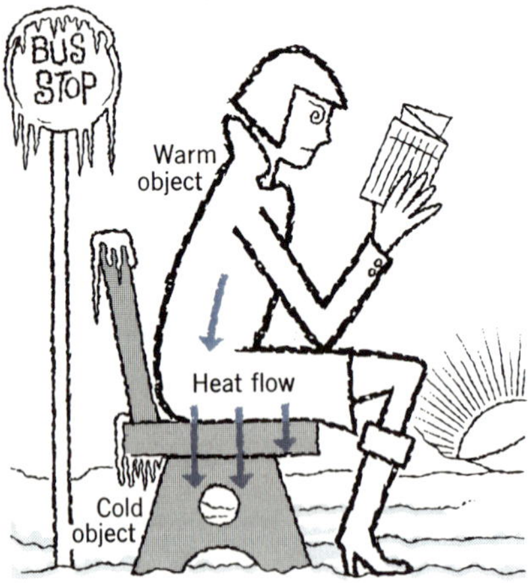

Fig. 6-6

force to start it rising; but once it is rising at a constant speed, a force equal to its weight is sufficient to keep it going upward.) If we lift it through a total height h, we have done some work on it since we have exerted a force equal to its weight w through a distance h. The total work done is

$$\text{Work} = wh = \text{energy transferred} \tag{20}$$

This quantity is equal to the gravitational potential energy the object gains by being raised to that height. Thus, in this example the energy put into the object by the work done appears as an increase in its gravitational potential energy.

Energy may be transformed as well as transferred by means of work. As an example, we need only let the object we lifted in the previous example fall freely back down through the height h. The constant force of gravitation acting on it causes it to acquire kinetic energy as it falls. The work done on this falling object by the force of gravitation transforms the gravitational potential energy into kinetic energy.

*Work Done by Forces not along the Direction of Motion

In this section we explain how to find the work done when the direction of the force is not parallel to the direction of motion. When this situation occurs, we use the component of force that is along the direction of motion and not the force itself:

$$\text{Work} = \begin{pmatrix}\text{The magnitude of the} \\ \text{component of force} \\ \text{along the direction} \\ \text{of motion}\end{pmatrix} \times (\text{distance}) \tag{21}$$

Figure 6-9 shows an example of this situation. The block is once again free to move on a frictionless surface, but it is being pulled by

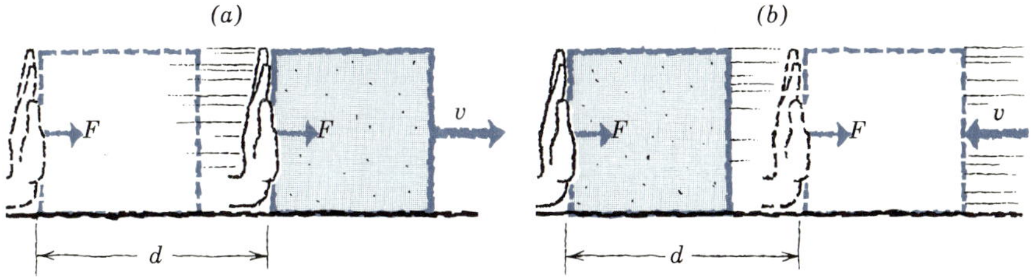

Fig. 6-7

a force at an angle to the direction of motion of the block. This force acts while the block moves a distance d. To calculate the work we do not use the force **F**, but rather the magnitude F_m of its component in the direction of motion.

$$\text{Work} = F_m d \tag{22}$$

Equation 22 correctly implies that there is no work done by a force on an object when the direction of that force is perpendicular to the direction of motion of the object.

*Power

In practical applications, engineers are interested not only in the total amount of work done but also in the rate at which it is being done. Therefore, we introduce the derived quantity *power*, which is defined as

$$\text{Power} = \text{the rate of doing work}$$

If the work is being done at a constant rate, then the power can be found easily by dividing the work done by the time required to do it:

$$\text{Power} = \frac{\text{work}}{\text{time}} \tag{23}$$

Sometimes the word *power* is used to refer to the rate of energy transfer or transformation when heat flow as well as work is involved.

Fig. 6-8

We have here used a more limited definition of the quantity power, restricting it to work alone.

The unit of power in the system of units we are using is called the *watt*, after James Watt, who helped to develop the steam engine. The watt can be expressed in terms of the joule and the second by inserting the units of power, work, and time into Equation 23. The unit of work is the same as the unit of energy (the joule) and therefore

$$\text{watt} = \frac{\text{joule}}{\text{second}} \tag{24}$$

The common English unit of power, the *horsepower*, is equal to about 746 watts.

APPLICATIONS

Binding Energy

Two objects held together by a force may be separated by a force. Work will be done in separating the two objects, since the force must be exerted through a distance. This means that energy must be supplied to the objects in order to separate them. The amount of energy needed to completely separate two objects is called their *binding energy*. The objects may actually have more energy when they are separated than when they are bound together because of the binding energy.

As an example of binding energy, consider the rock in Figure 6-10,

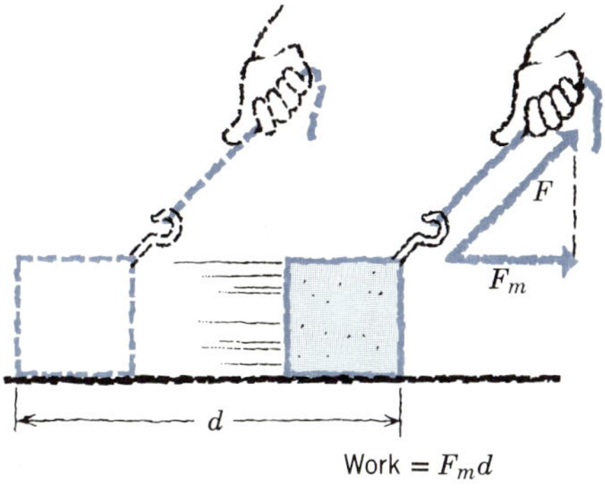

Fig. 6-9

which is held to the surface of the earth by gravitational forces. In order to remove this rock from the earth, we would have to do work on the rock to lift it so far from the earth that the remaining gravitational pull would be negligible. The energy that must be supplied in doing this work is equal to the binding energy. This energy used in lifting the rock increases the gravitational potential energy, so that the system actually has more energy when separated than before.

In Chapter 4, we mentioned that an object with sufficient speed could escape from the earth. We can now understand this "escape speed." To remove an object from the earth, we must supply energy equal to the binding energy. An object moving as fast as or faster than this escape speed already has the required extra energy in the form of kinetic energy, and therefore can escape from the earth.

Perpetual Motion

It is often said that perpetual-motion machines are impossible. Here we shall discuss one kind of perpetual-motion machine—a device that would be capable of putting out energy without taking any in (see Fig. 6-11). Many such devices have been designed, some of them, to judge from published diagrams, are marvels of engineering skill as far as workmanship and complexity of operation are concerned. Unfortunately, none of them has been demonstrated to work, despite the obvious ingenuity and effort that have evidently gone into their "discovery."

What principle would prevent a device from putting forth energy forever without ever taking any in? The principle, of course, is the law of conservation of energy. Energy cannot be created (or destroyed) by any machine or device. It can be stored in a device, of

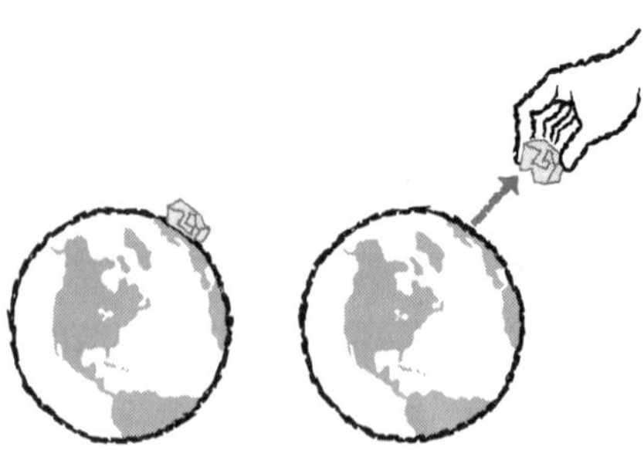

(a) (b)

Fig. 6-10

course, but the stored energy will eventually be used up unless it is replenished by energy from outside the device. Thus, according to the law of conservation of energy, devices which supply an inexhaustible flow of heat or do work forever without ever taking in any energy are impossible. If a successful perpetual-motion machine were

Fig. 6-11 Humorous design by M. C. Escher. *From the collection of C. V. S. Roosevelt, Washington, D.C. Used with permission.*

actually devised, we would have to abandon the law of conservation of energy. However, since this law has been verified in such a wide variety of phenomena and since all of the perpetual-motion machines produced so far have failed, physicists are very skeptical when a new perpetual-motion machine is proposed. We sometimes express the law of conservation of energy by saying that perpetual-motion machines of the kind discussed here are impossible.

*Simple Machines—The Lever

A machine is a device that we put energy into and get energy out of. The input and output energies may be of the same or different forms. Figure 6-12 illustrates the action of one simple machine, a lever. To operate the lever, we push down on the input end, putting energy into the machine by doing work. The other end does work in lifting the rock, thus putting energy out of the machine into the rock. Energy flows from us into the rock by way of the machine. Suppose that the rock has a weight w; how much force must we use to operate the machine and lift the rock? (We will disregard friction in the machine, which would ordinarily divert some of the energy from useful work into internal energy.)

We first push down on the lever with a force F and move the lever through a small distance d_1. Thus we have supplied an input energy equal to the work Fd_1. The other end of the lever will then lift the rock a small distance d_2. (The distances d_1 and d_2 are approximately vertical.) Thus there is an output energy equal to the work wd_2. Ignoring friction, the input energy must be equal to the output energy:

$$Fd_1 = wd_2 \tag{25}$$

Dividing both sides of this equation by d_1 we obtain

$$F = w(d_2/d_1) \tag{26}$$

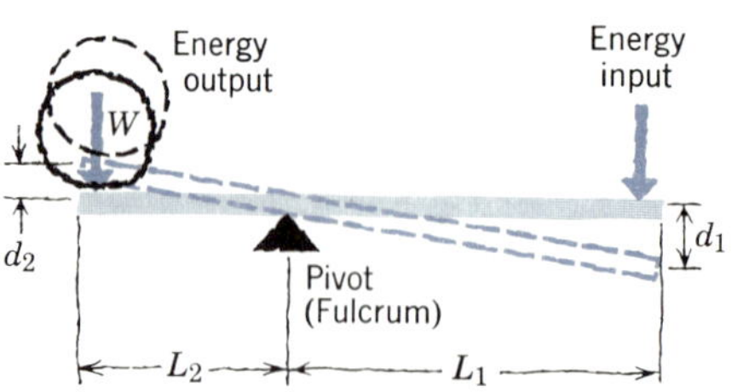

Fig. 6-12

The greater the distance of either end of the lever from the pivot, the farther that end will move when the lever turns about the pivot. In fact, the distance each end of the lever moves is proportional to the distance from the pivot to that end. We may write this proportionality as follows, letting L_1 and L_2 represent the distance of the two ends from the pivot:

$$(d_2/d_1) = (L_2/L_1) \tag{27}$$

Therefore, the force we must exert to lift the rock is

$$F = w(d_2/d_1) = W(L_2/L_1) \tag{28}$$

If the length of the output side of the lever L_2 is smaller than the length of the input side L_1, then we can use the lever to lift the rock by exerting a force smaller than the weight of the rock. This shows one use of machines—to increase the force that we can apply to an object. A small force on the input end is transformed by the machine into a large force on the output end. We had to exert our input force, however, over a larger distance than the rock itself moved. Therefore, we had to do the same amount of work as if we had moved the rock ourselves without the use of the machine. Machines cannot create energy, but they enable us to use it more effectively.

Problem solving

Most of the problems we take up now involve three forms of energy, and changes from one to another: kinetic energy, gravitational potential energy, and internal energy. The basic principle applied in solving conservation-of-energy problems is that the energy of a system before an interaction is equal to the energy after the interaction.

$$\text{Energy before} = \text{Energy after} \tag{29}$$

or, more generally, during a process

$$\text{Energy at one time} = \text{Energy at another time} \tag{30}$$

We must be sure to include all the important forms of energy at both times. Very often we shall be concerned with situations in which the only two important forms of energy are the following:

Kinetic energy: $1/2\ ms^2$ for each object
Gravitational potential energy: ma_gh for each object

These two forms of energy are usually the only important ones when we are considering objects falling or sliding without friction on smooth stationary surfaces, subject only to the pressure from the surfaces and the gravitational force. If there is friction between the object and the surface, some energy will be transformed into internal energy. Also, if there is an impact, some energy may be transformed into internal energy. For example, the ball rolling down the inclined plane in Figure 6-13*a* has an impact when it reaches the bottom of the plane because the surface forces the ball to change direction suddenly. This impact could cause the ball to lose kinetic energy and gain internal energy. An impact does not occur at the bottom of the plane in Figure 6-13*b* because the path curves gently, and the ball does not strike a corner. Idealized impacts which do not involve increases of internal energy are called *elastic* impacts.

In problems concerning energy or its conservation, we must bear in mind the ways in which the energy of a system of objects can be changed. A force arising from an interaction with an object outside the system can change the energy of the system if it acts through a distance on objects or substances inside the system. Objects sliding without friction along stationary surfaces are subject to a force exerted by the surface, but this force does not change their energy, since it is perpendicular to the direction of motion. A frictional force, however, will retard the object and change its energy. Heat flow to or from an object outside the system can also change the energy of a system. Such a change will occur, for example, if part of the system is in contact with an outside object or substance at a different temperature.

We can break down the solution of conservation-of-energy problems into several steps, similar to those presented for problems involving other conserved quantities. (See Chap. 3, p. 112.)

STEP 1 Draw a diagram including a reference frame that shows the condition of the object or objects both before and after the interac-

(*a*) (*b*)

Fig. 6-13

tion. Decide which objects are concerned in the conservation process you wish to describe. Choose the situations (1 and 2) which represent the "before" and "after" condition. The second situation will usually involve what you want to know. Subscripts 1 and 2 are used to identify the two situations in Equation 31 below.

STEP 2 Write the equation for conservation of energy, including every object whose energy changes and all the forms of energy that are important to the problem. We will often be dealing with motion under gravitational forces near the surface of the earth where there is no friction and no nonelastic impacts. For a single object this results in the following equation:

$$1/2\ ms_1^2 + ma_g h_1 = 1/2\ ms_2^2 + ma_g h_2 \tag{31}$$

Note that a_g is a positive quantity in this equation. If there are more objects than one, we must have additional terms on both sides of the equation for each form of energy for each object. If the problem involves gravitational potential energy, we will need to decide upon the level at which we want the potential energy to be zero. (Usually this will be the lowest level along an object's path.)

STEP 3 List the known and unknown quantities that appear in the equation(s) of Step 2.

STEP 4 Substitute the known quantities listed in Step 3 into the equation(s) in Step 2, and solve for the unknown quantities, if possible. If this is not possible, some other information or principle is necessary. It may be necessary to use another conservation law in order to obtain another relation between the quantities. In some of the more difficult problems both conservation of energy and conservation of momentum are necessary to determine the answer.

EXAMPLE 1

A flower pot of 5 kilograms mass falls off the top of a building of height 30 meters. A woman looks out her window and sees the flower pot going by at a speed of 20 meters per second. How high is her window?

Solution.

STEP 1 Figure 6-14 shows the situation and also our reference frame. Since the flower pot is falling freely under gravity, its energy is conserved if we disregard the friction of the air. The two situations are (1) the situation just as the pot begins to fall, before it has picked up any speed, and (2) the situation as it passes the window.

STEP 2 We choose the zero of potential energy at ground level.

There is just one object, and no friction or impact (yet), so we can use Equation 31:

$$1/2\ ms_1^{\ 2} + ma_g h_1 = 1/2\ ms_2^{\ 2} + ma_g h_2$$

STEP 3

m = 5 kilograms
s_1 = 0
s_2 = 20 meter/second
h_1 = height of building = 30 meters
h_2 = height of window = ?
a_g = 9.8 meter/second^2

STEP 4 We notice that the mass m will cancel out so we do not substitute its value into the equation. We do substitute the other values as follows:

$$1/2\ m\ (0)^2 + m\left(9.8\ \frac{\text{meter}}{\text{second}^2}\right)(30\ \text{meter}) = 1/2\ m\ (20)^2\ \frac{\text{meter}^2}{\text{second}^2} + m\left(9.8\ \frac{\text{meter}}{\text{second}^2}\right) h_2$$

$$294\ \frac{\text{meter}^2}{\text{second}^2} = 200\ \frac{\text{meter}^2}{\text{second}^2} + 9.8\ \frac{\text{meter}}{\text{second}^2} \cdot h_2$$

$$94\ \text{meter}^2 = (9.8\ \text{meter})\ h_2$$

$$h_2 = \frac{94\ \text{meter}^2}{9.8\ \text{meter}} = 9.6\ \text{meters}$$

EXAMPLE 2

A skier starts from rest and slides down a frictionless hill from a height of 10 meters. What is his speed at the bottom?

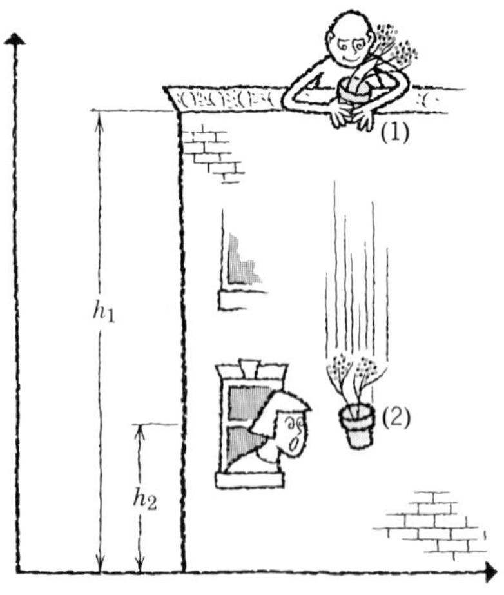

Fig. 6-14

Solution.

STEP 1 See Figure 6-15. Since there is no impact and no friction, we are concerned with only two forms of energy: kinetic and gravitational potential energy. We choose situation 1 at the start and situation 2 when the skier is at the bottom of the hill.

STEP 2 We write the equation as before:

$$1/2\ ms_1^2 + ma_g h_1 = 1/2\ ms_2^2 + ma_g h_2$$

STEP 3 m = unknown, and unnecessary

$s_1 = 0$, since the skier starts from rest at the top.
$s_2 = ?$
$a_g = 9.8$ meter/second²
$h_1 = 10$ meters
$h_2 = 0$

STEP 4 We substitute the values into the equation of Step 2:

$$1/2\ m\ (0)^2 + m\left(9.8\frac{\text{meter}}{\text{second}^2}\right)(10\ \text{meter}) = 1/2\ ms_2^2 + m\left(9.8\frac{\text{meter}}{\text{second}}\right)(0\ \text{meter})$$

$$(9.8 \times 10)\frac{\text{meter}^2}{\text{second}^2} = 1/2\ s_2^2$$

$$s_2^2 = 2\ (9.8 \times 10)\frac{\text{meter}^2}{\text{second}^2}$$

$$s_2 = \sqrt{2} \times \sqrt{9.8 \times 10} \times \frac{\text{meter}}{\text{second}}$$

$$s_2 = 10\sqrt{2}\frac{\text{meter}^2}{\text{second}^2}\ \text{approximately}$$

$$= 14\ \text{meters per second}$$

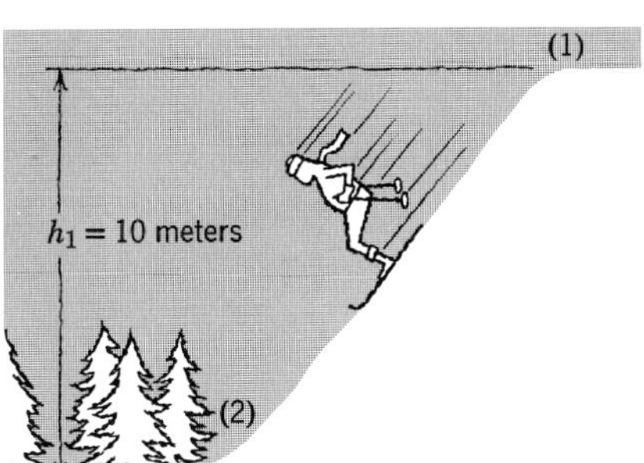

Fig. 6-15

EXAMPLE 3

A skier is approaching a smooth hill at 40 meters per second. How high up will he be when his speed has dropped to 20 meters per second? Disregard friction.

Solution.

STEP 1 See Figure 6-16. Once again we are concerned with only kinetic and gravitational potential energy. We establish situation 1 with the skier at the bottom and situation 2 when the speed has dropped to 20 meters per second.

STEP 2 Again $1/2\ ms_1^2 + ma_g h_1 = 1/2\ ms_2^2 + ma_g h_2$

We choose $h = 0$ at the bottom of the hill.

STEP 3

$$\begin{aligned} m &= \text{unknown and unnecessary} \\ s_1 &= 40 \text{ meter/second} \\ s_2 &= 20 \text{ meter/second} \\ a_g &= 9.8 \text{ meter/second}^2 \approx 10 \text{ meter/second}^2 \\ h_1 &= 0 \\ h_2 &= ? \end{aligned}$$

STEP 4 Once again the mass cancels out and we have

$$1/2\ m\left(40\frac{\text{meter}}{\text{second}}\right)^2 + m\left(10\frac{\text{meter}}{\text{second}^2}\right)(0) = 1/2\ m\left(20\frac{\text{meter}}{\text{second}}\right)^2 + m\left(10\frac{\text{meter}}{\text{second}^2}\right)h_2$$

$$1/2\ (1600)\frac{\text{meter}^2}{\text{second}^2} = 1/2\ (400)\frac{\text{meter}^2}{\text{second}^2} + 10\frac{\text{meter}}{\text{second}^2}h_2$$

$$10h_2 + 200 \text{ meter} = 800 \text{ meter}$$

$$10h_2 = 600 \text{ meter}$$

$$h_2 = \frac{600 \text{ meter}}{10} = 60 \text{ meters}$$

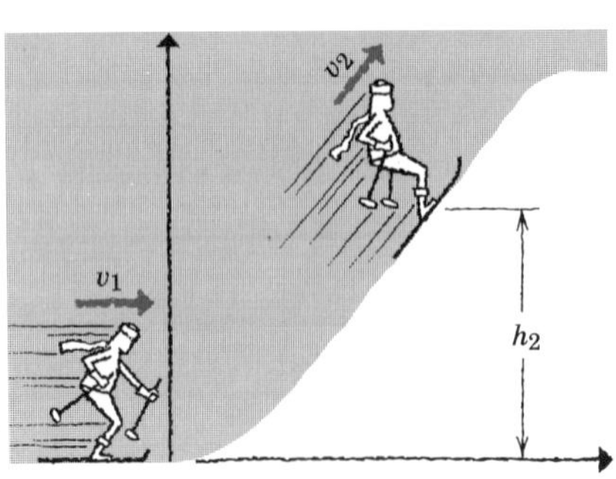

Fig. 6-16

EXAMPLE 4

Starting from the top of an inclined plane, a mass of 15 kilograms starts to slide down with friction at a speed of 10 meters per second and comes to rest at the bottom. The plane is 2 meters high at its highest point. How much has the internal energy of the mass and the plane together increased when the mass reaches the bottom?

Solution.

STEP 1 See Figure 6-17. Energy has been conserved if we take internal energy (U) into account. The first situation exists when the object is at the top of the plane. The second situation exists when the object is at rest at the bottom.

STEP 2 $U_1 + 1/2\ ms_1^2 + ma_g h_1 = U_2 + 1/2\ ms_2^2 + ma_g h_2$

We choose the bottom of the plane for $h = 0$.

STEP 3
$$\begin{aligned} U_1 &= ? \\ U_2 &= ? \\ m &= 15 \text{ kilograms} \\ s_1 &= 10 \text{ meter/second} \\ s_2 &= 0 \\ h_1 &= 2 \text{ meters} \\ h_2 &= 0 \\ a_g &= 9.8 \text{ meter/second}^2 \approx 10 \text{ meter/second}^2 \end{aligned}$$

STEP 4

$$U_1 + 1/2\ (15 \text{ kilogram}) \left(10 \frac{\text{meter}}{\text{second}}\right)^2 + (15 \text{ kilogram}) \left(10 \frac{\text{meter}}{\text{second}^2}\right) (2 \text{ meter})$$

$$= U_2 + 1/2\ (15 \text{ kilogram})\ (0) + (15 \text{ kilogram}) \left(10 \frac{\text{meter}}{\text{second}^2}\right) 0$$

$$U_1 + (1/2 \times 15 \times 100) \frac{\text{meter}^2 \text{ kilogram}}{\text{second}^2} + (15 \times 10 \times 2) \frac{\text{meter}^2 \text{ kilogram}}{\text{second}^2} = U_2$$

Thus

$$U_2 - U_1 = (750) \underbrace{\frac{\text{meter}^2 \text{ kilogram}}{\text{second}^2}}_{\text{joule}} + (300) \underbrace{\frac{\text{meter}^2 \text{ kilogram}}{\text{second}^2}}_{\text{joule}}$$

$$U_2 - U_1 = 1050 \text{ joules}$$

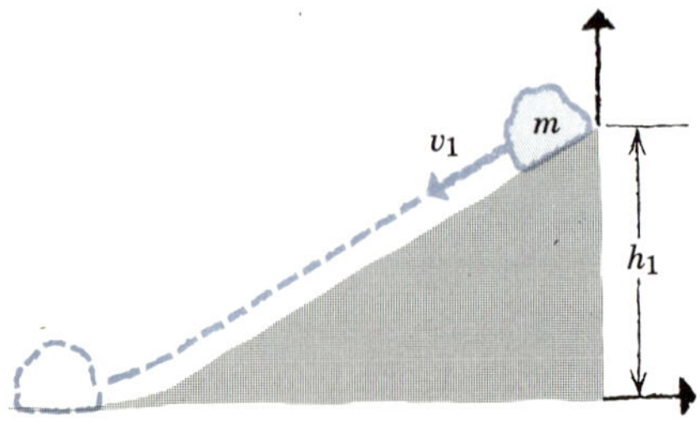

Fig. 6-17

The equations do not yield the value of either U_1 or U_2; however, $U_2 - U_1$, the increase in internal energy, is exactly the information we are looking for.

EXAMPLE 5

A constant force F is exerted on the block in Figure 6-7*a* while the block, starting from rest, moves a distance d. Show mathematically that the work done is equal to the change in the kinetic energy of the block.

Solution. The total work done is the force times the distance, Fd. The force is related to the acceleration and the acceleration to the distance by the following equations:

$$F = ma \tag{32}$$

$$s^2 = 2\,ad \tag{33}$$

We can use Equation 33 because the force on the block is constant, making the acceleration constant also, and the starting speed is zero. Dividing this equation by $2a$, we obtain the distance d:

$$\frac{s^2}{2a} = d \tag{34}$$

Taking F and d from Equations 32 and 33, we can write the work Fd as follows:

$$Fd = (ma)\left(\frac{s^2}{2a}\right) = (1/2)\ ms^2$$

This shows that the work Fd is equal to the kinetic energy $(1/2)\ ms^2$. Since the block started from rest, the change in kinetic energy is the total kinetic energy.

Summary

I. FUNDAMENTAL LAWS AND PRINCIPLES
   A. CONSERVATION OF ENERGY
II. QUANTITIES
   A. ENERGY: a derived scalar quantity. The unit of energy is the joule, which is equal to one kilogram meter^2/second^2. We have introduced three forms.
      1. Gravitational potential energy: ma_gh
      2. Kinetic energy: $1/2\ ms^2$
      3. Internal energy.
   B. QUANTITIES INVOLVED IN ENERGY TRANSFER AND TRANSFORMATION
      1. Heat flow.
      2. Work = Fd,*or F_md
III. APPLICATIONS
   A. ENERGY RELATIONS; INCLINES PLANES AND PENDULUMS
   B. BINDING ENERGY
   C. PERPETUAL MOTION
   *D. THE LEVER, A SIMPLE MACHINE
IV. PROBLEM SOLVING
   A. PROBLEMS INVOLVING THREE FORMS OF ENERGY AND CONSERVATION OF TOTAL ENERGY

Questions

1. How can the energy of two objects held together by a force be greater when they are separated than when they are together? binding energy?
2. What is "heat flow"?
3. What is "work"?
4. Why do we believe that there is more than one form of energy?
5. What kinds of energy transformation would you expect where friction occurs?
6. When a spring is stretched we say that it possesses potential energy, not gravitational potential energy, but another kind. Why do we say this?

7. Does the kinetic energy of an object depend on the direction of its velocity?
8. Would it be possible to project a snowball against a wall at such speed that it would entirely melt on impact? Explain.
9. A professor performed the pendulum demonstration as illustrated in Figure 6-2. In his enthusiasm to show it to his students, he gave the bob a shove as it left his nose. What happened and why?
10. What makes a swinging pendulum finally come to rest? Where does the energy go in this case?
11. Why does an automobile have to burn fuel continually just to keep going at a constant speed? What happens to the energy supplied by the burning fuel?
12. Suppose you were assigned the task of proving that "internal energy" really exists. What sort of experiment would you devise?
13. How do we know that energy can be transformed from one form to another by the process of work?
14. How do we know that energy can be transferred from one object to another by the process of work?
15. How do we know that energy can be transferred from one object to another by the process of heat flow?
16. Figure 6-18 shows two kinds of water wheel. Where does the energy come from to run each kind?
17. Why is energy an important quantity?
18. Give examples in which the various forms of energy are transformed into other forms or obtained from other forms.
19. Does a stretched rubber band contain energy? Explain.

(a) (b)

Fig. 6-18

20. It requires work to stir a pan of soup. What do you think happens to the energy thus transferred?

21. Why does a knife get hot when it is sharpened on a grinding wheel?

22. Figure 6-19 shows a proposed device for running a grinding wheel. The electric motor runs an electric generator, which generates enough electricity to run the motor and the grinding wheel. Without knowing much about the operation of electric generators and motors, why would you conclude that this device cannot work?

23. Artificial satellites often stay up for a long period of time and then eventually fall to earth. Why do they finally come down?

24. Why do the brakes on a car heat up when the car is stopping?

25. Figure 6-20 shows a device turned by escaping steam. Where does the energy come from? Follow the transformations of energy illustrated in this process.

26. What happens to the kinetic energy of a falling object when it hits the ground?

27. What is meant by the term "conservation"?

*28. Under what circumstances is the total energy of a system of objects conserved? Under what circumstances are various kinds of energy conserved for a set of objects?

*29. Would a constant-thrust rocket give more energy of motion per pound of fuel when it was just starting out or after it had been traveling for a while? Why?

*30. Explain from the standpoint of energy why a comet attains its greatest speed when it is nearest the sun.

*31. Do the earth and the moon have more or less energy at present than they would have if the moon were lying on the surface of the earth but things were otherwise the same? Explain.

32. Can you offer an explanation for the slowing down of the rotation of the earth?

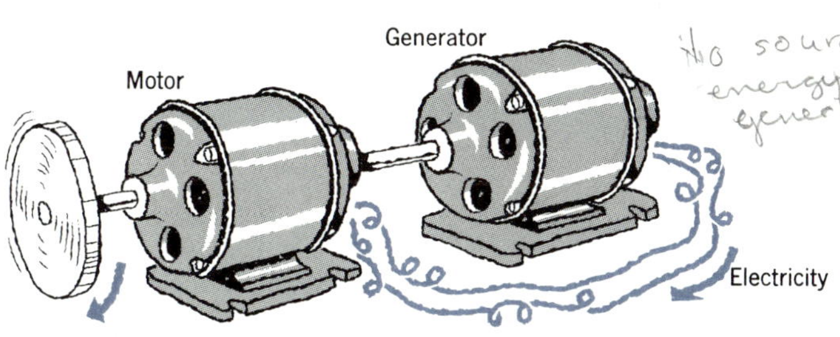

Fig. 6-19

33. Would a hard-boiled egg roll farther than a raw egg if they both started rolling at the same speed on a hard surface? Explain.[1]

34. What would convince you that sound involves a form of energy?

35. Why is it possible to obtain fire by friction?

*36. Design a machine that gives an output motion greater than the input motion. What must be sacrificed in order to obtain this result?

*37. Would a constant-thrust rocket develop more power when it was just starting out or after it had been traveling for a while? Explain.

*38. Does a more powerful automobile engine necessarily exert more force than a less powerful engine?

Problems

1. A skier is approaching a frictionless hill at a speed of 100 meters per second. How far up the hill does he get before he starts to slide back down again?

2. A skier traveling at 15 meters per second goes over a hill that is 5 meters high. What is his speed at the top of the hill?

3. A lead weight of mass 0.5 kilograms falls out of a window 30 meters high and goes through the roof of a car. How much is the internal energy of the car increased?

4. A man lifts a 10-kilogram weight one meter off the ground, holds it

[1]This question arises from a comment by John Perry, *Spinning Tops and Gyroscopic Motion*, New York: Dover, 1957, p. 54.

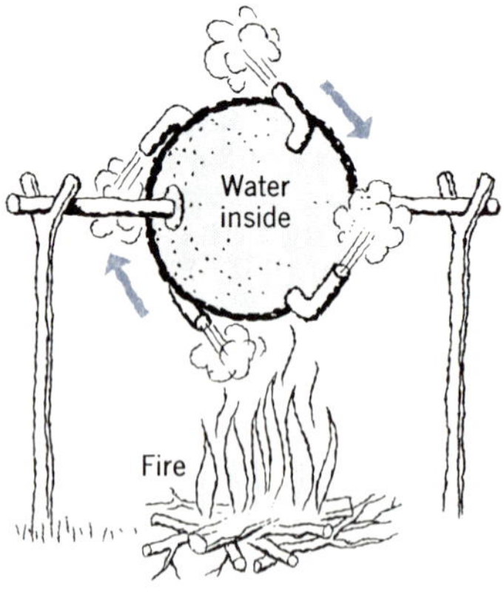

Fig. 6-20

there for 10 hours, and then gently sets it back down. How much work has he done?

5. A man lifts a 10-kilogram weight one meter off the ground and holds it there for 10 hours. How much work has he done?

6. A skier at a height h has a speed v. What will his speed be at height zero, disregarding friction?

7. A ball with a mass of 10 kilograms is thrown at a speed of 20 meters per second. What is its kinetic energy?

8. A wheelbarrow filled with cement is wheeled up a ramp to a height of 15 meters. If the full wheelbarrow has a mass of 100 kilograms, by how much is its potential energy increased?

9. If I shoot a cannonball straight up, how high will it rise as compared with another cannonball shot with half the energy?

10. On a frictionless surface, a body of mass $2m$ and speed v strikes a body of mass m at rest and sticks to it. How much kinetic energy is lost?

11. A pendulum is made by putting a bob of mass 2 kilograms at the end of a 1-meter cord. The bob is pulled out so that the cord is horizontal and released. What is the maximum speed of the bob?

12. In Example 1, p. 257, what is the speed of the flower pot just before it hits the ground? What happens to the energy when it hits the ground?

*13. How much energy is expended if a 60-watt bulb burns for 12 hours?

More Difficult Questions and Problems

1. If I drop a certain weight into a bucket of molasses from a height h, the temperature of the molasses rises by one degree. What happens if I drop the weight from a height of $2h$? What assumption are you making about energy, temperature, and molasses?

2. If I throw an object into a bucket of molasses at a speed v, the temperature rises by two degrees. What happens if I throw it at twice the speed?

*3. Suppose that I wanted to determine the work done by one body on another in a collision, and thus the energy transferred or transformed. Would I average the force over the distance or over the time? Why? Suppose that I wanted to determine the momentum transferred in the collision. How would I take the average?

*4. Show that the product Ft, the force times the time (called *impulse*), is related to transfer of momentum.

*5. A body is put into motion by a force F which acts through a distance d. The body is then brought to rest by a force equal in magnitude to F but opposite in direction, acting for a time t. What is the maximum speed of the body if it has mass m?

*6. Using Figure 6-21, show that the rule for components of forces follows from work and energy considerations. Assume that the surfaces are frictionless and the masses balance. Then move one mass and consider the work put into the machine as the mass is moved.

7. When a gun is fired, the momentum of the gun and the bullet are approximately equal in magnitude. Why, then, is it not just as dangerous to be hit by the gun as by the bullet?

8. If I shoot two cannonballs straight up, one with twice as much energy as the other, what will be their relative times of flight?

9. Of the two equal masses shown in Figure 6-22, one slips down the other from a height h, reaching a speed v (which is less than $\sqrt{a_g h}$). What is the velocity of the other mass? How do you know? Is there friction between the two blocks? How do you know? Assume that the horizontal surface is frictionless.

10. In Figure 6-23, the car (of mass m) starts from rest at the top of the slope on the left, and rolls without friction. What is its speed at the top of the loop?

11. In Problem 10, what is the force arising from the pressure between the car and the track at the top of the loop?

12. In Problem 10, find the radius of the loop so that the pressure referred to in Problem 11 will be zero.

13. In Problem 10, what is the force between the car and the track at the point labeled A?

14. A ball, starting from rest, rolls down an incline and bounces, as shown in Figure 6-24. The impact is fully elastic. When it bounces, does it come back to its original height? Explain.

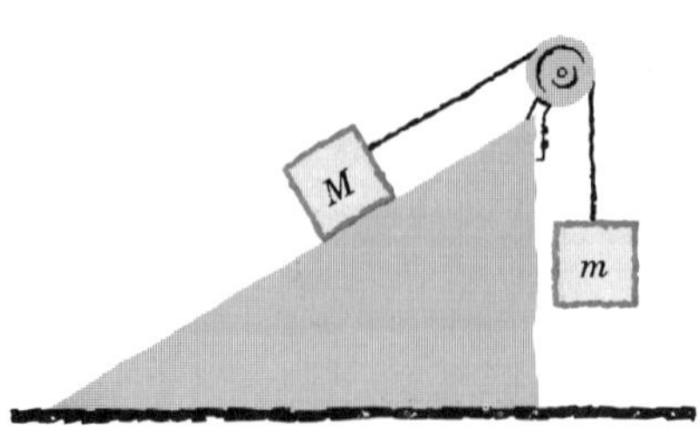

Fig. 6-21

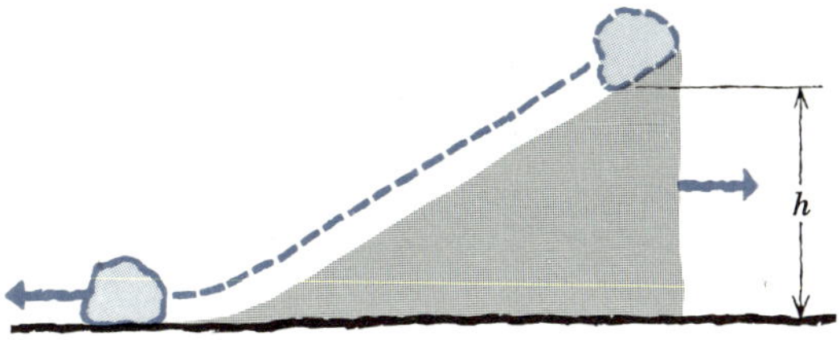

Fig. 6-22

15. A mass m moving at a speed v strikes a motionless mass $2m$ and sticks to it, then the two masses together slide up the incline as shown in Figure 6-25. If the surface is frictionless, how high up does the combined mass get?

16. What force F is needed to hold the pulley system shown in Figure 6-26 in equilibrium? (Disregard the mass of the pulleys and the rope.)

17. Two equal masses are free to slide on a frictionless surface. The first is initially moving at a velocity v. It strikes the second one (which is initially at rest), exploding a cap between them, and comes back at a velocity V in the opposite direction. Find the resulting velocity of the second mass, and also the energy obtained by both blocks together from the exploding cap. Consider the masses as known.

18. A ball is whirling around on the end of a string, as shown in Figure 6-27, at a speed v. We disregard effects of gravitation. If the string is pulled in to half its former length, what is the new speed of the ball? Where does the new energy come from?

***19.** In Problem 18, suppose that the cord is firmly attached to the end of a rod (see Fig. 6-28). As the ball whirls around, the cord becomes wound up on the rod. What is the speed of the ball when the string has been wound to half its original length? Why does this differ from the results of Problem 18? How has the angular momentum been changed?

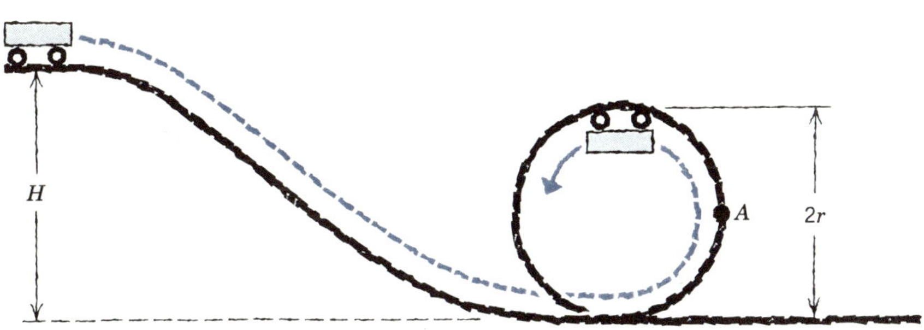

Fig. 6-23

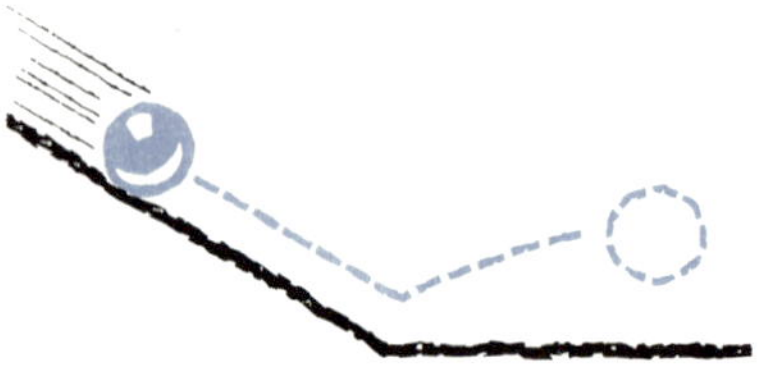

Fig. 6-24

20. Figure 6-29 shows three equal masses. Sliding mass A strikes masses B and C, which are in contact at rest and in line with the line of A's motion. Assuming that the masses are perfectly elastic, so that both energy and momentum are conserved, show why only one of the masses B and C will move after the impact.

21. A constant force F is exerted on the block in Figure 6-7a while the block moves a distance d, starting with a speed s_0. Show mathematically that the work done is equal to the change in the kinetic energy of the block.

*22. In the system illustrated in Figure 6-30, the area of the large piston is A and the area of the small piston is a. If the force exerted by the small piston on the fluid is f, find F, the force exerted by the fluid on the large piston.

*23. Show that the pressure in a fluid at rest must be the same in opposite directions at the same point. (In Figure 6-31, what happens if the force is not the same on both sides of the disc?)

*24. Show that the force on a surface in a fluid at rest must be perpendicular to that surface. (What happens to the small looped belt in Figure 6-32 if the forces are not perpendicular to the surface?)

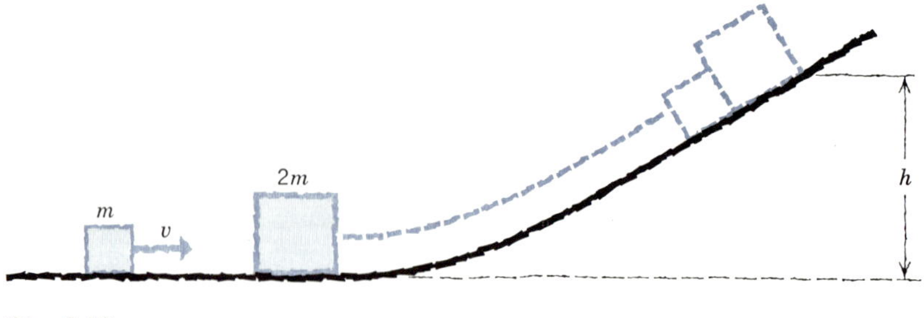

Fig. 6-25

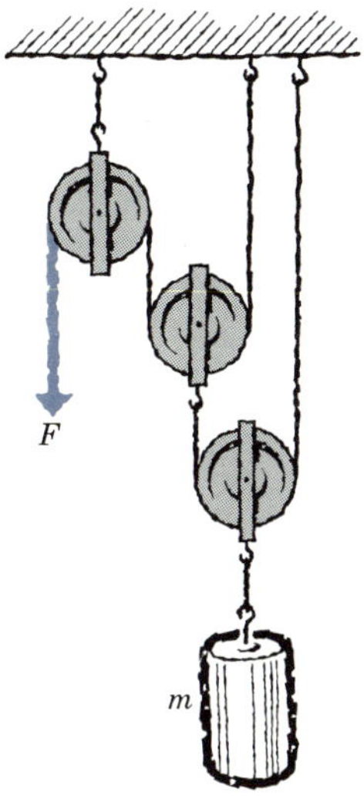

Fig. 6-26

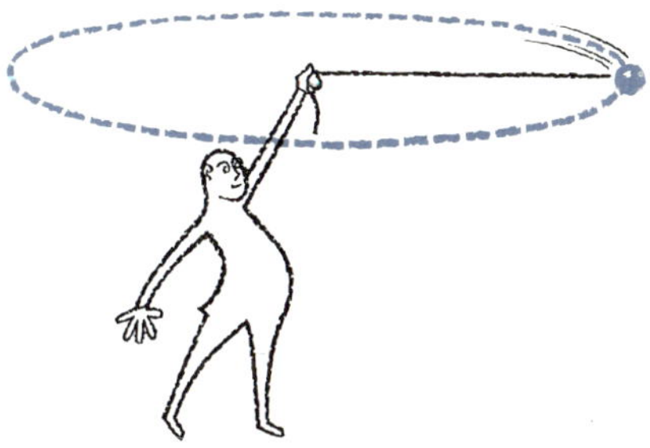

Fig. 6-27

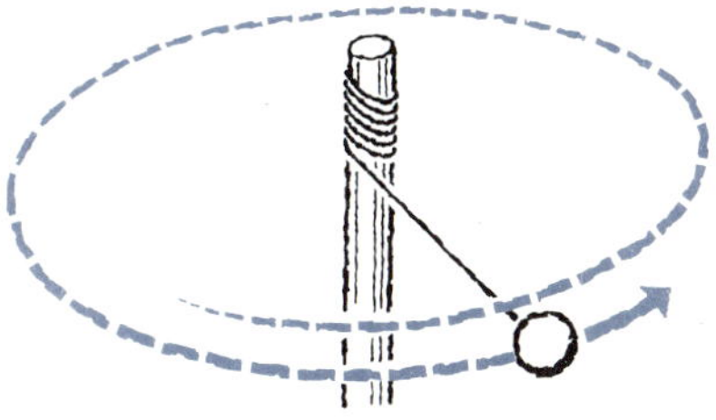

Fig. 6-28

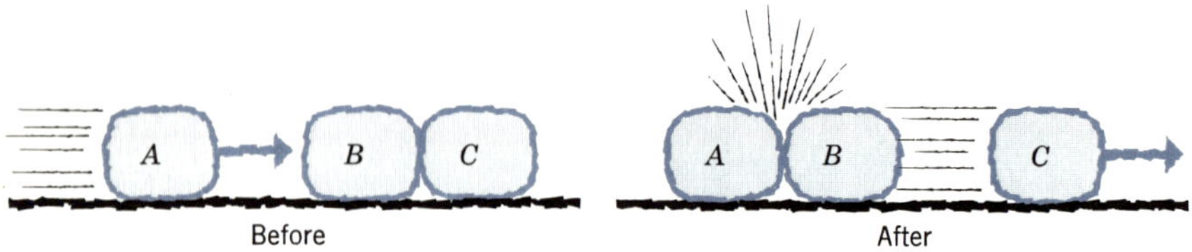

Fig. 6-29

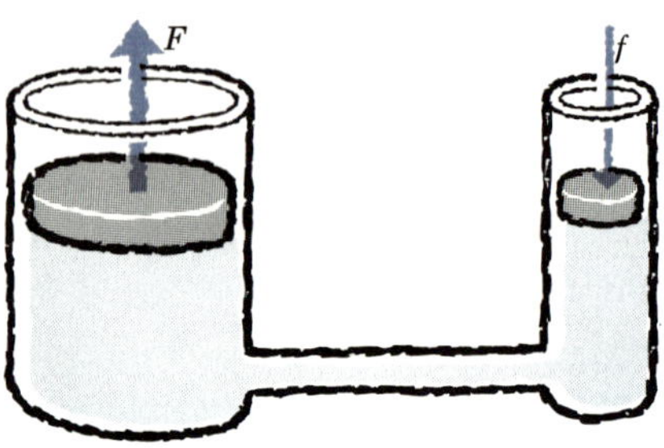

Fig. 6-30

Fig. 6-31

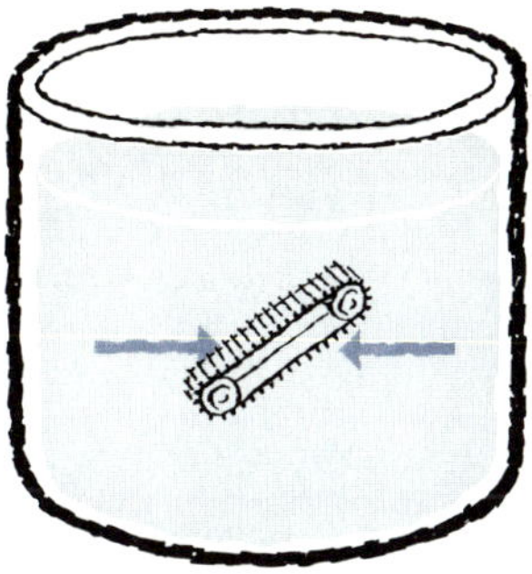

Fig. 6-32

7

MOLECULAR ORDER AND DISORDER

In the preceding chapters, we have presented the Newtonian model of the world, illustrating the laws that form the framework of this model by describing the behavior of matter in motion. We have assumed, provisionally, that matter is discrete rather than continuous—that is, made up of very small particles. In order to use the Newtonian model to explain a wider range of the phenomena of nature, it is convenient to assume also that these particles are generally in motion and obey the laws of motion that we have discussed. Although such particles are much too small to observe directly, we can justify this assumption by showing that it helps to explain the properties of matter. The theory that matter is made up of small particles in motion is called the *kinetic theory*. In this chapter we shall attempt to explain the behavior of matter using this theory. Since this subject is often difficult, we shall emphasize some basic and general ideas without attempting to give a complete explanation of all details and special cases. We shall give particular attention to irreversibility and the law of increasing entropy.

THE KINETIC THEORY OF MATTER

Molecules and Atoms

The particles of matter in which we are interested in this chapter are called *molecules*. Molecules themselves are made up of one or more particles called *atoms*. Each molecule of the gas hydrogen, for example, is made up of two atoms of hydrogen (see Fig. 7-1). Each molecule of the gas helium, however, consists of a single atom of helium. Each molecule of water consists of two atoms of hydrogen and one atom of oxygen. The various kinds of atom correspond to the various chemical elements. The laws of chemical combination by which atoms combine to form molecules are discussed briefly in Chapter 13.

Although molecules occur in various shapes and sizes, they are all extremely small, so small, in fact, that it is difficult to develop an appreciation for their size. To be told, for example, that a molecule of helium has a diameter of about 10^{-10} meters means very little if we have had no experience with anything else that small. The fact that a thimbleful of air near sea level contains 10^{19} molecules is also difficult for us to comprehend—this quantity is more than the number of all the people in the world multiplied by a billion. Later, in other chapters, we will discuss particles still smaller than atoms.

Hydrogen molecule
(2 atoms of hydrogen)

Helium molecule
(1 atom of helium)

Water molecule
(1 atom of oxygen,
2 atoms of hydrogen)

Fig. 7-1

Solids, Liquids, and Gases

We are familiar with matter in three common forms: solid, liquid, and gas. Since gases are ordinarily easily compressed, we think of the molecules of a gas as being far apart so that they can be pushed closer together (see Fig. 7-2). Since liquids are not easily compressed, we may think of them as being made up of molecules which are almost in contact with one another but are free to slide over one another so that the liquid can flow. Solids are more rigid, so we think of them as being made up of molecules that are held more firmly to one another by some force. Although these molecules can sometimes slip past each other, they are more commonly found vibrating in place. Since crystalline solids exhibit a regular shape, we consider that the molecules of these substances are arranged in an orderly fashion, stacked, for example, like oranges in a grocery-store display.

The Explanation of Gases in Terms of Moving Molecules

Many observable properties of matter are explained by the kinetic theory. To illustrate, we will discuss the internal energy, temperature, and pressure of a gas, and a few of the simpler aspects of the behavior of liquids and solids.

The basic picture of a gas, according to the kinetic theory, is one of molecules in motion, striking and rebounding from each other and from the walls of their container. These frequent collisions continually alter the speed and direction of individual molecules. As a result of these collisions, the molecules have a wide range of speeds, and there are about as many traveling in one direction as in another.

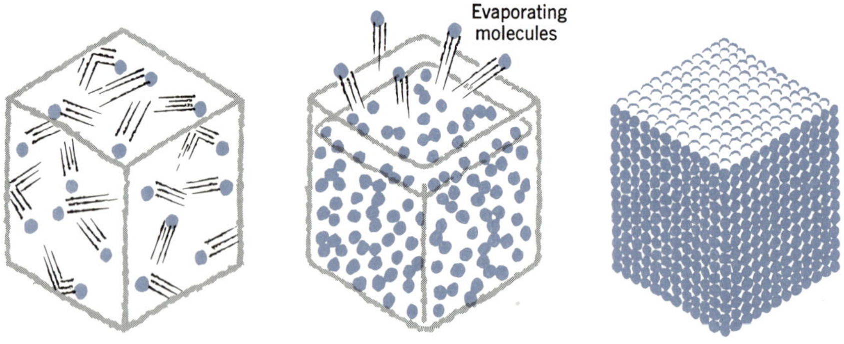

(a) Gas *(b)* Liquid *(c)* Solid

Fig. 7-2

PRESSURE. The kinetic theory enables us to account for the pressure that a gas exerts on the walls of its container (see Fig. 7-3). When a moving molecule strikes the wall of its container, a force is exerted on the wall during the impact. The continual rain of moving particles striking the walls continually produces these impact forces all over the walls of the container, accounting for the pressure of the gas.

INTERNAL ENERGY. The internal energy of a sample of a gas is the total energy of all of the molecules within it. In the simplest case, this is approximately equal to the sum of the kinetic energies of each of the molecules. For solids and liquids, and to a lesser extent for gases, the molecules also have nongravitational potential energy that arises from forces between molecules. In these cases, the internal energy is equal to the sum of total kinetic and potential energy of all the molecules. When a moving object slows down because of friction, the energy of the object is not lost but can be accounted for in the increased energy of the individual molecules of the object and its surroundings.

TEMPERATURE. We all have an intuitive understanding of temperature, derived from our sense of touch. Our sense of touch, however, is insufficient even to distinguish between very hot and very cold temperatures, as is illustrated by a well-known practical joke (see Fig. 7-4). The victim first sees a pot of water boiling on the stove and is then blindfolded. While he is blindfolded, his hand is plunged into a pan of ice water. His great distress at this event is due to his

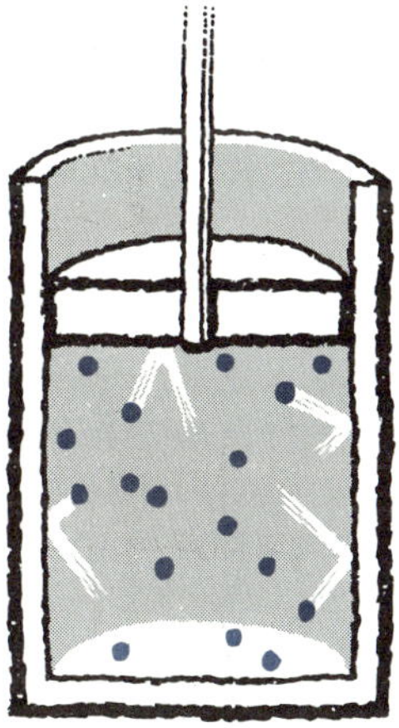

Fig. 7-3

inability to discriminate between the boiling and freezing temperatures of water.

Of course, we can use an ordinary mercury thermometer to measure temperature far more accurately than with our hands. As the temperature is increased, the mercury inside the thermometer expands and therefore rises in the glass tube. In fact, most substances expand when heated. However, we are interested in what temperature means in terms of the motion of the molecules.

According to the kinetic theory, the temperature is related to the *average* kinetic energy of the molecules. In a simple gas, in which the only important form of energy the molecules have is kinetic energy, the temperature is proportional to this average kinetic energy. When we heat a gas, the internal energy increases. The molecules start to move with greater speeds. Their average kinetic energy increases, and we observe a higher temperature. If the gas is not confined in a rigid container, it will expand as its temperature is increased. The faster-moving molecules collide, on the average, more frequently and with greater impact with each other and with the walls of the container. This results in a greater pressure on the walls, which can cause expansion.

Where the kinetic energy of the molecules is not the only form of energy, as in solids, liquids, and gases under some conditions, the simple proportionality between temperature and average kinetic energy does not hold. However, the temperature is still closely related to the average kinetic energy of the molecules and generally increases as their average speed increases.

The kinetic theory helps us to understand how heat flow occurs. Figure 7-5 shows a cold thermometer in hot air. Since the air is hot, its molecules will have high kinetic energies. Some of these molecules collide with the slower-moving molecules in the cold glass of the thermometer. In time, the molecules in the glass begin to move

Fig. 7-4

faster as a result of these collisions, and the temperature of the glass increases. These faster-moving molecules in the glass then speed up the mercury molecules as a result of collisions, and in this manner the temperature of the mercury is also increased. Finally, both glass and mercury are at the temperature of the air. There has been a heat flow into the thermometer. Thus the kinetic theory helps us to explain heat flow as a result of molecular collisions.

As we have said, most substances expand when their temperature rises. This is true for both the mercury and the glass of the thermometer. However, the mercury expands more than the glass, and therefore the mercury rises in the thermometer tube.

ABSOLUTE ZERO OF TEMPERATURE. If the temperature is closely related to the average kinetic energy of the molecules, then there should be some lowest possible temperature that would be attained when the molecules lose all their kinetic energy. There appears to be just such a temperature, called the *absolute zero*. Actually, even at absolute zero there is still a small amount of kinetic energy in the molecules, the so-called zero point energy. (The presence of this small zero-point energy can be explained by modern quantum theory.) We shall not explain the zero point energy here but shall simply point out that the absolute zero is the temperature at which the molecules have their minimum kinetic energy.

Most of us are used to measuring temperatures using the ordinary Fahrenheit scale, on which water freezes at a temperature of 32

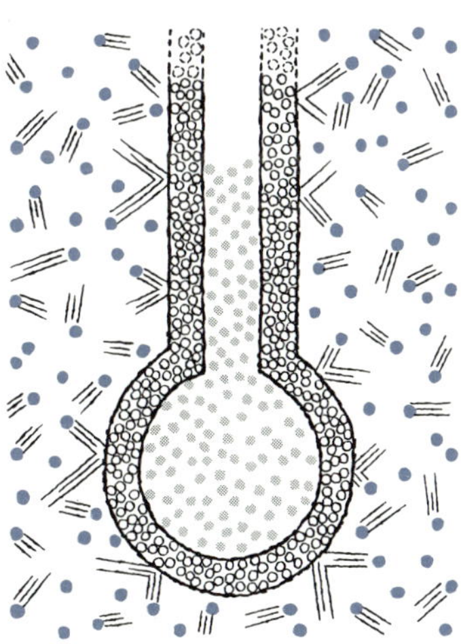

Fig. 7-5

degrees and boils at 212 degrees. The *absolute zero* of temperature corresponds to a temperature of about 459 degrees *below* zero on the Fahrenheit scale. Most scientists use the Centigrade scale of temperature. On this scale the freezing point of water is arbitrarily chosen to be zero degrees, and the boiling point is chosen to be 100 degrees. The absolute zero of temperature corresponds to a temperature of about 273 degrees *below* zero on the Centigrade scale. We can see that the absolute zero is far below the normal temperature of our environment. Physicists, however, have recently been able to obtain temperatures within 0.0001 of a degree of absolute zero in very small samples of matter in the laboratory.

Evaporation and Melting

According to the kinetic theory, matter is made up of molecules in motion. Because of this motion, the molecules of a small sample of matter will tend to disperse as moving molecules leave the sample, unless something prevents them from doing so. In a gas, the walls of the container prevent these moving molecules from escaping. In a liquid or a solid, there are forces of attraction (stronger than gravitational forces) that are able to hold the molecules together. However, the molecules of solids and liquids can escape in the process of evaporation.

Let us examine the process of evaporation in liquids—for example, water. (Evaporation in solids is similar.) In the continual jostling of molecules by other molecules that occurs in a sample of water, some of the molecules acquire, by collisions, more energy than others. For some of these molecules, this acquired energy will be greater than the binding energy that holds them to the other molecules. If these high-energy molecules are near the surface and moving toward it, they will escape from the liquid and form or join the gas above (see Fig. 7-6). In this way it is possible for a sample of water to evaporate completely and thus "disappear." Water molecules in the gas above, however, can also reenter the liquid. If there are the same number of molecules leaving and entering, on the average, in the same length of time, then the amount of water in the sample remains the same.

Under the right conditions, evaporation can take place within a liquid, forming bubbles of gas inside it which then rise to the surface. This process is known as boiling. The pressure in the fluid tends to collapse these bubbles of gas, however, and therefore they will not form unless there are enough evaporating molecules moving fast enough to exert a pressure that can withstand this surrounding pressure. As the temperature of a liquid increases, the pressure that can be exerted by the evaporating molecules increases because there are more high-speed molecules in the liquid. Finally, a temperature is reached at which bubbles can form. This temperature is the boiling temperature.

We can change the boiling temperature by changing the pressure on the liquid. In a pressure cooker, for example, the pressure rises higher than atmospheric pressure. In order for the evaporating molecules to form bubbles (boil) against this higher pressure, the temperature must be higher than the ordinary boiling temperature. This means that the water in the kettle can be raised to a higher temperature without boiling. This higher temperature causes the food to cook more quickly.

As we have seen, evaporating molecules have on the average more energy than the molecules left behind in the liquid. The loss of these high-energy molecules leaves the liquid colder than it was before. Therefore, evaporation cools the liquid. This effect explains why a wet hand feels colder than a dry one. The liquid on the wet hand is being cooled by evaporation. Some evaporated molecules will bounce back into the liquid. If a person blows on his wet hand so that these molecules are carried away and cannot return, evaporation is more rapid, and his hand feels even cooler.

Suppose that we increase the internal energy of some ice by letting heat flow into it. As we put energy into the ice, the temperature increases until it reaches the point at which some of the ice begins to melt. To melt the ice we must supply enough energy to overcome

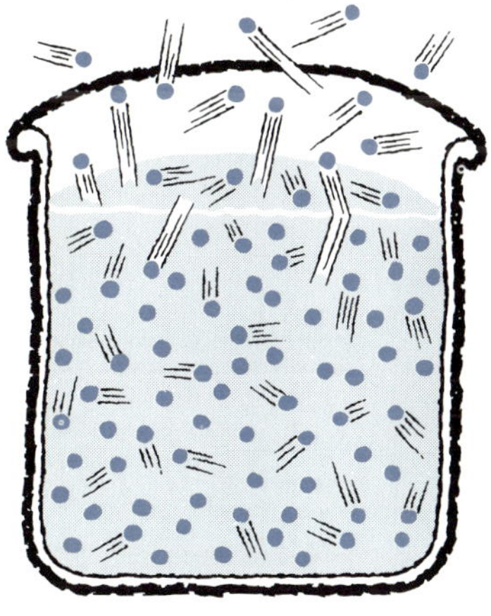

Fig. 7-6

that part of the binding energy that holds the molecules of the solid rigidly together and keeps them from sliding over one another. Even if we continue to supply energy, the temperature will not increase further until all the ice is melted, since all the energy we are supplying is being used to melt the ice. When all the ice has been transformed into water, then the energy that we are supplying raises the temperature of the water until the boiling point is reached. At this temperature the energy supplied is used to overcome the binding energy that keeps the molecules of the liquid together. Again, the temperature does not change until all the liquid has been transformed into a gas even though energy is still being supplied. Then the temperature again rises.

Evidence for the Kinetic Theory

The evidence for the kinetic theory of matter is its success in accurately explaining the behavior of matter. We have briefly outlined some of these explanations, and we shall present others in the remainder of this chapter. There are some phenomena that would be especially difficult to explain without the kinetic theory. We shall discuss two of these here—diffusion of gases, and Brownian motion.

The diffusion of gases is illustrated in Figure 7-7, which shows two chambers at the same temperature and pressure separated by a porous wall. This wall has tiny holes in it that will permit molecules to pass through. We start with helium gas on the right and a heavier gas, such as oxygen, on the left. (Oxygen molecules have eight times more mass than helium molecules.) The gases are at the same temperature, and therefore the average kinetic energy of the molecules is the same for both. Since the helium molecules have less mass than the oxygen molecules, they must have greater average speed in order to have as much energy. For this reason the helium

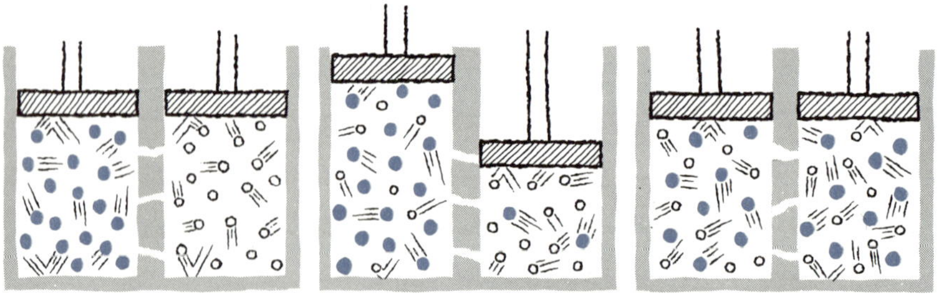

Fig. 7-7

molecules move through the small holes in the porous wall into the oxygen chamber more rapidly than the oxygen molecules diffuse into the helium chamber. Consequently, the pressure will rise in the oxygen chamber. After the diffusion is complete, half of the helium will have diffused into the oxygen chamber and half of the oxygen will have diffused into the helium chamber. At this time the pressure becomes and remains the same in the two chambers. In summary, first the pressure rises on the left, and then the two pressures become equal. This effect is experimentally observed and therefore supports the kinetic theory.

Brownian motion, discovered by Robert Brown in 1827, provides evidence for the kinetic theory which seems more direct. A particle in a liquid or gas is bombarded by individual molecules. The effect of this molecular bombardment is negligible for large particles: the mass of the particle is so much greater than the mass of the individual molecules that the effect of each collision is small, and the collisions occur on all sides of the particle so frequently that on the average the effects cancel out. However, particles of microscopic size may be seen to bounce around as a result of molecular collisions (see Fig. 7-8). Albert Einstein in 1905 calculated the expected rate of drift for such particles bouncing around in Brownian motion. The calculation agreed so well with the observed drift that it convinced the noted chemist Wilhelm Ostwald that atoms and molecules really existed.

THE LAW OF ENTROPY INCREASE

In this section we shall discuss time-irreversible processes and the problem that arises in trying to explain them in terms of molecules in

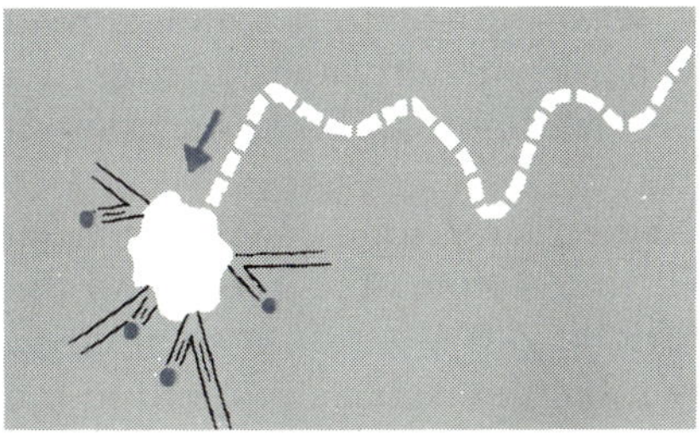

Fig. 7-8

motion. We shall introduce a scalar quantity, *entropy*, which is conserved in reversible processes but always increases in irreversible processes.

Time-Reversible Processes

All of the fundamental laws we have presented thus far are time-reversible, including, of course, those we have applied to the behavior of molecules. We introduced time-reversibility in Chapter 1 as a symmetry principle. The Newtonian model obeys this symmetry principle; that is, it would operate backward in time in the same way as it does forward. For example, a motion picture of a rock thrown into the air would show the rock obeying the same laws whether the film were run backward or forward. Reversed motion is a possible motion for the rock. The principle of time reversibility has recently been challenged for interactions of certain small particles; in this section, however, we shall assume that the principle of time-reversibility does hold for molecules.

Time-Irreversible Processes

Despite the fact that the behavior of molecules appears to be time reversible, there are certainly processes in nature which are not time-reversible. We list here several processes that do not occur in reverse.

1. When two objects at different temperatures are pressed together, heat flows from the hot object into the cooler one, and never in the opposite direction. You would not sit on a block of ice to get warm.
2. A ripe peach decays. You have never seen a rotten peach become whole and edible again.
3. When a man dives into a pool, waves are started which spread out and finally disappear. There are no reports of waves arising in a pool by themselves, converging on a swimmer, and tossing him out of the pool onto the diving board.
4. A block of wood on an inclined plane slides, with friction, down the plane and comes to rest at the bottom. The block and the plane are warmed as the result of friction. We have never noticed the block and plane start to cool down by themselves, propelling the block back up the plane.
5. A gas released from a small container under high pressure

expands to fill a large box (see Fig. 7-9). No one has ever seen the gas in a box (as, for example, a room) gather itself together like Aladdin's djinni and flow back into the bottle, building up pressure as it does so.

6. A layer of blue sand is carefully spread over a layer of white sand in a bottle. Vigorous shaking will mix the grains of blue and white sand together. Continued shaking has never been known to cause the blue and white grains of sand to separate themselves into two layers again (see Fig. 7-10).

All these processes are irreversible; each of them establishes a sequence of conditions in time that cannot be reversed. Most commonly observed processes are actually irreversible. Reversible processes (as, for example, motion without friction) are usually idealizations of some actually irreversible process.

The existence of irreversible processes presents us with two problems.

1. Objects are made of molecules in motion, which obey only reversible laws. How, then, can they undergo irreversible changes? This is one of the long-standing puzzles in physics. We shall consider this question again later.

2. What single law or principle can be devised to account for all irreversible changes? In the remainder of this chapter we shall search for such a principle. We shall find a scalar quantity which usually increases as time goes on and never decreases. This quantity is called *entropy.* Entropy is conserved in reversible processes, but it is *not* conserved in irreversible processes. After an irreversible process there is always more entropy than there was before. The principle that will enable us to account for irreversible processes is, therefore, the *Law of Entropy Increase.*[1]

[1]The law of entropy increase is generally known to scientists as the second law of thermodynamics. The first law of thermodynamics is the law of conservation of energy.

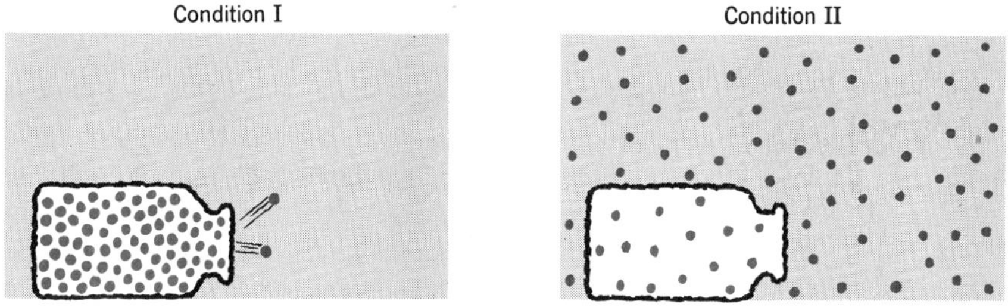

Fig. 7-9

The total entropy of the objects involved in a process never decreases. In irreversible processes the total entropy increases.

The entropy of a single object or system may well decrease. When we consider *all* the objects and systems involved in a process, however, it turns out that the total entropy always increases or remains the same. In later sections we shall attempt to explain what is meant by the quantity entropy, so that the law of entropy increase may become more meaningful.

Entropy as Disorder

In order to obtain our first clue to the nature of entropy, let us consider the previous example in which white and blue sand were mixed together. What was it that changed in the mixing process? There was order in the arrangement of the sand to start with; the blue and white grains were separated. In the process of mixing, this order was lost. *Disorder*, therefore, increased. Entropy also is supposed to increase. Perhaps this disorder is the entropy we are looking for. Indeed, this seems to be the case. The increase of the quantity entropy corresponds to an increase in disorder among the molecules that make up material objects and substances. We can check this interpretation by examining the other examples of the last section.

The decay of the peach is an obvious example of increasing disorder. It is a breakdown of the remarkable organization of molecules present in living cells. The waves coming from the diver show an ordered motion of the water molecules, which is broken up as the waves rebound against the sides of the pool and are dissipated. In

Fig. 7-10

this manner, cooperative motion of the molecules in a wave finally becomes part of the normal random motion of the molecules. In the fourth example, the ordered motion of the entire block moving down the plane is also reduced, by friction, to the increased random motion of the molecules in the block and plane. The molecules of the expanding gas were at first restricted to one small volume; after the expansion they are miscellaneously distributed over the whole volume of the box. Each of these irreversible processes can be thought of as an increase in disorder.

These examples strongly suggest that we might write the law of entropy increase in terms of disorder.

> The total amount of disorder never decreases, and it increases in irreversible processes.

Actually, this statement gives a good idea of the meaning of entropy increase.

The first example in the last section was concerned with the heat flow from a hot object to a cooler one. This process can also be understood as increasing disorder. The temperature of an object is related to the average kinetic energy of its molecules. The molecules of the hot object have, on the average, more kinetic energy than those of the cold object. To begin with, therefore, we have a partial separation between two classes of molecules—those of high kinetic energy and those of lower kinetic energy. After heat has flowed from one object to the other to make the temperatures of the two objects equal, the ordered arrangement of the molecules into these two classes is lost. The high-energy molecules are now mixed in with the low-energy molecules. Thus disorder increases here also. This example enables us to state the law of entropy increase in one of the ways in which it was first expressed.

> No process can occur in which the net effect is the flow of heat from one object into another at a higher temperature than the first.

The most complex and cunningly devised machines cannot make heat flow "uphill" against a higher temperature—unless some other effect also occurs at the same time. For example, a refrigerator can take energy from a cool system and expel it into a region of higher temperature. To do this, however, the refrigerator uses electrical energy or energy supplied by burning gas. It cannot make heat flow uphill without using some other energy.

Entropy as Probability

We can explain to some extent why disorder increases by introducing the concept of *probability.* Suppose that we tossed a large number of pennies. Would we expect to find them coming up either all heads or all tails? No; we would expect to find them in a more disordered and nonuniform condition, with about as many heads as there are tails. The most probable condition for these pennies is not the ordered state in which they are all heads or all tails but, rather, a disordered state. Indeed, for most systems a disordered condition is the most probable.

The mixing of the sand provides us with another example. The first condition, in which the blue and white grains of sand are all separated, is certainly possible. It is even possible that shaking the bottle after the sand has been mixed will happen, by chance, to separate the grains again into two groups, blue and white (see again Fig. 7-10). This event is so unlikely, however, that you would not believe anyone who said he had seen it happen. A condition in which the blue and white grains are all mixed together is so much more probable that it is the only one we expect to see as a result of random shaking. Thus, if a system of objects is tossed, bumped, jostled, or shuffled, there is a great probability that its order will be lost. The molecules of matter are constantly in motion, jostling, bumping, and disturbing each other. Thus it is not suprising that in irreversible changes their disorder increases. The less probable conditions are replaced by the more probable, disordered conditions. We may, then, state the law of entropy increase as follows.

> In an irreversible process the system changes to a more probable condition.

This statement also gives a good idea of what is meant by the law of entropy increase.

We can explain further why the more probable conditions of a system are usually disordered by considering what happens when a number of coins are tossed. If we toss two coins, there are four possible ways in which they can come up. Using T to stand for tails and H to stand for heads, we can list these four possibilities in a table (see Fig. 7-11). There are two conditions that might be called ordered. These are the conditions in which both coins come up heads or both come up tails. The other two conditions, each with one head and one tail, might be called mixed or disordered. If we increase the number of coins, we increase the number of possible disordered conditions. Figure 7-12 shows the possibilities for various numbers of coins. If we tossed a large number of coins, the overwhelming majority of possibilities would be mixed or disordered. Comparatively few would be completely ordered (all heads or all tails) or nearly ordered. If we should randomly disturb a collection of coins, we would be far more likely to obtain one of the many disordered conditions than one of the few ordered conditions. Each of the conditions would be considered equally likely, since we have no information that would lead us to choose one rather than another. The disordered conditions are so numerous, however, that we are much more likely to get one of them. On the basis of probability, then, we can explain why we will normally see an ordered condition change into a disordered condition, and not vice versa. Even a thimbleful of gas contains about 10^{19} molecules. With this many molecules involved, the ordered conditions will be extremely rare in comparison with the disordered conditions. In principle, however, it is possible for the law of entropy increase to be violated—one could by chance get a more ordered condition. The probability that this will happen, however, is ridiculously small.

The relation of entropy to probability enables us to understand how apparently irreversible processes can occur when all the molecules involved in these processes obey reversible laws. What we have called irreversible processes are not, strictly speaking, irreversible. The reversed process is not impossible; it is merely extremely improbable. In fact, it is so improbable that for practical purposes, and most theoretical purposes, it can be ignored.

In the above discussion we have not attempted to define order or probability precisely, and we have not discussed the effect of force

	Coin 1	Coin 2
1.	H	H
2.	H	T
3.	T	H
4.	T	T

Fig. 7-11

Case	Coin 5	Coin 4	Coin 3	Coin 2	Coin 1	
1	H	H	H	H	H	
2	H	H	H	H	T	1 Coin
3	H	H	H	T	H	
4	H	H	H	T	T	2 Coins
5	H	H	T	H	H	
6	H	H	T	H	T	
7	H	H	T	T	H	
8	H	H	T	T	T	3 Coins
9	H	T	H	H	H	
10	H	T	H	H	T	
11	H	T	H	T	H	
12	H	T	H	T	T	
13	H	T	T	H	H	
14	H	T	T	H	T	
15	H	T	T	T	H	
16	H	T	T	T	T	4 Coins
17	T	H	H	H	H	
18	T	H	H	H	T	
19	T	H	H	T	H	
20	T	H	H	T	T	
21	T	H	T	H	H	
22	T	H	T	H	T	
23	T	H	T	T	H	
24	T	H	T	T	T	
25	T	T	H	H	H	
26	T	T	H	H	T	
27	T	T	H	T	H	
28	T	T	H	T	T	
29	T	T	T	H	H	
30	T	T	T	H	T	
31	T	T	T	T	H	
32	T	T	T	T	T	5 Coins

Fig. 7-12

on disorder. The presence of forces can make it appear that order is increasing, not decreasing, in a process. In fact, where we see order in the behavior of particles, we usually assume that a force is acting, and we seek to discover this force. Also, our use of the term *order* in this discussion is actually rather curious. No matter how the coins fall when we toss them, they will have some kind of order. Similarly, the molecules of a gas, expanding into the box from the bottle in Figure 7-9, have a definite order whether they are cooped up in the bottle or moving around in the box. Why should we call one arrangement (or order) ordered, and another arrangement (or order) disordered? The answer seems to be that the arrangement is considered ordered if it means something to us, or gives us some information. We have more information about the location of the individual molecules of a gas when they are confined to a small volume than when they are allowed to move around in a large volume. Hence we consider the latter condition more disordered. Even in the small bottle, however, the molecules appear to be behaving as randomly as possible under the circumstances. Molecular behavior appears to be chaotic.

Perpetual Motion of the Second Kind

In Chapter 6 we discussed the impossibility of a perpetual-motion machine that would create energy and therefore violate the law of conservation of energy. This type of machine is called a perpetual-motion machine of the first kind. In this section we shall discuss another type of impossible machine—a perpetual-motion machine of the second kind. A perpetual-motion machine of the second kind violates the law of entropy increase. As an example, imagine a machine that can take in energy by heat flow and convert this energy entirely into work done by the machine, leaving the machine in its original condition. This process would violate the law of entropy increase; indeed, one of the early statements of this law stated the impossibility of such a process.

> No process can occur whose total overall effect is to take internal energy from a material by heat flow and convert this energy entirely into work done in the process.

We can see why such a process would violate the law of entropy increase by considering again the example of the block sliding with friction down an inclined plane (see p. 286, 289). As the block comes to rest, the energy is converted to internal energy, thus, as we have seen, increasing the disorder of the molecules. If the process occurred in reverse, the disorder of the molecules would have to decrease and would therefore violate the law of entropy increase. But this is just one example of the forbidden process described above: internal energy flowing out of the block and plane and being converted to the work needed to move the block back up the plane, with no other change to the system.

The law of entropy increase can be expressed as a principle of impossibility as follows.

Perpetual-motion machines of the second kind are impossible.

We have already stated the law of conservation of energy as a principle of impossibility—the impossibility of perpetual-motion machines of the first kind. It is often considered elegant to express physical principles as principles of impossibility.

If perpetual-motion machines of the second kind were possible, many convenient devices could be made. For example, it would theoretically be possible to make a machine which could extract internal energy out of sea water, thus lowering the temperature of the water, and use that energy to propel a ship. As long as the law of entropy increase holds, however, such a machine is impossible.

The energy in a heat flow can be partly converted into work, as occurs, for example, in a steam engine. There is, however, a heat flow out of the machine, to surroundings at a lower temperature than the source of the incoming heat flow. The steam engine cannot operate unless the surroundings are at a lower temperature than the boiler (the source of the incoming heat flow). Any engine that converts part of an incoming heat flow into work operates between at least two temperatures and has an outgoing heat flow at the lower temperature.

Free Energy and Heat Death

The fact that energy supplied as heat flow cannot be completely converted into work has some interesting consequences. Energy converted to internal energy in irreversible processes, by friction or any other means, can never be completely recovered and used to do work. Where differences of temperature or the means to produce them exist, we can use *some* of this internal energy to do work, but not all of it. Some of it flows into material at the lower temperature, tending to reduce the temperature difference. We can expect, therefore, that as time goes on, more and more energy will be irreversibly converted into disordered internal energy, and less and less will be available to do work. Although the energy is conserved, we say that the *free energy* is decreased. (*Free energy* may be defined roughly as the energy available to do work.) Another way of stating the law of entropy increase is to say:

> Free energy never increases, and it decreases in irreversible processes.

The continual increase of entropy or decrease of free energy within a system can eventually result in a total absence of free energy in the system. In this condition, unless more free energy is supplied from outside, nothing but random molecular motion can occur in the system. If we consider the whole universe as an isolated system, we can imagine a time in the future when the free energy has totally disappeared and the whole universe has reached the same temperature. In such a condition, often called the "heat death" of the universe, no work can be done.

Entropy as Lack of Information

A more recent point of view considers the law of entropy increase as a law of decreasing available information. Consider once again the expansion of a gas from a small bottle into a large box (see Fig. 7-9). We call the situation in which the gas is in the bottle, just ready to emerge, condition I. We call the situation in which the gas has emerged and fills the box, condition II. The irreversible process, therefore, may be called

$$\text{I} \rightarrow \text{II}$$

The hypothetical reversed process is

$$\text{II} \rightarrow \text{I}$$

We are curious as to why this last process is never observed. Since the laws of molecular interaction are assumed to be time-reversible, the process II → I should be possible.

Let us first see what we would have to do in order to reverse the process. If we were able to take all the particles in condition II and simultaneously reverse their velocities, we would expect them to retrace their paths and return to the bottle. The task of doing this for each of the 10^{22} molecules in the box, however, would be rather difficult; in fact, it might be impossible to reverse even one particle precisely without disturbing all the rest. We can, of course, prepare a situation that closely resembles condition II. That is, we can put the same amount of gas in a box, at the same pressure and temperature. However, we cannot expect to have all the molecules in the right place at the right time, going at the right speed in the right direction, to cause a transformation to condition I. "Condition II" is a general designation for a large number of possible conditions that look like condition II. Only an extremely small fraction of these will have the proper molecular motion to revert to condition I.

But the crucial problem is not the difficulty of preparation. Suppose that we had prepared or otherwise obtained a sample of gas in condition II which just happened to have the molecules behaving in exactly the right way to get a reversal. How would we know we had such a system? That is, how would we know we had it before it reversed to condition I and gave itself away? We would have to measure the speed and direction of all the molecules, which we could hardly do without hopelessly disturbing them. There does not appear to be any way of knowing in advance when we have the proper situation II. We cannot get enough information about the molecular motions of such a sample of gas to answer the question.

The real problem, then, is concerned with information. We cannot get enough information to tell which situation that looks like condition II is the right one to reverse. We *do* have sufficient information to tell which of the situations that look like condition I will go to condition II, because, practically speaking, they *all* will. We can therefore get I → II any time we like, but only by the most miraculous chance will we ever get II → I. Our knowledge of the temperature, pressure, and volume of condition I gives us more information about the condition of the molecules than similar knowledge gives us about condition II. Our information about condition I, for example, tells us that the position of each molecule must be within a volume the size of the bottle. This is more precise information than we have about condition II, in which we know only, for each molecule, that it is in the larger volume of the box.

In summary, this less precise information we have about condition II would not enable us to select or identify the few examples of condition II which could be reversed to condition I. Although essentially all of the systems in condition I will change to condition II, extremely few of the apparently identical systems in condition II will revert to condition I. This difference in behavior identifies a direction in time and thus helps explain time irreversibility.

This suggests the following statement of the law of increasing entropy.

> In an irreversible process, information about molecules decreases. (This applies to information available from the system itself and not its history.)

The restriction in the statement is necessary because if we know that a certain condition II has followed a condition I, we have additional information about the former condition (history) of the molecules. According to Newton's laws, the present condition follows precisely from the former condition; therefore, this history gives us some information about the present condition. Even if we have this historical information, however, it generally becomes lost because most systems are subject to the bombardment of outside molecules in an unknown way. Thus, even this historical information decreases because we do not know the exact position and velocity of these disturbing molecules.

The relation between the law of entropy increase and information has applications which extend beyond physics to such matters as the theory of information transmission in radio and telephone communications. We shall not discuss these applications here.

Entropy Changes Defined by Heat and Temperature

Let us look for a quantity that changes when heat flows from a hot object to a colder one, something we can identify with entropy (see Fig. 7-13). The quantity that might seem most real to us in this flow is the energy that is being transferred. However, the amount of energy that leaves the hot body is equal to the amount of energy that enters the cold body. Since the amount of energy is not changed in this process, it cannot be the quantity we are looking for. The *condition* of the energy has changed, however. At first it existed in a high-temperature condition; later it exists in a low-temperature condition. We need to devise a quantity whose change will depend on both the energy and the temperature. We can do this by dividing the amount of energy Q transferred by heat flow by the temperature

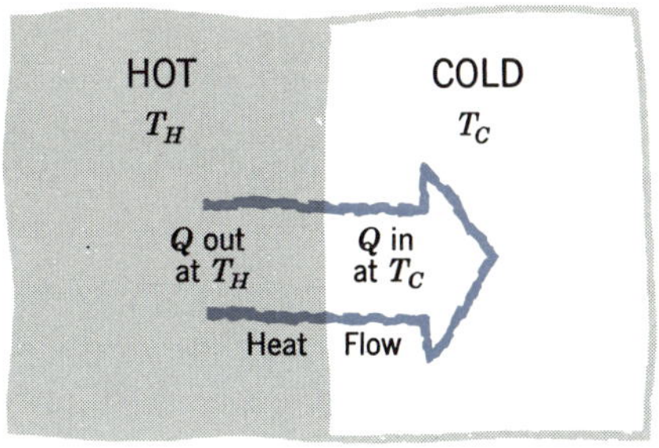

Fig. 7-13

T. This gives us a quantity that we shall identify as the *change of entropy*:

$$\text{Change of entropy} = Q/T \tag{1}$$

(This is the way in which entropy was defined before its relation to disorder was understood.)

We can now find the change of entropy that occurs when a total heat flow Q occurs between the objects in Figure 7-13. The object on the left has a temperature T_h, and the object on the right has a lower temperature T_c. The heat flowing into the object at the lower temperature causes a change of entropy Q/T_c for that object. The heat flowing out of the object at the higher temperature causes a change of entropy *minus* Q/T_h. (The entropy change is negative for this object since the heat is flowing *out* of the object.) The total entropy change for the system is the sum of the entropy changes for the two objects:

$$\text{Total entropy change} = Q/T_c - Q/T_h \tag{2}$$

Since the hotter object has a higher temperature than the colder one, the total entropy change is greater than zero. Therefore, the entropy increases. This suggests that we have indeed chosen the proper quantity for entropy. More careful analysis would confirm this choice of entropy for systems of objects in which heat and temperature are important.

This expression of entropy changes in terms of heat and temperature brings the problem of reversibility and irreversibility, which we have already discussed, into sharp focus. The law of increasing entropy is clearly not time-reversible. If time were reversed, the law would become a law of decreasing entropy, which is quite different from the law of increasing entropy we find in the real world. Yet, according to the kinetic theory, both heat and temperature, in terms of which entropy itself is defined, can be explained by the behavior of molecules in motion. These molecules are supposed to obey the time-reversible laws of physics.

The Carnot Cycle and the Definition of Temperature

A young French engineer, Sadi Carnot (1796–1832), laid the foundations for the laws governing heat exchanges and temperature. In his study of engines which convert internal energy to work, he devised what is called the *Carnot cycle*, which can be used to give a precise definition of temperature. To illustrate the Carnot cycle, we start with two sources of heat. We will refer to them as the hot stove, at a temperature T_1, and the cold stove, at a lower temperature, T_2. We also have some gas in a cylinder with a piston, as illustrated in Figure 7-14. The gas is originally at the same temperature as the hot stove.

There are four steps in the Carnot cycle.

1. We put the cylinder on the hot stove and let it take in energy by heat flow at that temperature. As the energy comes in, the gas expands, doing work.

2. After a time, we remove the cylinder from the stove but continue to let the gas expand, still doing work. Since work is being done, energy is leaving the gas. Thus the internal energy must be decreasing, and this means that the temperature is also decreasing.

3. As soon as the temperature of the gas has been reduced to that of the "cold" stove, we set the cylinder on the cold stove. We now do work on the gas by compressing it, thus putting energy into it. As we do this, we want the gas to maintain very nearly the same temperature as the cold stove, so there must be a heat flow into the cold stove from the gas.

4. After some time, we take the cylinder off the cold stove, but continue to compress the gas. During the compression, of course, we are still doing work on the gas, and so energy is entering the gas. This

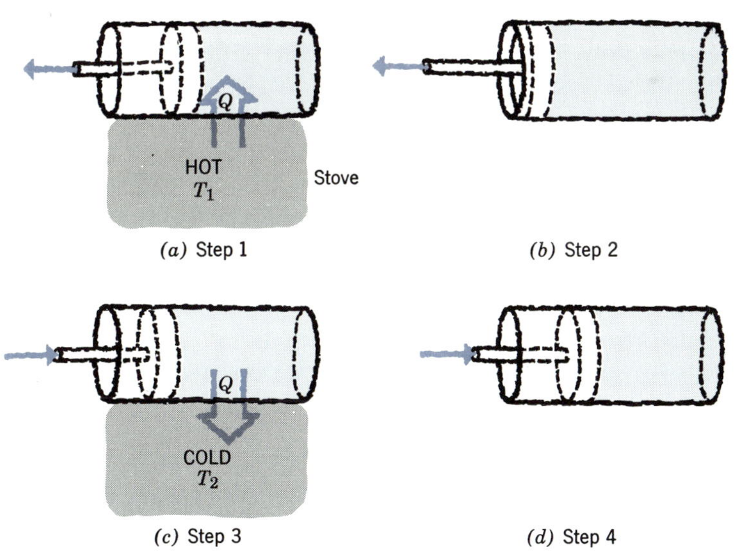

Fig. 7-14 Carnot cycle.

means that the internal energy, and thus the temperature of the gas, must rise. We allow the temperature of the gas to rise until it is once again equal to that of the hot stove. If we have been clever about how long we continue each process, the final state of the gas will be exactly the same as the original state, and thus we can repeat the whole cycle again and again.

We note that in Steps 1 and 2 the gas is doing work on its surroundings. In Steps 3 and 4, we are doing work on the gas. It turns out that the work done on the gas in Steps 1 and 2 is greater than the work done on the gas in Steps 3 and 4. In Step 1, heat is flowing into the piston; in Step 3, heat is flowing out of the piston. The work done comes from the heat added, and more heat is added in the first process than is lost in the third. (Actually, the entropy increase of the gas in Step 1 is the same in magnitude as the entropy decrease in Step 3.) The total effect of this cycle is that part of the energy taken in by heat flow has been put out as work. Thus the cycle converts internal energy partly into work.

The Carnot cycle is reversible. This claim may seem surprising, since the cycle involves the flow of heat between the stove and the gas in the cylinder and we have noted that heat flows from a hot object to a cooler one, and that such a flow is irreversible. However, in the Carnot cycle, we think of the stove and the cylinder of gas as being at very nearly the same temperature when they are in contact. Therefore, the heat flow is very nearly reversible, since the heat is leaving one object at almost the same temperature as it is entering the other object. Although the two objects cannot be at exactly the same temperature, or heat flow would cease, we can think of the temperature difference between the two objects as being as small as we please. Thus the Carnot cycle, in an idealized sense, is reversible, or perhaps we should say, "as reversible as we please."

Carnot argued that a reversible engine using the ideal reversible Carnot cycle would be the most efficient heat engine that could work between the same two temperature levels. That is to say, no other engine could convert a greater fraction of internal energy into work, and of course no real engine could do so well (see Fig. 7-15). If there were indeed another engine that could do better, then we could use it to drive the reversible engine backward. (Driven backward, the engine acts like a refrigerator—we put work into it and it pumps heat out of the cold stove into the hot stove.) Since the more efficient engine would use less internal energy than the reversible engine, there would be a greater heat flow to the high temperature from the reversible engine running backward than is required to run the more efficient engine forward. Thus the net effect would be to have a heat flow from a low temperature to a high one. This is impossible because it would violate the law of increasing entropy. In fact, the law of increasing entropy is equivalent to the statement that a Carnot cycle furnishes the most efficient heat engine operating between any two given temperatures.

Lord Kelvin (1824–1907) did much to put the laws of heat and temperature in their modern form. He used the Carnot cycle to give a more exact theoretical definition of absolute temperature than was known before. We already mentioned that the Carnot cycle is the most efficient heat engine working between two temperatures. By the same reasoning as before, it follows that all Carnot cycles operating between the same two temperatures must be equally efficient. Indeed, when operating between the same temperatures, all *reversible* cycles must be equally efficient. Otherwise, we could use one engine to run another backward, as we have already discussed, and thus would violate the law of increasing entropy. The efficiency of a reversible engine, therefore, depends only on the two temperatures between which it operates. This fact enables us to define temperature in terms of the efficiency of a Carnot cycle. We can show this quantitatively as follows.

Since the Carnot cycle is reversible, total entropy can neither increase nor decrease in the process. The gas must therefore lose the entropy on the cold stove that it acquires on the hot stove. If Q_1 is the

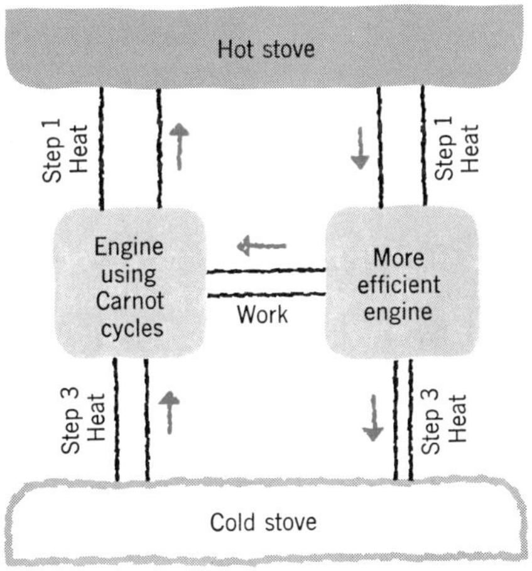

Fig. 7-15

heat gained on the hot stove, and Q_2 is the heat lost on the cold stove, then

$$\frac{Q_1}{T_1} = \frac{Q_2}{T_2} \tag{3}$$

Multiplying both sides of the equation by T_2 and dividing both sides by Q_1 we obtain

$$\frac{T_2}{T_1} = \frac{Q_2}{Q_1} \tag{4}$$

Subtracting 1 from both sides,

$$\frac{T_2}{T_1} - \frac{T_1}{T_1} = \frac{Q_2}{Q_1} - \frac{Q_1}{Q_1} \tag{5}$$

or

$$\frac{T_2 - T_1}{T_1} = \frac{Q_2 - Q_1}{Q_1} \tag{6}$$

Now Q_1 is the total energy put in, and $Q_2 - Q_1$ is equal to the total work done, since it is the net energy which has been supplied by heat flow. Therefore:

$$\text{Efficiency} = \frac{\text{work done}}{\text{energy put in}} = \frac{Q_2 - Q_1}{Q_1} \tag{7}$$

Now, of course, $T_2 - T_1$ = the temperature difference, so

$$\frac{\text{Temperature difference}}{T_1} = \text{efficiency} \tag{8}$$

and

$$T_1 = \frac{\text{temperature difference}}{\text{efficiency}} \tag{9}$$

Equation 9 defines the absolute temperature T_1 in terms of the efficiency of the ideal Carnot cycle and the temperature difference.

To relate the temperature defined by the Carnot cycle with the temperature of known phenomena, such as the boiling and freezing points of water, we can proceed as follows. Suppose that we have a Carnot cycle operating between the freezing and boiling points of water. We can arbitrarily choose the temperature difference between these two temperatures to be 100 degrees. Then we can use Equation 9 to express the upper temperature T_1 (the boiling point) in terms of this temperature difference and the efficiency of the Carnot cycle. T_2 is just 100 degrees lower. The rest of the temperature scale can be found by using Carnot cycles which operate between one of these two temperatures and other temperatures. Suppose, for example, that we had a Carnot cycle operating between T_1, which we consider known, and an unknown temperature. Knowing T_1 and the efficiency of the Carnot cycle, we can use Equation 9 to find the temperature difference between T_1 and the unknown temperature. From this we can find the unknown temperature. The absolute zero of temperature turns out to be about 273 degrees below the freezing point of water. The temperature scale defined in this way is called the *Kelvin* scale, after Lord Kelvin. The zero point is at absolute zero, and the size of the degree is determined by the fact that there are 100 degrees between the freezing and boiling points of water.

It would be very difficult to use the Carnot cycle in practice to measure temperature; however, it does enable us to form a precise definition of temperature that is very satisfactory from a theoretical point of view. It is not entirely satisfactory to attempt to define temperature by using the expansion of a particular substance, as, for example, the mercury in a mercury thermometer, because different substances expand differently. The efficiency of the Carnot cycle, however, does not depend on the gas used; indeed, the substance need not even be a gas. Thus the temperature defined by the Carnot cycle does not depend on the properties of any particular substance.

Summary

I. FUNDAMENTAL LAWS AND PRINCIPLES

- **A.** THE LAW OF ENTROPY INCREASE. Entropy never decreases, and it increases in irreversible processes.
  or: Heat never flows from an object into a hotter object;
  or: Perpetual motion machines of the second kind are impossible.

II. QUANTITIES

- **A.** ENTROPY, a derived scalar quantity.
  1. Entropy as disorder.
  2. Entropy as probability.
  - *3. Entropy as lack of information.
  - *4. Entropy in terms of heat and temperature.
- **B.** TEMPERATURE. From the standpoint of the kinetic theory we may regard temperature as a derived quantity. Originally it was thought to be fundamental.

1. Temperature is related to the average kinetic energy of molecules.
2. Absolute zero of temperature.
*3. Temperature as defined by the Carnot cycle.

III. MODELS

A. THE KINETIC THEORY. Assumes that matter is made up of particles subject to the laws of motion. Used to explain:
1. Temperature of a gas.
2. Pressure of a gas.
3. Internal energy.
4. Heat flow.
5. Evaporation and melting.
6. Irreversible processes, in terms of reversible molecular motion.

B. EVIDENCES FOR THE KINETIC THEORY
1. Success of explanations for items 1–6, III-A, above, and others not covered in this book.
2. Diffusion phenomena.
3. Brownian motion.

IV. PARTICLES

A. ATOMS

B. MOLECULES

Questions and Problems

1. Why does the pressure on a gas enclosed in a rigid container increase as the temperature increases?
2. Why does it take longer to boil an egg on top of a mountain than at sea level?
3. Why are conservation laws not adequate for explaining irreversible processes?
4. Why does the temperature of boiling water remain the same as long as the boiling continues?
5. Would it be possible theoretically for water to boil and freeze at the same temperature?
6. Why do we believe that matter is made up of small particles in motion?
7. Distinguish between atoms and molecules.

8. How does the kinetic theory account for the temperature and pressure of a gas?
9. How does the kinetic theory account for heat flow?
10. Explain the absolute zero of temperature.
11. Why does evaporation occur?
12. How does evaporation cause cooling?
13. What causes Brownian motion?
14. Give an example of a time-reversible process.
15. Give examples of time-irreversible processes.
16. How do you know that the processes in Question 15 are not time-reversible?
17. Why do we sometimes hold up a wet finger in trying to find out the direction of the wind?
18. Suppose that our discussion of evaporation and melting (pp. 282–284) had been entitled "condensation and freezing" instead. How would the discussion have been changed?
19. Some water is put into a bottle, air is pumped out of the top, and the bottle is sealed. It is then observed that evaporation takes place inside the bottle for a while, and then apparently stops. Why?
20. Give some examples not mentioned in the text in which irreversible processes cause increased disorder.
21. Why are some hands in a card game more unlikely than others?
22. Does it make sense to try to cool yourself by using an electric fan in a closed room? Explain.
23. List all the possible results of an experiment in which four coins are tossed.
24. Give an example of a situation in which the presence of a force makes a system appear to be moving toward a condition of increasing order, in apparent contradiction to the law of increasing entropy.
25. If an astronaut put a conventional thermometer out of the hatch of his orbiting spacecraft, would it record the temperature? Explain.
26. The quantity entropy has been called "times arrow." Can you imagine why?
27. A water molecule has a mass of about 3×10^{-27} kilograms. How many molecules of water would be contained in a thimbleful? (Water has a density of 1000 kilograms per cubic meter.) Estimate how many it would take to fill up the ocean.

28. From your answers to Question 27, estimate how many of the same molecules you would get back again if you emptied your thimbleful into the ocean, mixed thoroughly, and then took out another thimbleful.

29. From the information in Question 27, estimate the size of a molecule of water. Suppose the information had been given for a gas. Would that enable you to estimate the size of the gas molecule?

30. Explain entropy in terms of order and probability.

31. Why can't a machine take in sea water, put out sea water at a lower temperature, and do work with the energy thus obtained?

32. Express the other symmetry principles and conservation laws as principles of impossibility.

33. Suppose that a person claims that living processes do not obey the law of increasing entropy because living matter grows, reproduces, and thus increases the amount of ordered material. Comment on this claim.

34. Suppose that another person claims that evolution is impossible, because the appearance of higher forms of life implies an increase in order, which would contradict the law of increasing entropy. Comment on this claim.

35. How does the kinetic theory of matter seem to contradict the idea of irreversible processes?

36. Distinguish between the two kinds of perpetual motion.

37. What advantage would there be in a perpetual-motion machine of the second kind?

38. What is meant by the "heat death" of the universe, and what would cause it?

39. In what sense can it be said that temperature arises from the energy of the molecules and pressure from their momentum?

40. Does a steam engine obey the law of increasing entropy? Explain.

41. Give as many reasons as you can to explain why heat passes from a hot body to a cold body instead of vice versa.

42. In Question 31, why doesn't the machine, by ejecting the water at a lower temperature, obey the law of increasing entropy?

43. Can the entropy of a material system decrease? Explain.

44. Can the entropy of an isolated material system decrease? Explain.

45. Why might an astronaut's blood boil if he stepped out of his capsule without his space suit?

*46. Show how one could produce a refrigerator by operating the Carnot cycle in reverse.

*47. What advantage is gained by using the Carnot cycle to help define temperature?

*48. What does information have to do with irreversibility? How can what we know about a process have anything to do with whether or not it is reversible?

*49. In Figure 7-9, why do we know more about Condition I than Condition II?

*50. Do you think we could obtain information about one situation without giving up the possibility of having or getting information about another situation? Explain.

*51. Why would an engine using the Carnot cycle be at least as efficient as any other heat engine operating between the same two temperatures?

*52. Show that all engines operating between the same two temperatures using the ideal Carnot cycle would have to be equally efficient.

*53. Show that all reversible engines operating between the same two temperatures would have to be equally efficient.

8

ELECTRIC CHARGE AND FIELDS

In 1600, Sir William Gilbert (1540–1603), contemporary of Galileo and physician to Queen Elizabeth I, presented to the world his famous book *De Magnete*. This volume, a synthesis of his life's work, was sufficiently accurate to bring the study of electricity and magnetism into the realm of modern science. The practical use of the magnetic compass in navigation caused him to expend great effort in the study of magnetism, and not without results. For example, he investigated the region around spheres of magnetite (a magnetic iron ore) with a compass needle and correctly concluded by analogy that the earth too was a magnet. Still, for a long while after Gilbert's time, electrical and magnetic phenomena were considered by many to be manifestations of supernatural powers. John Wesley, for example, called electricity the "soul of the universe."[1]

Benjamin Franklin dramatically dispelled the superstitions of the public when he collected the electricity of lightning through a kite string and showed it to be the same type of stuff that is in the sparks produced by scuffing shoes on a rug. (Incidentally, we do not advise our readers to risk trying to repeat this experiment of Franklin's.) Schneer describes the impact of the experiment as follows:

> The effects of Franklin's discovery were popularly electrifying. His identification of the lightning and thunder universally associated with supernatural power as the corporeal, mechanical phenomenon, electricity, was in itself of the character of miraculous revelation. Franklin seemed like

[1]Cecil J. Schneer, *The Search for Order*, Harper & Brothers (now Harper & Row), New York, 1960, reprinted as *The Evolution of Physical Science*, Grove Press, Inc., New York, pp. 221–223.

a sage from a nobler world, soothing the fears of mankind. To the philosophers of the Age of Reason it was one more blow, as great perhaps as had ever been struck, against the forces of ignorance and superstition which they thought of as an enemy. In Newtonian mechanics, the rationalists found the ordered world-system to fit their ideas, but the *Principia* was after all a remote and forbidding territory of hostile theorems and implacable geometry. Franklin's lightning was immediate to the perception. Franklin had not only rationalized the lightning but with his invention of the lightning rod he had rendered it powerless. To the more literal minded, both in the camp of the clericals who deplored Franklin's impiety and the anticlericals who hailed the backwoods savant, it appeared that Franklin had disarmed the Deity and rendered the awful visitation of divine chastisement an impotent if dramatic display of frustrated rage. In France, Buffon, who had already been forced to retract his naturalistic theory of the origin of the earth, arranged for the translation of Franklin's letters, while the Abbe Nollet, who had scourged the Carthusian monks with his Leyden jar, indignantly wrote that he would prefer church bells to lightning rods.[2]

Over the years, scientists discovered many facts and principles about electricity and magnetism. The Scottish physicist James Clerk Maxwell (1831–1879) added to this work and gathered these principles together into a few mathematical laws. Maxwell's laws do for electricity and magnetism what Newton's laws do for motion and gravitation. They give us a few precise laws which describe a wide range of electrical and magnetic phenomena. In this chapter, we will discuss some of these facts and principles.

[2]*Ibid.*

ELECTRICITY

Most people are familiar with electrical effects produced by friction in many everyday situations. When you scuff your shoes on a rug and then reach for the door knob, you sometimes see or feel a spark between your fingers and the knob. If you comb or brush your hair vigorously you sometimes hear sparks crackle. Similarly, you may produce sparks if you slide across a plastic car seat and then reach for the door handle. Electrical effects similar to these have been known for ages.

It has long been known that if a glass rod is rubbed with silk (or plastic) it can be used to illustrate electrical effects. Figure 8-1 shows a rod that has been rubbed with silk and suspended by a thread. When a second rod is also rubbed with silk and brought near the first, the suspended rod is pushed away. A force of repulsion is produced by the electricity of the two rods. Electricity can also produce attractive forces. Suppose that the second rod is made of hard rubber and is rubbed with fur. When this rod is brought near the silk-rubbed glass rod, the two rods are attracted rather than repelled.

Three significant facts are evident in these simple experiments. First, the forces of attraction and repulsion occur even though the rods do not touch one another. Second, the force is extremely large compared to the gravitational force between the same objects. Third, there is something intrinsically different between a rubber rod (rubbed with fur) and a glass rod (rubbed with silk) since they be-

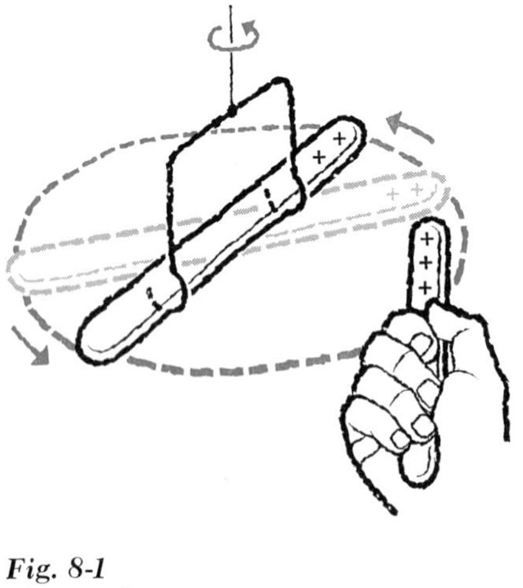

Fig. 8-1

have differently in the presence of the suspended glass rod (rubbed with silk).

Electric Charge

We can explain the electrical effects discussed above by introducing a new fundamental, scalar quantity: *electric charge*. Rods and other objects can acquire an electric charge by friction when they are rubbed. Because we obtained a different result with the rubber rod than with the glass rod, there must be two kinds of electric charge. The electric charge we obtained on the glass rod is arbitrarily called a positive charge, and the electric charge we obtained on the rubber rod is called a negative charge. In the experiment we have described, the positive charge on the glass rod produced a repulsive force when brought near a similarly charged rod. The negative charge on the rubber rod produced an attractive force when brought near the positively charged rod. These two examples illustrate a general principle:

Like charges repel each other, and unlike charges attract each other.

Two Fundamental Particles

The fact that objects can acquire an electrical charge by friction strongly suggests that matter has an electrical nature. In fact, today we believe that atoms are made up of particles with electrical properties. Two of the *fundamental particles* that make up the atom are the *electron* and the *proton*. For example, every atom of hydrogen is made up of one electron and one proton. The electron has a mass of 9.1×10^{-31} kilograms. The proton has a mass of 1.7×10^{-27} kilograms. These two masses are fundamental constants, since they are the masses of fundamental particles. The proton supplies most of the mass of the hydrogen atom since it is about 1840 times more massive than the electron. Other atoms contain different numbers of fundamental particles.

Although the electron and proton are different in mass, they have exactly the same amount of electric charge. Their charges, however, are not of the same kind. The electron has a negative charge, and the proton has a positive charge. Most objects are electrically neutral: that is, they have the same number of electrons as protons, and therefore have no excess, or net, electric charge. Objects are *charged* when they do not have the same number of electrons and protons.

Such an imbalance is ordinarily a result of the gain or loss of electrons by the object. When two objects acquire an electric charge by being rubbed together, the charge is acquired because electrons from one object are transferred to the other.

Conservation of Charge

Charge is a conserved quantity; that is, it cannot be created or destroyed. If an object loses charge, it is because the charge has been transferred somewhere else. For example, we can give the glass rod a positive charge by rubbing it with silk. At the same time, however, the silk acquires a negative charge of the same amount. Some of the electrons in the glass simply move from the glass to the silk, carrying their negative charge with them.

The Unit of Charge

A charge may be measured by the electrical force it exerts on another charge. The unit of charge is called a *coulomb*, after the French physicist Charles Augustus Coulomb (1736–1806). This amount of charge may be defined as follows:

> One coulomb of charge is the amount of charge that would exert a force of 8.99×10^9 newtons on an equal charge one meter away.

The direction of the force, of course, depends on whether or not the two charges are of like or unlike kinds, but the magnitude does not. Because 8.99×10^9 newtons is a very large force, the coulomb is a very large charge compared to the amounts we normally deal with on charged glass rods, etc.

The Law of Electrical Force for Charges at Rest

Coulomb discovered the exact law that gives the magnitude of the force between any two charged objects at rest. This law is expressed as follows:

$$F_e = K_e \frac{q_1 q_2}{d^2} \tag{1}$$

In this equation, q_1 is the charge on one object, q_2 is the charge on the other object, d is the distance between the two objects, and F_e is the magnitude of the electrical force between them. The electrical force depends on the charges of the two objects in the same way that the gravitational force depends on the masses. Both the electrical and gravitational force depend on the distance d in the same way. The quantity K_e is a fundamental constant that is fixed by the strength of the electrical force. K_e has a value of 9.0×10^9 newton meter^2/coulomb^2. This value is very large compared to the gravitational constant G. The electrical forces between electrons and protons are much larger than the gravitational forces between them. The direction of the electrical force whose magnitude is given in Equation 1 is along the straight line between the two charges. This force is a force of repulsion for like charges and attraction for unlike charges.

Electric Currents

Some materials, such as most metals, allow electrons to move freely from place to place on and inside them. Such materials are called electrical *conductors*. Other materials do not allow electrons to move freely on or inside them. Such materials are called *insulators*. The glass rod and the rubber rod we discussed above were both insulators. Therefore the electric charge placed on them by friction tended to stay on them, since it could not easily move. If we had used metal rods instead, the electrons would have been able to move out of the rod, into our hands, and perhaps into the ground. (The flow of electrons could also have been in the opposite direction.)

A flow of electric charge is called an electric current. This is usually a flow of electrons through a conductor. The amount of current in a conductor, such as a copper wire, is determined by the rate at which charge is passing through. The unit of current is the *ampere*. A current of one ampere is a current in which one coulomb flows by in one second.

Fields

It may seem strange that a charged object has an electrical force on it because of another nearby charged object, when they are not even touching. Our previous study of gravitation, however, has already given us an example in which one object is influenced by another across a distance without being touched. Thus the falling apple does not stop to ask for directions; it heads for the center of the earth. Often it is convenient to assume that the apple is in a *field* of influence that exists around the earth, and that it falls because it is in that field. We may think of the earth, because of its mass, as setting up a gravitational field of force around it. The apple, because it is *in* this field of force, is accelerated toward the earth *by* this field of force. In this section we shall discuss gravitational fields (which are set up by masses) and electrical fields (which are set up by "charges").

GRAVITATIONAL FIELDS. We would like to use a picture to represent the force exerted by the gravitational pull of the earth. We want this picture to indicate the force exerted on any object placed in any location with respect to the earth. This picture will represent the gravitational field of the earth, because it will enable us to tell how the presence of the earth affects other objects.

We draw this picture of the gravitational field of the earth by drawing arrows at various points in the space around the earth (see Fig. 8-2*a*). These arrows represent the gravitational force that would be exerted on an object if it were placed at those points. Since objects of different mass would have different forces on them, we agree to use an object of one-kilogram mass. The arrows, then, represent the force that would be exerted on a one-kilogram mass. The one-kilogram mass is pulled harder when it is near the earth than when it is farther away. For this reason the arrows are longer nearer the earth and become shorter as we go farther out into space. The length of the arrows represents the magnitude of the force, and the direction of the arrows gives the direction of the force. The arrows point toward the center of the earth, since gravitational forces pull objects in that direction. Ideally we can think of an arrow drawn at every point in space. This collection of arrows could be used to represent the gravitational field of the earth, since it tells us the gravitational force that would be exerted on one kilogram of mass at any point. If we

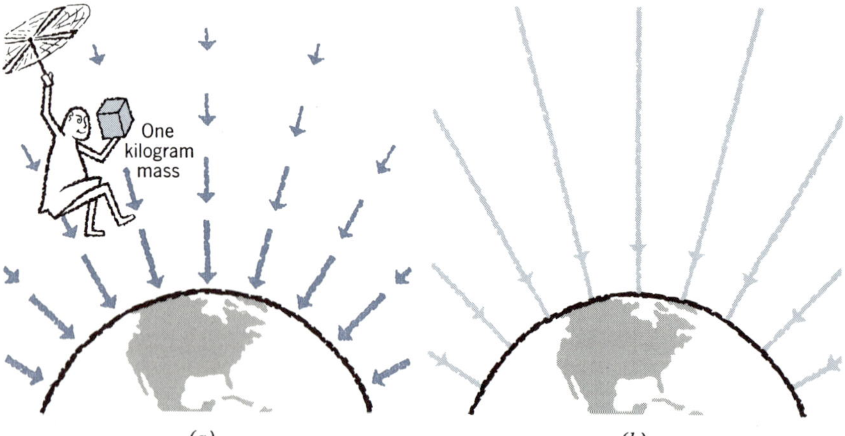

(*a*) (*b*)

Fig. 8-2

want the force that would be exerted on an object of different mass, we simply multiply the mass of the object by the force on the one-kilogram mass. This allows us to find the force that would be exerted on any object at any point in the gravitational field of the earth.

We can represent the gravitational field more conveniently by lines than by arrows. Figure 8-2*b* shows how these lines look. Each line is drawn by following the arrows from point to point. Therefore, at each point the lines are in the direction of the arrows that would be drawn at that point. Where the force is the strongest, the lines are closest together. Where the force is weak, the lines are far apart. These lines, called field lines, give us the picture that we desired. The lines of the gravitational field do not end except on a mass. (Of course, this is not obvious from what we have said, but like a similar statement to be made later about electric field lines, it is true.)

THE ELECTRIC FIELD. We can picture the electric field around a charge just as we have pictured the gravitational field around the earth. Figures 8-3*a* and 8-3*b* show the arrows drawn at various points, and also the field lines. The arrows at a given point represent the force that would be exerted on one coulomb of positive charge if it were placed at that point. (We use a one-coulomb charge here rather than the one-kilogram mass because, while the gravitational force acts on masses, the electric force acts on charges.) The direction of the field lines (or arrows) is the direction of the force that would be exerted on a positive charge. The force on a negative charge would be in the opposite direction. Once again, the closer the lines of force are to each other, the stronger the force that would be

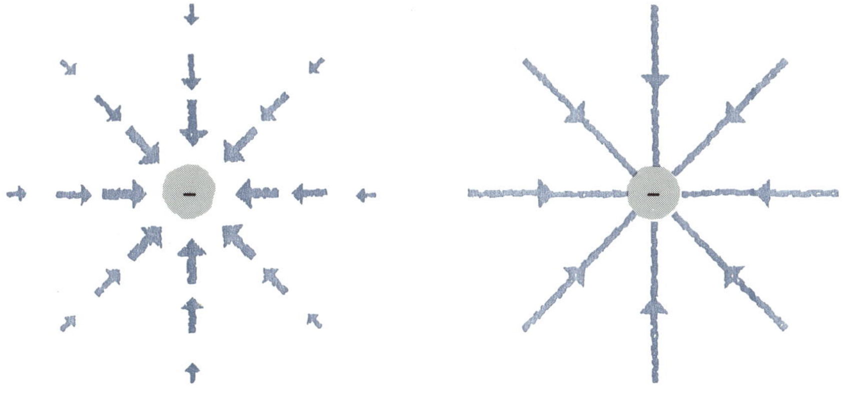

(*a*) (*b*)

Fig. 8-3

exerted. In Figure 8-4 we show the field lines for several arrangements of charged objects. In each case, a small positive charge would be impelled in the direction of the field lines. A small negative charge, such as an electron, would be impelled in the opposite direction. The field lines of the electric field do not end except on a negative charge, and do not begin except on a positive charge.

It may seem from the above discussion that the field concept is an unnecessary fabrication. In fact, in this discussion, we have simply been looking at the forces between objects in a different way. Instead of thinking of objects as exerting forces on each other at a distance, we have decided to consider that each one of them has set up a field and that it is this field that exerts the force on the other object. Thus far, there is no reason to choose one approach over the other, except convenience. In Chapter 9, we shall find reason to believe that the field has a real existence.

The English physicist Michael Faraday (1791–1867) used the concept of field lines in his research in electric and magnetic phenomena. Although Maxwell later put these ideas in a more elegant mathematical form, Faraday was able to understand fields sufficiently well to make important discoveries using field lines.

Electric Potential Energy

We have seen in Chapter 6 that an object may have gravitational potential energy because of its position in a gravitational field. A *charged* object can also have potential energy because of its position in an electric field. This must be true because a charged object in an electric field will experience a force and, if released, will accelerate, acquiring kinetic energy. This energy must come from somewhere. In fact, it comes from the *electric potential energy* the charged object had in its starting position.

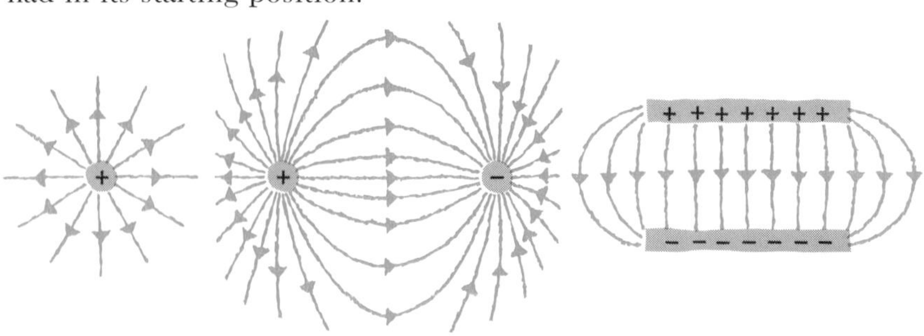

Fig. 8-4

Another way to recognize the existence of electric potential energy is to move a charged object in an electric field. We recall that we had to exert a force to lift a mass in a gravitational field. In so doing, we transferred energy to the object by doing work. The gravitational potential energy of the object was increased as we lifted it against the gravitational force. In the same way we increase the electric potential energy of an object when we move it against an electric force. The man shown in Figure 8-5*a* is doing work against the electric force and thus is increasing the electric potential energy of the charged object. The potential energy is transformed into kinetic energy when the charged object is released (Fig. 8-5*b*). It is then accelerated in the electric field by the electric force.

When we discussed gravitational potential energy we found that it was not the potential energy itself that was important but the change in an object's potential energy as it moves from one position to another. Thus we were really interested in the difference between the potential energies of the object at two points in the gravitational field. Similarly, here we are interested in the difference between the potential energies of a charged object at two points in the electric field. We can choose one point to have zero potential energy and use that point as a reference point for the potential energies of other points.

Since the force on a negative charge in an electric field is opposite to that on a positive charge, the two types of charge will have different potential energies at the same point in the same field. The potential energy of a negative charge in an electric field will be the negative of the potential energy that a positive charge of the same amount would have if it were at the same point.

Often in discussing an object in an electric field we speak of its

(*a*) (*b*)

Fig. 8-5

electric potential instead of its electrical potential energy. The electric potential of a charged object is simply its electric potential energy divided by its electric charge. The common unit of electric potential is the *volt*, which is defined as follows:

$$\text{volt} = \frac{\text{joule}}{\text{coulomb}}$$

*Illustrations of the Operation of Electrical Forces and Fields

INDUCED CHARGES. In the experiment illustrated in Figure 8-6, we have suspended a light cotton ball wrapped in tin foil by a thread. The thread acts as an insulator to prevent the "leaking" of electric charge from the ball. The ball is electrically neutral to begin with—that is, it is not charged. We now charge a glass rod by rubbing it with silk. As we bring the glass rod up to the ball, the ball is attracted to the rod, despite the fact that the ball has no net charge (see Fig. 8-6*b*). When the ball touches the rod, it suddenly jumps away. After this, the ball is repelled by the rod (see Fig. 8-6*c*). How are we to explain this behavior?

First we must explain how the ball can be attracted by the positively charged rod although the ball itself has no net charge. The reason is that the positive charge on the rod sets up an electric field. The negative charges in this field are attracted toward the positive rod. This attraction causes many of the electrons in the tinfoil covering the ball to move over onto the side of the ball nearest the glass rod. This sets up a negative charge on that side of the ball and leaves an equal positive charge on the other side. Thus one side of the ball is attracted to the glass rod and the other side is repelled. The two charges on the ball are equal in magnitude, but they are at

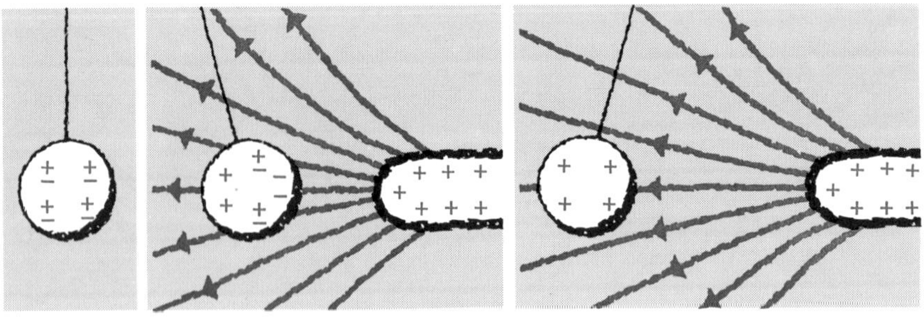

(*a*) (*b*) (*c*)

Fig. 8-6

different distances from the rod. The electric field of the rod is stronger closer to the rod. Therefore, the attraction of the negative charge on the ball overcomes the repulsion of the positive charge, and the ball is pulled toward the rod. We say that the field of the rod *induces* a charge separation in the conducting metal that covers the ball. This causes an induced negative charge on one side of the ball, and an induced positive charge on the other.

When the rod actually touches the ball, some of the electrons flow off onto the glass rod, attracted by the positive charge. Since the rod is an insulator, the electrons from the ball tend to neutralize the positive charge of the rod only in the region where the rod touches the ball. Therefore, the rod still retains part of its positive charge. Now, however, with the loss of these electrons from the ball, the ball *itself* is positively charged. In this condition, the positively charged ball will be repelled by the positively charged rod.

THE ELECTROSCOPE. The behavior of an electroscope may also be understood in terms of induced charges. The electroscope is a device used to detect charge (see Fig. 8-7). When there is a charge on the electroscope, there will be a repulsion between the two metal leaves so that they are held apart. Without this force, the leaves fall together.

In Figure 8-7*a*, the leaves of the electroscope are together, indicating that the electroscope is uncharged. If we bring a charged glass rod up to, but not touching, the top of the electroscope (see Fig. 8-7*b*), the positive charge of the glass rod pulls some of the electrons out of the leaves of the electroscope into the ball on the top. The

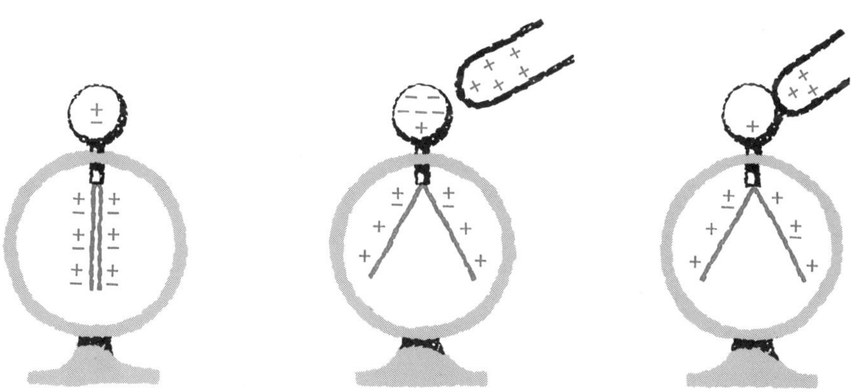

(*a*) (*b*) (*c*)

Fig. 8-7

leaves of the electroscope are left with a positive charge. This charge makes the leaves separate because like charges repel each other. When the rod is withdrawn, the electrons return to the leaves and the leaves come together again. If we touch the rod to the top of the electroscope, however, some of the electrons are pulled into the charged glass rod (Fig. 8-7*c*). Then, when the rod is removed, the leaves will remain apart, since the electroscope now has a positive charge. The electroscope appears to behave in the same way if we do the same experiment using a charged rubber rod (negative charge).

The Charge on the Electron

Although the electron is extremely small, its charge can be measured by a rather direct experiment, conceived and executed by Robert Andrews Millikan (1868–1953) and his students. This experiment is called the "oil-drop experiment" because it measures the charge on small drops of oil. The experimental apparatus is diagramed in Figure 8-8. Droplets of oil from the fine mist sprayed from an atomizer are allowed to fall into the electric field between two charged plates. The droplets are charged as they come out of the atomizer, by friction or some other means. When illuminated by a strong light, they can be seen individually as tiny specks in a microscope.

As illustrated, there is a positive charge on the upper plate and a negative charge on the lower. This creates an electric field that tends to lift negatively charged objects. It is possible to adjust or reverse

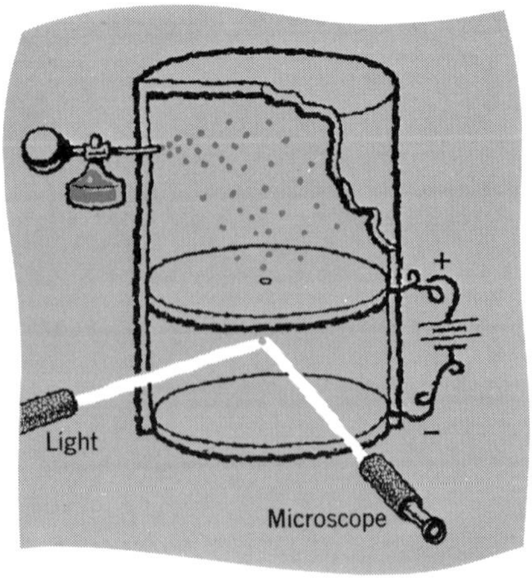

Fig. 8-8

the field by changing the charge on the plates, or to eliminate the field altogether by discharging the plates. Suppose, for example, that we have a negatively charged droplet. We can adjust the field so that the electric force pulling the droplet upward is exactly balanced by the weight of the droplet pulling downward. If we know the weight of the droplet we can determine the electric force and thus the charge on it.

Actually, in practice, we find the weight of the droplet by observing how fast it falls against air resistance when no electric field is present. We then compare this speed with its speed as it rises in a known electric field. This method is more convenient than trying to balance the droplet perfectly, but the principle is the same: we compare the gravitational force with the electric force on the same drop. The electric force depends on the charge on the droplet and therefore allows us to measure that charge.

In this way the charge, both positive and negative, on many droplets can be measured. The smallest amount of charge ever observed is 1.60×10^{-19} coulomb. Every other observed charge is some whole number times this smallest charge—either two, three, four, etc., times 1.60×10^{-19} coulomb. Thus charge occurs in small discrete "bundles." Each electron has one such "bundle" of negative charge. A droplet may gain or lose electrons, and thus have a negative or positive charge. The charge, however, always corresponds to the gain or loss of a whole number of electrons.

The amount of charge on a proton is the same as that on an electron, except that the proton's charge is positive rather than negative. Therefore, the charge on a proton is *plus* 1.60×10^{-19} coulomb. This charge is represented by the letter e. Wherever charge is found, it occurs in

$$e = 1.60 \times 10^{-19} \text{ coulomb}$$

discrete bundles of this amount; therefore, this quantity e is one of the fundamental constants of nature. (Recently there has been speculation that particles of one third or two thirds this amount of charge may exist, but this is still unverified.)

*Electrical Conditions in Conductors

Here we shall discuss several facts about solid conductors in which the charges are at rest. These facts apply to the solid material of the conductor and not necessarily to holes within it. If currents are flowing, then some of these conditions may not hold perfectly. The facts are as follows.

1. There is no electric field inside a conductor.
2. The excess charge on a conductor is on its surface.
3. The concentration of charge on the surface of a conductor is greater where the surface is more outwardly curved or pointed. The field just outside the surface is also greater in these regions. These are the regions at which charge can most easily leave or enter the conductor.

The first of the above facts may be explained as follows. If there were an electric field inside a conductor, the field would cause the electrons of the conductor to move. The electrons would rearrange themselves until the field disappeared.

Since there is no field inside a conductor, there can be no net charge there either. A charge, as, for example, the single charge in Figure 8-9, would have lines of field issuing from it (or toward it). These field lines would thus constitute a field inside the conductor, which is impossible. One might think that other nearby charges might supply field lines that would balance out this field, but this does not occur. We cannot go into the mathematical details that prove this, but we can indicate the principles that are important.

All field lines must start on positive charges and end on negative charges, if they start or end at all. Conversely, all fundamental charged particles must have field lines starting or ending on them; in fact, this is true of all charge. We can see this by examining the electric force very close to a charged particle. According to Equation 1, the electric force increases very rapidly as we near a charge and decreases very rapidly as we move away from it. Therefore, there is a small region around a small charged particle in which the electric force it causes is much larger than the force from any other charge. Each small particle is surrounded by a cluster of field lines which begin at the particle if it is positive and end on the particle if it is negative, as, for example, the small negative particle in Figure 8-9. Therefore, any charged particle in the middle of a conductor would cause a field there, a field which, as we have seen, does not exist. This field can be eliminated by removing the charge, or by placing

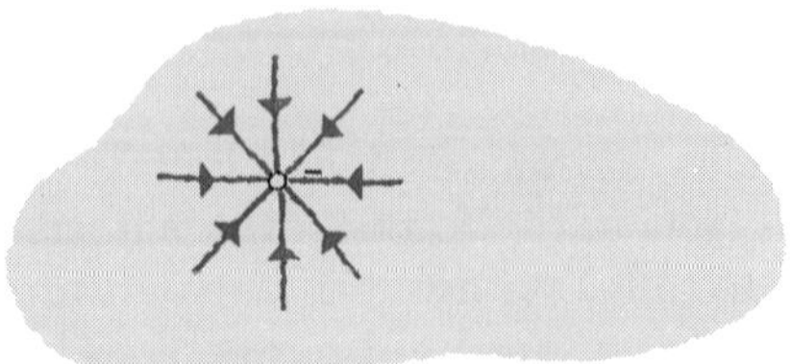

Fig. 8-9

on top of it an opposite charge of equal strength. In either case there is no excess charge in the interior of the conductor. (Actually, of course, the material of the conductor is made up of positive and negative charges that are not exactly on top of each other but are so arranged close to each other that the argument given here is not invalidated.)

Since charge cannot remain inside a conductor it will move to the outside and remain there. This charge will be more concentrated where the conductor is more curved or pointed, provided that these curves or points are outward (convex). If they point or curve inward they will be among the *least* charged regions of a conductor. The field lines coming from these charges will be most closely grouped where the charge is most concentrated. Therefore, the field just outside the surface will be strongest where the charge concentration is greatest. Where the field is greatest, there is the greatest tendency for charge to leave, or opposite charge to enter, the conductor. We shall not attempt to prove all these related facts, but we will give a partial explanation of why charge tends to leave or enter these pointed regions.

On a flat surface, the force on a charge from neighboring charges is directly along the surface and thus does not tend to push the charge off the surface (see Fig. 8-10*a*). Where the surface is pointed outward, however, the forces on a charge from neighboring charges are not in opposite directions. They have a resultant that tends to push the charge off the surface (Fig. 8-10*b*). The more sharply pointed the surface, the more pronounced this effect.

A lightning rod acts as a conducting point on the surface of the conducting earth. The electric charge in lightning tends to enter these high conducting points rather than the nearby structures that the rod is designed to protect.

During the American revolutionary war, King George III evidently decided that the pointed lightning rods invented by the rebellious Benjamin Franklin were inferior to the blunt rods championed by a loyal British subject. He therefore removed the pointed rods from powder magazine and palace. As far as we are able to determine, the

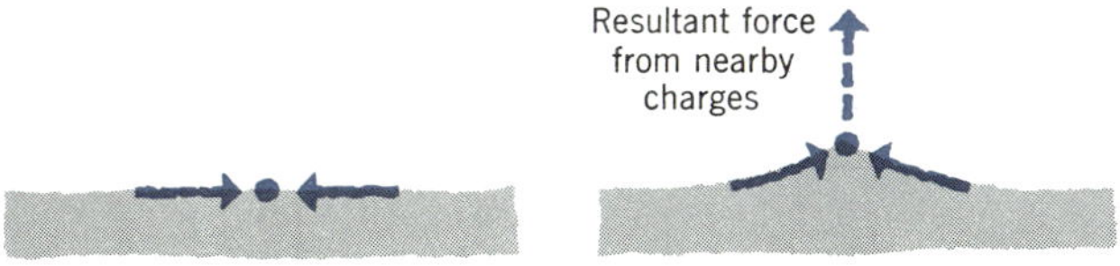

(*a*) (*b*)

Fig. 8-10

laws of electricity and magnetism were not affected by this royal decision. The renowned American scientist and statesman took this decision with characteristic good humor, suggesting that his only regret was that the king did not give up lightning rods altogether.[3]

*The Van de Graaff Generator

The Van de Graaff generator is a laboratory device used to collect charge on a sphere at a very high electric potential energy. This potential energy can be used to accelerate charged particles to high kinetic energies for various experiments. As we have seen, there is no field or excess charge inside a conductor, and the charge stays on the outside. This situation is not changed if part of the material inside is removed to form a hollow conductor, since there is no excess charge in the material removed. There will be no field or charge inside this hollow region unless some charge is brought in and held there. Even if there is a small hole leading into the hollow region from outside the conductor the field inside will be very small.

The top of the Van de Graaff generator (Fig. 8-11*b*) is a large hollow spherical conductor, with a hole at the bottom through which charge can be inserted. Since there is very little field inside this sphere, this charge is easily released onto the inside of the sphere. The charge does not remain there but moves to the outside of the sphere. By bringing in more charge, we are able to build up a large charge on the outside of the sphere. We could not do this by bringing the charge to the outside of the sphere because there is a high electric field building up there as the charge is increased. Charge would flow off the outside of the sphere, rather than onto it.

The charge is brought into the sphere on a belt made of insulating material. It is collected on the sharp points at the top, which are connected to the inside of the conducting sphere. Charge may be put on the belt at the bottom in several ways—for example, by friction.

MAGNETIC FORCES AND FIELDS

Magnetic forces were first studied in connection with magnets. Magnets are generally made of some combination of iron and nickel.

[3]James Parton, *The Life of Benjamin Franklin*, Vol. II, Boston: Houghton, Mifflin, and Cambridge: The Riverdale Press, 1864, 1892, pp. 216–217. Also see Carl Van Doren, *Benjamin Franklin*, New York: Viking Press, 1938, p. 658.

(a)

Courtesy of the High Voltage Engineering Corporation

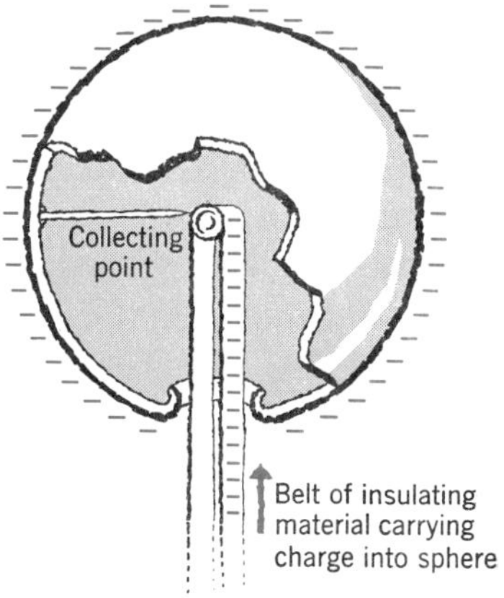

(b)

Fig. 8-11

They will attract each other or other objects made of iron, nickel, and certain other metals. Most magnets have two regions in which the magnetic forces are especially powerful, called the *poles* of the magnet. The earth itself is a magnet, with two magnetic poles close to the ordinary North and South poles but not quite in the same places. The two poles of a magnet are not alike. If a magnet is suspended by a thread, one of its poles will point to the north and the other to the south. A compass needle is such a magnet. The pole that points north is called the north-seeking, or north, pole of the magnet. The other is called the south-seeking, or south, pole. All north magnetic poles repel each other. All south magnetic poles repel each other. All north magnetic poles attract all south magnetic poles, and vice versa. In other words, like poles repel, opposite poles attract.

Cutting a magnet in two will not produce a free (isolated) north or south pole (see Fig. 8-12). If we tried this experiment we would find that another pole automatically appears in each half, so that each piece is a complete magnet with two poles. There are no free magnetic poles—magnetic poles always occur in pairs.

The Magnetic Field Around Magnets and Currents

The concept of the magnetic field helps us to describe the influence of a magnet in the space around it. As in Figure 8-13, we can use a compass needle to map out the field lines. The north pole of the compass needle will point in the direction of the magnetic force on it. The force on the south pole is in the opposite direction. Like the electric field, the magnetic field is strongest where the field lines are closest together and becomes weaker as the lines spread apart.

During a class demonstration, the Danish physicist Hans Christian Oersted (1771–1851) discovered a remarkable connection between electricity and magnetism. Happening to have a compass near a wire, he was surprised to see the compass needle move when

Fig. 8-12

an electric current flowed through the wire. Although the compass needle is uncharged, it will turn at right angles to the wire when a current flows, reversing itself when the current flows in the opposite direction (see Fig. 8-14). This means that an electric current, or indeed any moving electric charge, produces a magnetic field. We can use a compass to map out this field just as we did for the permanent magnet. Figure 8-15*a* shows the magnetic field around the current in a straight wire. Figure 8-15*b* shows the magnetic field around an electric current in a loop. The more current flowing in the wire, the stronger the magnetic field it produces. In fact, the field at any point in space is proportional to the current that produces it. The magnetic field lines do not end anywhere. If there were such a thing as a free magnetic pole, similar to a free electric charge, then the magnetic field lines could end on the poles. So far, however, no free magnetic poles have ever been discovered.[4]

The current loop in Figure 8-15*b* has a magnetic field very much like the magnet in Figure 8-13 (see also Fig. 8-16). One face of the loop is a north pole, and the other face is a south pole. If we make the current in the loop flow in the other direction, then the poles will be reversed.

Field lines are imaginary lines drawn in space to indicate the direc-

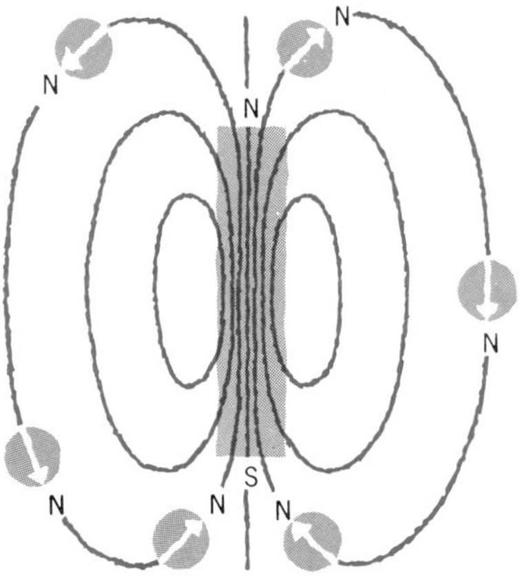

Fig. 8-13

[4]In gravitational, electric, and magnetic fields, there can be points in space at which the field has a magnitude of zero. At such points, field lines can appear either to cross or to come to an end without ending on a mass or a charge. These points are possible exceptions to the general rules that field lines do not cross, that gravitational field lines end only on masses, that electric field lines begin only on positive charges and end only on negative charges, and that magnetic field lines do not begin or end anywhere.

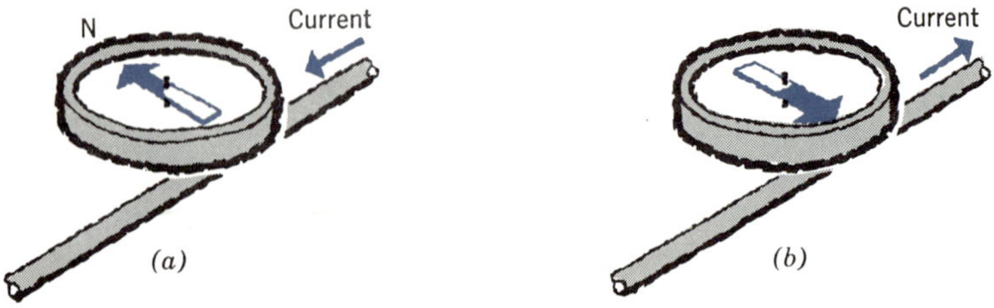

Fig. 8-14

tion and magnitude of the field. The direction of the field, as we have seen, is given by the direction of the field lines. Figure 8-17*a* and *b* show how the magnitude of the field is related to the spacing of the field lines. The small square is on a surface perpendicular to the field lines. In a strong field (Fig. 8-17*a*) many field lines go through the square. In a weak field fewer field lines go through it. The strength of the field is proportional to the number of field lines going through a given small area (such as the square) on a surface at right angles to the field lines. (We imagine this area to be small enough so that the strength of the field does not vary much from one part of the area to another.)

Figure 8-17*c* and *d* show what happens to the magnetic field lines when the current through a loop is increased. If the current is doubled, then the number of field lines passing through the coil is also doubled. In fact, the strength of the field caused by the loop is doubled everywhere in space around the wire of the loop.

Scientists believe that the magnetic properties of magnets and

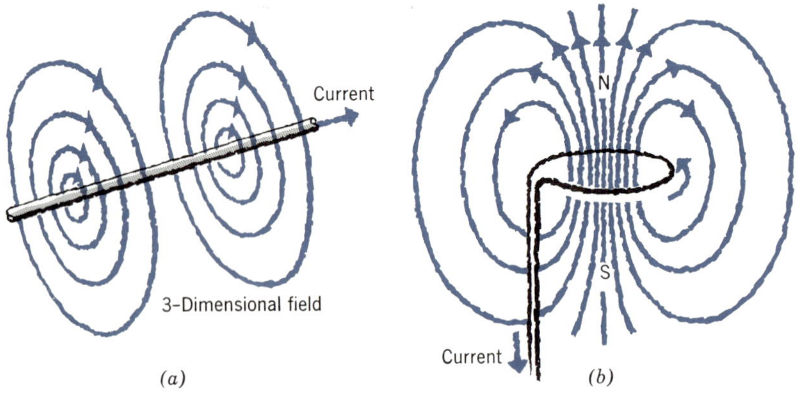

Fig. 8-15

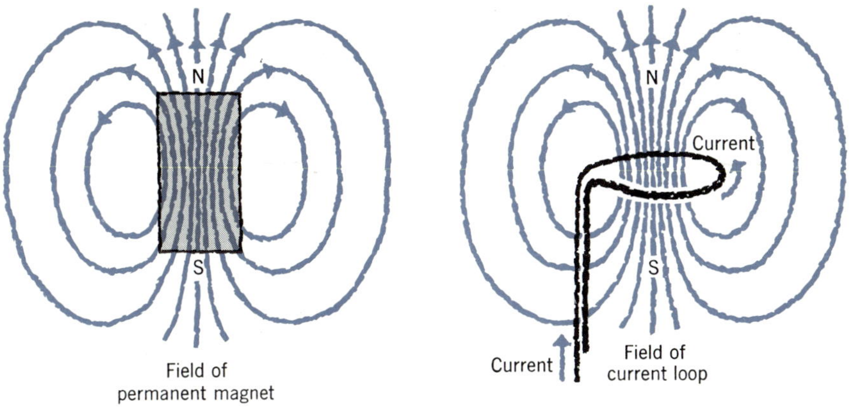

Fig. 8-16

other types of matter are actually caused by electric currents moving within the atoms. Atoms contain charged electrons moving around in circles or spinning, thus producing tiny current loops which, either individually or in groups, act as small magnets and produce magnetic fields. In some elements, such as iron, the magnetic fields of these tiny current loops added together cause the entire atom to act like a small magnet. In unmagnetized iron, these small magnets are disordered and act in different directions. Therefore, their magnetic forces cancel each other out. In magnetized iron, these small magnets are more nearly lined up, so that they cooperate to produce a large magnetic field.

The small magnets in a bar of iron can be lined up by the influence of the magnetic field from a permanent magnet, just as the compass needles are lined up by the magnetic fields in Figure 8-13. This is known as the process of magnetization. In the influence of this field, the bar itself acts as a magnet and is therefore attracted by the perma-

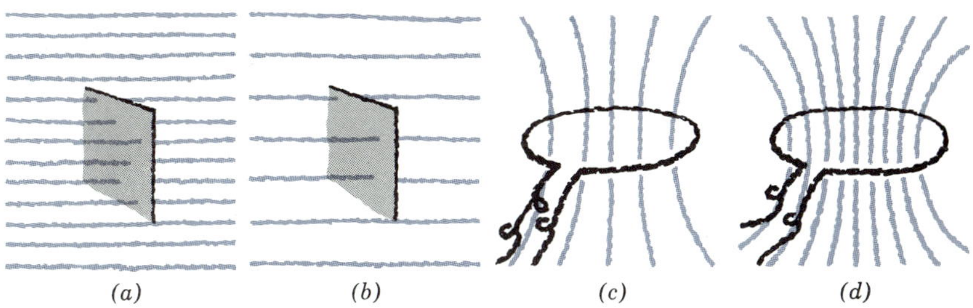

Fig. 8-17

nent magnet (see Fig. 8-18). This is the process by which magnets are able to attract other objects made of magnetic materials. The object itself becomes an induced magnet, with its north and south poles oriented so that attraction occurs.

A magnetized bar will remain magnetized to some extent even after it has been removed from the influence of the permanent magnet. Magnetic materials differ greatly, both in their ability to become magnetized in the presence of another magnet and in their ability to retain their magnetization after the other magnet has been removed.

That the orderly arrangement of these small magnets in matter is partly responsible for magnetic effects is shown by the experiment illustrated in Figure 8-19. The suspended iron block is attracted to the permanent magnet. When the block is heated over a burner, it eventually loses its attraction to the magnet and swings away. As it cools, it is once more attracted. The heat flow into the block of metal increases its internal energy and, therefore, the random motion of molecules and electrons inside the block. This vigorous random motion destroys the alignment of the small magnets in the block that has been set up by the magnetic field of the permanent magnet.

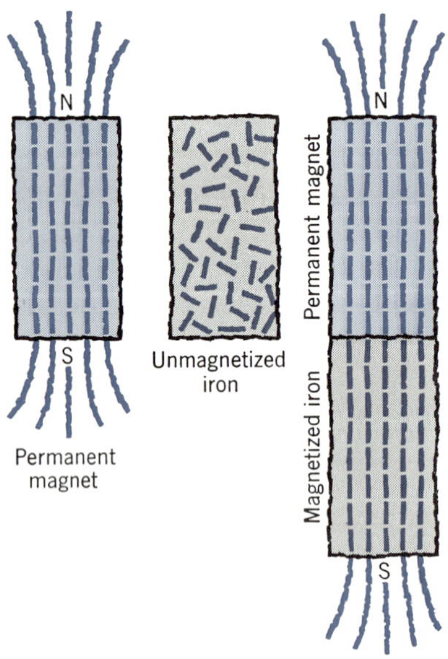

Fig. 8-18

Without this order the block cannot act as a magnet and be attracted to the permanent magnet.

An electromagnet can be made from current-carrying loops of wire and an iron core. When current flows through the wire, the current loops act as a magnet, setting up a magnetic field. This field aligns the small magnets in the iron core, which then greatly strengthen the total field. When the current is turned off, these small magnets tend to become disordered again. In electromagnets, the core is made of a material that is easily magnetized by a magnetic field but retains little of this magnetization when the field is turned off.

We have illustrated in this section the principle that *all* magnetic fields arise from moving charges—whether they are moving in the ordinary electric currents or the hidden electric currents inside matter. The magnetic force actually arises from an interaction between moving charges. Wherever there are moving charges there are magnetic effects.

The Magnetic Force on Moving Charges

All magnetic effects are electrical in origin. A magnetic field is not the same as the electric field produced by a stationary charge but is an *additional* field produced whenever the charge is in motion. As we pointed out, a current loop and a bar magnet are surrounded by similar magnetic fields. For both these situations, the magnetic force resulting from the field is easily described

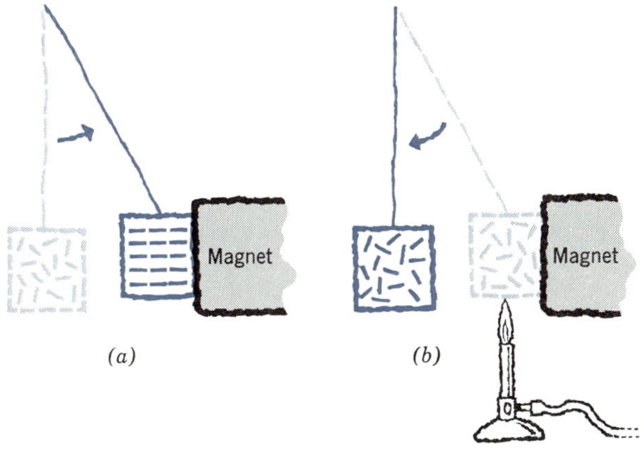

Fig. 8-19

in terms of the attraction and repulsion of magnetic poles. Much of the time, however, we are interested in magnetic forces on streams of charged particles or on current-carrying conductors of a variety of shapes. The pole concept does not help us much in these situations.

The general form of the law of magnetic force is complex. Two simple facts which come from this law are that currents in the same direction attract each other and that currents in opposite directions repel. We can get a more general feeling for these forces by seeing how currents (moving charges) set up magnetic fields, and then by studying the effect of magnetic fields on the behavior of other currents (moving charges). Accordingly, we describe the basic character of the magnetic force as follows.

With one exception, whenever a charge is moving in an external magnetic field it will experience a force at right angles to both the direction of its motion and the direction of the external field. That is, a moving charge (or a current in which many charges are moving through a conductor) has a magnetic field around it. When it passes through the magnetic field of some other system, it experiences a deflecting force at right angles to its velocity. The exception occurs when the charge is moving parallel to the direction of the magnetic field. In this one case, there is no deflecting force.

We can illustrate this basic description of the magnetic force in several ways. For example, suppose that we have a beam of electrons moving in a straight line (see Fig. 8-20). The electrons pass through a tube and strike a fluorescent screen, such as is used in an ordinary television set. The electrons produce a luminous spot on this screen when they strike it. When a magnet is brought near, the electrons are deflected, and they strike a different spot on the screen.

The motion of an electron, or other charged particle, in an unchanging magnetic field is actually rather curious. (We assume that there are no electric or gravitational fields present.) The particle

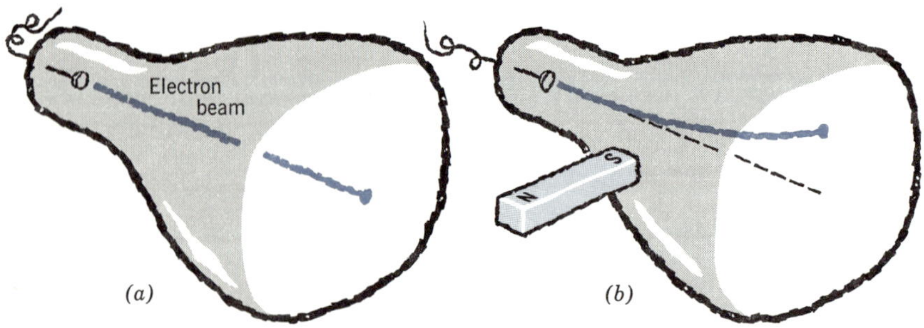

Fig. 8-20

curves around the field lines, changing its direction but not its speed (see Fig. 8-21*a*). The magnetic field alone does not increase the speed of the moving charged particle because it acts at right angles to its velocity. A simple kind of motion occurs when a particle is moving at right angles to a constant uniform magnetic field. Such a field is not changing with time, and it is the same in magnitude and direction over the region of space in which we are interested. In this case the particle moves in a circle (see Fig. 8-21*b*). The magnetic force supplies the centripetal force necessary to maintain this circular motion.

Applications of the Magnetic Forces

THE ELECTRIC MOTOR. The magnetic force can be used to make a current-carrying wire move or accelerate. Figure 8-22 shows a simple conducting wire wrapped around a frame and connected to a battery. (As far as we are concerned here, the battery is simply a device which keeps a current flowing through the wire.) The wire runs between the poles of a magnet and is therefore in a magnetic field. When the switch is closed so that a current flows through the wire, the wire moves, either up or down depending on the direction of the current. The electrons moving through the wire are deflected at right angles to their motion. This makes the whole wire move perpendicular to the magnetic field lines and to itself.

The force illustrated in the above experiment can be used to operate an electric motor. Figure 8-23 shows one kind of electric motor, in which electrons flow through a loop or coil of wire in a magnetic field. The force on the electrons flowing in the coil causes the coil to turn, and thus turns the motor. A drive shaft attached to the coil can be used to operate an egg beater, a grinding wheel, a vacuum cleaner, and other devices.

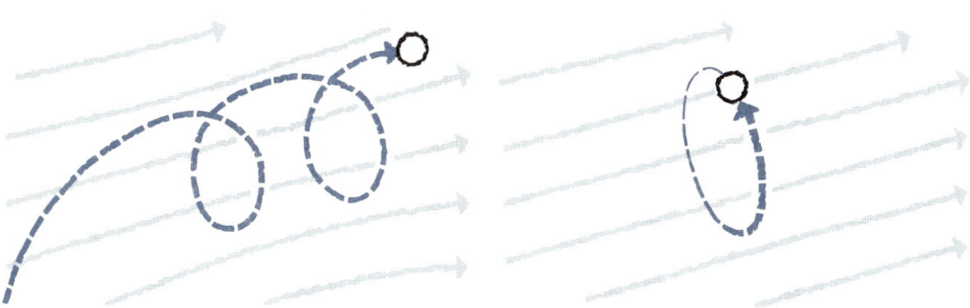

(a) (b)

Fig. 8-21

If the current were to flow in the opposite direction through the loop, the motor would turn the other way, since the force would be in the other direction. To prevent this we set up the two curved plates on the ends of the wire of the loop as shown. This arrangement insures that the incoming current in the wire always enters whichever side of the loop happens to be close to the north pole of the magnet as the loop rotates. Thus the motor will keep running in the same direction. Of course, practical motors are more elaborately designed so that they run more smoothly and efficiently.

THE CYCLOTRON. The American physicist Ernest O. Lawrence (1902–1958) invented a device which uses a combination of electric and magnetic fields to accelerate charged particles, such as protons. This device, called a cyclotron, is illustrated in Figure 8-24. Two hollow D-shaped conductors are oppositely charged, producing an electric field between them. There is also a magnetic field going through the two D's, perpendicular to them and to the electric field. The proton starts out near the center of the space between the two conductors and is accelerated by the electric field, from one D to the other. Inside the conducting D's the electric field is weak. Because of the magnetic field the proton turns in a semi-circle and comes back out into the space between the two D's. When it emerges into this space it finds that the electric field has been reversed while it was inside the D. Therefore it is accelerated again, enters the other D, and turns around again—but this time it has more speed and turns in a larger circle. This procedure continues, with the electric field increasing the speed of the particle again and again while the magnetic field turns the particle around and keeps it from getting out of the cyclotron. When the particle reaches the outer edge of the D's it is deflected out with its acquired high energy (Fig. 8-24*b*). A wide variety of similar accelerators has been designed to obtain high-energy particles for use in industry and research. These devices use

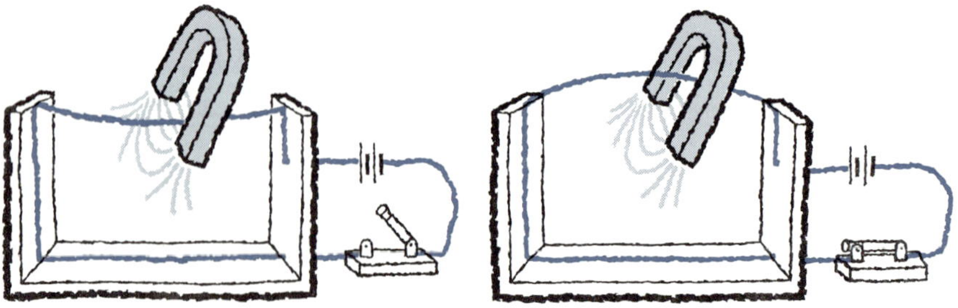

Fig. 8-22

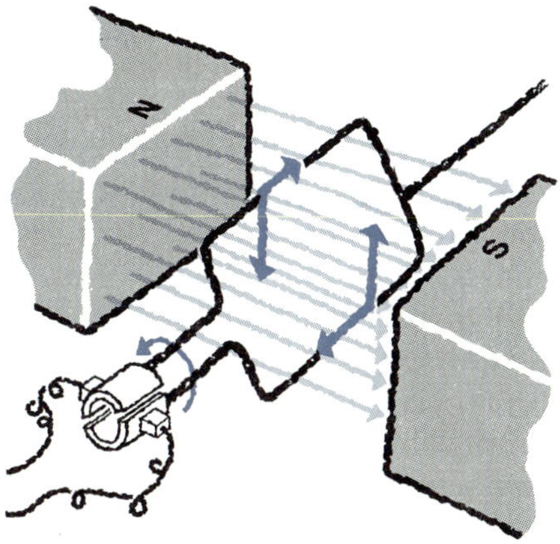

Fig. 8-23

magnetic fields to keep the particles moving in a circle or close to it so that electric fields may be used repeatedly to increase the particles' energy.

**Directions and Magnitudes of Magnetic Forces and Fields*

Figure 8-25*a* illustrates a "right-hand rule" for determining the direction of the magnetic field around an electric current or a moving charge. If the thumb of the right hand points in the direction of the current, then the bent fingers of the right hand point in the direction of the magnetic field. The direction of the current is the direction in which (positive) charge is moved. If the current is caused by the

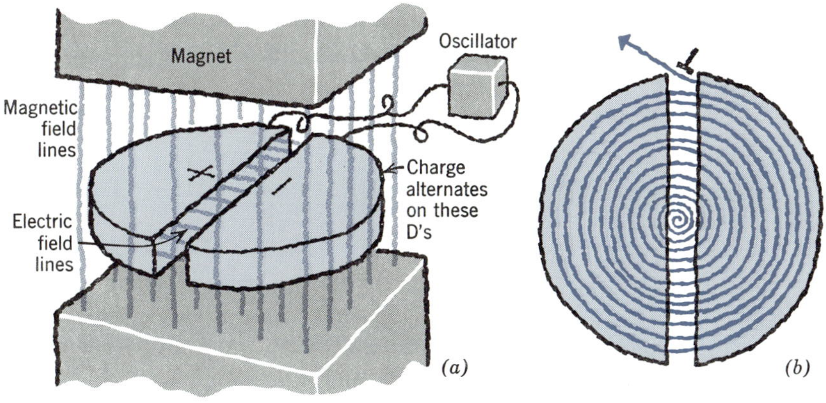

Fig. 8-24

motion of negative charges, such as electrons, then the direction of the current is opposite to the direction of motion of the negative charges (see Fig. 8-25*b* and *c*).

The direction of the field in a current loop can also be found by this rule. However, we can use a different, more convenient, rule for this special case (see Fig. 8-26.) The direction of the current in the loop is indicated by the fingers of the right hand, and the direction of the magnetic field lines through the middle of the loop is in the direction of the thumb. The north magnetic pole of this current loop is in the direction of the thumb.

The next rule tells us the direction of the force on a moving charge in a magnetic field. We shall consider a moving positive charge; the force on a moving negative charge will be in the opposite direction. Figure 8-27 shows how the directions are related. We point the thumb of the right hand in the direction of motion of the positive charge. We point the first finger in the direction of the magnetic field lines. Then the middle finger will be in the direction of the force on the charge. For a negative charge, one can use the left hand.

The simplest kind of current would seem to be a single charge in motion. However, the magnetic field around such a charge is not very simple. The direction of the magnetic field lines around a moving charge is shown in Figures 8-25*a* and 8-25*b*. The magnetic field lines are in planes perpendicular to the direction of motion.

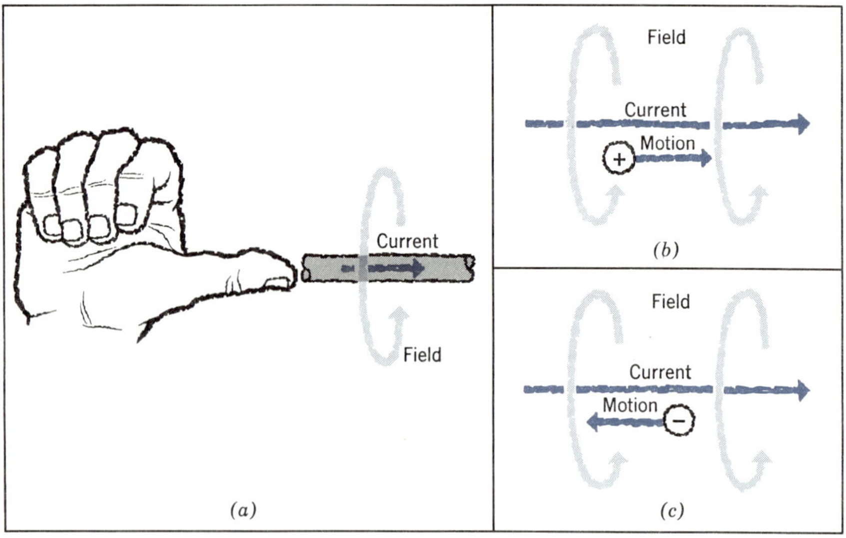

Fig. 8-25

They form complete circles whose centers are on the line of motion of the particle. The magnetic field lines of a moving charge are perpendicular to the electric field lines of the same charge. Like the magnitude of the electric field of a single charge the magnitude of the magnetic field of a moving charge falls off with distance in such a way that the field is inversely proportional to the square of the distance from the charge. Unlike the electric field of a stationary charge, however, the magnetic field at a given point of space caused by a moving charge depends on both the distance and direction of the point from the charge. The magnetic field is essentially zero in the line of motion directly behind or in front of the moving charge.

The magnitude, B, of the magnetic field caused by the current in a long wire is given by a simple equation

$$B = \frac{2K_e I}{c^2 r} \tag{3}$$

(See Fig. 8-25*a*.) The constant K_e in this equation is the same constant that appears in the law of electric force (Equation 1); c is the speed of light, 2.99793×10^8 meters per second; I is the current in the wire; and r is the distance from the wire to the point at which B is the magnitude of the magnetic field. This equation technically holds for a straight wire of infinite length, but it is very accurate for long wires, and even curved wires at points close to the wire. It gives the magnetic field outside (not inside) the wire.

The magnitude, F, of the force on a charge moving in a magnetic field perpendicular to the direction of the field is given by the equation

$$F = qsB \tag{4}$$

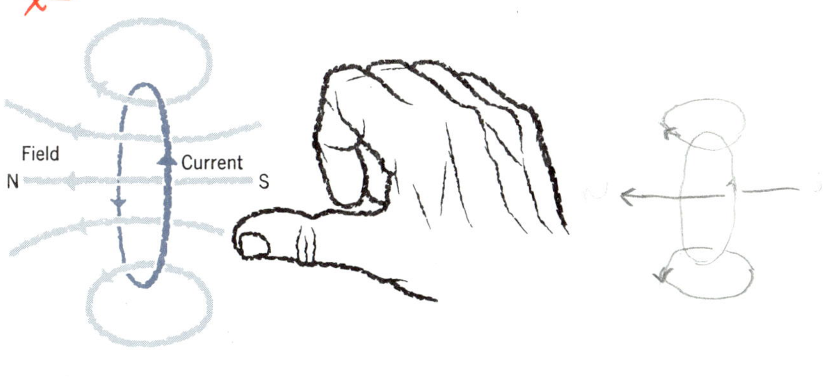

Fig. 8-26

In this equation, the force is exerted on a charge, q, moving with a speed, s, in a magnetic field of magnitude B (see Fig. 8-21*b*). If the direction of motion is not perpendicular to the field, the equation can still be used if s represents the component of the velocity of the particle that is perpendicular to the direction of the field.

INDUCED ELECTRIC CURRENTS

The Law of Induced Currents

We have seen that an electric current produces a magnetic field. The English physicist Michael Faraday (1791–1867) and his contemporaries thought that a magnetic field in turn ought to be able to produce an electric current in a conductor. Various scientists tried putting various coils in various positions in various magnetic fields. Faraday, convinced of the symmetry of nature, finally discovered that he could get such a current but only when the magnetic field was changing or when the conductor was moving in certain ways in the field. These currents are called *induced* currents.

Figure 8-28 shows one way in which the motion of a conductor in a magnetic field can cause an electric current to flow. The wire from the coil is hooked up to the light bulb so that any current in the coil must go through the light bulb also. A magnet is placed in the coil and then quickly pulled out. When the magnet and the coil are

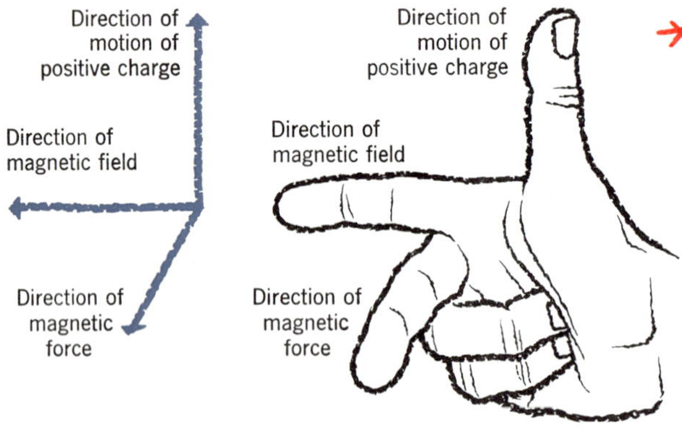

Fig. 8-27

moving relative to each other, the light glows, showing that current is flowing in the coil. When the relative motion stops, the light goes out. The faster the relative motion, the brighter the light, showing that the quicker the motion, the larger the current induced in the coil. Whenever the total number of magnetic field lines through a coil is changing, a current is induced in the coil. The rate at which this change takes place is proportional to the current induced.

The Electric Generator

Now that we have discussed induced currents, we can explain how an electric generator works. Figure 8-29 is a diagram of such a device. As the water wheel turns, the coil moves in the magnetic field. This changes the total number of magnetic field lines through the coil. As long as the coil turns, the field through the coil is changing. This change generates an induced current in the coil, which flows out through the terminals and causes the bulb to light. This is a simple device for obtaining an electric current from mechanical motion. It is very similar to the electric motor, in which electric current generates mechanical motion. These are opposite processes, but the devices which accomplish them are very similar.

We know that electrical machinery is capable of doing work. Therefore there is energy associated with electric currents. The motor and the generator are devices that can be used for transforming electric energy into some other forms or obtaining it from other forms. The energy is transformed by means of work, either done on the generator or done by the motor.

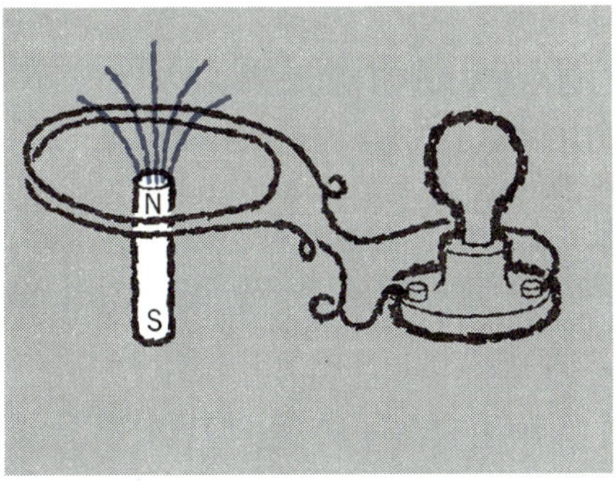

(a)

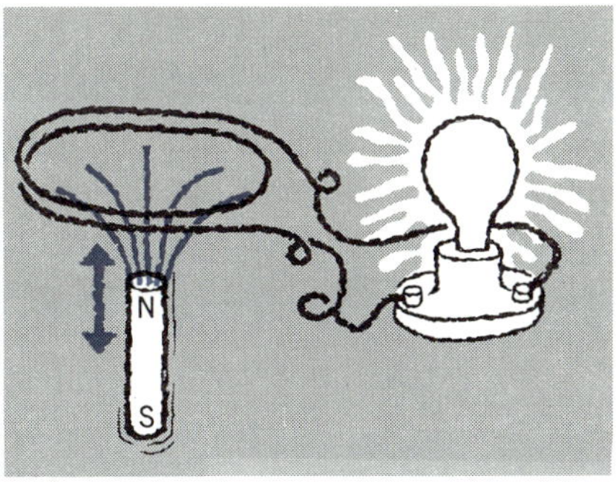

(b)

Fig. 8-28

*The Direction of Induced Currents

An induced current has a magnetic field of its own. This field opposes whatever change induced the current in the first place. If it was the motion of a conductor that caused the change, the induced current will be in such a direction that the force on it opposes the motion. This enables us to state a general law that determines the direction of induced currents.

The direction of the induced current is such that this current opposes the change that caused it.

THE ELECTROMAGNETIC INTERACTION

Although we have discussed electric and magnetic forces separately in this chapter, we have seen that these two kinds of force are very intimately associated with each other. In fact, together they account for that interaction between two charged objects which is due to their charge. Therefore, instead of considering the forces separately, we should properly speak of the *electromagnetic interaction*, which includes them both. If the two charged objects are at rest, then the magnetic forces are zero, and only the electric forces need be considered. If the charged objects are in motion, however, then both electric and magnetic forces should be considered.

Fig. 8-29

Since the electric and magnetic forces are intimately related, their fields are also intimately related, and we should properly speak of the *electromagnetic field*. A charge at rest gives rise to an electromagnetic field comprised of an electric field alone. A charge in motion gives rise to an electromagnetic field which includes both an electric and a magnetic field. According to the special principle of relativity, a charge can be considered either at rest or in motion, depending on the reference frame, despite the fact that the fields are different in the two cases. The two parts of the electromagnetic field are very intimately associated indeed.

Thus far we have introduced two interactions, the gravitational interaction and the electromagnetic interaction, and their fields. These two interactions account for nearly all the forces we are ordinarily aware of in daily life. The gravitational interaction keeps us on the earth. Nearly all other forces we are aware of arise (directly or indirectly) from the electromagnetic interaction. We know, of course, that much of our machinery and instrumentation is electrical. However, even the ordinary pulls and pushes of one object against another involve electrical forces. The atoms are held together by the electric attraction between the positive protons and the negative electrons. An understanding of the behavior of these forces on the atomic level, however, is impossible with only the laws discussed in this chapter. The principles of quantum physics, which are discussed in Chapter 11, will help us to understand better how the electric forces hold the atom together.

Summary

I. FUNDAMENTAL LAWS AND PRINCIPLES

A. THE LAW OF ELECTRIC FORCE. The magnitude is given by the formula $f = K_e q_1 q_2 / d^2$. The direction is along the line between the two charges, and is given by the rule that like charges repel and unlike charges attract. The electric force and the magnetic forces mentioned below comprise the electromagnetic interaction, the second fundamental interaction we have discussed.

B. THERE ARE NO FREE MAGNETIC POLES

C. MAGNETIC FORCES EXIST BETWEEN CHARGES IN MOTION. We do not give the exact law but rather discuss it in terms of the origin and action of magnetic fields.

1. The magnetic field produced by moving charges.

a. Straight current.

b. Current loop.

2. The magnetic force on moving charges.

D. THE LAW OF INDUCED ELECTRIC CURRENTS

E. CONSERVATION OF CHARGE

II. QUANTITIES.

A. FUNDAMENTAL QUANTITIES

1. Charge, a scalar. Two kinds, positive and negative. The unit of charge is the coulomb.

B. DERIVED QUANTITIES

1. Current. The amount of charge that flows by per second. The unit is the ampere, which is one coulomb per second.

*2. Voltage. Electric potential energy per charge, measured in volts, or joules per coulomb.

III. FUNDAMENTAL PARTICLES

A. ELECTRON. Negatively charged, 1.6×10^{-19} coulomb. Mass 9.1×10^{-31} kilogram.

B. PROTON. Positively charged, 1.6×10^{-19} coulomb. Mass 1.7×10^{-27} kilogram.

IV. FUNDAMENTAL CONSTANTS

A. FUNDAMENTAL CHARGE. the amount of charge on the proton: $e = 1.6 \times 10^{-19}$ coulomb.

B. MASS OF THE PROTON: $m_p = 1.7 \times 10^{-27}$ kilogram.

C. MASS OF THE ELECTRON: $m_e = 9.1 \times 10^{-31}$ kilogram.

D. THE ELECTRIC FORCE CONSTANT: $K_e = 9.0 \times 10^9$ newton meter^2/coulomb^2.

V. FIELDS

A. THE GRAVITATIONAL FIELD

B. THE ELECTROMAGNETIC FIELD: comprising the Electric Field and the Magnetic Field.

VI. APPLICATIONS

A. APPLICATIONS OF THE ELECTRIC FORCE AND FIELD

*1. Induced Charges.

*2. The Electroscope.

3. The Oil Drop Experiment.

*4. Electrical Conditions in Conductors: relating to charge and field distribution in conductors, lightning rods.

*5. The Van de Graff Generator.

B. APPLICATIONS OF MAGNETIC FORCES AND FIELDS

1. Model. Use of the magnetic forces between currents to explain the behavior of permanent magnets.
2. The Electric Motor.
3. The Cyclotron.

C. APPLICATIONS OF INDUCED CURRENTS

1. The Electric Generator.

Questions

1. How do we know there is only one kind of mass but two kinds of electricity?

***2.** Does a cotton ball wrapped in tinfoil "remember" whether or not it has touched a charged object? Explain.

3. What is a "field"? How do we know that a field exists?

4. What was measured in the oil-drop experiment? How were these results interpreted?

5. Is a wire with a current in it necessarily "charged"?

6. Is there any net charge on a permanent magnet?

7. Is there any detectable difference between the magnetic fields produced by a permanent magnet and by moving charges?

8. Can you explain the existence of a permanent magnet in terms of moving charges?

9. What is the difference between an electric field and a magnetic field?

10. What is the difference between an electric motor and an electric generator?

11. What is the difference between the electric force and the magnetic force?

12. State what happens in each of the following demonstrations and then explain it in terms of fundamental principles.

(a) A coil hooked to a light globe is pulled through a magnetic field.
(b) An electric egg beater is plugged in.
(c) A magnet is brought near a beam of electrons.
(d) A proton is released in a cyclotron.

13. A magnet can be used to distort a television picture. What can you deduce from this fact?

14. What is charge?

15. Why might magnetism in an iron bar be reduced by heating the bar?

16. Design a device that obtains electricity from the turning of a crank.

17. Design an electric motor.

*18. A negative charge is propelled toward a current-carrying wire on the opposite side of which is a positive charge. What kinds of field are present, and what sort of motion is expected of the negative charge?

*19. A current is passed through the lower wire in Figure 8-30. What effects would you expect to observe in terms of electric and magnetic fields and currents in wires and space, and what would happen to the compass needle?

20. How can objects be set in motion by using electrical and magnetic forces?

21. How may electricity be obtained?

22. You are stranded on a deserted island that has forests, streams, waterfalls, outcroppings of magnetite (a magnetic iron ore), native copper, etc. How would you go about building a modern technical industrial civilization?

*23. How does a Van de Graaff generator work?

24. How does a cyclotron work?

*25. If I put a piece of fur on top of a Van de Graaff generator while it is being charged, what will happen? Will it be more or less difficult to charge the machine? (*Hint.* Consider the fur as a conductor in these high fields.)

*26. What would happen to an electroscope if the experiment described on pp. 321–322 were repeated using a negatively charged rod? Describe what would happen to the charge on the various parts of the electroscope at each step.

27. How do we know that the fur used to rub a hard rubber rod has acquired a positive charge?

28. What is the electrical force between an electron and a proton one meter away?

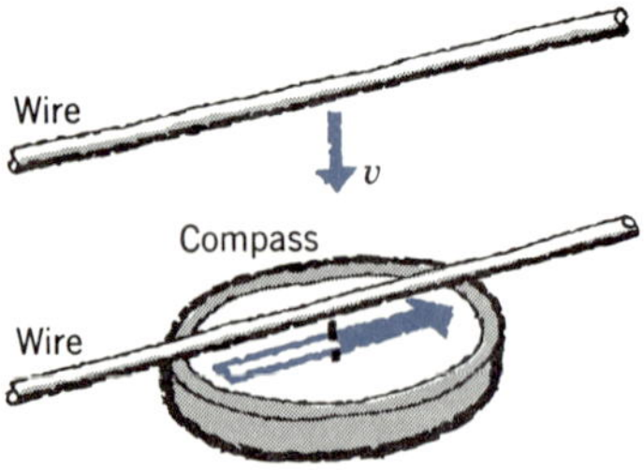

Fig. 8-30

29. What is the gravitational force between an electron and a proton one meter away?

30. A current of two amperes is flowing in a wire. How many electrons pass a given place in the wire each second?

31. Can a charge of 2.9×10^{-19} coulomb exist?

32. Could mass be said to be gravitational charge? Explain.

33. What are the sources of the gravitational and electromagnetic fields?

34. List and explain all the interconnections you can think of to justify the grouping together of electric and magnetic forces into the electromagnetic interaction and of electric and magnetic fields into the electromagnetic field.

35. Is the north magnetic pole of the earth near the North Pole?

36. Why is a core of magnetic metal used in electromagnets?

37. If the aurora were caused by charged particles from the sun, why would they be found near the North and South Poles?

38. We have three identical iron rods. Two of them are magnetized, with poles at the ends. How would you find out which one was not magnetized without using any other objects?

39. Answer question 38 for the case in which only one of the rods is magnetized.

More Difficult Questions and Problems

1. In what sense is the electric field like a flow of water?

*2. Figure 8-31 shows a curious device known as the "flying doughnut

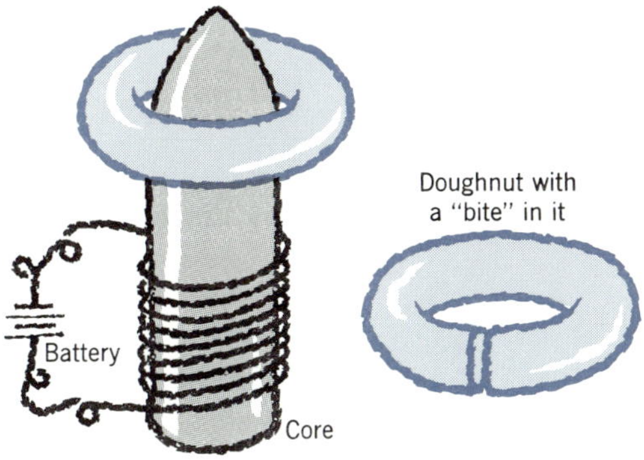

Fig. 8-31

machine." The "doughnut" is made of a good electrical conductor. When we close the switch in the coil of the electromagnet, the doughnut flies up and off. Why? If the doughnut is cut as shown it does not fly off. Explain.

*3. In Figure 8-32 I allow the paddle to swing down between the poles of the magnet. If I use a solid paddle, it stops between the poles. If I use a paddle with slots in it, it swings right through. Both paddles are made of a nonmagnetic material that is, however, a good electrical conductor. Explain the behavior of the two paddles.

4. The conducting rod shown in Figure 8-33 is pushed between the poles of a magnet. What forces act on the electrons in the rod?

*5. How do we know that the field just outside a conductor is perpendicular to its surface?

*6. Show that the rule for the direction of induced currents is necessary in order to conserve energy. (Assume that it does not hold, and see what happens.)

*7. Using the information about the forces on moving charges in magnetic fields, show that currents in the same direction attract each other and that currents in opposite directions repel each other.

*8. Show that the forces between currents mentioned in Problem 7 account for the attractions and repulsions of magnetic poles.

*9. Two positive charges are moving in straight lines parallel to each other. What kinds of force are acting on them? Show that no matter how fast they are moving, the forces between them which attract them to each other can never be strong enough to overcome the electric repulsion between them.

10. Electric and magnetic fields obey a principle called the principle of superposition. According to this principle, the field at a given point due to a number of charges is equal to the sum of the fields at that point due

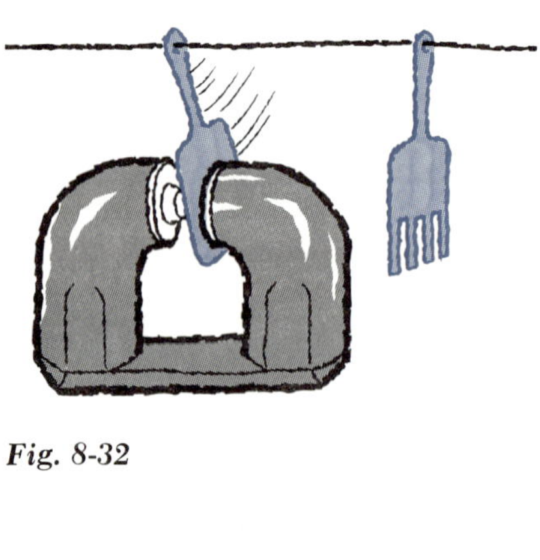

Fig. 8-32

to each charge separately. Since the field has a magnitude and a direction at each point, the fields are to be added by vectors. Show that a field of the form shown in Figure 8-4 around two opposite charges (equal in magnitude) separated by a short distance can be obtained by adding together the fields for each charge separately. What would the field look like if the charges were equal?

11. What are the units of the electric field?

*12. What are the units of the magnetic field?

*13. What is the radius of the circle that describes the path of an electron moving with a speed s perpendicular to a magnetic field of magnitude B?

*14. Using Equation 4 find the force on a short length d of a wire carrying a current I in a magnetic field.

*15. Using Equations 3 and 4 find the force between two wires of length L separated by a distance d and carrying currents i and I. (Assume that d is much less than L.)

16. Explain why field lines can never cross.

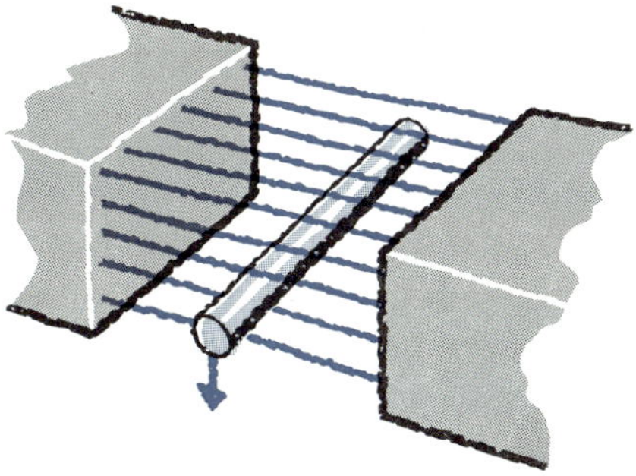

Fig. 8-33

9

WAVES

Nearly everyone is familiar with waves in water and in other materials. If a rock is tossed into still water, the rock creates a disturbance at the point where it hits the water. However, this disturbance does not remain at that point but spreads out in waves. In this chapter we shall discuss the general behavior of waves, both in matter and in the electromagnetic field.

It is convenient to begin by describing the waves in a long spring (see Fig. 9-1). One end of the spring is fastened to a support, and the other is held in the man's hand. When the man moves his hand up and down, he disturbs the spring. The disturbance does not remain in place but travels away from his hand to the other end of the spring and is reflected back. Of course the metal in the various parts of the spring does not move from one end of the spring to the other; it merely participates in the motion caused by the wave as the wave passes by, and then returns to rest. The wave is a disturbance, with some kind of form, which passes along the spring. In this example, the motion of each small part of the spring is mainly in a different

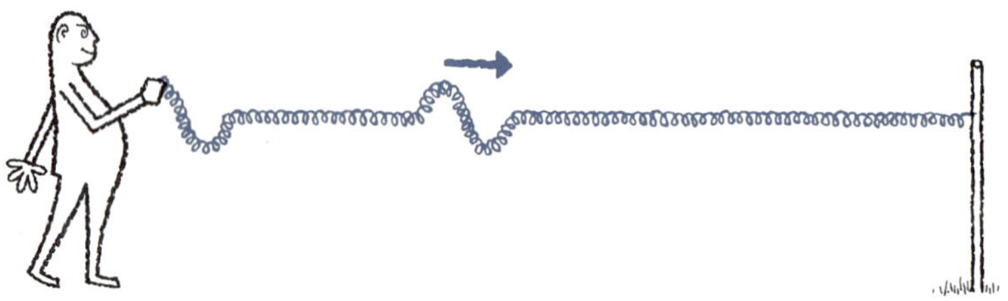

Fig. 9-1

direction from the direction of the wave. The wave moves along the spring horizontally, but each small region of the spring moves mainly up and down as the wave goes by. This behavior is similar to the behavior of water (or floating objects) when water waves go by. As a smooth wave goes by in water, floating objects bob around but remain more or less in place.

These examples illustrate the behavior of waves. First we have a medium, such as water or a spring. Second, we have some device that creates a disturbance in that medium, such as a rock striking the water, or a hand moving the spring. This disturbance may act once, sending out a pulse or two, or it may act continually, in a more or less regular manner. The source of the disturbance is usually called an *oscillator*, particularly if it is capable of acting continually. For example, we can refer to the man's hand in Figure 9-2 as an oscillator. If he moves his hand up and down continually, he will send out, not just one or two waves, but a whole series of waves along the spring. We call such a series of waves a *wave train.* The disturbance

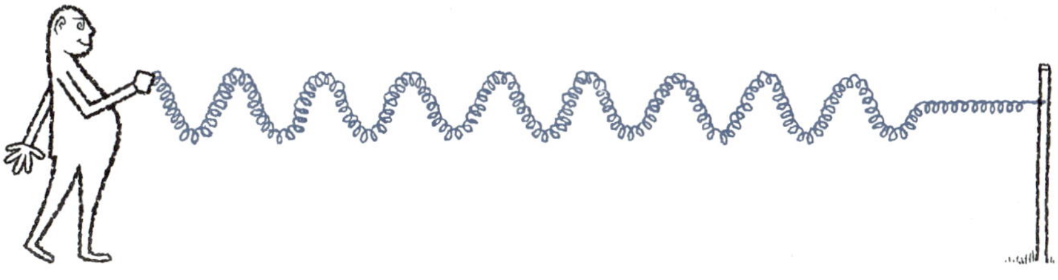

Fig. 9-2

created by the oscillator does not remain near the oscillator but moves out through the medium in wave form. The particles of matter that make up the medium move in their own neighborhood when the wave passes by, helping to support and propagate the disturbance.

The wave disturbance must carry energy with it, since it causes the matter of the medium to move wherever it goes. The wave receives energy from the oscillator and transmits this energy through the medium. This energy may be given up by the wave elsewhere or slowly dissipated by friction in the medium.

WAVE DESCRIPTION

Properties of Waves

The form and motion of simple waves can be described in terms of their speed, frequency, wavelength, and amplitude.

Each vibration, or back and forth movement, of the oscillator sends a wave down the spring. The speed at which the waves travel down the spring is the *speed*, *s*, of the wave.

The series of waves in Figure 9-3 is made up of repeated waves, each of the same form. The length of each of these repeated wave forms is called the *wavelength*, *L*. The wavelength is the distance between similar points on successive waves. Within each wavelength of these simple waves there is an upward pulse (or crest) and a downward pulse (or trough). Each complete vibration of the oscillator sends a wave, one wavelength long, down the spring.

The *frequency*, *f*, of the series of waves is the number of complete waves that pass a fixed point in each second. The oscillator also has a frequency, which is the number of complete vibrations the oscillator

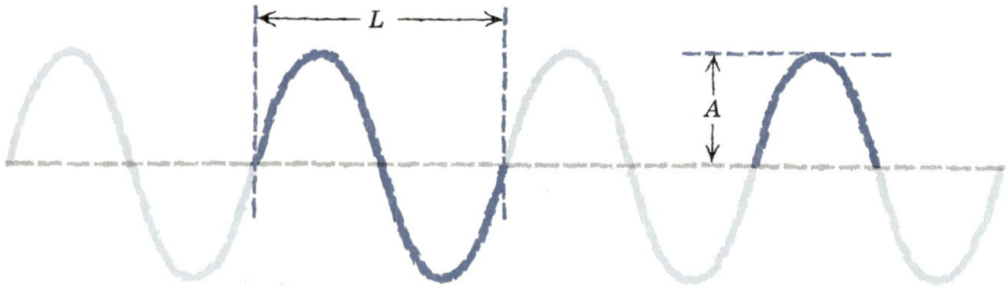

Fig. 9-3

makes in each second. If the location of the oscillator does not change, this frequency is the same as the frequency of the wave.

There is a very simple relationship that holds among the frequency, speed, and wavelength of a series of waves, illustrated in Figure 9-4. Since the frequency is equal to the number of waves that pass a certain point during each second, it is also equal to the number of wavelengths that a wave travels along the spring during each second. For example, in Figure 9-4 the frequency of the wave is three vibrations per second. This means that three complete waves pass point A during one second. If the wave train starts at A, as in Figure 9-4*a*, one second later three waves will have gone by, as in Figure 9-4*b*. During this second, the beginning of the wave train has moved three wavelengths to the right. Thus, as we have stated, the number of wavelengths the wave has traveled in one second is equal to the number of vibrations per second, which is the frequency. Therefore the frequency gives us the speed of the wave in wavelengths per second. To obtain the speed in meters per second, we simply multiply the frequency by the number of meters in the length of a wave, which is just the wavelength itself:

$$\text{speed} = \text{frequency times wavelength} \tag{1}$$

In symbols, this is

$$s = fL \tag{2}$$

The simple waves we have discussed in this section are periodic—that is, the wave form repeats itself. It is, of course, possible for the wave form to be very irregular. In such cases the idea of a single frequency and a single wavelength does not apply.

The *amplitude*, A, of a wave is the maximum distance through which the material of the medium is displaced from its rest position when the wave passes by. Figure 9-3 shows the amplitude of a wave on the spring. The maximum distance the spring rises as each part of the wave goes by, the high point of the wave, gives the amplitude. Of course we could just as well obtain the amplitude by taking the

Fig. 9-4

maximum distance that each part of the spring drops. The energy in a wave is generally proportional to the square of the amplitude.

Types of Waves

There are two types of waves that we wish to describe: *transverse* and *longitudinal.* We have already described one kind of transverse wave, the wave in the spring in Figures 9-1 and 9-2. We have seen that the individual parts of the spring in Figure 9-1 move up and down as the wave goes by horizontally; such waves are transverse. The motion of the medium as a transverse wave passes is at right angles to the motion of the wave.

When a longitudinal wave passes, the motion of the medium is along the same line of direction that the wave travels. Figure 9-5 shows a longitudinal wave, using the same spring as before. This time we start the wave by pushing and pulling on the spring along its own direction. The push puts a small region of compression into the spring, which continues as a wave down the spring. The pull stretches a small region, and a region of "stretch" travels down the spring. In this type of wave, the motion of the metal of the spring is back and forth along the direction of the spring. Thus the wave and the material of the spring move along the same line, and we have a longitudinal wave. For the longitudinal wave the amplitude is still given by the maximum distance each part of the spring moves away from the rest position as the wave goes by. In this case, however, this distance is along the length of the spring.

Some waves, such as waves in water, are partly transverse and partly longitudinal. When a large water wave goes by, the molecules of water move not only up and down but also back and forth along the direction of the wave.

Wave Phenomena

There are certain characteristic kinds of behavior that are common to nearly all waves. Since sound and light exhibit this behavior,

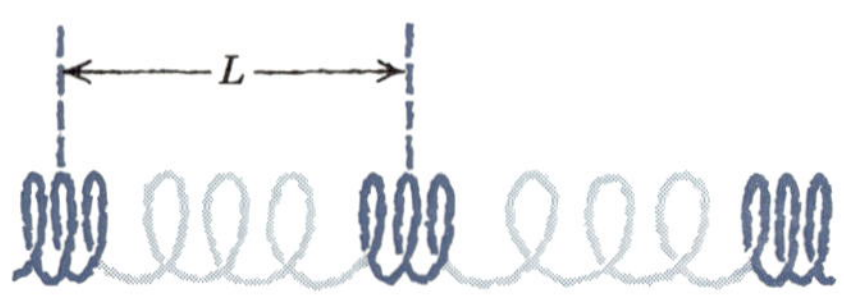

Fig. 9-5

scientists have concluded that sound is transmitted as waves in the air and other material, and that light is transmitted as waves in the electromagnetic field. Light and sound will be discussed in more detail later in this chapter; here we give some examples of the characteristic behavior of waves.

REFLECTION. When a wave reaches the edge of the transmitting medium, or indeed, any abrupt change in the medium, part or all of the wave is reflected. We have already seen that a wave on a spring is reflected back from the end of the spring. Reflected sound waves are heard as echoes. Reflected light waves are responsible for one's reflection in a mirror.

REFRACTION. When waves travel from one region into another in which their speed is different, refraction occurs. This involves the slowing down (or speeding up) of the wave and generally a change in its direction of travel. If there is a definite boundary between the two regions, the direction of travel will change at the boundary unless the direction of the wave is perpendicular to the boundary. Very often, in such a case, there will be a wave reflected from the boundary and also a wave refracted at the boundary entering the new region. One example of the refraction of light waves is shown in Figure 9-6. Light waves from the partly submerged pencil are bent, or refracted, as they leave the water. This causes the pencil itself to appear bent.

DIFFRACTION. A very important property of waves is that they bend around corners. This is obvious in the case of sound, since we can hear around corners (see Fig. 9-7). This property is illustrated for

Fig. 9-6

Fig. 9-7

water waves in Figure 9-8. The waves bend to some extent around the edges of the barrier. Notice that this diffraction (or bending around corners) is less obvious in Figure 9-8*a*, where the hole in the barrier is much larger than the wavelength of the wave. Where the hole is not much bigger than the wavelength, as in Figures 9-8*b* and 9-8*c*, diffraction effects become very important. We can make diffraction effects more important either by increasing the wave-length, as in Figure 9-8*c*, or by decreasing the size of the aperture through which the waves move, as in Figure 9-8*b*.

INTERFERENCE. What happens when more than one wave train is moving through the same medium? To investigate this question, we set up in water two wave trains that move through intersecting channels (see Fig. 9-9). In the center region, where the waves are together, the wave pattern appears to be disrupted. However, the waves reappear in their original form after passing through this

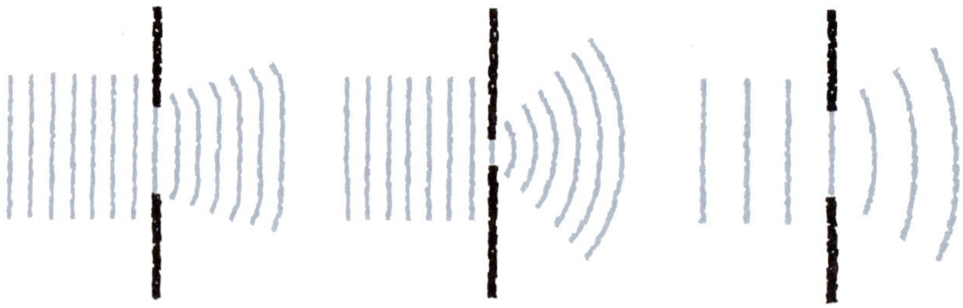

(*a*) (*b*) (*c*)

Fig. 9-8

region. We conclude that waves can travel right through each other without permanent distortion.

What is happening in the region where the waves are passing through each other? Even though both wave trains are present, we cannot identify either wave train separately. All we observe is the net cooperative effect of the combined, or superimposed, waves. We call the wave pattern in this region the *interference pattern* set up by the two wave trains. (It would be just as appropriate to call this a *cooperative* pattern.)

To see what happens when two waves come together to form an interference pattern, let us look at Figure 9-10*a*. This figure shows two pulses moving toward each other on a spring. When the pulses come together, they momentarily add up to create a large pulse; then each pulse reappears in its original form on the other side of this region. We call this pattern *constructive interference*, because the two small pulses build up to (construct) one large one. Similarly, if two downward pulses come together they result in one large downward pulse; this is also called constructive interference.

Figure 9-10*b* shows what happens when an upward pulse and a downward pulse come together. Where the pulses meet, they momentarily cancel each other out. If the pulses were not of the same size, they would be only partly cancelled out. This type of interference is called *destructive interference*.

We can summarize all this by saying that at each point where two waves come together, the resulting disturbance is found simply by adding the disturbance caused by one wave to that caused by the other. In the region where the two wave trains of Figure 9-9 come

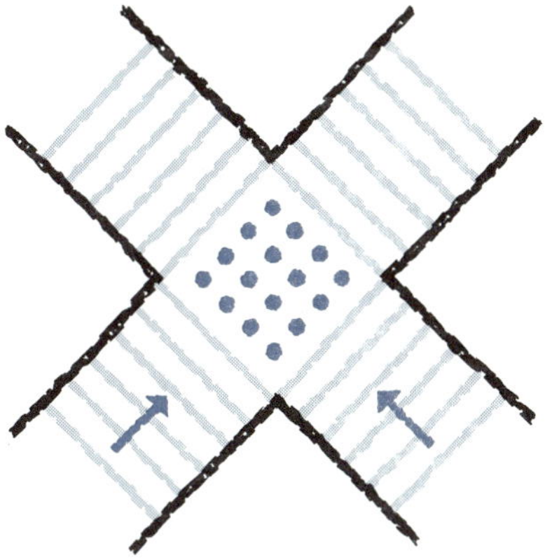

Fig. 9-9

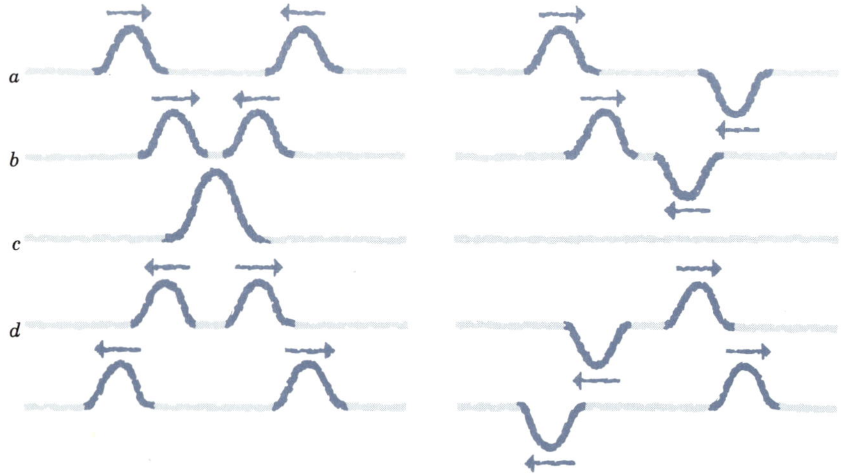

(*a*) Constructive interference (*b*) Destructive interference

Fig. 9-10

together, therefore, the interference pattern is caused by a combination of constructive and destructive interferences occurring at different points.

A very important type of interference pattern is produced by two oscillators that are sending out simple waves of the same frequency and the same amplitude (see Fig. 9-11a and b). In the vicinity of certain lines, as indicated in the figure, there is destructive interference at all times. In these positions the waves from the two oscillators always cancel each other out. Between these lines the waves coming from the two oscillators interfere to produce waves moving outward.

Sound waves

The Nature of Sound

Sound is transmitted by waves through solids, liquids, and gases. In this section we are interested in sound waves moving through the air. These are longitudinal waves that are started in much the same way as the longitudinal waves in the spring illustrated in Figure 9-5. Figure 9-12 shows how a vibrating tuning fork can cause sound waves in air. As the prong of the fork moves, it momentarily com-

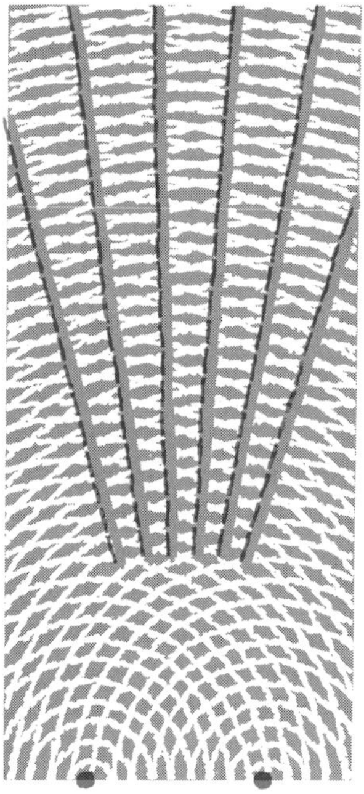

(a)

Fig. 9-11

presses the air in front of it, causing a small region of high pressure. This compression causes a small pulse of high pressure to move outward in a wave. As the prong of the tuning fork moves back, it leaves a region of rarefied air behind it, causing a pulse of low pressure to move outward in a wave. In this way we get a series of pulses of compressed and rarefied air.

Sound waves move through the air at a speed of 331 meters per second when the air temperature is at the freezing point of water. This speed increases with temperature but does not change very much with air density. It is nearly the same for all audible frequencies. (If this were not true, you can imagine how an orchestra would sound at the back of an auditorium: the notes of different frequency would arrive at our ears after different time lags.)

That sound is indeed composed of pressure waves is strongly suggested by a simple experiment using the wheel illustrated in Figure 9-13. As the wheel is rotated, the jet of air blows against it and through the holes as they come by. As each hole passes in front of the jet, the air jet passing through the hole causes a pressure pulse

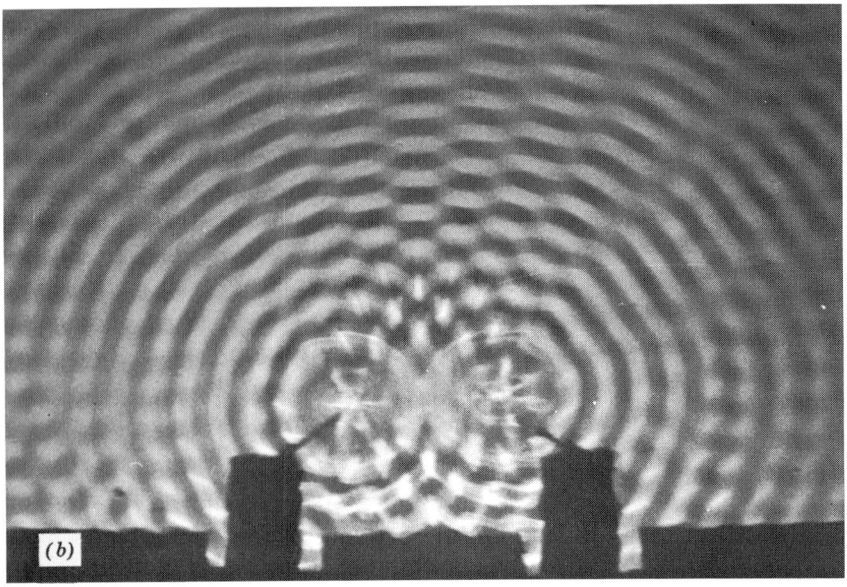

Fig. 9-11b Two-source water wave interference pattern. From PSSC PSYSICS, D.C. Heath and Company, Boston 1965. Used with permission.

in the air on the other side of the hole. If the holes are equally spaced we will obtain a wave with a definite frequency, which sounds to us like a musical note with a definite pitch. If we increase the frequency of the wave by speeding up the wheel or by spacing the holes closer together, we hear a higher pitch. The pitch of a sound wave depends on the frequency of the sound wave. The higher the frequency, the higher the pitch. If the holes are irregularly spaced on the wheel, we no longer hear a definite musical note, but what we would call noise instead.

Musical instruments are oscillators that are designed to vibrate in a regular way to produce musical notes and not noise. The tuning fork, in particular, is designed to produce a simple wave of definite frequency. Such devices as the reeds of woodwind instruments, the air columns of flutes and organ pipes, and the strings of pianos and violins are all oscillators that vibrate in a somewhat more complex manner.

Interference of Sound Waves

BEATS. When two musical tones of nearly the same pitch are sounded together, *beats* are heard. These beats may be heard as a

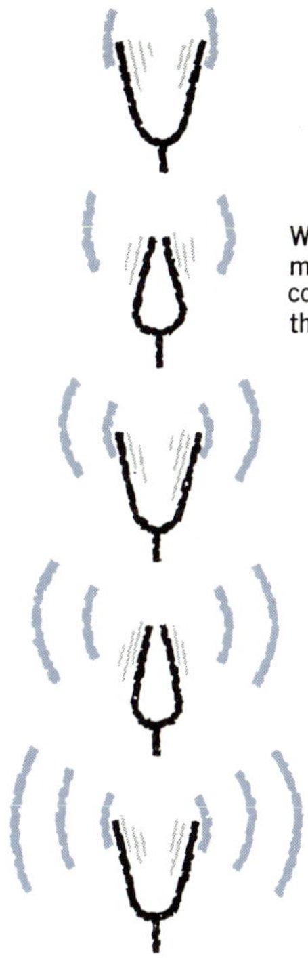

Fig. 9-12

rhythmic rise and fall in the loudness of the sound. If the two tones are far enough apart in frequency, the beats may occur in such rapid succession that they are not heard individually but as an additional musical tone. Figure 9-14 shows how these beats arise. Actually the waves are longitudinal, but we have drawn them as transverse in order to show the effect more clearly. These waves might arise from two tuning forks of slightly different frequencies that are sounded at the same time. Figure 9-14*a* shows two waves of slightly different frequencies. Their wavelengths are also slightly different, because the speed is the same for both waves. The actual wave that we hear is the sum of these two waves, as shown in Figure 9-14*b*. When the two waves interfere constructively, the tone is louder. When they interfere destructively, the tone is diminished. When the wave of Figure 9-14*b* comes by the ear, the ear hears these alternating loud and soft tones, or beats.

STANDING WAVES. The principle of wave interference helps us to understand why musical instruments (as oscillators) produce musical

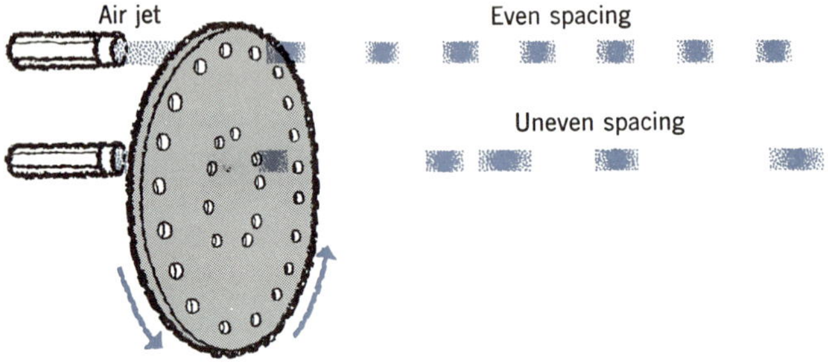

Fig. 9-13

tones instead of noise when they are plucked, struck, blown, or scraped with a bow. We shall confine our discussion to the musical tones produced by a stretched string and a metal plate, but the principles are widely applicable.

Figure 9-15*a* shows part of a stretched string of infinite length. On this string two trains of transverse waves of equal amplitude are traveling at the same speed in opposite directions with the same frequency and, of course, the same wavelength. The pattern formed when these two trains come together and interfere is illustrated in Figure 9-15*b*. The waves in this pattern appear to vibrate in place. For this reason they are called *standing waves*. At certain points, called *nodes*, the string does not move up and down at all. Between these nodes, the string vibrates up and down with the same frequency as the original waves. The nodes are seen to be one half wavelength apart.

The strings used in musical instruments are not infinite, so we shall discuss what happens when the stretched string is of some definite length, fixed at the ends, as in Figure 9-16. Since any waves

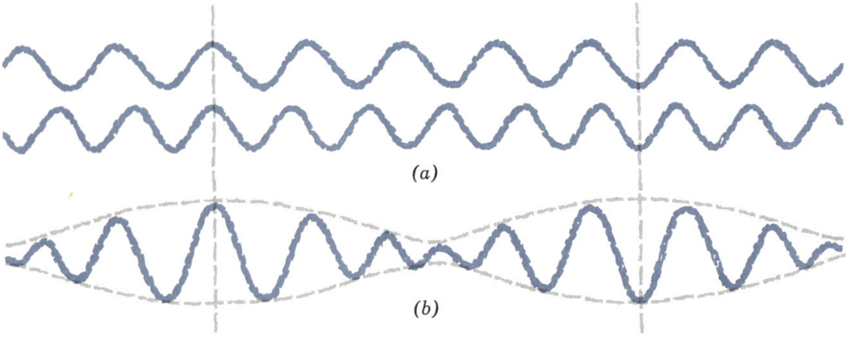

Fig. 9-14

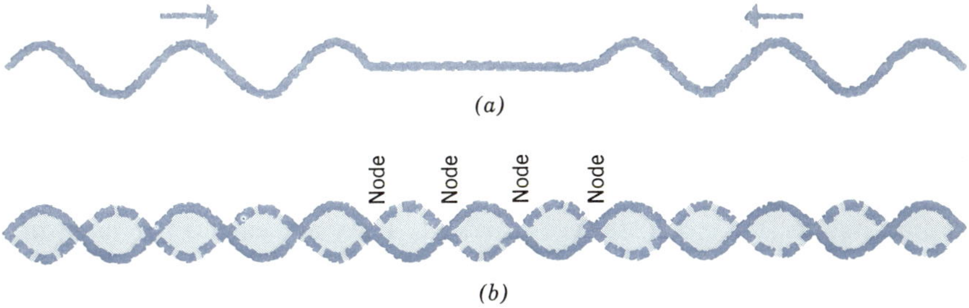

Fig. 9-15

traveling on the string are reflected at the ends, a disturbance will eventually produce waves traveling in both directions. It is possible, therefore, to produce standing waves on such a string. However, as we shall see, standing waves will not be produced for all frequencies on a given stretched string. Rather, a particular stretched string will have standing waves of certain characteristic frequencies.

In order for standing waves to occur on a stretched string, there must be a node at each end of the string, since each end is fixed and cannot move. There may or may not be nodes elsewhere on the string. Since the nodes, as we have seen, must be a half wavelength apart, they must be equally spaced along the string. Figure 9-16*a-c* shows some of the possibilities. There can be no nodes, two nodes, three nodes, indeed, any number of nodes, equally spaced along the string in addition to the two nodes at the ends of the string. Figure 9-16*d*, on the other hand, shows a wave that *cannot* occur. The wavelength of this wave is too long to form two nodes (in addition to the endpoints of the string) and too short to form just one. With such a wavelength, it is impossible to form nodes *equally* spaced along the string. Therefore, standing waves on a particular stretched string can have only certain wavelengths. Since the wavelength is related

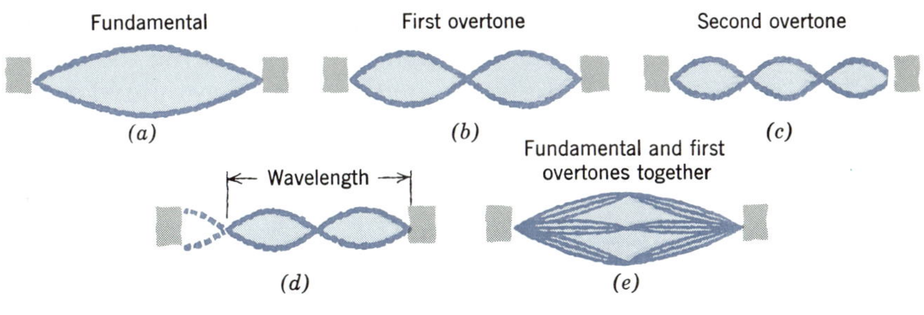

Fig. 9-16

to the frequency of the wave through Equation 2, there will be certain permitted frequencies also, corresponding to the permitted wavelengths. (Waves of different wavelength have very nearly the same speed along the string. This speed may be changed by tightening or loosening the string, as in tuning a guitar or a violin.)

These permitted frequencies are the frequencies heard when guitar strings, for example, are plucked. They are called the *natural*, or *characteristic*, frequencies of the stretched string, and they include the *fundamental* frequency and various *overtones*. The *fundamental* frequency is heard when the string vibrates as a whole, as in Figure 9-16*a*. The first *overtone* is heard when the string vibrates as in Figure 9-16*b*. The second overtone is heard when the string vibrates as in Figure 9-16*c*. In general, the number that identifies the overtone (first, second, etc.) corresponds to the number of nodes on the vibrating string (not counting the ends) that produces the overtone. For example, the seventh overtone is produced by a vibrating string with seven nodes, excluding the end nodes, and so on. The string may vibrate in such a way as to produce more than one overtone at once. For example, Figure 9-16*e* shows a string vibrating to produce both the fundamental and the first overtone at the same time.

The vibrating string is an oscillator that produces sound waves in the air. The wavelength of the standing wave on the string is not necessarily the same as the wavelength of the sound wave it produces, however, since the speed of the waves on the string is not necessarily the same as the speed of the sound wave in the air. However, the frequency of the standing wave will be the same as the frequency of the sound wave it produces, since each complete vibration of the string causes one complete wave in the air. We have already mentioned that the frequency of a wave is the same as that of the oscillator which produces it, assuming that the location of the oscillator is not being changed.

We can easily find the relation between the various natural frequencies of a stretched string of length d by first finding the wavelengths L on the string. These wavelengths can be determined easily if we recall that the distance between nodes is one half wavelength:

Fundamental $L = 2d = 1 \times (2d)$ (since d is a half wave length; see Fig. 9-16*a*)

First Overtone $L = d = 1/2 \times 2d$ (since d includes two half wavelengths; see Fig. 9-16*b*)

Second Overtone $L = (1/3)\, 2d$ (see Fig. 9-16*c*)

Third Overtone $L = (1/4)\, 2d$

And, in general, we can see that the series goes as follows: 1, 1/2, 1/3, 1/4, 1/5, 1/6, 1/7, etc. To find the frequencies from the wavelengths, we use Equation 2:

$$fL = s$$

If we divide both sides of this equation by L, we obtain

$$f = s/L \tag{3}$$

We now substitute the values of L found above to get the values of f.

Fundamental $f = s/2d$

First Overtone $f = s/d = 2(s/2d)$

Second Overtone $f = \frac{s}{\frac{1}{3}(2d)} = 3\left(\frac{s}{2d}\right)$

Third Overtone $f = \frac{s}{\frac{1}{4}(2d)} = 4\left(\frac{s}{2d}\right)$

Fourth Overtone $f = \frac{s}{\frac{1}{5}(2d)} = 5\left(\frac{s}{2d}\right)$

The first overtone, therefore, has a frequency twice that of the fundamental. The second overtone has a frequency three times that of the fundamental, etc. The frequencies thus form a series that goes 1, 2, 3, 4, 5, etc.

Standing waves can also occur in organ pipes, drum heads, and other objects. Figure 9-17 shows standing waves in a metal plate. The plate is held by a clamp in the center and is set into vibration by a violin bow. The waves caused by the bow spread out across the plate, are reflected at the edges and at the center, and combine to produce standing-wave patterns. Standing waves for two of the natural frequencies of the plate are shown in the figure. On a surface, such as this plate, the nodes are *lines* where no vibration occurs, rather than points as on a string. To help us locate the nodes, salt has been spread over the plate. It is shaken off by the vibration of the plate, except at the nodes.

The simple numerical relations among the natural frequencies of the string make it pleasing to the ear when more than one frequency sounds at once. The natural frequencies of the plate, however, are not so simply related, and it does not sound "harmonic" when more than one overtone is heard at the same time. This same harsh lack of harmony is heard in the clash of cymbals.

Circumstances often occur in which waves or other disturbances strike an oscillator, setting it in motion. Some of the energy in the waves is transferred to the oscillator, and it begins to vibrate. Energy is more readily absorbed from the wave by the oscillator if the frequency of the wave is the same as, or very close to, one of the natural frequencies of the oscillator. For example, a violin string will begin to vibrate if sound waves reach it that have the same frequency as one of its natural vibrations. Each oscillator, in this way, tends to "pick up" vibrations at certain frequencies. This effect can be observed by using two tuning forks, as shown in Figure 9-18. If one of the tuning forks is set into vibration, it will produce sound waves that will travel to the other tuning fork and start it vibrating. The effect will be hardly noticeable for forks of different pitch, but for those of the same pitch (frequency) it will be marked.

The reason an oscillator can pick up vibrations of its own natural frequencies better than vibrations of other frequencies may be indicated by using an ordinary swing (see Fig. 9-19). The swing is a pendulum, a very simple oscillator, with its own natural frequency. If a swing is set in motion, it will swing back and forth with that natural frequency. (As the total angle through which the swing is swinging changes, this natural frequency changes also; however this frequency change is so small that we can ignore it here.)

The man in Figure 9-19*a* gives the swing a push each time it comes

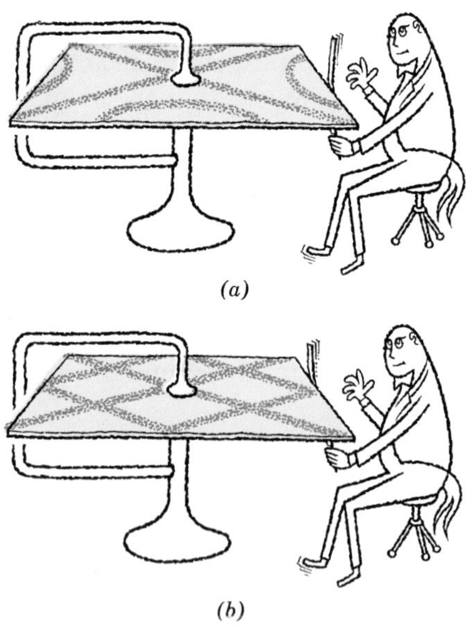

Fig. 9-17

back to his position. He pushes the swing just after it starts back on its downward path, pushing it in the direction in which it is moving. Thus, as he pushes it, he puts energy into the swing; this causes the swing to go higher and higher. The important point is that his pushes come at the same frequency as the natural frequency of the swing. Each time the swing makes one trip back and forth it receives one push. Furthermore, the pushes come at just the right time to increase the motion of the swing rather than oppose it.

Suppose, however, that the pushes were at a higher frequency than the frequency of the swing (see Fig. 9-19*b*). Then a push would sometimes increase the speed of the swing and sometimes oppose it. Sometimes the push would put energy into the swing and at other times it would take energy out. Therefore, the pushing would be about as likely to retard the motion as to increase it. The most efficient way to get energy into the swing is to time the pushes so that they occur at the same frequency as the oscillations of the swing.

Of course, if the pushes come every second swing, or every third swing, energy will also be transferred into the swing, but not so quickly. Thus the oscillator will also respond to half, one third, one fourth, and so on of its natural frequency. Energy can also be taken *out* of the swing very effectively by a series of pushes at its natural frequency. If, instead of pushing on the swing just as it is going away, the man pushes on it just as it is approaching him, energy will be removed from the swing each time.

The example of the swing illustrates the fact that an oscillator can most effectively acquire energy from an outside disturbance if the disturbance has the same frequency as one of the natural frequencies of the oscillator. It will not acquire energy so efficiently from disturbances of other frequencies. Therefore, if the disturbance contains several frequencies, the oscillator will respond most vigorously to its own natural frequency, thus appearing to select that frequency from the others.

Fig. 9-18

The Doppler Effect

The sound of a low-flying jet airplane changes in pitch as it passes overhead. This change cannot be accounted for by the change in the plane's distance from us, nor does it indicate any change in the performance of the engine. Other sources of sound, such as racing cars and train whistles, give the same effect. The pitch of the sound is higher when the source is approaching us than when it is not moving. When it is going away from us, the pitch is lower than when it is not moving. The greater the speed at which the source is moving toward or away from us, the greater this change in pitch. Similar changes of pitch occur when it is the listener and not the source that is in motion. When the source and the listener are approaching each other, the pitch is higher. When they are moving apart, the pitch is lower. This phenomenon is called the Doppler effect, after the Austrian physicist who described it.

We can explain the changes in pitch (frequency) in the Doppler effect by analogy with the situation shown in Figure 9-20. Instead of a source emitting waves which move through the air, we have a source of pies that move along a belt. The man at the left is putting pies on the belt at a certain frequency—say, one every five seconds. The man at the right is trying to remove them at the same frequency. If the first man moves to the left along the belt as he puts down the pies (Fig. 9-20*b*), the pies will be more widely spaced along the belt, and the man at the other end will receive them at a lower frequency. (The speed of the pies has not changed, of course.) If the man at the left moves in the opposite direction (Fig. 9-20*c*), still putting down

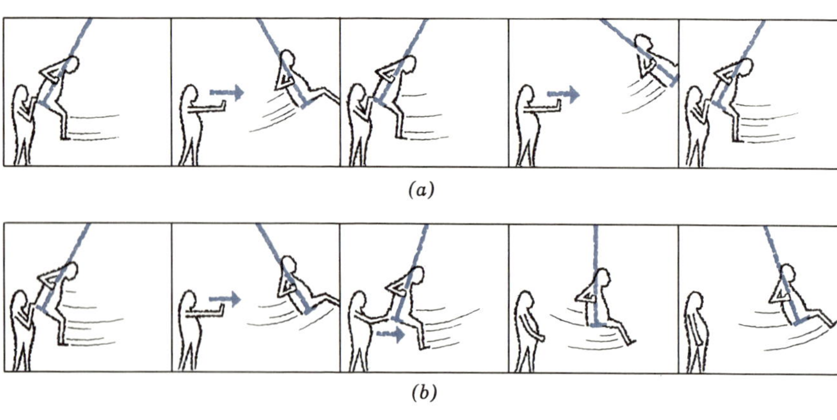

(a)

(b)

Fig. 9-19

pies at the same frequency, the pies will be closer together on the belt, and the receiver will receive them at a higher frequency. In the Doppler effect, it is waves rather than pies that are being sent out and received, but the principle is the same. The Doppler effect occurs for light waves also, although the theory is somewhat different.

Earlier in this chapter we emphasized the fact that an oscillator and the waves it emits have the same frequency if the oscillator is stationary. Because of the Doppler effect, the frequency of the waves may not be the same as that of the oscillator if the oscillator is in motion. Of course, if the person who observes the waves is moving at the same speed and in the same direction as the oscillator he will receive the waves at the same frequency as the oscillator. If there is relative motion between oscillator and the observer, however, the frequency at which the waves will be received in the reference frame fixed to the observer will be different from the rate at which they are emitted in a reference frame in which the oscillator merely oscillates and has no net or average velocity.

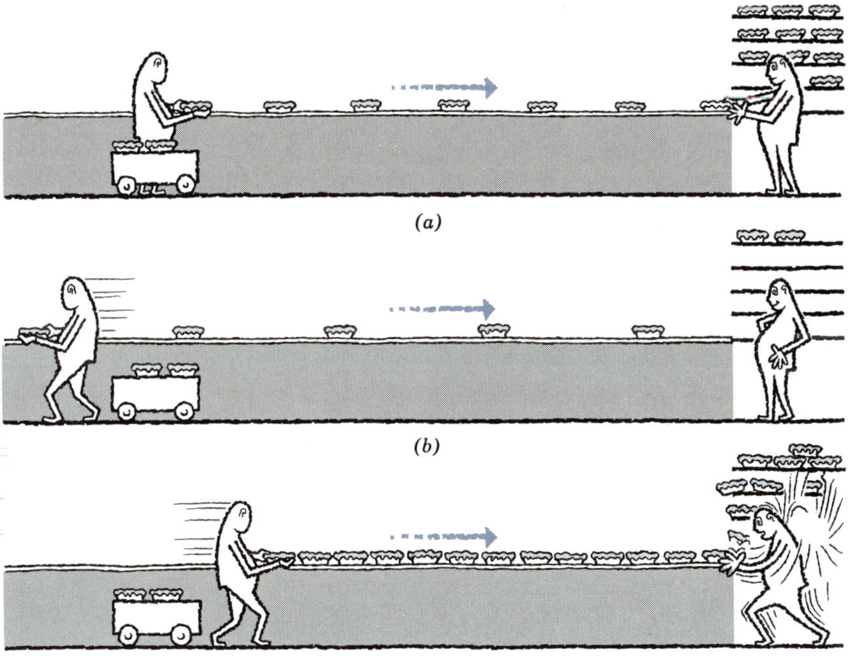

(a) *(b)* *(c)*

Fig. 9-20

Light as a wave

The nature of light has long posed a fundamental problem for physicists. Many scientists, influenced by Newton, pictured light as a beam of tiny particles. The Dutch physicist Christian Huygens, a contemporary of Newton, strongly supported the idea that light was essentially wavelike.

An Italian Jesuit, Francesco Maria Grimaldi (1618–1663), had already performed an important experiment (see Fig. 9-21). He made a tiny hole in a window shutter facing the sun and allowed the light coming through to fall on a second screen that also had a small hole in it. Beyond this second hole he placed a white paper. He found that a region larger than the second hole was illuminated on the white paper. In fact, the smaller the second hole, the larger the illuminated region on the paper. In a second experiment, Grimaldi removed the second screen and placed a small object in the beam of light coming through the hole in the first screen. He found that the shadow this object cast was smaller than the object itself. This showed that light does not travel in straight lines and cast sharp shadows but diffracts, or bends around corners.

Although Newton knew of this experiment, he did not believe that it necessarily disproved his own particle theory of light. Huygens, although he believed in the wave nature of light, could not account for diffraction because of errors in his theory, and he did not accept Grimaldi's results. Later, after interference as well as diffraction was

Fig. 9-21

observed by Thomas Young (1773–1829), the wave theory of light propagation became well established.

The Speed of Light

Galileo made an attempt to measure the speed of light in an experiment performed with lanterns from two hilltops (see Fig. 9-22). Galileo first uncovered his lantern. As soon as his assistant on the nearby hill saw the light, he uncovered his lantern. The time between the signal and the return signal was noted. Actually, as Galileo realized, light moves so fast that it was only the observers' reaction time that was measured. Nevertheless, the experiment drew attention to the idea that light has a finite speed. A Danish astronomer, Olaf Roemer, obtained a value of the speed of light from astronomical data in about 1676. More direct methods of measurement, similar to that illustrated in Figure 9-23, were used by later experimenters. As indicated in the figure, the light beam passes through one notch of the wheel and returns through another when the wheel is at rest. When the wheel is turning, the light that has been sent through the first notch may be blocked when it returns because the second notch has moved. By knowing how fast the wheel must rotate in order to block or admit the returning light, we can find the total time it takes for the light pulse to travel back and forth. From this and the total distance we can find the speed of light.

The speed of light in a vacuum is one of the fundamental constants of nature. It is usually represented by the letter c and has a value of

$$c = 2.99793 \times 10^8 \text{ meter/second}$$

This is a very high speed, and it is no wonder that Galileo failed to measure it with the primitive instruments available to him. As far as

Fig. 9-22

we know, no material object, no energy, and no information can travel faster than this speed.

Interference and Diffraction of Light

THE TWO-SLIT EXPERIMENT. About 1798 an English physician, Thomas Young, performed the experiment diagramed in Figure 9-24. Young used two pinholes in a card, but we shall use two slits that are very close together. The light passing through the slits falls on a screen. If the light beam directed at the two slits were composed of very small particles, we would expect to see just two illuminated regions where the particles of light coming through the slits struck the screen. Actually, however, the pattern observed is very similar to the interference pattern seen in Figure 9-11. As the reader will recall, this pattern was caused when water waves were emitted at the same time by two oscillators of the same frequency.

Of course, in this two-slit experiment we do not see the whole wave pattern in space, but merely the illumination caused by these waves when they strike the screen. We see only a series of light and dark strips on the screen. The brightest parts of the pattern on the screen are about in the middle of the light strips. These are the regions of maximum intensity. The darkest parts of the pattern are about in the middle of the dark strips. These are the regions of minimum intensity. The pattern is made up of a series of these maximums and minimums, one after the other. The closer the slits are to each other, the greater the distance between successive maximums and minimums. This is a general rule for this type of pattern.

Figure 9-24*b* allows us to compare the water-wave interference pattern from Figure 9-11 with the light-wave interference pattern caused by light coming through two slits. In the earlier experiment

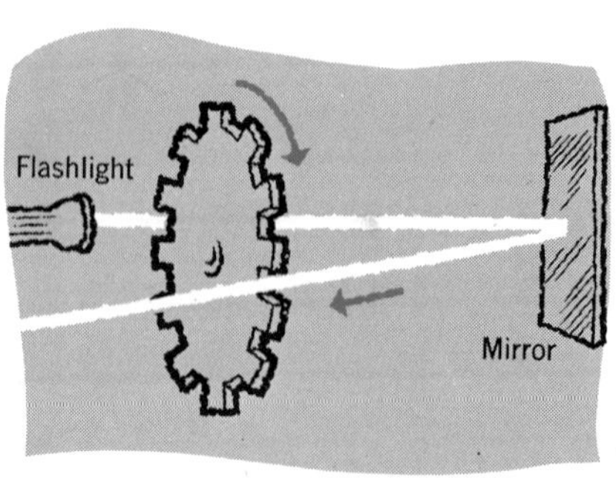

Fig. 9-23

we noticed that there were lines in the pattern along which there was always destructive interference. In Figure 9-24*b* we see that the regions of darkness occur where these lines reach the screen. Between them, the waves are flowing outward. These waves cause the bright maximums where they reach the screen. This similarity between the behavior of light and the behavior of water waves is strong evidence that light indeed travels as a wave. At least, it certainly exhibits wave like behavior. As we shall see, light of different frequencies (or wavelengths) appears to the eye as light of different colors. The wavelengths of visible light range from about 7.4×10^{-7} meter for very red light to about 4.1×10^{-7} meter for violet light.

If we know the distance between the maximums and the minimums on the screen, the distance between the slits, and the distance to the screen, we can actually find the wavelength of the light. Figure 9-25 illustrates how this can be done. The light from the two slits must interfere constructively at each maximum on the screen. We shall discuss one of the maximums nearest the center maximum on the screen.

In order to have constructive interference at this maximum, the light from one of the slits must travel a path that is a full wavelength longer than the path of the light from the other slit. This ensures that the two light waves will combine at that point on the screen to form a

(*a*)

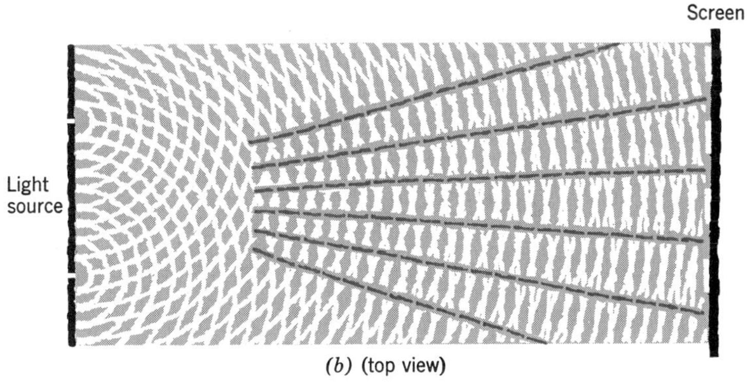

(*b*) (top view)

Fig. 9-24

maximum, and not a minimum or something in between. Therefore, the wavelength of the light can be found from the difference between these two path lengths. This difference is hard to measure since light has a very small wavelength, but it can be calculated from the other quantities mentioned above.

The fact that the wavelength of the light can be calculated from the position of the first maximum in a two-slit interference pattern means that the position of this maximum depends on the wavelength. Different wavelengths (colors) of light will have their interference maximums at different positions on the screen. This fact enables us to separate the wavelengths (colors) in a beam of light. Actually the two-slit arrangement is not very effective for this purpose; we usually use a *diffraction grating*, a device with about 15,000 slits per inch. The interference pattern of this device is much the same as that of the two-slit arrangement, except that the additional slits allow more light through and also, it turns out, produce some very narrow maximums. Since these maximums are so narrow, we can more easily separate the maximums caused by the different wavelengths. These separated maximums appear as different colors on the screen.

The diffraction grating is a very useful instrument because it enables us to determine the wavelengths (colors) present in a beam of light. Figure 9-26 illustrates the use of such a grating to analyze the light emitted by an ordinary light bulb. The central maximum of the interference pattern does not separate the colors because it is in the same position regardless of wavelength. The first maximum on each

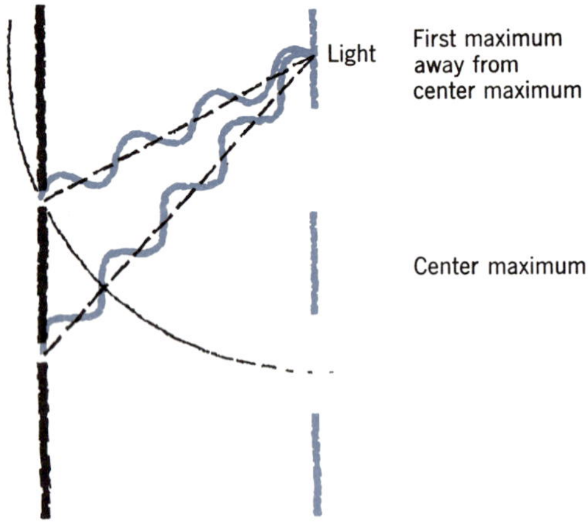

Fig. 9-25

side of the center, however, is spread out to form what looks like a rainbow. Since the position of this maximum is different for different colors, light of each color strikes the screen at a different place. This rainbow is called a *spectrum.* The fact that the spectrum of the light bulb resembles a rainbow proves that the light from the bulb contains all visible colors. In later chapters we shall have occasion to discuss light that does not contain all colors and that therefore has a spectrum in which certain colors are missing.

Our discussion here has focused on the first maximum away from the center of the interference pattern. Spectra are also obtained from the maximums farther from the center.

THE ONE-SLIT EXPERIMENT. In Figure 9-8 we saw that diffraction occurs when a water wave passes through an aperture between two barriers. We can perform a similar experiment for light. We simply use one slit instead of the two we used before. Figure 9-27 shows the diffraction pattern we obtain when we pass light through a single slit. The presence of maximums and minimums in this pattern shows that we are getting interference effects from the light that comes through different parts of the same slit. We shall not discuss these interference effects further, except to note that they do exist. As the slit is made more narrow, the light pattern on the screen retains the same form, but becomes broader. This is a general property of patterns caused by diffraction. The smaller the object or aperture causing the diffraction, the larger the actual diffraction pattern. Of course, if the aperture is so big that diffraction is unimportant, this rule does not

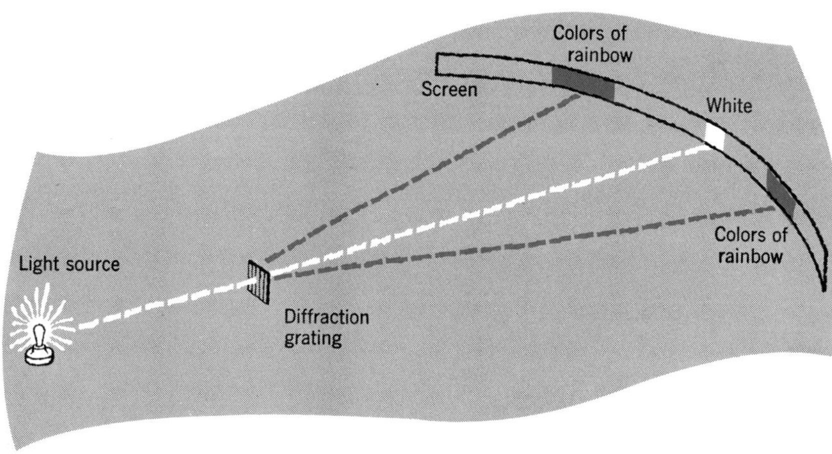

Fig. 9-26

apply. A slit six meters wide will permit a wider beam of light to pass through than a slit a half meter wide; but for these wide slits, diffraction is not the important consideration.

We have omitted one effect from our discussion of the two-slit interference pattern. Diffraction, as well as interference, occurs in the two-slit experiment as well as the one-slit experiment. In fact, waves from both slits must diffract—that is, bend around corners—in order to be present in regions away from the center of the pattern. The light from each slit separately would form its own one-slit diffraction pattern. Because the two slits are usually very close together, these two patterns are nearly on top of each other and may be represented together as the one-slit pattern in Figure 9-28. The two-slit interference pattern cannot occur except where light from both slits reaches the screen, and this will only be in the maximums of the one-slit pattern. Therefore, the actual two-slit pattern is as shown at the bottom of Figure 9-28. The two-slit interference maximums appear only within the broader regions where the one-slit pattern has its maximums.

The one-slit diffraction pattern, with its alternating light and dark regions, is similar to the diffraction patterns obtained when light is allowed to pass through small holes or around objects. For example, Figure 9-29 shows the type of pattern that Grimaldi (see Fig. 9-21) should have obtained when he let light pass in a straight beam through a small round hole. Once again we have the bright central maximum, which, as Grimaldi noticed, was actually larger than the hole through which the light passed. In addition, however, there are a series of concentric maximums and minimums, which were probably too dim for him to detect.

THE MICHELSON INTERFEROMETER. The principle of light interference is used in a very accurate instrument, the Michelson inter-

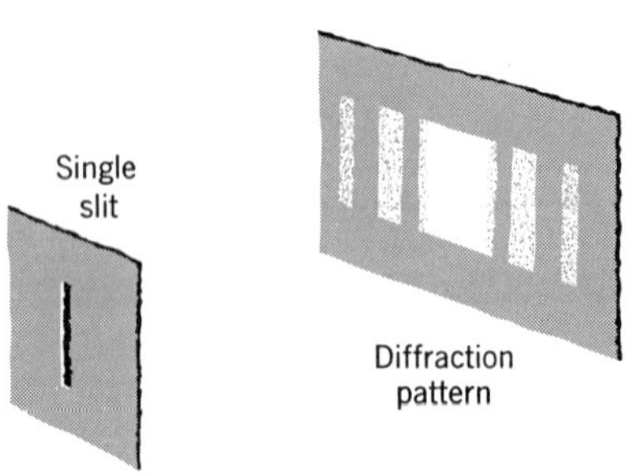

Fig. 9-27

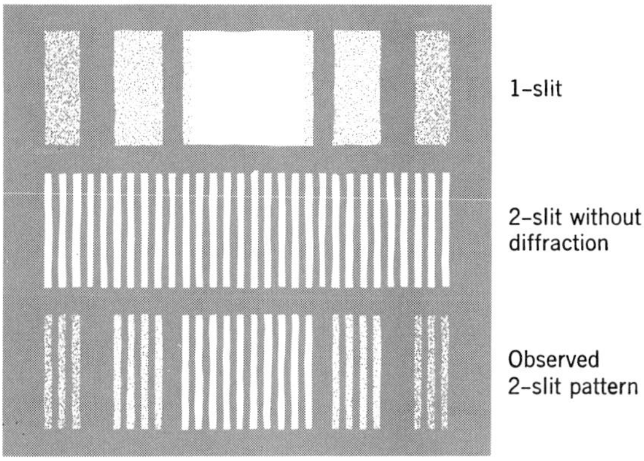

Fig. 9-28

ferometer, to measure either distance or wavelengths. The instrument is diagramed in Figure 9-30*a*. The light beam is split by a partially reflecting mirror. Half the beam continues through the mirror, and half is reflected. Both of these beams are reflected back to the partially reflecting mirror, and part of each of these reflected beams arrives at the point *A*. Thus the instrument takes a beam of light from a single source, divides it into two parts, sends each of these parts along a different path, and then reunites a portion of each of these two parts. If these separated beams have traveled the same distance at the same speed, then there will be constructive interfer-

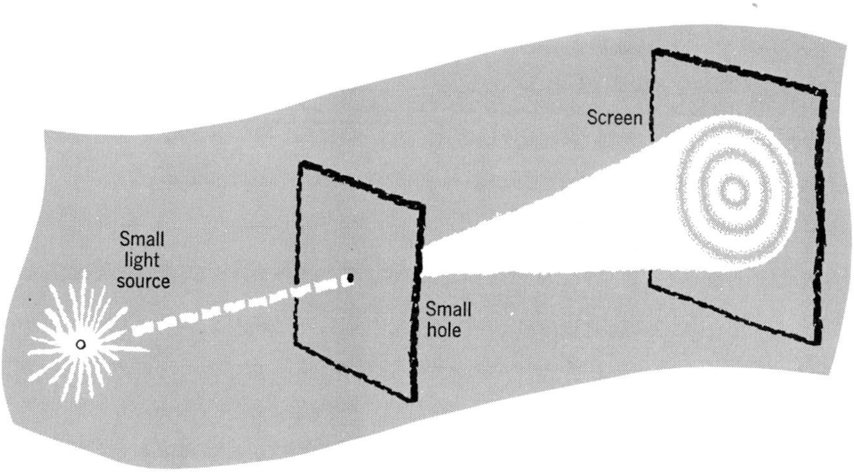

Fig. 9-29

ence when they come back together again, and a bright spot will appear in the center of the viewer located at A (see Fig. 9-30*b*). This will also happen if one wave returns a full wavelength behind the other, or even two or several whole wavelengths behind. Suppose, however, that the two beams have not traveled the same distance and that one beam therefore came back half a wavelength behind the other. Then the two waves would interfere destructively, and a dark spot would appear in the center of the viewer at A (see Fig. 9-30*b*). This will also happen if one wave returns 3/2, or 5/2, or 7/2, etc., wavelengths behind the other.

If we move the mirror at B, we change the distance that one of the beams travels. If we watch the viewer at A while slowing moving the mirror at B, we will see alternating dark and bright spots appearing one after another in the center of the viewer. By counting these changes, we can determine by how many wavelengths we have changed the path length. From this information we can relate the wavelength to the distance the mirror has been moved. If we know either of these lengths, we can then find the other.

*Polarization

One reason for believing that light waves are transverse is that they can be polarized. To explain what is meant by polarization, let

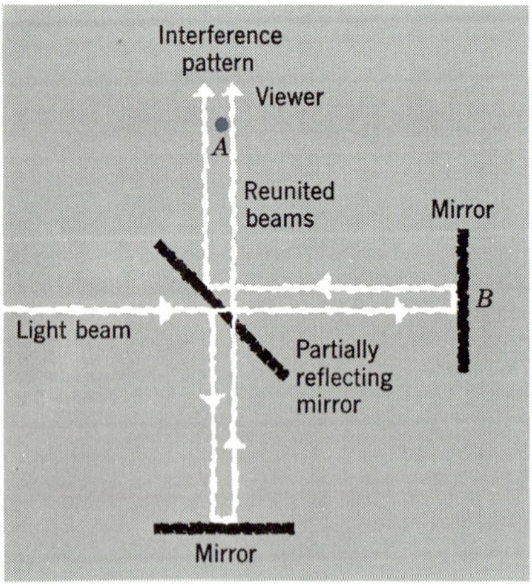

(*a*)

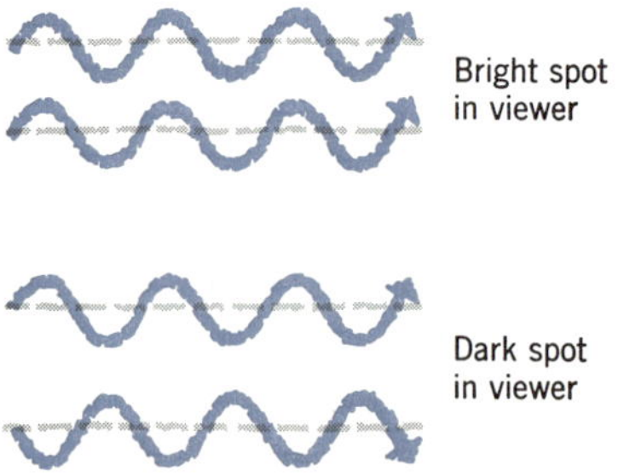

(*b*)

Fig. 9-30

us examine transverse waves more closely. Figure 9-31 shows transverse waves in two springs. In Figure 9-31*a* the spring moves up and down as the wave goes by. In Figure 9-31*b*, however, the spring moves in a horizontal direction as the wave goes by. Both waves are transverse, but they are not the same. The first one is said to be polarized in a vertical plane; the second, in a horizontal plane. Transverse waves may be polarized in other planes, since the oscillation that starts the wave can be in any number of different directions. "Ordinary" light (from sunlight and ordinary lamps) may be thought of as containing many waves, polarized in different planes.

Light polarized in only one plane can be obtained from ordinary light by using a polaroid sheet (see Fig. 9-32). The polaroid material absorbs the light that is polarized in one plane while allowing the light to pass through that is polarized in a perpendicular plane. For example, the polaroid sheet at the left in Figure 9-32*a* allows the light that is polarized in the vertical plane to pass through but stops light polarized in a horizontal plane. Thus, the light that emerges from this polaroid sheet is dimmed, since some of the light has been absorbed. This light is not dimmed to any great extent by the second polaroid sheet, because the light is already polarized in the proper plane to pass through this sheet. If this second sheet of polaroid is turned through a right angle, as shown in Figure 9-32*b*, then the light coming from the first polaroid will not pass through it. Thus, by assuming that light consists of transverse waves, we are able to explain the effect of the polaroid sheets. Purely longitudinal waves could not account for this effect.

Light can also be polarized by other methods. Light reflected from a surface is often strongly polarized. Polaroid sheets are often used in sunglasses and oriented to absorb much of this polarized reflected glare. Of course, they absorb some of the direct light also.

A rough analogy of the action of polaroid sheets in polarizing a beam of light is shown in Figure 9-33. This figure shows a wood

Fig. 9-31

gatherer striding through a forest of tall straight trees. The long brittle poles he holds in his hand are oriented in all directions. If the trees are closely spaced with no low branches, all the poles he is carrying will be broken off (absorbed) except those that are vertical. When he emerges from the forest he has, in effect, a handful of polarized poles.

Light as an Electromagnetic Wave

Electromagnetic Waves

On the basis of his formulation of the laws of electricity and magnetism, Maxwell predicted the existence of waves in the electromagnetic field. Furthermore, the predicted speed of these waves turned out to be the same as the already measured speed of light. Therefore it seemed evident that light itself must be a wave in the electromagnetic field. The theory of light became part of the theory of electricity and magnetism. Light waves, ultra-violet waves, infrared waves, X rays, radio-broadcasting waves, radar waves, and others are all

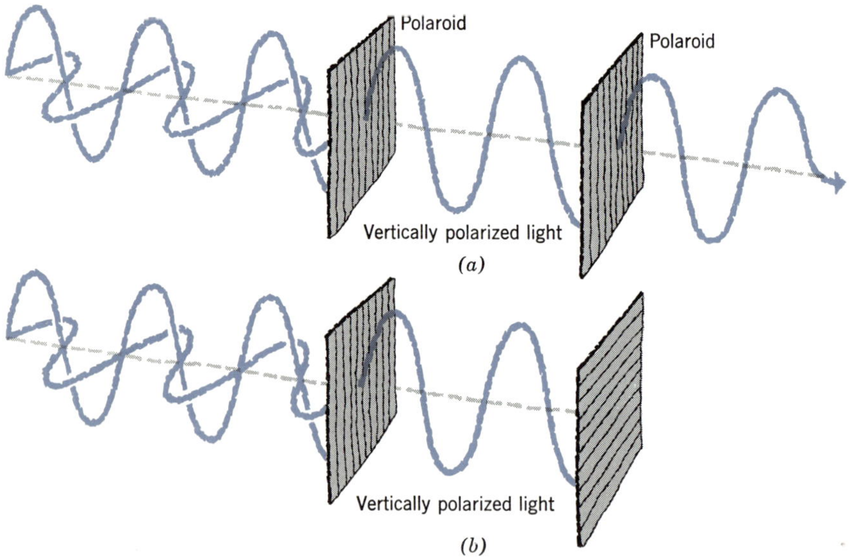

Fig. 9-32

waves of different frequencies in the electromagnetic field (see Fig. 9-34). It was indeed an impressive intellectual achievement to include so much of nature in one organized theory.

Electromagnetic waves such as light involve both the electric and the magnetic field in combination. We can see this by asking how an electromagnetic wave can get started. Suppose that we have an electron or some other charge sitting in space. We are going to use this electron as an oscillator. This means that we will make the electron move back and forth. When it moves, its electric field must move also. Thus we have created a disturbance in the electric field. Moving the electron also causes a current, which sets up a magnetic field. As the electron is accelerated, its velocity, and therefore the current changes, changing the magnetic field. Therefore, by means of this oscillating charge, we have created a disturbance in both the electric and the magnetic fields. This disturbance produces electromagnetic waves moving outward at the speed of light. All of this is in accordance with the laws of electricity and magnetism which state that an *accelerating* electric charge should radiate electromagnetic waves, but a stationary or uniformly moving charge should not. This radiation occurs in many situations. The oscillator can consist of the electrons moving back and forth in a radio-broadcasting antenna, emitting radio waves. Or the oscillator can be an electron in an atom, emitting light and other waves. Or it can be an electron accelerated by crashing into the metal target of an X-ray tube, to emit X rays.

What happens when these electromagnetic waves run into another charged particle, as, for example, another electron? The electron could be in a radio antenna, an atom, or elsewhere. The electron will feel the accelerating effects of the oscillating electric and mag-

Fig. 9-33

netic fields in the wave. Therefore it will be set into an oscillating motion by the wave it receives.

Energy in the Electromagnetic Wave

We have seen that an accelerating charge can emit an electromagnetic wave, such as light, and that this wave can accelerate another charge some distance away. This means that energy must be transmitted in the process. The natural assumption is that this energy is transmitted in the wave. There is, of course, a delay while the wave is traveling between the two charges. This delay can be appreciable in some cases; light from the sun takes eight minutes to reach the earth. Where is the energy during those eight minutes? It would seem obvious that the energy is in transit, carried in the light wave. Momentum is also carried with the wave. Light can exert pressure on an object and cause it to move.

These conclusions may seem easily acceptable. Yet we must recall that light is a wave in the field—the field that was invented to explain magnetic and electric effects. The field did not itself appear to be material. Now we find that it can support waves, and that it contains energy and momentum. This makes it seem a little more real. Both particles and fields seem to be fundamental realities.

The Ether

Many people felt that some material medium was required to support the field and its waves. After all, who had ever heard of a wave without a material medium to wave in? What is waving, if there is no medium? Therefore, the "ether" was devised to support the field and its waves. This ether, of course, has nothing to do with the ether of medical practice; rather, it was supposed to be a material

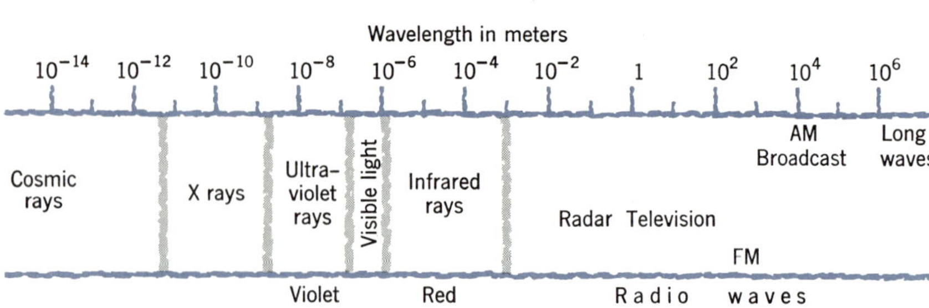

Fig. 9-34

medium which filled all space. The planets had to move through it without appreciable resistance. Furthermore, it had to transport transverse waves of various frequencies. The ether could hardly be a normal liquid or gas and allow transverse waves to pass through it. In fact, even steel is not rigid enough to support a transverse wave moving as fast as light. Therefore, the ether, although apparently necessary to support electromagnetic waves, presented several difficulties. We shall consider some of them in Chapter 10.

Summary

I. THE NATURE OF WAVES
   A. PROPERTIES OF WAVES
      1. Wavelength.
      2. Frequency.
      3. Speed: $fL = s$
      4. Amplitude.
   B. TYPES OF WAVES
      1. Transverse.
      2. Longitudinal.

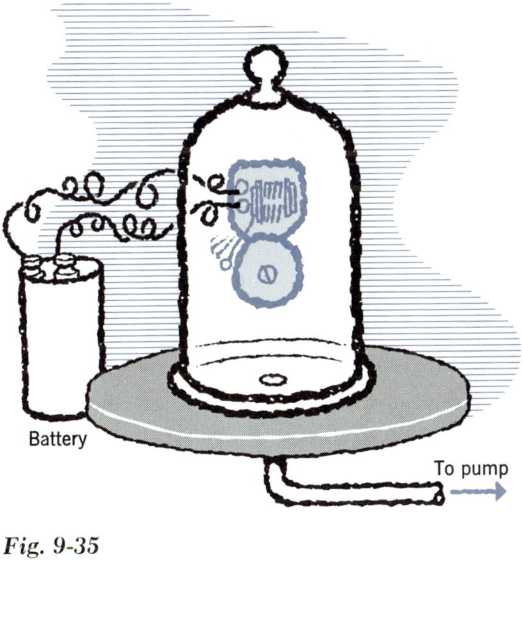

Fig. 9-35

C. BEHAVIOR OF WAVES
   1. Reflection.
   2. Refraction.
   3. Diffraction.
   4. Interference
      a. One-slit.
      b. Two-slit.
      c. Standing waves.
      d. Beats.
   5. Polarization of transverse waves.
   6. Doppler effect.

II. FUNDAMENTAL CONSTANT

A. The speed, c, of light or of any electromagnetic wave in a vacuum. $c = 2.99793 \times 10^8$

III. FIELDS

A. WAVES, including light, in the electromagnetic field.

B. ENERGY in the electromagnetic field.

C. ETHER to support the electromagnetic field and the waves traveling in it.

IV. APPLICATIONS: The wave model explains behavior of:

A. SOUND

B. LIGHT

C. WAVES in springs, water, and stretched strings.

Questions and Problems

***1.** Would you be surprised if you found a polarized beam of sound waves? Why, or why not?

2. Why does a violin string have certain natural tones and not others? How do the natural frequencies relate to each other?

3. Why do we think that light waves exist? How do these reasons differ from our reasons for believing that other kinds of wave exist?

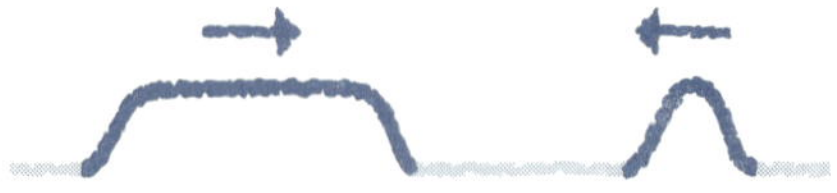

Fig. 9-36

4. Why do we think that sound waves exist?
5. What is an oscillator?
6. What is the difference between a transverse wave and a longitudinal wave? Which kind would you expect to find in solids, liquids, and gases? Explain.
7. How do we know that waves carry energy?
8. Why does the frequency of an oscillator equal the frequency of the waves it emits? Are there any apparent exceptions? Explain.
9. Why, in each wavelength of the simple waves in Figure 9-2, is there one upward pulse and one downward pulse? Why not one upward pulse in one wavelength and one downward pulse in the next?
10. What is the frequency of an electromagnetic wave traveling in a vacuum if the wavelength is (a) 1.54×10^{-10} meter (x-ray); or (b) 3×10^{-2} meter (radar)?
11. What is the wavelength of a sound wave with a frequency of 256 vibrations per second (middle C)?
12. Air is evacuated from the bell jar in Figure 9-35 while the bell is ringing. Why can we no longer hear the bell after the air is pumped out?
13. Explain in your own words what is meant by the terms reflection, refraction, diffraction, and interference.
*14. How would you try to convince someone that light is a transverse wave?
*15. What is meant by polarization?
16. What is the reason for the dark places in the two-slit interference pattern?
17. Give a reason for believing that the electromagnetic field can contain energy.
18. What is a wave?
19. Why was the ether invented?
20. What difficulties does the ether present?
21. Explain why the Doppler effect occurs when the source of waves is stationary and the listener (receiver) is moving.
22. Does the Doppler effect hold for water waves? Explain.
23. Why can we call a musical instrument an oscillator?
24. Two wave forms are moving toward each other on a spring as shown in Figure 9-36. What happens when they meet?
25. Why are the nodes on a vibrating string one half wavelength apart?

26. Why, if light is a wave, do we speak of light *rays*?
27. Why did Galileo's attempt to measure the speed of light fail? What was accomplished by this experiment?
28. Why does a radio or television set use an antenna?
29. How many overtones does a stretched string have?
30. When the two wave pulses in Figure 9-10*b* cancel each other out, what happens to their energy?
31. Describe the interference and diffraction patterns characteristic of one-slit and two-slit experiments.
32. Since waves are coming from both oscillators, how does it happen that there are lines in Figure 9-11 where there appear to be no waves?
33. Why do we believe that the two pulses in Figure 9-10*a* actually pass through each other rather than just bounce back off each other?
34. What is a shadow?
35. Why does the speed of sound in air increase with temperature?
36. Why does the speed of sound in air *not* vary much with pressure, for normal pressure changes?
37. Why does a tuning fork stop vibrating?
38. If I hold the handle of a tuning fork solidly against a table, the sound from the tuning fork becomes louder. Why? How will this affect the length of time the fork keeps vibrating? Explain.
39. How can the result of Grimaldi's experiment be explained by the wave theory of light?
40. Give a reason for believing that electromagnetic waves involve both the electric and the magnetic fields.
41. What information must we have in order to specify a wave completely?

More Difficult Questions and Problems

1. By what combination of experimental results and reasoning can we show that the color of light depends on its frequency?
2. Draw a diagram approximately full scale showing where the first minimum will fall for a two-slit interference pattern with slits one inch apart, a screen three inches away, and waves of wavelength one inch.
3. What method or device can be used to separate a white beam of light into its colors?
4. In terms of the wavelength, what is the difference in the distances

traveled to the second maximum from each slit in the two-slit experiment? To the third maximum?

5. In the two-slit experiment, what is the path difference in terms of the wavelength for waves from the two slits to the first minimum? To the second minimum?

6. Show that there will be a beat frequency equal to the difference between two given frequencies if the given frequencies are sounded together.

7. Why does a violin have more wood in it than is necessary to keep the strings supported and stretched?

8. In some auditoriums there are spots in which it is hard to hear the music, although there are spots all around it where it is easy to hear. Explain.

9. In Figure 9-30, if I move the mirror at B a distance of 1.5×10^{-5} meters, and at the same time I notice that the bright spot in the middle of the viewer at A changes to dark and back to bright again 100 times, what wavelength of light am I using?

10. Suppose that an oscillator is emitting sound waves of frequency f and wavelength L. I am approaching the oscillator at a speed v. What is the change of frequency I hear as a result of the Doppler effect?

11. We mentioned in the text that the first maximum away from the center maximum will be at different positions for different colors of light. Will this be true of the next maximum out also? The next one after that? The center maximum? Explain.

12. Knowing that light waves are transverse and that they involve both magnetic and electric fields that are perpendicular in direction to the wave velocity, can you think of a way to account for light pressure?

10

MODERN RELATIVITY

The luminiferous ether, that is the only substance we are confident of in dynamics. . . . One thing we are sure of, and that is the reality and substantiality of the luminiferous ether. . . . You may regard the existence of the luminiferous ether as a reality of science. *Lord Kelvin*[1]

Einstein's special theory of relativity

Ether and the Field

The existence of the electromagnetic field raised interesting problems in physics. The field, as well as particles of matter in motion, was necessary to describe the physical world. It was natural to think that waves in the field, such as light, would need some sort of material medium in which to travel, and that indeed the field itself must have some material basis. Thus the idea of the ether, an unobserved substance filling all space, was introduced. The ether was thought to be a form of matter, probably composed of particles. Light waves, coming through the void from the remote stars, were believed to travel through the ether. The ether filled the void.

The concept of the ether raised at least two problems. First, a transverse wave such as light seemed to require a solid substance through which to travel. However, the earth and the planets appeared to travel through the ether without the slightest resistance.

[1]Lord Kelvin, *Popular Lectures and Addresses* (1891), Vol. I, pp. 317, 334, 354.

The second problem involved the special theory of relativity. Ordinary objects can appear to be moving at different speeds if they are observed from reference frames in relative motion. As we pointed out in Chapter 1, for example, the stewardess of an airplane walking down the aisle appears to be moving at a considerably different speed if she is observed from the ground than from a seat in the airplane. Naturally it was assumed that the speed of a beam of light in a vacuum could also appear to be different in different frames. However, Maxwell's laws of electricity and magnetism predicted a single definite value for the speed of light in a vacuum; if this speed were actually different in different reference frames, then Maxwell's laws themselves would have to be different in different frames. This conclusion, of course, would appear to violate the special principle of relativity. Strictly speaking, the problem could be resolved, if somewhat uneasily, by assuming the existence of a material ether. Then the speed of light predicted by the laws of electricity and magnetism could be the speed observed in a reference frame at rest with respect to the ether. These and other considerations convinced many scientists of the existence of the ether and may help to account for the now obsolete claim expressed in the quotation from Lord Kelvin at the beginning of this chapter.

If the ether really exists, then the earth should be moving through it at a definite speed. The American scientists A. A. Michelson and E. W. Morley tried to measure that speed in 1887, using the Michelson interferometer (see p. 378). If the earth is moving through the ether, then from the earth the ether should appear to be moving past the earth. By the same principle, if we are riding in an open car on a calm day, a wind appears to be blowing by the car. Actually it is the car, not the air, that is "moving," but we feel the relative motion as a wind. Similarly, there should be an ether "wind" blowing by the earth. Michelson and Morley attempted to detect this "wind."

To explain the principle on which the Michelson-Morley experiment depends, we refer to Figure 10-1, which shows a boat race. The boats are going to travel the same total distance relative to the ground, but under different conditions. One boat goes a certain distance upstream and then back downstream to the starting point. The other boat goes across stream the same distance and then back. Both boats move with the same speed with respect to the water in which they are traveling. They will, however, have different speeds with respect to the river bank. The first boat will be slowed by the motion of the water when it is going upstream, and it will be speeded up by the water as it goes downstream. The second boat, in going across stream and back, will be slowed a little in both directions because it must be aimed a little upstream in order to keep the current from pulling it off course. It turns out that the boat going across stream wins every time. The total result of the changes in speed caused by the moving water is that the time for the round trip is less for the crossstream journey.

The Michelson-Morley experiment is diagrammed in Figure 10-2. Here we have a situation similar to the boat race. Instead of the flowing water, we assume that we have the ether flowing by the

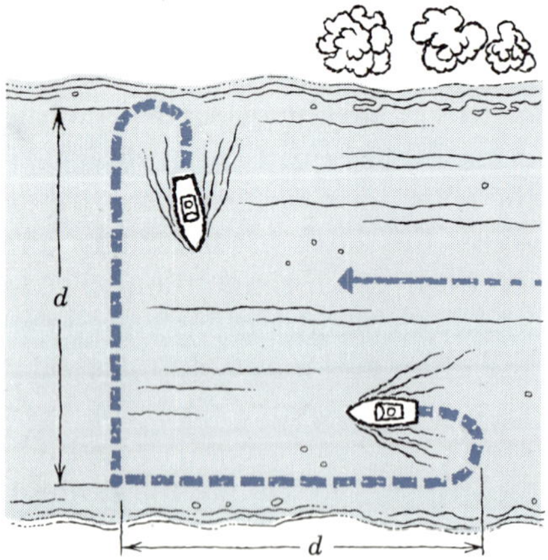

Fig. 10-1

earth. Instead of the boats, we have the waves in two light beams that have been split from the same beam by a partially reflecting mirror. One of these goes across the ether stream and back, and the other goes upstream and down. If the ether is really flowing by, and the speed of light is constant with respect to the ether, then the waves going across and back should return before the waves going upstream and back. That is, the waves in beam B should return before the waves in beam A.

How can we tell that one set of waves gets back before the other? We know that we get interference effects when the two beams of light come back together. Whether there is an ether there or not, some pattern will be set up; for example, there might be a bright spot in the viewer. Let us now rotate the instrument until it is in the position shown in Figure 10-2*b*. In this situation, it is now beam *A* that goes across the stream and beam *B* that goes downstream and back. Therefore the waves in beam *A* will now return before the waves in beam *B*. Because of this shift, the interference pattern should change as we rotate the interferometer. The bright spot in the viewer could become dark and perhaps bright again. This result would signify that this ether "wind" really existed.

In fact, no significant result was observed; this, of course, raised a problem. Unless the earth were sitting motionless in the ether, there should be some effect. Perhaps the earth just happened to be moving with the ether on that particular day, so that there was temporarily no relative motion between the earth and the ether. However, the earth, in traveling around the sun, is always changing the direction of its

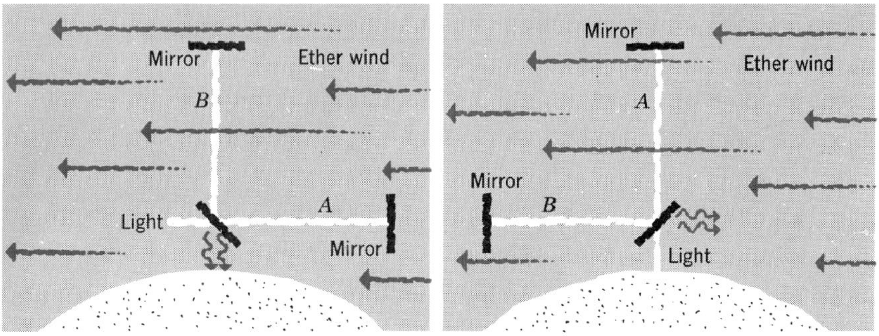

(*a*) (*b*)

Fig. 10-2

motion. Six months later the earth would be traveling in the opposite direction and should then have a speed with respect to the ether. Experiments were performed to check this possibility, but again there was no effect.

Perhaps the earth dragged some of the ether with it through space, so that the ether near the earth was at rest with respect to the earth. If this were true, however, how could the aberration of starlight be explained? The explanation given in Chapter 1 (see p. 26) assumed that light traveled in straight lines from the star to the earth. If the earth were carrying the nearby ether with it, then the light from the star would start out in stationary ether and enter the moving ether near the earth. This would be like a water wave starting in a quiet pool and then moving into a flowing stream. The wave would be distorted, and its direction of motion would change.

Perhaps, as a desperate measure, we should consider light as moving with its characteristic speed, not with respect to the ether, but with respect to the object that emitted it. Then light emitted forward from a moving object would travel faster than light emitted backward from the same object. This possibility appeared to contradict certain astronomical observations. In the heavens there are many stars that occur in pairs, revolving around each other. If the light from the star moving toward the earth traveled faster than the light from the star going away from the earth, we would know it. We would not observe the two stars to move around each other in the uniform way they do. For this and other reasons, this explanation was abandoned.

Another suggested explanation was that the arm of the interferometer in the direction of the ether flow was somehow automatically shortened by its motion through the ether, so that the beam going up and downstream actually had a shorter distance to travel. Even if this contraction were large, it could not be measured by a ruler, since the ruler would also contract. According to this assumption, however, another interferometer experiment using an interferometer with arms of different length should give some positive result. Such an experiment was performed, but still no positive result was obtained. This could be explained if one assumed that time in a moving system was somehow altered also. In fact, H. A. Lorentz (1853–1928) obtained correct equations with this kind of explanation.

No one was able to measure the speed of any reference frame relative to this supposed ether. Nor did the measured speed of light seem to be influenced by such an ether. At this point Albert Einstein proposed a solution.

Einstein's Two Postulates

As a boy Albert Einstein (1879–1955) was informed by one of his teachers that he would never amount to anything. Of studying for examinations he himself said:

> It had such a deterring effect upon me that, after I had passed the final examination, I found the consideration of any scientific problems distasteful to me for an entire year.[2]

About five years later, however, after managing to find a steady job as a patent clerk, he published a series of scientific papers. One, on relativity, created the subject from a modern point of view. Another, on Brownian motion, we mentioned in Chapter 7. A third, which we shall discuss in the next chapter, helped lay the foundation for quantum physics.

We have seen how the special principle of relativity seemed to be challenged by the laws of the electromagnetic field. Yet experiments had been unable to measure the earth's speed with respect to the ether. Nor could experiments find any speed for light in a vacuum other than the value predicted by Maxwell's electromagnetic theory. Einstein decided that the facts should be accepted at face value; why not assume them to be true instead of trying to explain them away by some theory? The basic reality was the observed phenomenon; the theory was made up to make things simple or calculable. Einstein therefore chose the following postulates as the basis for his special theory of relativity.

1. The velocity of light in a vacuum is the same in all reference frames moving uniformly, . . .
2. All laws of nature are the same in all reference frames moving uniformly,. . .[3]

(Notice that the special *theory* of relativity and the special *principle* of relativity are not the same. The special *theory* of relativity includes two postulates, *one* of which is the special *principle* of relativity (see Chap. 1, pp. 21–25); the other postulate concerns the constant speed of light in a vacuum.)

These two postulates simply recognize the continued failure of

[2]Quoted in Martin J. Klein, "Einstein and Some Civilized Discontents," *Physics Today*, **18** (January 1965), pp. 38–44.

[3]Albert Einstein and Leopold Infeld, *The Evolution of Physics*, New York: Simon & Schuster, 1938, p. 186. We have substituted "reference frames" for "CS" in the original, and "a vacuum" for "vacuo."

certain experiments and state those failures as principles. No one had been able to measure a speed relative to any "preferred" reference frame, in the ether or elsewhere, so the special principle of relativity must hold. The second postulate, therefore, is just the special principle of relativity. Only one speed had been found for light in a vacuum, regardless of the reference frame, so the first postulate states that the speed is always the same. The reference frames mentioned in Einstein's two postulates do not include reference frames that are being accelerated; for example, a reference frame fixed to a merry-go-round would be excluded. In fact, the theory is called the *special* theory rather than the *general* theory because it excludes such reference frames. The postulates in the special theory refer only to reference frames that are either at rest or in uniform straight-line motion. According to the special principle of relativity, the laws of physics are the same in all such reference frames. In accelerated reference frames, however, the laws of physics may appear to be different.

The first law of motion, which is not altered by Einstein's special theory of relativity, affords a test as to whether a given reference frame is or is not accelerating. The first law of motion is the same in all nonaccelerating reference frames, as indeed it must be according to the special principle of relativity. In an accelerating reference frame, however, the first law of motion will appear to be violated. For example, a man standing still, watching a merry-go-round, is obeying the first law of motion. However, if we watch him from a reference frame fixed to the merry-go-round, he will appear to be going around and around, without benefit of any centripetal force. Thus he will appear to be violating the first law of motion. Actually, of course, this apparent violation is because of our choice of an accelerating reference frame. It is our reference frame that is going around, not the man, (in the special theory of relativity).

Einstein's two postulates may seem innocent enough, but they have some very surprising consequences. For example, imagine a space station out in space, far away from any other object (see Fig. 10-3*a*). We may consider it as stationary if we please. Suppose that there is a rocket moving past the station at half the speed of light —that is, 1.5×10^8 meters per second—and sending out a forward beam of light waves. These light waves will have the speed of light, about 3×10^8 meters per second, with respect to the rocket. What speed will they have as observed from the space station? One might think that the answer should be the sum of the two speeds, or 4.5×10^8 meters per second, but it is not. It is 3×10^8 meters per second, just the same speed as if the waves had been sent out by the station and not by the rocket. This is one of the curious results of modern relativity theory.

The example we have just discussed is an extreme case of the more general situation illustrated in Figure 10-3*b*. In this figure the rocket fires a projectile straight ahead, at some speed less than that of light. Ordinarily, to find the speed of the projectile with respect to the space station, we would simply add the speed of the rocket with respect to the station to the speed of the projectile with respect to the rocket. As an example of this rule, suppose that the rocket is traveling four meters per second with respect to the station and that the projectile is moving 30 meters per second with respect to the rocket. Then, according to our rule, the speed of the projectile with respect to the station should be the sum of these two speeds, or 34 meters per second. According to the special theory of relativity, however, our rule does *not* give the correct answer. Actually, the projectile will be going less than 34 meters per second, by an extremely small amount. For such low speeds, our rule is very nearly true, but it becomes more and more inaccurate as the speeds approach the speed of light. If it were possible for the projectile to reach the speed of light, we would have the same situation as we did for the light beam in Figure 10-3*a*. The projectile would then have the speed of light in any reference frame. The speed of light acts as a natural speed limit for material objects. In fact, as far as we know, the speed of light in a vacuum is the maximum speed at which matter, energy, or information can be transferred from one place to another.

As we have illustrated, we cannot obtain a relative speed for an object greater than the speed of light by adding up relative speeds

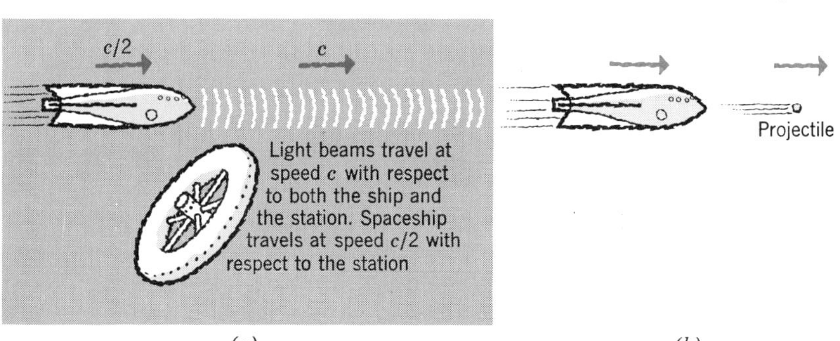

Fig. 10-3

that are less than or equal to the speed of light. If this is true, and we believe that it is, then something must happen to ordinary time and space measurements when we compare two systems in relative motion. In fact something *does* happen, as we shall now see.

Some Consequences of Einstein's Two Postulates

TIME DILATION. Time and space are abstract notions. Our understanding of them depends on our measurement of time and space. Einstein therefore concentrated on the behavior of clocks and measuring rods. In investigating the behavior of clocks in moving systems, we shall see that Einstein's postulates lead us to some remarkable conclusions.

Let us first perform a thought experiment to see how a clock keeps time when it is moving past us at a high speed. We shall choose a very simple, but impractical, kind of clock—the so-called "light clock," illustrated in Figure 10-4*a*. In this clock we do not have to worry about the springs, gears, etc., that we find in ordinary timepieces. The light clock consists of two parallel mirrors with a pulse of light bouncing back and forth between them. Each time the light

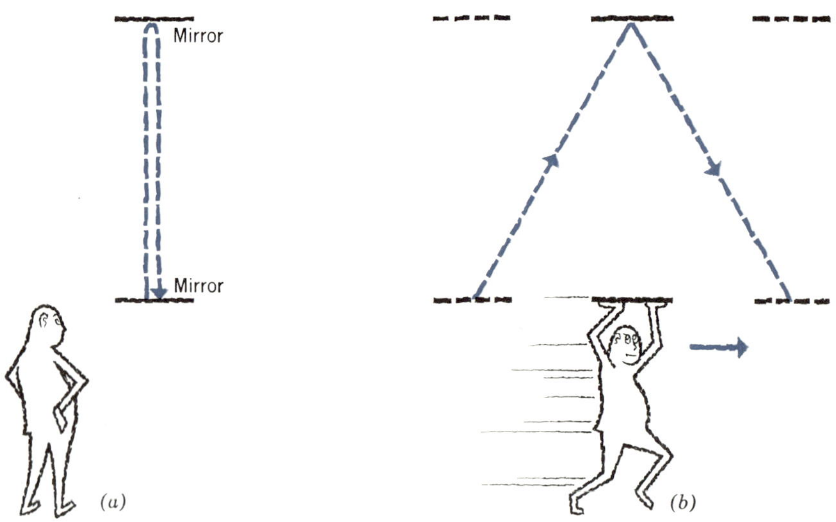

Fig. 10-4

pulse is reflected from one of the two mirrors, we shall imagine that one tick of the light clock has occurred.

Now we hand the light clock to someone else, and have him race by us at a high speed. (Since this is only a thought experiment, we might as well have him moving at about half the speed of light.) As he races by, he holds the clock as shown in Figure 10-4*b*. As far as he is concerned, in his reference frame the clock is behaving quite normally. The light pulse is simply bouncing straight back and forth between the two mirrors. It looks the same to him as it did to us before we gave it to him. He thinks it is still keeping perfectly good time, and in his reference frame it is.

From our reference frame, however, things are different. According to our observation, the light bouncing between the mirrors does not move vertically as the runner moves but travels the diagonal path shown in Figure 10-4*b*. Because the mirrors are moving to the right in our reference frame the light pulse must move to the right also as it bounces back and forth so as to keep up with the mirrors. The pulse actually travels a longer distance as it moves between the mirrors in our reference frame than in that of the runner. Therefore, the pulse also travels for a longer time between the mirrors in our reference frame than in his, since the speed of the pulse of light is the same in both reference frames. As we see it from our reference frame, there is a longer time lag between ticks when the clock is in motion than when it is stationary. Therefore, the moving clock appears to run slow. This is what is meant by the term *time dilation*. Clocks in motion run slow. The faster they move, the slower they run. If it were possible to accelerate a clock so that it reached the speed of light, it would appear to stop completely.

The person moving with the clock, however, does not observe any slowing down of the clock. On the contrary, he thinks his clock is operating normally. In his reference frame the light clock is stationary, with the beam of light bouncing back and forth between the mirrors at its characteristic speed, *the same for all reference frames*. However, if he checked a clock in our reference frame, he would find our clock to be running slow.

Although this has only been a thought experiment, it correctly describes the behavior of any kind of clock, as far as is known—all moving clocks run slow. This curious fact is, of course, the result of the assumption that the speed of light is the same in all reference frames. We were led to this assumption by the fact that the speed of light in a vacuum appeared to be the same in all experiments.

From Figure 10-4, using simple geometry and algebra, we can calculate the relation between a time duration, t_0, registered on the moving clock, and the time duration t, which we think should have been registered on it. (See Example 4, p. 437.) The result is:

$$t = \frac{t_0}{\sqrt{1 - s^2/c^2}} \tag{1}$$

in which s represents the speed of the moving clock as observed from our reference frame, and c represents the speed of light in a vacuum. The equation indicates that t is greater than t_0, as should be the case if the moving clock is running slow.

Several experiments have been performed that show the time dilation predicted by Einstein's special theory of relativity; we shall discuss one of them here. The factor $1 - s^2/c^2$ is very small when the speed is much smaller than the speed of light; therefore, the time dilation is hard to observe for ordinary objects. We usually do not have clocks whipping through the laboratory at half the speed of light or thereabouts. However, there are many small particles which do indeed move at speeds comparable to that of light. We have discussed two such particles, the electron and the proton. There is another fundamental particle, called the muon, which is almost exactly like the electron but has much larger mass. Muons are formed high above the earth when cosmic rays, made up of high-energy protons and other particles, crash into the atmosphere. These muons are continually being formed at about the same rate in the upper atmosphere. They are thus continually descending through the atmosphere to the surface of the earth, at speeds nearly that of light. While they are descending, some of them change into other particles. There is no way of telling how long an individual muon will last before changing. However, for a large group of muons, there is a definite average lifetime, which is characteristic of muons. By studying this average lifetime, Bruno Rossi and his collaborators obtained the first experimental evidence for the time dilation.

Rossi's muon experiment may be performed using the equipment illustrated in Figure 10-5, located on the top of a mountain.[4] The shield illustrated in the figure is so constructed that only muons traveling at about 95 percent the speed of light or faster can pass through it. All others are absorbed. Detector 1 will count all muons entering the can. Muons with speeds greater than 95 percent the

[4] *Time Dilation: An Experiment with Mu-Mesons*, a motion picture distributed by Educational Services, Inc., 39 Chappell St., Newton, Mass., 1963.

speed of light will not stop in the can, so detector 2 is needed to count them. The difference between the numbers of muons counted by the two detectors will tell us exactly how many muons with speeds very close to 95 percent the speed of light are caught in the can. A third detector, which is not shown, is able to measure the time between the entry of a muon into the can and its change into other particles. In this way we can determine the lifetime of these muons. From the lifetimes we can predict how many of them would have lived long enough to reach the bottom of the mountain had they been allowed to do so. These lifetimes are so short that we would expect very few muons to reach sea level.

Next we take our equipment down to sea level and adjust the shield to account for the fact that the additional air between sea level and the mountain top would absorb some of the muons. Then we repeat our experiment and count the muons that would have had 95 percent of the speed of light at the mountain top. We find that we get almost as many as before. In other words, the muons lasted longer in flight than they did at rest in the laboratory. When we measure how much longer, we get a value which agrees with the time-dilation formula we have just obtained from Einstein's postulates. Thus it appears that Einstein's theory is correct, or at least on the right track.

THE APPARENT PARADOX OF THE TWINS. The time dilation predicted by the special theory of relativity has given rise to a paradox that is still discussed among physicists. It seems paradoxical that each of two observers in relative motion will claim that the other's

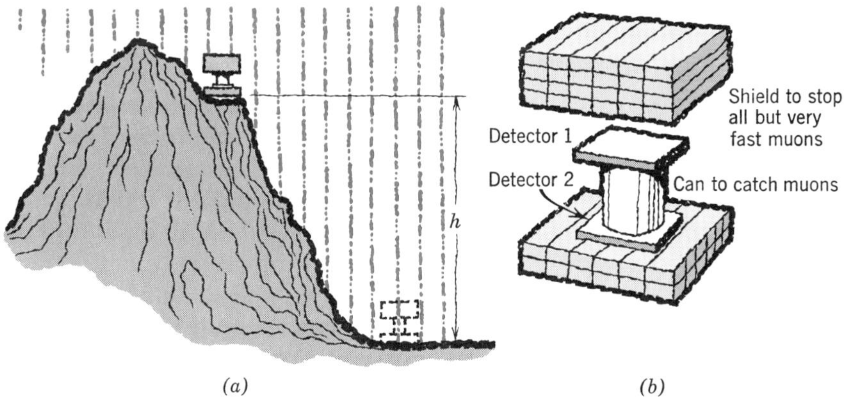

Fig. 10-5

clocks are running slow. However, no contradiction arises as long as each maintains his velocity, since they will meet at most once, and then recede from each other forevermore. What would happen, however, if some means were found to bring them back together again? Whose clocks would really have been running slow?

Putting the paradox in its usually stated form, suppose that there are two twins, one of whom remains home on earth while the other takes a high-speed journey to a star (see Fig. 10-6). After arriving near this star, the traveling twin reverses his direction and returns to earth. During his outward journey, and during his return journey as well, his clocks will appear to be running slow as observed from the reference frame of his stay-at-home brother. Since this slow down should apply to all processes, including biochemical processes, the traveler will have aged less than his brother, and therefore should appear to be the younger of the two when he returns, even though they were both the same age when he left. This would be a remarkable result, but in itself it contains no contradiction.

However, we might also look at the matter from the reference frame of the traveler. Since he is at rest in his own reference frame, he notices no slowdown of his clocks. It appears to him, rather that the earth and his brother are moving; therefore he observes that his brother's clocks are running slow. When he returns to earth, then, he will conclude that he is the elder, that it is his brother who has aged less than might be expected. This result is directly contradictory to the result we obtained when we looked at the matter from the reference frame of the other twin. Thus we have an apparent paradox—which twin really would be the elder?

The traveler, of course, cannot have been traveling at a constant

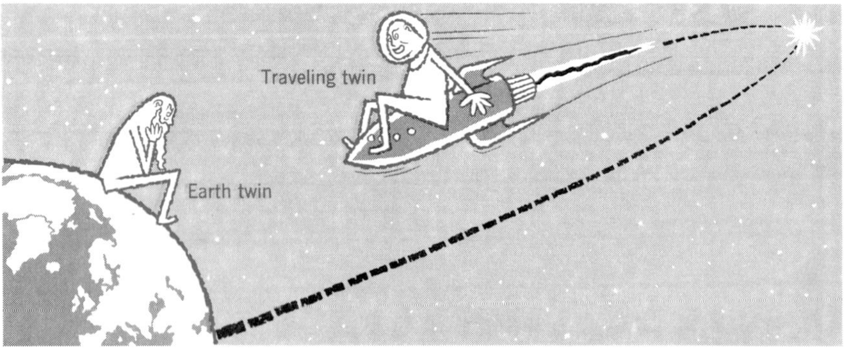

Fig. 10-6

speed in a straight line during his entire journey. Therefore his reference frame is not one of those referred to in the postulates of the special theory of relativity. The stay-at-home's reference frame has not been accelerated and therefore is acceptable, but the traveler's frame is not acceptable. The traveler must have been accelerated to start on his journey, he must have been accelerated to turn around and start home, and he must have been accelerated (negatively) in order to stop when he finally returned home. According to the special theory of relativity, the laws of physics need not be the same in an accelerated reference frame; therefore we need not give the same status to the traveler's view that we do to that of his brother, and the contradiction expressed in the last paragraph was not validly obtained.

Even if we restrict ourselves to the reference frame of the stay-at-home twin, however, we still need to consider the possible effect of acceleration on the clocks of the traveler. The theory needed for such an analysis would take us beyond the level of our present discussion; however, we can say that according to currently accepted theory the acceleration apparently does not eliminate the effect of time dilation. The traveling twin would indeed return to earth younger than his brother. Moreover, a recent experiment seems to justify this conclusion.

Atoms in crystals are in continual vibration, because of the kinetic energy associated with the temperature of the crystal. Measurements on small particles within these atoms indicate that the time-dilation formula holds for them regardless of their acceleration as they vibrate back and forth. Therefore the traveler, despite his acceleration, finds when he returns home that he has aged less than his brother. It is to be expected that there will be further discussion among physicists (and others) on this subject. It is indeed curious that the stay-at-home spends more time watching the journey than the traveler spends taking it.

LORENTZ CONTRACTION. Let us return to the problem of the muons, discussed on page 402. According to the special principle of relativity, we should be able to consider the whole process from the muon's reference frame as well as from our own. In our reference frame, the muon is moving down to earth at a certain speed. The muon, in its reference frame, sees the earth moving up at the same speed. We have already found that the "normal" lifetime of the muon is not long enough to allow very many muons to travel the vertical distance from the top of the mountain to sea level, even though they are moving at nearly the speed of light. We have also seen that in fact the muons *do* successfully cover this distance. We solved this problem in our own reference frame by showing that the lifetimes of the moving muons are stretched out because of time dilation. This solution will not work, however, if we look at the problem from the reference frame of the muon.

The muon is at rest in its own frame and would observe no time dilation. Therefore, the muon will not survive the trip to sea level if it has to travel the whole distance. The only solution is that the muon must see the distance from the mountain top to sea level as shorter than we see it. It must see the distance shortened in exactly the same way that we see the time stretched:

$$L_0 = \frac{L}{\sqrt{1 - s^2/c^2}} \tag{2}$$

In this formula, L is the height of the mountain that the muons see, and L_0 is the height of the mountain that we see. The muons see a contracted length, a length that is shorter along the direction of motion than the length we see. This shortening, called the *Lorentz contraction*, is another result of Einstein's two postulates. The length of any moving object should be shortened along its direction of motion. The faster the object moves, the shorter it becomes. If it were possible for the object to be accelerated to the speed of light, its length along the direction of its motion would collapse to zero.

In order to clarify the Lorentz contraction, we refer to Figure 10-7. Here we have two measuring rods moving at a high speed. The one that is parallel to the direction of motion is shortened by the Lorentz contraction. The other, which is moving sideways, is not shortened by this contraction. However, it *is* made thinner. The Lorentz contraction alters the length *along* the direction of motion, and not perpendicular to it. Of course, someone moving with the rod would not notice any shortening of the rod, since the rod is at rest with respect to him. Instead, he would think the rod in *our* system was shortened by this contraction.

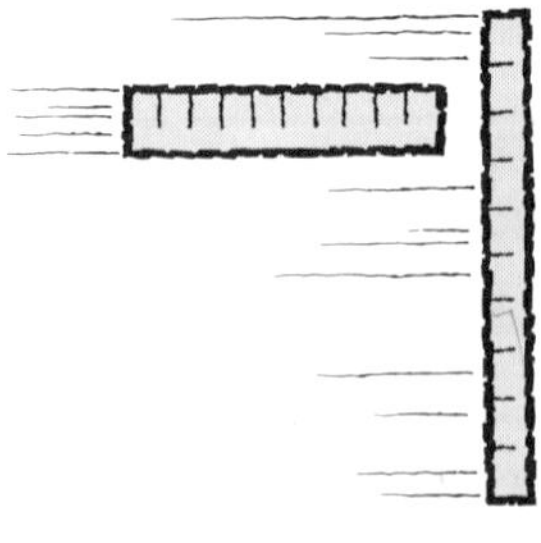

Fig. 10-7 Lorentz contraction.

The Lorentz contraction and the time dilation are very small at ordinary speeds. Therefore, they may be ignored for most of our ordinary activities in everyday life. They become very important, however, for objects moving at nearly the speed of light.

Our discussion has centered on the behavior of clocks and measuring rods. Yet we have spoken of time and distance. Because the time dilation and the Lorentz contraction appear to affect all clocks and all measuring rods according to the same principles, we consider these principles as properties of time and space, and not just as properties of objects. After all, in physics, our precise knowledge of time and space is dependent on the devices we use for measurement. For this reason some physicists would claim that the properties of space and time are the properties of our measuring instruments.

THE INCREASE OF MASS WITH SPEED. When we discussed the example of the rocket and the space station, we noticed that even a light wave emitted from the front of a rocket cannot go faster than light in the reference frame of the space station. This is a general principle. We cannot accelerate any object enough for it to reach the speed of light. We can, of course, apply a force to an object for a long time. As a result of this force, the object will continue to accelerate. That is, its speed will continue to increase as long as the force is applied. However, this increase, or even a series of such increases, cannot cause the object to reach the speed of light.

What can prevent a moving object from reaching the speed of light when a constant force is exerted on it? If the speed were to continue increasing at a constant rate, it would eventually equal and even surpass the speed of light. The acceleration of an object, however, depends not only on the force but on the mass as well. If the mass is increasing with speed, then the force will have less and less effect as the speed increases. In fact this is what happens, according to the theory of relativity. The mass actually depends on the speed, according to the equation

$$m = \frac{m_0}{\sqrt{1 - s^2/c^2}} \tag{3}$$

In this equation, m is the mass of a particular object at any speed s. The letter m_0 represents the mass the object would have at rest, the so-called *rest mass*. The increase of mass with speed is not important for low speeds. It is important, however, for electrons in a cyclotron. The increasing mass due to the increasing speed of the electrons causes the paths of the electrons to change in the cyclotron and spoils the operation of the device. Other devices must be used to accelerate electrons to very high energies. This circumstance, although a practical nuisance in accelerators, is another experimental fact which supports Einstein's theory.

The momentum of a moving particle, according to the special theory of relativity, is still equal to the mass of the particle m multiplied by the velocity v. The mass, however, is the mass m given in Equation 3. Therefore, the momentum p is given by the equation

$$p = mv = \frac{m_0 v}{\sqrt{1 - s^2/c^2}} \tag{4}$$

E = MC². The results of the last section also have another consequence. In trying to accelerate a very fast particle to the speed of light, we are pushing on it with a force while it is traveling through space. Thus, we are applying force through a distance, or doing work on the particle; therefore its energy should increase. But the speed cannot increase very much since the particle cannot go faster than light. The energy must show up in some other way. At the same time, we know that the mass of the particle is also increasing. This leads us to suspect that the mass itself must represent energy, and it turns out that this is correct. In fact, the total energy of the particle is

$$E = mc^2 \tag{5}$$

The particle still has some mass when it is at rest—its rest mass. If Equation 5 is correct, then the particle should have some energy when it is at rest, because of this rest mass. In fact, this energy should be

$$E_0 = m_0 c^2 \tag{6}$$

This form of energy, rest energy, is at present mostly inaccessible in most objects. However, in atomic-bomb explosions and similar processes, rest energy can be transformed into other forms which are "useful." Such processes, therefore, provide experimental evidence that rest energy really exists.

In fact, according to all relevant experiments that have been performed, the equation $E = mc^2$ appears to hold for all forms of energy and all mass. Energy, in whatever form, has mass. Mass, wherever it may be, represents energy. It appears that mass and energy are really the same thing but are measured as different quantities in different units. It is incorrect, then, to say that mass is a form of energy. "Mass" and "energy" relate to the same real entity, sometimes called mass-energy. These two great central concepts of nineteenth-century physics turn out to refer to the same basic thing. This fact is often called "the equivalence of mass and energy." The mass-energy equivalence combines two of the great conservation laws. Both mass and energy are still conserved. Since they basically represent the same thing, however, we often speak of one conservation law—the conservation of mass-energy.

What we are accustomed to regard as a very large amount of energy does not have very much mass. This explains why the equivalence between mass and energy was not discovered earlier. As Equation 5 indicates, the energy is equal to the mass multiplied by c^2, and c^2 is very large. Even a small mass multiplied by the very large c^2 can yield an enormous amount of energy. For example, in the atomic bomb and other processes, a rather small change in rest mass releases tremendous amounts of energy.

It should be clear that the theoretical and experimental results of the theory of relativity have greatly modified our ideas. Since the ether had no observable properties, the idea of the existence of the ether has largely been abandoned. Waves in the field travel through space without its help. Because of the equivalence of mass and energy, the electromagnetic field and its waves, which we know contain energy, must also contain mass. Thus we have two types of "matter" in motion—the fundamental particles and also the field with its waves. Mass-energy and momentum are conserved. The mass of an object changes with speed. Our ideas of space, time, and matter are greatly changed, and thus nearly all of physics is also changed. When these changes are made, it turns out that Maxwell's laws in their original form are still correct. It is Newton's law of gravitation that must be modified, and the second law of motion requires some attention also.

Further Implications of Einstein's Special Relativity

LIGHT VELOCITY, INFORMATION, AND CAUSE AND EFFECT. We have pointed out that the fundamental constant c, the speed of light in a vacuum, acts as nature's speed limit for the transfer of energy, information, and causes. This means that material objects as we know them cannot travel faster than this speed. Suppose that there is an explosion on the sun, and then a disturbance of compass needles all over the earth. Could the explosion on the sun be the cause of the compass disturbance? Only if the time between these events is as long as or longer than the time required for light to travel from the sun to the earth. If the time is not this long, then the explosion could not be the cause of the disturbance. A "cause" cannot be propagated through space faster than light. Even the "news" of the explosion on the sun could not get to the earth any faster than light. The maximum transmission speed of signals, causes, information, energy, and material objects is the speed of light in a vacuum, according to Einstein's special theory of relativity.

*__THE FOURTH DIMENSION.__ The theory of relativity is nearly always associated with the so-called "fourth dimension," which is supposed to be *time*. In physics, the number of dimensions is the number of pieces of information necessary to specify a position, a condition, or an event. In Chapter 1 we pointed out that three pieces of information are needed in order to locate a position in space with respect to a reference frame. For this reason we regard a point in space as a three-dimensional entity. For simplicity, the following discussion refers to points on a surface, which are two-dimensional entities. We need only two pieces of information to locate a point in a reference frame on a surface, as Figure 10-8*a* shows. The point may be specified by giving its perpendicular distance from each of the axes of the reference frame—the x and y distances in Figure 10-8*a*. These two pieces of information determine the point. (This is not the only way the point can be located; however, other ways also require two pieces of information.) To fix our attention on a definite object we assume that there is a rod extending between the origin of the reference frame and the point P. The location and length of the rod are completely specified by the location of its end points.

In Figure 10-8*b* we see what happens when we rotate the reference frame, leaving the rod unaltered. Its position is still the same, since it has not been moved, and its length is still the same also. However, we obtain new distances from the new axes in the new reference frame, as indicated. The rod has not changed, but the pieces of information necessary to locate it have. They have both been changed together at the same time by a single change of the

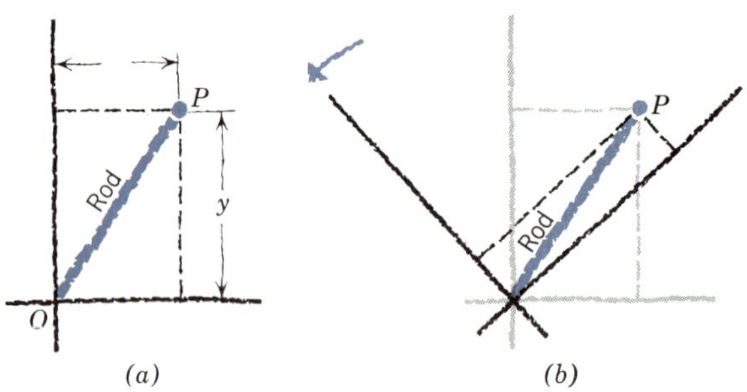

Fig. 10-8

reference frame. Both items change together in a regular manner. These related distances that locate the rod depend on the reference frame and, as we know, we can choose the reference frame in any orientation because of one of our symmetry principles. In this paragraph we have restricted our discussion to a surface. However, the same type of situation occurs in three-dimensional space. The three distances that specify the location of a point in three-dimensional space are also very much like each other. They also can change together in a regular manner when the reference frame is rotated.

It is easy to add a new dimension. For example, if we were interested in the temperature at a given point in space, we would be dealing with a four-dimensional entity. Four items of information would be required—three to specify the point in space, and one to specify the temperature. However, the temperature at a point does not depend in any way on the orientation of the axes of the reference frame, although the other three pieces of information do. As we have seen, these three distances are very much like each other and can be altered together by a rotation of the reference frame. But this would not change the temperature. Thus, the temperature is in a sense independent of the three distances.

An event may be specified by giving the position in space at which the event occurred, and also the time at which it took place. Therefore, an event is a four-dimensional entity. It requires three items of information to determine the position in space, and another to determine the time it happened. Before Einstein, it was thought that time, like temperature, was independent of the three dimensions of space, that we could change the reference frame as we pleased, but time measurements would not be altered.

Now suppose that we take the rod in Figure 10-8 and put it in a moving reference frame, along with a clock, as in Figure 10-9. Not only will the time dilation change the time measurement, but the Lorentz contraction will change the length of the rod. Both the time measurement and the distance measurements will be altered if we change from a reference frame at rest to the one in motion. We have therefore found a way to change the reference frame so that the item of information necessary to specify time and those which specify distance have been altered together. There is no way to cause these particular changes separately; both time dilation and the Lorentz contraction are necessary consequences of the moving reference frame. The time of occurrence needed to specify an event is not independent of the distances needed to locate its position because it changes together with one or more of them when we change to a reference frame with a different uniform motion.

Time measurements appear to be unaltered by changes in the position or orientation of the reference frames. When we switch to a reference frame moving with respect to the first, however, both time and space measurements change. Because the time and space measurements change together in this way when the reference frame is changed, we speak of four-dimensional space-time, rather than three-dimensional space plus one-dimensional time, as it was understood before Einstein. The changes in time and space when we change the motion of the reference frame are similar (but not identical) to the changes in the two items of space information that occured when we rotated the reference frame in space (as in Fig. 10-8*b*). These changes in the time measurement can actually alter the order of events in restricted cases, as we shall see in the next section. This means that the sequence in time can be altered along with directions in space in a manner similar (but not identical) to the change of direction of the reference frame in Figure 10-8*b*.

We must emphasize here, however, that there are strict limitations on the amount of changing of space and time measurements that is possible in relativity. No change in the motion of an observer can cause time to run backward for him. No change of this kind can change the speed of light or the laws of physics. Nor can cause-and-effect relations be reversed; the effect is not changed into the cause or vice versa by a change in the motion of an observer. Thus, the theory of relativity includes some common-sense limitations on the idea of time as the fourth dimension. Nevertheless, this "mixing" of time and space was sufficiently impressive to cause H. Minkowski (who pointed out this way of looking at Einstein's theory) to say:

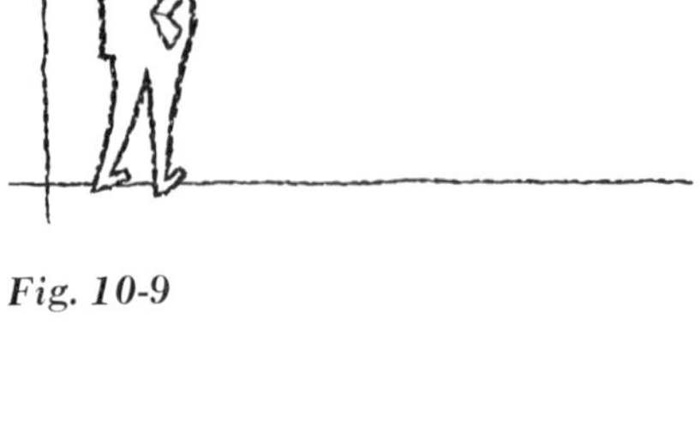

Fig. 10-9

> The views of time and space which I have set forth have their foundations in experimental physics. Therein is their strength. Their tendency is revolutionary. From henceforth space in itself and time in itself sink to mere shadows, and only a kind of union of the two preserves an independent existence.[5]

The theory of relativity causes a revolution in our ideas of time and space, a revolution based on theoretical insight combined with the success of careful experiments. The experimental effects of relativity are extremely small in ordinary motions familiar to us in daily life. In this respect they are like the experimental evidences for the rotation and revolution of the earth—small in many practical effects, but of deep philosophical import.

In the first chapter of this book we showed how the principle of relativity figured in the seventeenth-century revolution in science, even though it was not completely understood. The new physics of Newton, with the laws of motion and the forces arising from the gravitational interaction, actually obeyed this principle. Again, at the beginning of the twentieth century, in what must be considered another revolution in physics, we find the special principle of relativity once more deeply involved in a change in our basic ideas of the physical universe. This time it is the electromagnetic interaction, with its finite light speed, which has to be reconciled with the principle.

We may form some idea of the importance of Einstein's theory by reconsidering the revolution of ideas that was necessitated when it was discovered that the earth was not stationary but was actually rotating and revolving. As long as the earth was thought to be at rest in the center of the universe, the "up" and "down" directions had to be regarded as very different from any sideways direction. "Down" was the direction along which objects were pulled toward the center of the earth, which was also the center of the universe. With a moving earth, however, what was up or down at one time would not be in the same direction after the earth itself had turned and moved. Thus, there was no absolute "up" or "down"; the direction was relative to the position of the earth. Under the right circumstances, any direction could be up or down. Thus the abandonment of a stationary earth in favor of a moving earth destroyed the difference among the three dimensions of space. Similarly, the abandonment of the station-

[5] H. Minkowski, *Space and Time*, translated from an address delivered at the 80th Assembly of German Natural Scientists and Physicians, Cologne, September 21, 1908, as published in A. Einstein, H. A. Lorentz, H. Weyl, and H. Minkowski, *The Principle of Relativity*, New York: Dover, 1923, p. 75.

ary ether destroyed part of the difference between the three dimensions of space and the one dimension of time. So revolutionary were this and later developments that when we speak of the new physics today, we are usually referring to relativity and quantum physics rather than to Galileo, Kepler, and Newton.

*SIMULTANEOUS EVENTS.** Two events that happen at the same time when observed in one reference frame do not necessarily happen at the same time when observed in another reference frame moving with respect to the first. We can illustrate how this follows from Einstein's postulates by an example taken from one of Einstein's books.[6]

Suppose that there is an observer on the ground near a railroad track watching a train go by (see Fig. 10-10). At the instant that the train is halfway past the observer, it is struck by lightning at both ends. The light from each lightning flash travels to him, and he sees both flashes at the same time. Since he knows that his distances from the two flashes are the same, he concludes that the two flashes occurred at the same time.

Now suppose that a passenger on the train is located exactly halfway between the two ends of the train. He would not see the two flashes at the same time. He would see the light flash coming from the front of the train a short time before he sees the flash coming from the rear of the train, since he is moving toward the light wave coming from the front of the train and away from the light wave coming from the rear of the train. Therefore the light wave from the front of the train reaches him first and he sees it first.

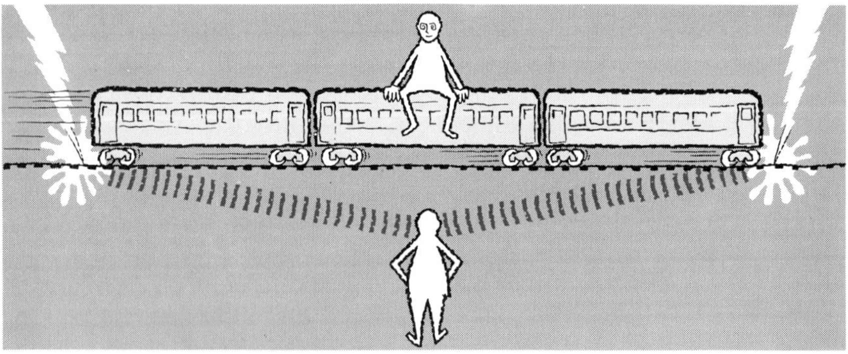

Fig. 10-10

[6]See Albert Einstein, *Relativity, the Special and General Theory*, trans. by Robert W. Lawson, Gloucester, Mass.: Peter Smith, 1931, and London: Methuen and Co., Ltd.

What will the passenger conclude about the two flashes of lightning, as he observes them from his reference frame fixed to the train? He will decide that the flash he saw first happened first. Even if he takes into account the time the light takes to reach him, he will still decide that the one he saw first happened first. After all, both ends of the train are an equal distance away from him. Einstein's first postulate tells him that the speed of light is the same in his reference frame as it is in the other. Therefore he will conclude that the waves from the two flashes of light took the same time to reach him. Therefore, the flash he saw first occurred first.

This example shows that two events which happened at the same time for the observer on the ground did not happen at the same time for the observer on the train. The idea of simultaneity depends to a certain extent on the reference frame. This is rather disturbing; we would prefer to think of two events as simultaneous, or not simultaneous, in an absolute sense. The fact that we cannot do this without specifying a particular reference frame is a consequence of Einstein's special theory of relativity. Of course, if we refer all events to the same reference frame, we can avoid talking about differences in simultaneity. But we have no justification for giving one particular reference frame this special treatment. Which reference frame should we choose? Einstein's second postulate is the special principle of relativity. According to this principle, one reference frame is no better than any other that is moving uniformly. Until some reason is found to prefer one of these reference frames to another and abandon the special principle of relativity, simultaneity must be considered as relative to the reference frame.

The results of the preceding paragraphs may seem startling, but they really do not violate common sense as much as may first appear. For example, if two events occur at the same time and at the same place, they will be observed to be simultaneous in all reference frames. There will appear to be a time interval between two "simultaneous" events in some reference frames only when they are a distance apart. In this case, the observer can be present at only one of the events and must obtain his information about the other from a signal. Light is the fastest signal, but it still takes time to travel from one place to another. Observers in relative motion will disagree as to the distance between events and therefore as to the time it takes light to travel between them.

***FURTHER ANALYSIS OF SIMULTANEOUS EVENTS.** Further analysis is necessary in order to understand what is meant by simultaneity. How can we find out whether or not two events actually did occur at the same time at different places? In the train example above we had light signals from the two events and used the known speed of light to determine how long the signals had been traveling. Another method would be to put clocks in each of the two places. The two clocks would have to be synchronized; then we could use each clock to tell us when the events in each place occurred. Although these methods (the use of signals and the use of clocks) may appear to be different, they are actually very closely related, as we shall see.

Suppose that we are on a long plank somewhere out in space, and that we have a clock at each end of the plank (see Fig. 10-11).[7] We wish to set the clocks so that they both read the same time. We assume that they are already going at the same rate, and that we need merely set one to agree with the other. We also assume that we do not know how the operation of a clock is disturbed if we accelerate it. Therefore we do not take a clock already set from one end of the plank to the other. The only way we can set the clocks, therefore, is to use a signal to carry information about the setting on the first clock over to the second clock. For example, we could use a runner to go back and forth carrying information. Let us send him out when the first clock reads twelve o'clock. When he gets to the second clock, he cannot simply set it at twelve o'clock; he must take into account the time he spent traveling between the two clocks. In order to do this he must know how long it took him to get from one clock to the other; therefore, he must know his speed. In order to measure his speed, however, he must have one clock at his starting point to tell when he left, and another clock at the finish line to tell when he arrived (see Fig. 10-12). Thus he would have to have two clocks, some distance apart, already set. But this is just the problem we are trying to solve. We are therefore in a dilemma. We cannot know the one-way speed of our signal until the clocks are set, and we cannot set the clocks until we know the one-way speed of our signal. What can we do?

Fig. 10-11

[7]Part of this discussion is similar to a treatment by Hans Reichenbach in *From Copernicus to Einstein*, trans. by Ralph B. Winn, New York: Philosophical Library, 1942, pp. 46–47.

There is nothing to stop us from measuring the two-way speed of the signal. We can time our messenger's trip across the plank and back again by using just one clock, and thus obtain his average speed for the entire two-way trip. But this method does not give us his one-way speed. We can improve matters by using the fastest signal at our disposal, to minimize the travel time of the signal. No information can travel faster than light, and therefore we use light signals instead of a runner. However, even light signals take some time to travel between the clocks, so we are still faced with the original problem: we cannot measure the one-way speed of our light signal without having the clocks alreay set. Since there appears to be no experimental way to measure the one-way speed of light, we shall obtain it by a definition. We *define* the one-way speed of light to be equal to the measured two-way speed. In a sense, then, we have *created* the one-way speed of light by our definition. The two-way speed can be obtained by a physical measurement; the one-way speed requires a definition. We could define it differently, but the definition we have used preserves the symmetry of the two directions. In fact, we define the speed of light (in a vacuum) to be the same in all directions.

By this definition of the speed of light, we have also defined what we mean by simultaneity. Knowing the one-way speed of light, we can proceed to set our clocks. Two events are then simultaneous if the clocks near each event read the same when the events occur. Within the limits given by the time it takes for light to go across our plank and back, we could have set our clocks any way we pleased. We have chosen to split the difference—that is, we assume that the light signal took the same time going each way.

In our previous example of the train hit by lightning (p. 414), we used this definition of simultaneity when we assumed that the light traveled at the same speed in both directions in both frames of reference. We saw, however, that this definition led the two observers to disagree about the timing of the two events—events that were simul-

To measure the one way speed of the signal we need two clocks

Fig. 10-12

taneous to one observer were not simultaneous to the other. Although the definition is consistent in a single reference frame, it leads to disagreement between observers who are moving relative to each other. The disagreement arises because each defines the speed of light to be the same in each direction in his own reference frame.

How can this difference be resolved? If there were some experimental way to show that there was just one special reference frame in which light had the same speed in all directions, then all our setting of clocks could be done in this special reference frame. Such a reference frame, however, would be the special reference frame at rest in the ether that the Michelson-Morley experiment failed to find. It was, after all, the idea that light would have a different speed on a "cross stream" journey than on a journey "up and downstream" that motivated the experiment. This experimental failure and others led to the adoption of Einstein's two postulates. If these postulates are true, there is no reason to choose the clock settings of the passenger on the train over those of the man on the ground, or vice versa. Therefore, no absolute answer can be given to the question as to which flash occurred first. Both reference frames are equally valid, and both time measurements are equally valid also.

*The General Theory of Relativity

The General Principle of Relativity

The special theory of relativity states that the laws of physics should be the same in all reference frames that are not accelerating. We are naturally tempted to generalize this principle so that it will apply to all reference frames, however they move or accelerate. We would like to state a *general* principle of relativity.

The laws of physics are the same in any reference frame.

Einstein's general theory of relativity is an attempt to establish this principle. After all, why should nature care what reference frame we are using? It is the same universe, regardless of the reference frame from which we view it; we ought then, to be able to express the laws of physics in a way that will be the same in any reference frame.

In our discussion of this principle of relativity, we shall make use of two particular accelerating reference frames. The first, illustrated in Figure 10-13, is fixed to a large rotating disk. We set up a laboratory on the disk and try to discover the laws of physics. The second reference frame, illustrated in Figure 10-14, is fixed to an elevator, which is being accelerated through space at a constant rate by a constant outside force exerted upward on the bottom of the elevator. We also set up a laboratory inside this elevator and try to determine the laws of physics. According to our new principle, we would like the laws to be the same even if they are determined in these strangely moving reference frames.

Let us discuss why these accelerated reference frames were excluded before. If we are on the rotating disk and release an object, it will fly off the disk (see Fig. 10-15). Without the centripetal force to keep it going around with the disk, the object will leave the disk. To us on the disk, it will appear as if the motion and acceleration of the object occurred without any force. Indeed, it requires a centripetal force to keep an object in place *on* the disk as it rotates. Without this force things start to move relative to the disk. Therefore Newton's laws of motion apparently do not hold in the reference frame fixed to the disc. If we were to release an elephant and a feather at the same time and at the same place, they would both fly away together along the same path. From the standpoint of an outside observer, they just keep going in a straight line. From a reference frame fixed to the disk, however, they appear to be pulled off the disk as if by some unknown force. To those on the disk, this acceleration without force appears to violate Newton's laws.

Now let us look at the reference frame fixed to the elevator. If we were to release an object in the elevator, it would appear to drop to the floor of the elevator (see Fig. 10-16*a*). Although the object would continue to move at a constant velocity, the elevator with us in it

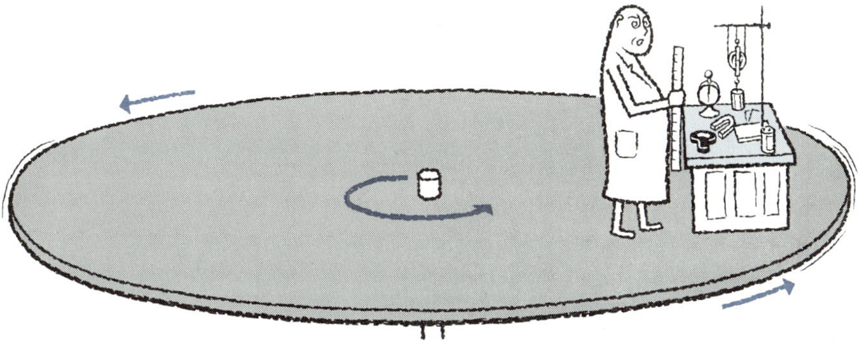

Fig. 10-13

Fig. 10-14

would keep accelerating. Thus the floor of the elevator would catch up with the released object, making it appear as though the object had fallen to the floor. If an elephant and a feather were released in the elevator at the same time, they would appear to fall together as the floor caught up with them. Thus, in this reference frame also we apparently have acceleration without force, in contradiction to the laws of motion. This explains why we did not allow accelerating reference frames earlier. The laws of motion as we have given them are not valid in accelerating reference frames.

In order to establish the general principle of relativity, we must

Fig. 10-15

somehow account for this apparent acceleration without force. We have to invent a force to account for this type of acceleration, one that affects objects in such a way that objects of different mass follow the same trajectorys. We have to invent a force that would cause an elephant and a feather to "fly" together off a rotating disk. We need a force that would cause an elephant and a feather to appear to fall together, as in the elevator. But just such a force has already been investigated—the force of gravitation.

The Principle of Equivalence

Let us set up another elevator, similar to that in Figure 10-16*a*, but resting on the earth (Fig. 10-16*b*). A ball will drop in this elevator much as it appears to in the accelerated elevator. Now, however, we can account for this fall in terms of gravitational force. If the man could not see outside the elevator in Figure 10-16a, he might not know whether the elevator was resting on some planet, and therefore was in a gravitational field, or was being accelerated through space by some constant force. The effect of the acceleration appears to be *equivalent* to the effect of the gravitational field. This equivalence of the two effects is called the *principle of equivalence.*

The reason the gravitational field appears to be equivalent to an acceleration of the reference frame is that all objects fall at the same acceleration in a gravitational field. As we have seen (Chap. 4), this equality is dependent on the equality of gravitational mass and ordinary mass. The principle of equivalence may be said to arise from the equality of the two kinds of mass. If gravitational mass were

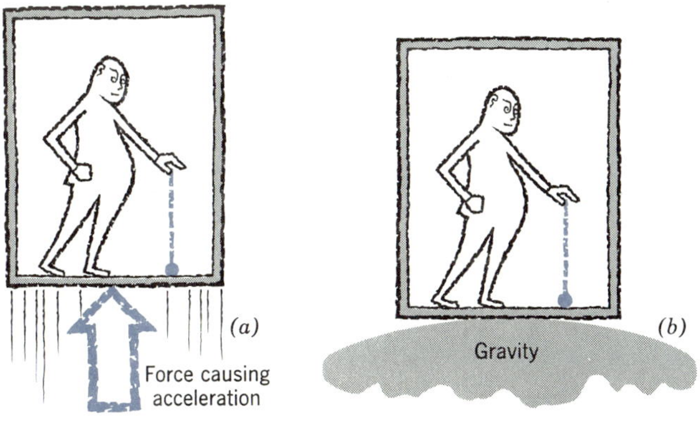

Fig. 10-16

different from ordinary mass, objects would not fall together in a gravitational field, and the principle of equivalence would not hold.

It seems a remarkable coincidence that gravitational and ordinary mass are exactly the same; there appears, at first, to be no reason why they should be. Thus the principle of equivalence also appears to be a great coincidence. Whenever we have precise coincidences in physics, we try to arrange the laws of physics so that the "coincidence" actually follows from basic principles. The principle of equivalence is one of the basic principles of the general theory of relativity.

We can express the principle of equivalence in a more precise form.

> There is no locally measurable difference between the effects of a gravitational field and the effects that arise from the acceleration of the reference frame.

Notice that there is one reservation in this statement—we restrict the equivalence between gravitational and accelerational effects to *local* effects. The need for this reservation can be explained by referring to Figure 10-17. There we see that in the accelerated elevator the balls appear to fall parallel to each other. In the other elevator, however, the gravitational force pulls the balls toward the center of the earth; therefore the paths are not parallel but intersect at the center of the earth. If we restrict ourselves to a very small region, the paths are very nearly parallel. Thus we are talking about a local equivalence that applies in very small regions.

The forces on the disk that appear to pull outward and the forces in the accelerating elevator that appear to pull downward might be called "fake" forces: they arise only because we wish to pretend

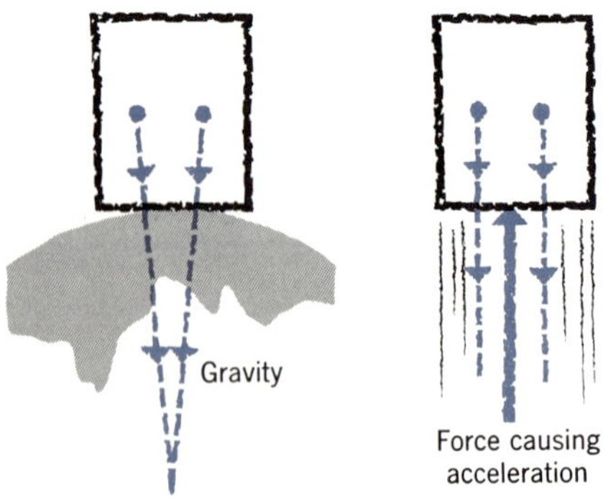

Fig. 10-17

that our reference frame is not accelerating. Nevertheless, they act locally in the same way that gravitational forces do. Could it be that gravitation itself is a "fake" force arising from the nature of space and time in our reference frame? Do we ever actually feel the gravitational force as a push or a pull? A little reflection convinces us that we do not. As we stand upon the earth we feel pressure between our feet and the ground, but that is the ground pushing up on us. If we jump off the diving board we no longer feel this pressure; essentially we feel no forces until we hit the water. The astronaut in his satellite is under the influence of the gravitational force, but he feels weightless.

Why, then, do we believe in this gravitational force that we cannot feel? We need the gravitational force to account for the gravitational acceleration, which we observe in the motion of falling bodies, projectiles, and orbiting satellites. Where there is an acceleration, according to the second law of motion, there must be a force *causing* the acceleration. Where no force is acting, according to the first law of motion, we have motion in a straight line at constant speed. If we abandon gravitation as a force, we must also abandon the laws of motion in their present form. We must allow motion in a curved line without force, and there must be something in addition to force that determines the path in which an object moves.

If the gravitational force is indeed like the "fake" forces we have discussed, then it too should depend upon our reference frame. In fact, by properly choosing our reference frame we can make the gravitational acceleration disappear locally—that is, in small regions. Consider again the astronaut in his orbiting satellite. In a reference frame fixed to the earth, he is accelerating with the gravitational acceleration. We normally consider this gravitational acceleration to be the result of the gravitational force. Let us consider, however, the reference frame fixed to the astronaut himself and accelerating with him. The astronaut experiences no acceleration in this reference frame—indeed, he is not even moving in it. Nor does he feel any gravitational force—he feels weightless. The effect we are describing is a local one, however. Although this choice of reference frame greatly simplifies the motion of the astronaut, it certainly does not simplify the motion of the earth, the moon, satellites in other orbits, or most other falling objects. Other "fake" forces would have to be assumed or discovered to account for the accelerations of these other objects.

This discussion gives us the clue to how a general theory of relativity should treat these apparent forces found in accelerated reference frames. It should treat them as forces of "gravitation." On the rotating disk, therefore, we shall try to assume that the disk is at rest and that there is some "gravitational" force pulling the objects outward. Will this trick work? The only way to find out is to work out the consequences mathematically and then compare these consequences with experiment. Einstein and others successfully accomplished this.

Because of the principle of equivalence, the general theory of relativity can treat the apparent forces in accelerated systems as "gravitational" forces and can also treat gravitational forces, in a sense, as apparent. Thus, the general theory of relativity is also a new theory of gravitation, replacing Newton's theory.

Curved Space

The foregoing argument has an interesting implication: the general theory of relativity leads us to believe that space (and space-time) is curved. It is difficult to visualize curvature in the three-dimensional space in which we live, but it is easy to visualize curvature of a surface. Therefore we shall start by discussing a curved surface—the surface of a sphere.

Figure 10-18 shows an object that we shall assume is free to move without friction on the inside surface of a stationary sphere but is not allowed to leave that surface. We imagine the sphere to be far out in space, so that gravitational forces on it and the object are negligible. Once set into motion on the sphere, the object will continue to move and never stop, since there is no friction. Its motion will not be in a straight line, however, because it cannot leave the sphere. Its path will be what is called a *great circle.* A great circle is a circle on a

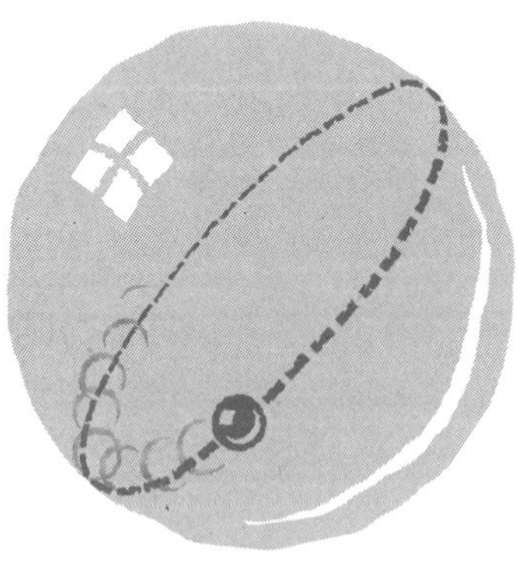

Fig. 10-18

sphere whose center is the same as the center of the sphere. For example, the lines of longitude (meridians) and the equator are along great circles on the earth. Except for the equator, the "parallels" of latitude are not great circles. The object must move in a great circle because, in the absence of friction, the only forces on it are perpendicular to the surface of the sphere, and therefore act on the object as centripetal forces. The center of the object's path must be the point to which these centripetal forces are directed—the center of the sphere.

Great circles have properties on the sphere that are very similar to the properties of straight lines on a flat surface. If, in the previous example, the surface had been flat instead of spherical, the object would have moved in a straight line instead of a great circle. Also, the shortest path between two points on a flat surface is along a straight line. The shortest path between two points on a sphere lies along a great circle—assuming that the path remains on the surface of the sphere. This is why ships and airplanes travel along great circles on the earth—provided that weather, natural obstacles, and unfriendly nations permit it.

The shortest path between two points on any surface is called a *geodesic*. A straight line, then, is a geodesic on a flat surface, and a great circle is a geodesic on a sphere. Other surfaces also have geodesics. Figure 10-19 shows one way of visualizing the geodesics on a surface. Let us imagine a frictionless rubber band stretched between two points, free to adjust itself to the shortest length but forced to stay on the surface. This rubber band will lie along the geodesic between the two points. In three-dimensional space geodesics have the same property they have on a surface: they are the paths of shortest length between any two points. Ordinarily we think of straight

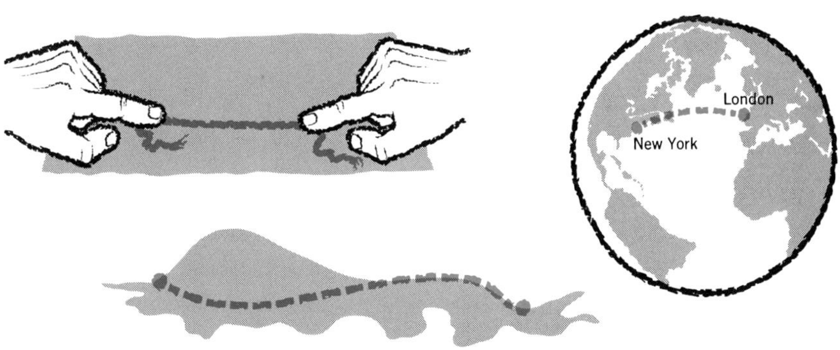

Fig. 10-19

lines as the shortest distance between two points. If space is curved, however, then the shortest path is not necessarily along a straight line.

The centripetal forces are the only forces on the object in Figure 10-18, and they have no component along the surface of the sphere. Let us now suppose that the surface of the sphere is the only space that there is, that we live on the sphere, and that only the surface of the sphere is accessible to our observations. Then these centripetal forces would be invisible or inaccessible to us. We would simply observe the object taking the straightest path available (a great circle), and it would probably not occur to us to look for any forces.

The above example suggests, by analogy, how a curved space might cause an object to move "naturally" in a curved path (a geodesic) without any apparent force. We can extend this analogy to show that curved space may cause objects to be accelerated toward one another without any force of attraction between them. Suppose that there are two objects moving without friction on the surface of a sphere. If they are moving parallel to each other at the outset, in the positions indicated in Figure 10-20, they will end up striking each other where the two great circles cross. To an observer on the sphere, however, it might appear as though the two objects were subject to a force of attraction that pulled them together.

Our analogies here are far too simple, but they do show that the effects demanded by a general theory of relativity might be supplied by a suitably curved space. The general theory of relativity involves not merely the curvature of a surface but the curvature of three-dimensional space, and of space-time itself. According to the general theory of relativity, there is a curvature surrounding objects that have

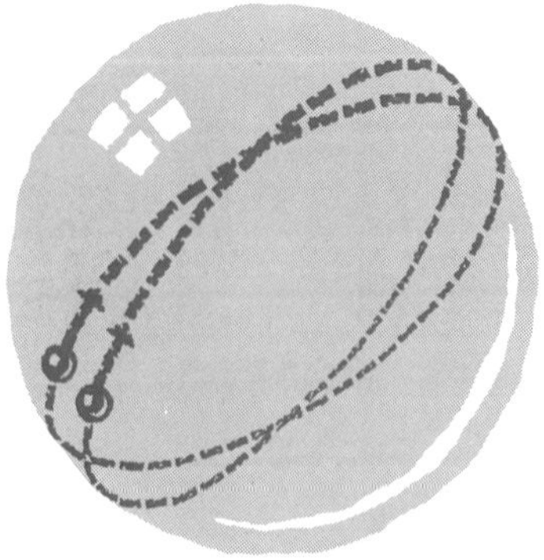

Fig. 10-20

mass. This curvature accounts for the gravitational attraction ascribed to these objects. The objects, actually following geodesics, appear to be attracted to each other.

Up to now, we have been using reference frames with straight-line axes. On a sphere, however, there are no straight lines; indeed, in curved spaces we may have to use curved axes. We therefore extend our idea of reference frame as far as the general principle of relativity is concerned. We allow the reference frames mentioned in that principle to include frames with curved axes. Figure 10-21 shows how curved axes can be used on a sphere. This reference frame uses lines like the ordinary lines of latitude and longitude. The laws of nature should not depend on the method we use to identify points in space and time.

Suppose that we are beings living on a spherical surface (such as the earth). How can we be sure that it is spherical without seeing the whole surface at once? To answer this question, we draw two large circles, one on a spherical surface and the other on a flat surface, as in Figure 10-22. We measure the circumference and diameter of each circle. If we divide the circumference of the circle on the flat surface by its diameter we obtain a constant number, usually represented by the Greek letter *pi* (π), and approximately equal to 3.14. When we perform the same operation on the sphere, however, we get a different result. The "true" diameter cannot be measured, because it does not lie on the surface; therefore, we must measure the curved "diameter," as indicated in the figure. With the same circumference, we have a different "diameter" and therefore we obtain a different number when we perform the division. The fact that we obtain a number different from pi indicates that the surface is curved. The

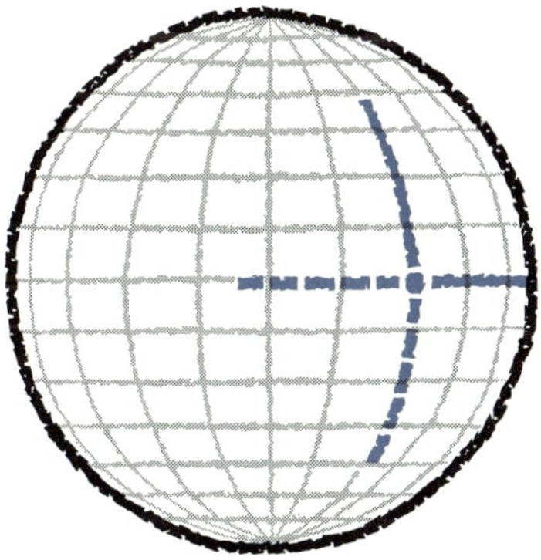

Fig. 10-21

same principle applies to circles drawn in space. (It is, of course, possible for space to be curved in one region and flat in another.)

The technique described above, although not sufficiently accurate to determine the curvature of space in our vicinity, enables us to discuss a thought experiment that gives us one more reason to think that a general theory of relativity would involve curved spaces. Let us suppose that a man on the rotating disk of Figure 10-23 is trying to measure the circumference and diameter of the disk. If we check his work from a reference frame that is not rotating with the disk, we will agree with his measurement of the diameter but not with his measurement of the circumference. When he measures the circumference his ruler will be oriented along the direction of motion and therefore shortened by the Lorentz contraction. If space is not curved, as viewed from our reference frame, then we will obtain the number pi when we divide the circumference by the diameter. Because of the contraction of his ruler, however, he will not obtain pi, and therefore he will conclude that space, viewed from his reference frame, is curved.

This analysis, of course, is given to show that we may expect curved spaces when we deal with rotating reference frames; it is *not* intended as a rigorous analysis of the rotating disk. We must be careful in applying the results of the special theory of relativity to accelerating reference frames. However, our analysis again suggests that we may expect curved spaces in the general theory of relativity, which *does* include accelerating reference frames.

Many people are puzzled by the idea of curved space; they find it hard to visualize curved space in three dimensions. So do we. But we are concerned, not with the visualization of the space, but with

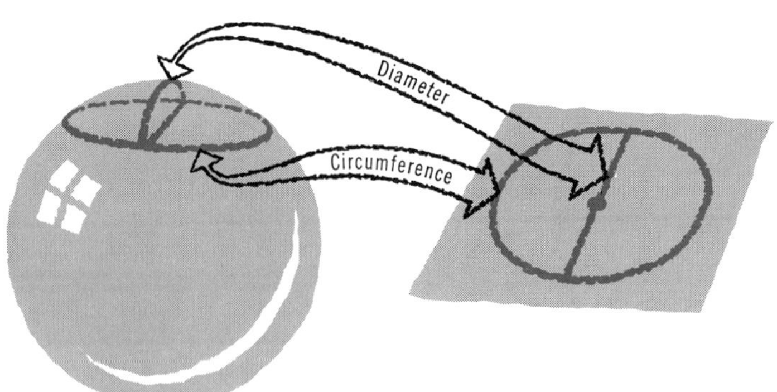

Fig. 10-22

its properties. If we measure a real circle and find that the ratio of the circumference to the diameter is the familiar value pi, then we conclude that the space is flat, and the ordinary flat "high school" geometry of Euclid applies. If this value turns out to be different from pi, then we have two possibilities. We can say that space is curved, since it behaves as we would expect it to behave if it were curved. On the other hand, we could try to save the situation by saying that something went wrong with our measuring instruments. If all our instruments always go wrong in precisely the same way, however, we might consider it simpler just to admit that space is curved. For example, we might insist that the earth is flat. When two airplanes fly directly north, however, they will still appear to be getting close together as they approach the North Pole. To explain this, we could assume that our measuring devices were affected as we went north so as to give this "illusion." It is much easier just to admit that the earth is curved. Of course, we can save almost *any* theory if we are willing to make enough extra assumptions and explanations. We can insist that nature is doing anything our hearts desire if only we postulate enough unobservable mechanisms to explain why it appears to be doing something entirely different. As physicists, however, we shall try to accept the geometry that our physical observations suggest to us.

Einstein's General Theory of Relativity

As we have seen, there are two main principles to be included in a general theory of relativity.

1. *The (General) Principle of Relativity.* The laws of physics may

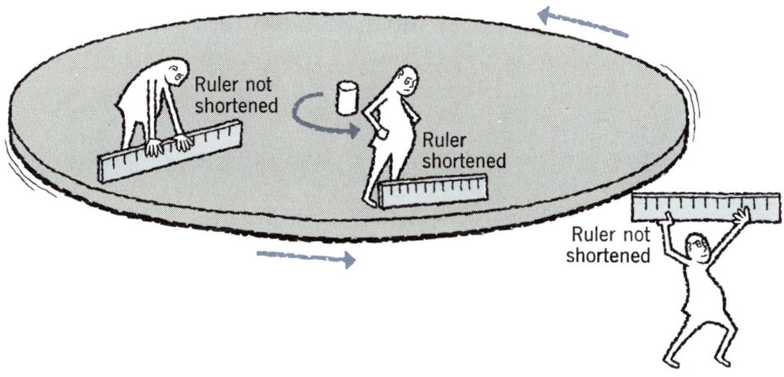

Fig. 10-23

be expressed in the same form in any reference frame, whether it is accelerating or not, whether its axes are curved or straight.

2. *The Principle of Equivalence.* There is no locally measurable difference between the effects of a gravitational field and the effects that arise from the acceleration of the reference frame.

The principle of equivalence makes it clear that the general theory of relativity is a theory of gravitation. As we have seen, we expect the general theory to lead to curved space and curved space-time, and it does. Gravitational effects arise from the curvature of space and space-time. In regions of small curvature and small gravitational fields, we expect that the physics of Einstein's special theory, rather than Newtonian physics, will be approximately valid.

Albert Einstein was the first to formulate a successful general theory of relativity which turned out to have the consequences which we have discussed. Ten years after his first publication on the special theory of relativity, he published his first paper on the general theory. In this theory both space and space-time are curved, especially in the vicinity of material objects, causing gravitational effects. Far away from material objects, space-time may be flat, or it may continue to curve.

Experimental Evidence for the General Theory

Is the general theory of relativity really correct? There is some evidence that it is superior to Newton's theory of gravitation in predicting certain very small effects. Since it takes into consideration Einstein's modern theory of special relativity, while Newton's theory of gravitation does not, the general theory is logically superior to Newton's theory. Nevertheless, it must be admitted that the theory has not been tested adequately. Experiments performed to date do not fully test the theory, and the effects that verify it are small and difficult to detect.

The general theory of relativity predicts that light will be deflected as it passes by a large body, such as the sun. This deflection may be considered as made up of two parts: first, a deflection resulting from the falling of the light toward the sun that can be accounted for either by Einstein's or Newton's theory; and second, an additional, equal deflection arising from the curvature of space that is not explained by Newton's theory. (The first deflection can be thought of in Newtonian terms as a result of the gravitational pull of the sun on the mass-energy in the light waves.) As far as we can tell, the actual deflection of starlight as it passes by the sun is in better accord with the new theory than with Newton's. This deflection can be observed in the shift of star positions during a total eclipse of the sun (see Fig. 10-24).

Mercury, in its orbit, comes closer to the sun than any of the other planets. Its orbit is sufficiently different from a circle to permit us to determine the point of closest approach (called the perihelion). If we consider all the forces on Mercury from the sun and the planets, we can account for almost all of the changes in the position of the perihelion as time goes on. There is left unaccounted for a change of 43 seconds of arc per century. This very small effect is not accounted for by Newton's theory, but it was accounted for by Einstein's theory. Just recently, however, R. H. Dicke discovered an oblateness of the sun. The fact that the sun is not a perfect sphere could account for about four seconds of the change in the position of Mercury's perihelion, possibly requiring some revision of the general theory of relativity.

According to the general theory of relativity, a clock in the gravitational field of a large mass should appear to run slow if observed from a point further away from the mass. This means that light emitted from stars and from the sun should appear to us to be of lower frequency, and therefore should look redder than light emitted from the same kind of atoms in a laboratory on earth. This effect appears to be observable in sunlight and is observable in the light from white "dwarf" stars (stars that are very small, dense, and hot). We call this effect the *gravitational red shift.* The red shift of other stars may be explained by the Doppler effect as well as by the general theory, so that it is still doubtful whether the gravitational red shift is actually observed in light from stars other than white dwarfs.

Recently it has been possible, by very accurate methods, to measure

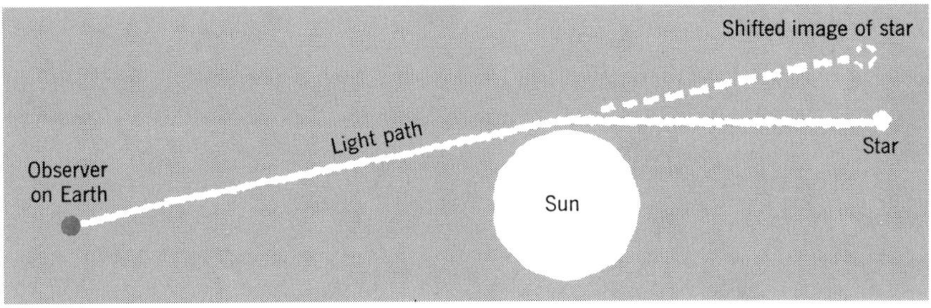

Fig. 10-24

the change in the frequency of light falling in the earth's gravitational field, an effect which tests the same principle as the gravitational red shift. Light is emitted at the top of a tower and detected at the bottom, where the change in frequency is observed. The increase in frequency corresponds to what would be expected from the general theory of relativity. The results of this experiment and the gravitational red shift are predicted by the general theory of relativity and are therefore consistent with it; however, they do not constitute a strenuous test of that theory because they can be explained without it and are consistent with other theories of gravitation.

There have been other tests suggested and at least one that has been performed, but we shall not discuss them here. We can say that the general theory of relativity is correct as far as we know, but more stringent tests would be very welcome.

Mach's Principle

To explore a very interesting, unanswered question pertaining to the general theory of relativity, let us imagine that we have the water pails shown in Figure 10-25. In Figure 10-25*a* the pail is at rest, and the surface of the water is flat. In Figure 10-25*b*, however, the pail is spinning around a vertical axis, and the water appears to be pulled out toward the edges to some extent. This effect arises from the same kind of apparent outward pull that is exerted on an object on a rotating disk. The fact that the surface of the water in the rotating pail is not flat suggested to Newton that the pail (and the water in it) was indeed rotating, not at rest. He believed, therefore, that accelerated

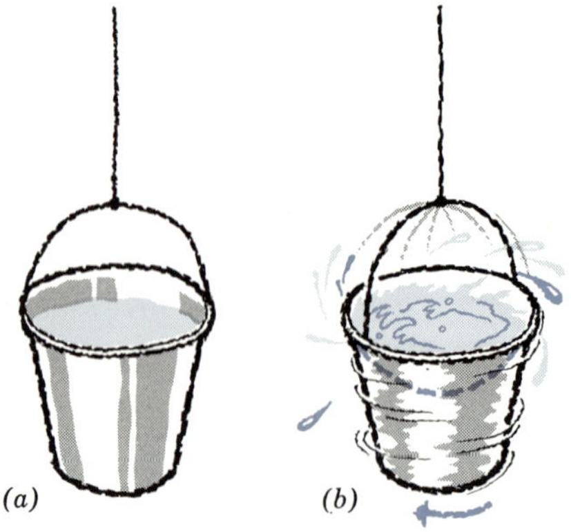

(a) (b)

Fig. 10-25

motion was an *absolute* motion, not merely motion relative to another object. Ernst Mach (1838–1916) disagreed with this conclusion, pointing out that the spinning pail is in relative rotation with respect to the stars but that the stationary pail is not. If it were possible to keep the pail "stationary" and move the stars and the rest of the universe around it, he believed, the surface of the water would once again be curved. To him it was the relative motion of the spinning pail that mattered, not its assumed absolute motion.

In order for the stars, in hypothetical motion around the pail, to exert force on the water in the pail, they would have to exert some other force beside their gravitational attraction. Some additional force would have to arise from their motion. Now we know that electric charges do indeed have an additional force between them when they are in relative motion—the magnetic forces. If moving masses had a similar force between them, of the right strength and direction, then Mach's belief might be proved true. So far, however, there is no experimental evidence to prove such an effect. True, according to the general theory of relativity there is an added gravitational force between objects in motion. However, it is unknown whether or not this force would cause the effect Mach predicted. Thus we do not yet know whether or not the general theory of relativity agrees with Mach's belief. The general theory of relativity is, in fact, so complex that it is often very difficult to find out what consequences follow from it.

Another aspect of the problem Mach considered is suggested by the elevator in Figure 10-14 (p. 420). Suppose that the man in the elevator looked out a window. Then he might find out that the elevator was being pushed from outside. If he were a determined relativist, however, he could still insist on considering himself at rest. But he would then need to find another force on the elevator that he could think of as counteracting the force pushing up on it from below. Then he could consider the elevator to be in equilibrium. The general theory of relativity, as we have seen, allows him to invent a force, similar to gravity, pulling the elevator so as to oppose the force from below. This same invented force of gravitation is the one that would account for the apparent fall of the objects dropped in the elevator. In the general theory of relativity, however, the force of gravitation arises in the change of reference frame, and no other cause or source for it is necessarily given. If the man in the elevator looked out at the stars, however, they should appear to be accelerating past his elevator. Perhaps, he might think, the accelerating stars cause a force on his elevator. Perhaps they cause the "gravitational" force he needs. Whether or not this effect would really occur is unknown at present. If these proposed effects of the accelerating stars prove to be a reality, they would explain why there is such a difference between accelerating reference frames and frames moving at uniform velocities. In this event, the accelerated frames would not be in some absolute acceleration but would merely be accelerated with respect to the average motion of the stars. This would be a complete vindication of the principle of relativity of motion.

In both of the examples discussed in this section we have suggested that the "fake" or apparent forces observed in accelerating reference frames can be ascribed to the motion of other objects, and therefore might be thought of, not as apparent, but as real—at least as real as the gravitational forces which can be ascribed to the position of other objects. This suggestion is often called Mach's principle, since it is a generalization of Mach's point of view. Mach's principle is sometimes regarded as one of the basic postulates of the general theory of realtivity, along with the general principle of relativity and the principle of equivalence.

Problem solving

We are concerned in this chapter with the use of the following equations (In this section, v is used in place of s to represent the speed.):

$$E = mc^2 \tag{6}$$

$$t = \frac{t_0}{\sqrt{1 - v^2/c^2}} \tag{7}$$

$$L = L_0 \sqrt{1 - v^2/c^2} \tag{8}$$

$$m = \frac{m_0}{\sqrt{1 - v^2/c^2}} \tag{9}$$

Example 1

A 10-kilogram ball is traveling at 1000 meters per second. How much mass does the kinetic energy have?

Solution.

Step 1 This problem requires us to find the amount of mass involved in a certain amount of energy; therefore, the proper formula is

$$E = mc^2$$

STEP 2 The known and unknown quantities are as follows:

$$E = ?$$
$$m = ? \text{ (Note that this is not the mass of the ball.)}$$
$$c = 3 \times 10^8 \text{ meter/second}$$

STEP 2a Look for more information or equations; we cannot solve with what we have. Since the speed is low compared to the speed of light, we can use the old formula for kinetic energy:

$$E = 1/2\ Mv^2$$

in which

$$M = 10 \text{ kilograms (the mass of the ball)}$$
$$v = 1000 \text{ meters/second}$$

then

$$E = 1/2 \times 10 \times (1000)^2 \text{ kilograms } \frac{\text{meter}^2}{\text{second}^2}$$
$$= 5 \times 10^6 \text{ kilogram } \frac{\text{meter}^2}{\text{second}^2}$$

STEP 3 Substitute the quantities in Steps 2 and $2a$ into the equation $E = mc^2$ and solve:

$$5 \times 10^6 \frac{\text{kilogram meter}^2}{\text{second}^2} = m(3 \times 10^8)^2 \frac{\text{meter}^2}{\text{second}^2}$$
$$5 \times 10^6 \text{ kilogram} = m \times 9 \times 10^{16}$$
$$m = \frac{5}{9} \times \frac{10^6}{10^{16}} \text{ kilogram}$$
$$= 5.6 \times 10^{-10} \text{ kilogram}$$

This result shows that the increase of mass caused by a velocity of 1000 meters per second is extremely small. (Could Equation 9 have been used to obtain this result?)

EXAMPLE 2

How fast will a ruler have to move in the direction of its length in order to appear to a stationary onlooker to be half its proper length?

Solution.

STEP 1 We are dealing with the Lorentz-contraction equation:

$$L = L_0 \sqrt{1 - \frac{v^2}{c^2}}$$

STEP 2 The known and unknown quantities are as follows:

L_0 = proper length at rest
$L = L_0/2$ = shortened length as observed from stationary system
$v = ?$
$c = 3 \times 10^8$ meter/second

STEP 3 Substitute and solve.

$$\frac{L_0}{2} = L_0 \sqrt{1 - \frac{v^2}{c^2}}$$

$$1/2 = \sqrt{1 - \frac{v^2}{c^2}}$$

Square both sides:

$$1/4 = 1 - \frac{v^2}{c^2}$$

$$\frac{v^2}{c^2} = 1 - 1/4 = 3/4$$

$$\frac{v}{c} = \sqrt{\frac{3}{4}} = \frac{\sqrt{3}}{2}$$

$$v = 1/2\sqrt{3}$$

$$= 1/2\,(1.73)$$

$$= .87c = .87 \times 3 \times 10^8 \frac{\text{meter}}{\text{second}}$$

$$= 2.6 \times 10^8 \frac{\text{meters}}{\text{second}}$$

Notice that the ruler is close to the speed of light when it shows this much contraction.

EXAMPLE 3

For an object with the speed found in Example 2, what is the total mass in terms of the rest mass, and what is the kinetic energy in terms of the rest mass?

Solution. STEP 1 We will be using the following equations:

$$E = mc^2$$

$$m = \frac{m_0}{\sqrt{1 - v^2/c^2}}$$

STEP 2 We list the known and unknown information:

$$c = 3 \times 10^8 \text{ meter/second}$$
$$v = \frac{\sqrt{3}}{2} c$$
$$m_0 = m_0, \text{ "known"}$$
$$m = ?$$
$$E = ?$$

We often find it convenient in relativity problems to express v in terms of c. We consider m_0 known because we are allowed to use it in our answer.

STEP 3 Substitute and solve:

$$m = \frac{m_0}{\sqrt{1 - v^2/c^2}} = \frac{m_0}{\sqrt{1 - 3/4c^2/c^2}} = \frac{m_0}{1 - 3/4}$$
$$= \frac{m_0}{\sqrt{1/4}} = \frac{m_0}{1/2} = 2\ m_0$$

Thus the total mass is twice the rest mass. This means that the excess mass arising from the speed is equal to the rest mass. This excess mass, equal to m_0, must represent the kinetic energy. To see how much energy the mass m_0 represents, we must use the equation $E = mc^2$. Thus this kinetic energy is m_0c^2.

EXAMPLE 4

Show that the time-dilation equation (Equation 1) is correct.

Solution. Figure 10–26 shows the moving light clock in three successive positions. The diagonal lines represent the path of the light pulse as it starts at the upper mirror, moves to the lower mirror, and then moves back to the upper mirror. We have marked distances ct_0, ct, and vt on the diagram. Each of these distances is obtained from the fact that the distance traveled by a uniformly moving object is equal to its speed multiplied by the time.

The time t is the time it takes the light pulse to travel between the moving mirrors as observed from the "stationary" reference frame.

Since the pulse moves at the speed c, it covers a distance ct in this time.

At the same time the clock is moving to the right with a speed v. Therefore it travels a distance vt while the pulse is traveling a distance ct.

The distance between the mirrors is the same in either frame, since the relative motion is horizontal, not vertical. The symbol t_0 represents the time for the pulse to move from one mirror to the other as observed in the reference frame fixed to the light clock. In that reference frame, light also moves with the speed c, and therefore the distance between the mirrors is ct_0.

We have a right triangle in Figure 10–26, in which ct is the length of the hypotenuse, and ct_0 and vt are the lengths of the legs. Therefore, we apply the Pythagorean theorem (see Appendix II-D) and obtain:

$$c^2t^2 = c^2t_0^{\,2} + v^2t^2$$
$$c^2t^2 - v^2t^2 = c^2t_0^{\,2}$$
$$t^2(1 - v^2/c^2) = t_0^{\,2}$$
$$t^2 = \frac{t_0^{\,2}}{(1 - v^2/c^2)}$$
$$t = \frac{t_0}{\sqrt{1 - v^2/c^2}}$$

Summary

I. FUNDAMENTAL LAWS AND PRINCIPLES

A. EINSTEIN'S TWO POSTULATES for the special theory of relativity.

1. The special principle of relativity.

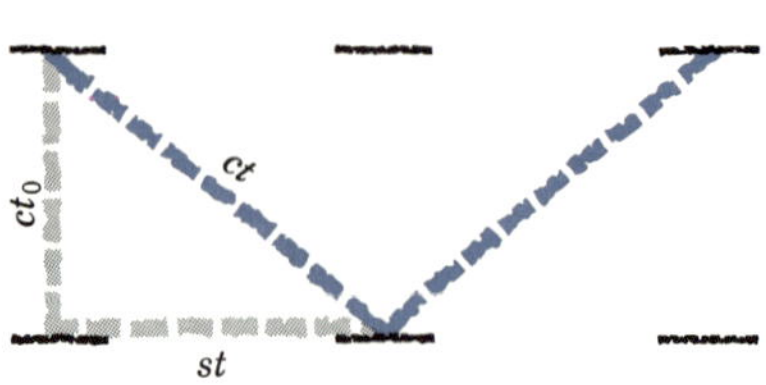

Fig. 10-26

2. The constancy of the speed of light, regardless of the reference frame.

*B. PRINCIPLES IN GENERAL RELATIVITY
1. The general principle of relativity.
2. The principle of equivalence.
3. Mach's principle.

II. CONSEQUENCES AND EVIDENCES OF RELATIVITY
A. THE SPECIAL THEORY
1. Consequences.
a. Time Dilation.
b. Lorentz contraction.
c. Increase of mass with speed.
d. $E = mc^2$
e. Limiting nature of c.
f. The fourth dimension.
g. Problem of simultaneity.
2. Evidences.
a. Michelson-Morley experiment.
b. Negative evidence from aberration of starlight and binary stars.
c. Lifetime of moving muons.
d. Change of electron mass in a cyclotron.
e. Atomic bomb and similar processes.

B. THE GENERAL THEORY
1. Consequences.
a. Curved space.
b. New theory of gravitation.
2. Evidences.
a. Deflection of starlight near the sun.
b. Change in perihelion of Mercury.
c. Gravitational red shift.
d. Dropping of light.

Questions

1. Would the mass of the earth-moon system be greater or less if the moon and the earth were pulled closer together?

2. Do two marbles weigh more or less when they are apart than when they are together?

3. Is a hot biscuit heavier than a cold one? Would you expect to be able to detect the difference in weight?
4. A die-hard flat-earther claims that the earth is stationary, citing as evidence the results of the Michelson-Morley experiment. How might he argue, and how would you reply?
5. Why was the ether invented, and why was it abandoned?
6. Suppose that the speed of light were 100 miles per hour.
   (a) Could you drive a distance of 500 miles in less than five hours? Explain.
   (b) Could you use the Lorentz contraction to help reduce? Explain.
   (c) Would the effects of car accidents at comparable speeds be worse or better than they are now?
   (d) Would it still be practical to send out time signals over the radio?
   (e) Would radar be obsolete?
   (f) Would it be necessary to have speed limits? If the speed limit were close to 100 mph, how would you suggest that the police measure speed?
7. A moving ruler of rest length one meter is passing through the laboratory and is shortened by the Lorentz contraction so that it loses 10 percent of its length. It is caught in a sidewise swipe, as shown in Figure 10–27, and thus is imprisoned in a box of total length less than one meter. If we look at the process in a reference frame at rest with respect to the ruler, however, the ruler is its proper length and the box is even shorter, because of the Lorentz contraction. Thus, in that system it appears that the ruler has been caught in a box smaller than it is. How is this possible?
8. A certain star is 1000 light years away. Could a human being live long

Fig. 10-27

enough to make a journey there, according to the laws of physics? (A light year is the distance that light travels in one year.)

9. A teenager claims to be the mother of a grayhaired old lady. Is this physically possible? Explain.

10. Which of the following processes increase the mass of a body of matter and which ones decrease it: boiling, freezing, shining sunlight on the object, starting it in motion, cutting it in two, stretching it, bringing it to rest, cooling it, expanding it, contracting it, twisting it, spinning it?

11. Compare the results of the "boat race" and the Michelson-Morley "light race." Which contestant won each race and why?

12. If a hunter riding in an open jeep stands up and fires his gun in the direction in which the jeep is traveling, will the bullet travel faster than it would if he had fired when standing on the ground? Explain.

13. If a hunter riding in an open jeep at night stands up and directs a powerful searchlight in the direction in which the jeep is traveling, will the light travel faster across the ground than it would if he were standing on the ground with the searchlight? Explain.

14. Explain the operation of a light clock when it is at rest and when it is moving.

15. Could time dilation occur without Lorentz contraction? Explain.

16. What was measured in the muon experiment, and how does it support the predicted time dilation of special relativity?

17. What is the difference between rest mass and ordinary mass?

18. Why did people think there was an ether "wind" blowing by the earth?

19. What does the aberration of starlight have to do with the theory of relativity?

20. What do double stars have to do with the theory of relativity?

21. Why, when we are interested in time and space, do we speak of clocks and measuring rods instead?

22. How can we make a rocket move faster than light?

23. Why can we use a light clock in our thought experiment instead of an ordinary clock? Why should this special type of clock give the same results as any other clock? For that matter, why is this thought experiment valid—isn't an actual experiment required to prove something?

24. How can we be sure of the results of the muon experiment since the same muons are not used in both parts of the experiment?

25. Explain what is meant by the statement, "Time is a property of clocks."

26. If we keep applying a force on an object, why won't it eventually go faster than light?

27. Why aren't people blown off the earth by the ether wind? (Give a reason other than the fact that there isn't any ether.)
28. Why did many scientists believe in an ether? Why did Einstein adopt his two postulates.
29. Why is it difficult to measure the changes of mass resulting from the ordinary changes in an object's energy?
30. What has happened to the conservation laws as a result of the theory of relativity?
31. What is space? What is time?
32. Is the field material?
*33. What is the fourth dimension?
34. The speed of light c is often called a limiting speed. What does it limit?
*35. Why is space three-dimensional?
*36. Why is time called the fourth dimension?
*37. What is space-time?
*38. Why will problems result from defining light speed to be the same in both directions in each of two reference frames moving with respect to each other?
*39. What does the principle of equivalence have to do with the equality of gravitational and ordinary mass?
*40. Does the principle of equivalence state that a gravitational field is the same as the result of acceleration?
*41. How can you tell whether or not space is curved?
*42. How would you go about defending the thesis that the earth is flat?
*43. Is there such a thing as gravitation? Is there such a thing as gravitational force?
*44. How do we know that the general theory of relativity is valid?
*45. State Mach's principle in your own words.
*46. What is a geodesic?

Problems

1. What will the length of a meter stick appear to be if the meter stick is traveling in a lengthwise direction at 3/5 the speed of light?
2. How much will a clock be slowed down if it is moving at 4/5 the speed of light?
3. How much will the mass of a body be increased if it is moving at 1/13 the speed of light?

4. How much energy is represented by one kilogram of mass?
5. A ball has a mass of one kilogram. If you could exert a force of one newton, how far would you have to push it before it had twice that mass?
6. An electron has a mass of 9×10^{-31} kilogram. How fast would it have to move in order to have the mass of the earth? (The mass of the earth is 6×10^{24} kilograms.)
7. Two bodies, each of rest mass one kilogram, are traveling straight at each other. They collide and stick together. They are each going at 3/5 the speed of light at the time of the collision. What is the rest mass of the final stuck-together mass?
8. If we give one of the masses in Problem 7 twice the energy it had in that problem and allow it to strike the other one at rest, will the rest energy of the stuck-together mass be the same as before? Why or why not?
9. One particle has twice the rest mass of another. Does it still have twice the mass of the other particle when they are both moving at 9/10 the speed of light?

More Difficult Questions and Problems

1. Show that the boat that goes across stream wins the race.

*2. Show that it is possible for event *A* to precede event *B* in one system while the order is reversed in another system.

*3. Show that the results of Problem 2 cannot make a cause occur after an effect.

*4. Devise a surface on which the ratio of the circumference of a circle to its diameter is greater than pi.

*5. We have suggested that space is a property of our measuring rods. We also know that heat will cause rulers to expand. Why do we not consider, then, that heat contracts space?

6. A body is moving at 9/10 the speed of light. If its energy is doubled, how fast will it be going?

*7. Make the best case you can to show that the force of gravitation is a fake. Why is it more difficult to make the same claim for the electromagnetic force?

*8. If instead of gravitational and inertial mass always being equal, it turned out that the first was always twice the second, would this destroy the principle of equivalence? Explain.

*9. Why was the notion of absolute simultaneity abandoned?

*10. What difficulty arises in trying to measure the one-way speed of light?

*11. How many dimensions are required by the following quantities or combinations of quantities: (a) temperature, (b) the direction of an arrow, (c) the direction and position of an arrow in space, (d) the state of motion of a particle, (e) the state of motion of two particles, (f) the state of motion of two particles connected by a rigid rod, (g) the position and motion of a single particle.

*12. If clocks are set as Einstein's theory requires in two systems moving uniformly in a straight line with respect to each other, show why two events that are simultaneous in one system may not be simultaneous in the other.

*13. We (a) define or (b) assume the one-way speed of light in a vacuum to be equal to the two-way speed. Which word is correct and why? Why is the other choice incorrect?

*14. Are astronauts in orbit truly weightless? Consider both Newton's theory and Einstein's. Which do you think corresponds better to appearances?

15. A clock in either of two reference frames in relative motion appears to be running slow as observed from the other frame. Thus each of two clocks could appear to be running slower than the other. Can you resolve the apparent paradox?

16. A ruler in either of two reference frames in relative motion appears to be shortened when observed from the other frame. Thus each of two rulers could appear to be shorter than the other. Is this possible? Explain.

17. How can we be sure that the speed of light is really the same in a moving system as in a stationary system, in view of the fact that the clocks and rulers we use to measure it may be affected by time dilation and the Lorentz contraction?

18. We have discussed time in terms of the behavior of clocks. However, a pendulum clock will also run slow if it is immersed in a vat of molasses. Does this mean that time slows down in molasses? Explain.

19. A source of light emits a pulse of light. Show that the source remains in the center of the pulse of light if it is observed in a reference frame moving with the source (Fig. 10-28*a*) but not if it is observed from other reference frames.

*20. Does Newton's theory of gravitation account for the deflection of starlight passing near the sun? Why do we need Einstein's theory?

*21. Why do we think of three-dimensional space as curved when we cannot visualize another dimension of space into which it can curve?

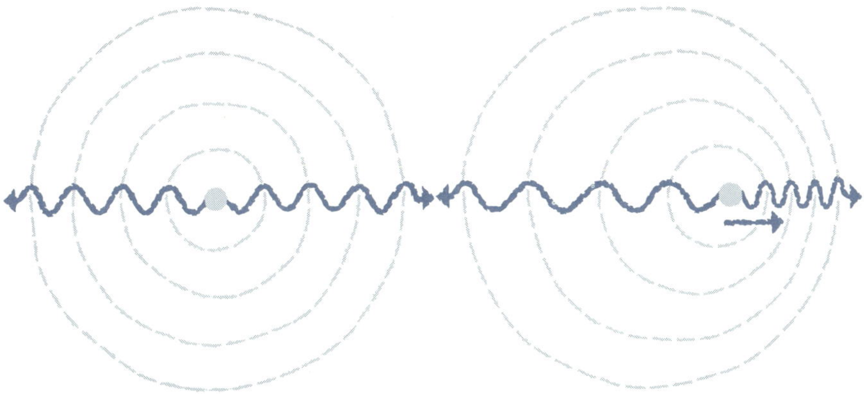

(a) *(b)*

Fig. 10-28

11

QUANTUM PHYSICS

. . . There was a time when the newspapers said that only twelve men understood the theory of relativity. There might have been a time when only one man did, because he was the only guy who caught on, before he wrote his paper. But after people read the paper a lot of people understood the theory of relativity in some way or other, certainly more than twelve. On the other hand, I think I can safely say that nobody understands quantum mechanics. So do not take the lecture too seriously, feeling that you really have to understand in terms of some model what I am going to describe, but just relax and enjoy it. I am going to tell you what nature behaves like. If you will simply admit that maybe she does behave like this, you will find her a delightful, entrancing thing. Do not keep saying to yourself, if you can possibly avoid it, "But how can it be like that?" because you will get . . . into a blind alley from which nobody has yet escaped. Nobody knows how it can be like that.

Richard P. Feynman[1]

Particles of Light

Planck's Constant

In 1900 Max Planck (1858–1947) used a revolutionary trick to solve a difficult problem concerning electromagnetic waves. The problem had to do with energy contained in the waves that should be found in

[1]Richard Feynman, *The Character of Physical Law*, Cambridge: Massachusetts Institute of Technology Press, 1967, p. 129.

an enclosed cavity in a material object at a given temperature (see Fig. 11-1). Since the object has internal energy, its molecules are vibrating and hence accelerating. Charged particles in the molecules also are accelerating and cause waves in the electromagnetic field. Therefore, in a cavity (with a small hole for observation) there should be electromagnetic waves bouncing around, being absorbed, reflected, and emitted by the walls of the cavity. The attempt to account theoretically for the energy contained in these waves according to the known laws of physics was unsuccessful—it kept turning out that there should be an infinite amount of energy in the waves, which of course was impossible. This was a theoretical catastrophe.

Planck avoided the catastrophe by assuming that the waves could be emitted only in discrete "bundles" of energy. The higher the frequency, the more energy there was in the bundle that had to be emit-

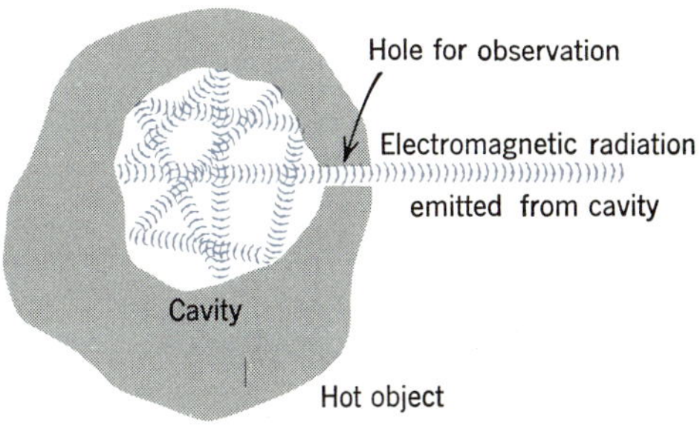

Fig. 11-1

ted. According to his calculations, this made it harder to emit at high frequency, so hard that one no longer got an infinite amount of energy in the waves. This assumption corrected the situation, and Planck was able to obtain a realistic law for cavity radiation.

For our purposes, it is not the problem of the cavity radiation itself that is important (we have discussed it only briefly and incompletely) but the fact that its solution led Planck to "discover" the discrete aspect of the electromagnetic field. Planck stated that each of the bundles or packages of radiated waves had an energy given by the formula.

$$E = hf \qquad (1)$$

E is the energy of the bundle and f is the frequency of the emitted wave. The constant h, called Planck's constant, has a value of 6.63×10^{-34} joule second. This new fundamental constant was destined to be as important in physics as the constant c, the speed of light in a vacuum. Today we call such a bundle a *quantum* of light or a *photon*. The photon is a very important fundamental particle.

The Photoelectric Effect

Planck's constant was used by Einstein when he explained the photoelectric effect, which is the emission of electrons from a material surface, such as a metal, when light is allowed to shine on the surface. If a beam of light strikes the surface of a metal the energy of the light is transferred to some of the electrons in the metal, and they pop out (see Fig. 11-2). Now, light waves are waves in the electromagnetic field. As we have seen, these waves can disturb electrons or any charged particle. Therefore, it is entirely reasonable that these waves should disturb the electrons in a metal surface, and even disturb them so much that they are tossed out. However, P. Lenard (1867–1947)

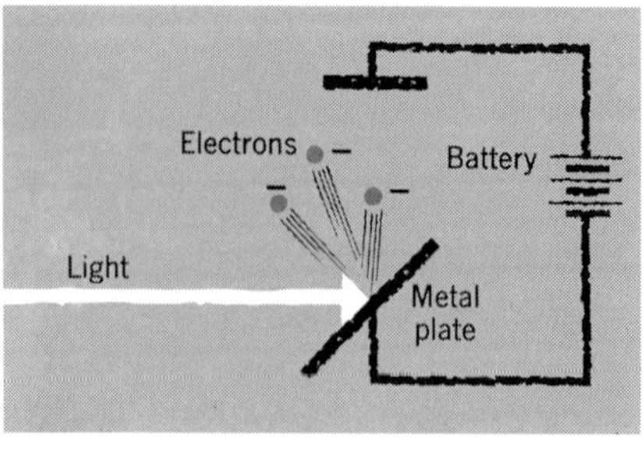

Fig. 11-2

discovered two things about the way in which these electrons are tossed out that did not seem reasonable according to known physical laws. First, let us shine a light of a single color on the metal surface and measure the speed of the electrons that come out of the surface. Then let us make the light brighter, without changing its color. The frequency of the light is therefore still the same, although we would expect the amplitude of its wave to be greater. As a result of this brighter light, more electrons escape from the metal surface. However, the maximum speed with which the electrons leave the metal does not change. This result is rather surprising, since we would think that the more energetic wave (higher amplitude) would give the electrons a stronger toss. As an analogy, consider the bird on the clothesline shown in Figure 11-3. The bird could probably ride out a small-amplitude wave without being thrown off the rope. The big wave, however, should throw him off violently. Electromagnetic waves evidently do not have quite this effect on electrons. The amplitude of the wave apparently does not affect the speed with which the electrons leave the surface.

If we now change the color of the light, so that waves of higher frequency are striking the metal, we *do* increase the maximum speed with which the electrons leave the metal. The greater the frequency, the greater this speed. Below a certain frequency, the electrons do not have enough energy to leave the metal at all.

Fig. 11-3

The known laws of the electromagnetic field were insufficient to give a convincing explanation of these two experimental facts. Einstein, in one of his three famous papers of 1905, suggested that the light waves were absorbed by the electrons in bundles of energy. The amount of energy in the bundle (or photon) depended on the frequency according to Planck's formula, $E = hf$. An electron got either the whole photon or nothing at all. If the photon had too little energy, the electron did not get enough energy to break free from the metal surface. If, however, the frequency were above some minimum, the electron would have enough energy to get out of the metal and still have some energy left as kinetic energy.

Increasing the brightness of the light did not increase the energy of the photons but merely supplied more photons. Therefore it did not increase the maximum energy with which the electrons were emitted. However, it *did* increase the *number* of electrons being emitted since there were more photons striking the metal. If, on the other hand, the frequency of the light was increased, the energy in each photon was greater. The more energetic photons supplied more energy to the electrons, and the electrons were emitted with a higher maximum energy.

We use the term *maximum energy* because some electrons lose more energy than others on their way out of the metal, by colliding with atoms or in other ways, and therefore come out with less kinetic energy than the maximum. They cannot come out with more than the maximum, however. When the electron is expelled with the maximum kinetic energy, the energy of the photon, hf, is accounted for as follows:

$$hf = 1/2\ mv^2\ \text{(maximum)} + W \tag{2}$$

W is called the *work function* of the metal surface and represents the minimum energy an electron must have in order to escape from the metal. Later work showed that Einstein's equation described the photoelectric effect precisely.

The Compton Effect

Another experiment, performed with X rays by Arthur H. Compton (1892–1962), confirmed the existence of photons. X rays are high-energy electromagnetic waves, much more energetic than ordinary light waves. Their wave lengths range from 10^{-8} to 10^{-11} meter (the size of an atom is about 10^{-10} meter). One of the effects that occurs when X rays strike matter is the so-called Compton effect (see Fig.

11-4): A beam of X-ray waves of the same frequency (and wavelength) is allowed to fall upon a sample of matter. The X-ray waves interact with the individual electrons in the matter and are scattered. The frequency in the scattered waves is not the same as the frequency of the original wave but depends on the direction of the scattering.

Compton found that he could predict the correct value for the frequency of the scattered waves if he assumed that the X-ray beam was a beam of particles (photons) instead of waves. He thought of the interaction of the wave with the electron as a collision between two particles, the photon and the electron. By applying the laws of conservation of energy and momentum to this collision, he could find the energy of a photon scattered in any direction. From this energy he could find the frequency of the wave by using Planck's equation, $E = hf$.

In order to apply this method, Compton also used Planck's equation to determine the energy of the photons in the incoming beam. The momentum of these photons could be found from a fact predicted by the laws of electricity and magnetism. The magnitude of the momentum in a series of electromagnetic waves can be obtained by dividing the energy they contain by the speed of light, c (see Example 4, pp. 486).

From these relationships, Compton was able to assign both an energy and a momentum to photons and to treat the interaction be-

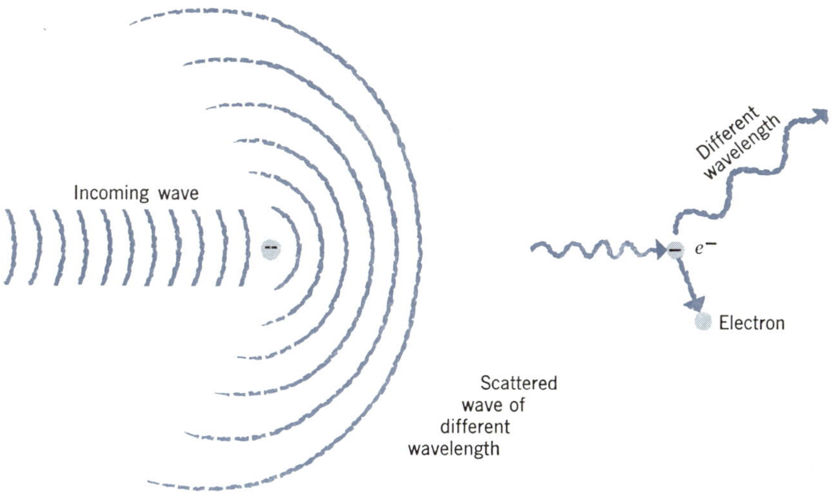

Fig. 11-4

tween the wave and the electron as if it were an ordinary collision between two particles, the photon and the electron. This success supplied more evidence that electromagmetic waves often acted as if they were made up of the particles we call photons.

Waves in the Field as Particles—Photons

The experiments described above established the fact that electromagnetic waves, such as light, had a particle aspect. These waves act in many circumstances as if they were particles (photons) with definite energy and momentum. Yet light unquestionably behaved like a wave in such phenomena as diffraction and interference. Thus there was a problem—the wave-particle duality problem. The character of light (and the field) had these two proved (yet apparently contradictory) traits.

WAVES OF MATTER

The Bohr Atom

An atom is made up of a very small, positively charged nucleus, with a cloud of electrons around it. (The discovery and composition of the nucleus will be discussed in Chapter 12.) The atom is held together by the forces of electrical attraction between the nucleus and the electrons. This fact raised two problems about the structure of the atom that needed to be solved.

First, the electrical forces between the nucleus and the electrons should cause the atom to collapse. The electrical forces would cause the electrons to be pulled straight to the nucleus if they were not moving. But there is good experimental evidence to show that they *are* moving. The magnetic properties of matter are explained by the moving charges within matter, and there are more direct experiments[2] as well. The atom, then, is pictured as a central nucleus with the electrons revolving around it (see Fig. 11-5). However, a revolving electron is an accelerating electron, and an accelerating electron should radiate according to the laws of electricity and mag-

[2]The details of these experiments are beyond the level of this book. Briefly, an experiment of J. W. M. DuMond and P. Kirkpatrick showed that an observed variation in the Compton frequency shift matched that predicted by the expected momentum of orbiting electrons in the Schrodinger theory.

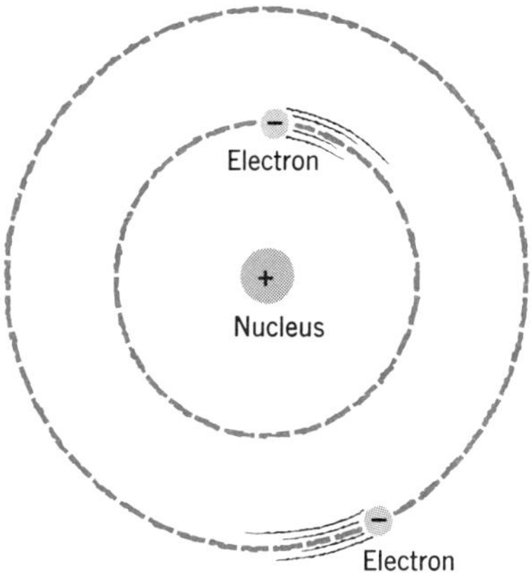

Fig. 11-5

netism, sending out electromagnetic waves. As it sends out these waves, it should lose energy and spiral into the nucleus, causing the atom to collapse. However, atoms do not collapse. They appear to be very stable structures.

When atoms do radiate, after various kinds of disturbances, the electromagnetic waves they send out are of certain definite frequencies. Each kind of atom has its own set of definite frequencies. Figure 11-6 shows an example of the frequencies emitted by one kind of atom. Each line represents a definite frequency of emitted light. One can identify the kind of atom, be it hydrogen, helium, oxygen, etc., by the characteristic frequencies of the waves it sends out. These frequencies must arise from the structure of the atom because they are the same for the same kind of atoms, and different for different kinds of atoms. How is it possible to discover this structure and explain the reason for these frequencies?

Neils Bohr (1885–1962), a Danish physicist, attacked this problem

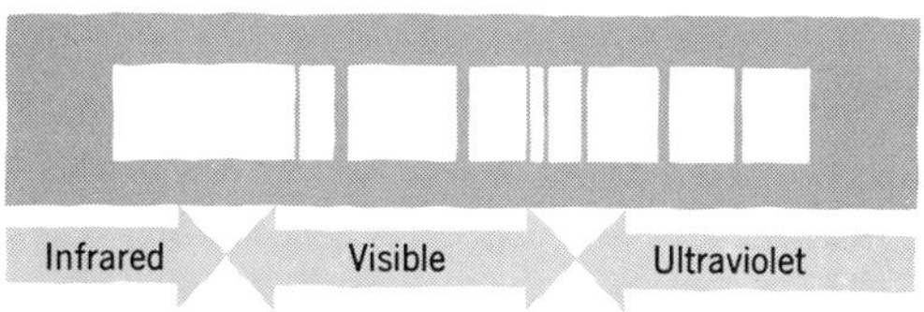

Fig. 11-6 *The light emitted from the Spectrum of mercury atoms. This spectrum can be obtained by allowing light from a sample of glowing, gaseous mercury to pass through a diffraction grating and strike a photographic film (see Chap. 9, p. 376).*

by constructing a theory that modified the laws of physics. According to this theory, the revolving electrons in the atom did not emit electromagnetic waves as long as they stayed in certain definite orbits (see Fig. 11-7). Each orbit corresponded to a certain energy that the atom could have. In order for an electron to move to a higher orbit, farther from the nucleus, energy would have to be supplied to the atom. When an electron dropped from one orbit to a lower orbit, energy would be emitted from the atom, perhaps as a photon.

Now we are in a position to understand why an atom emits only certain definite frequencies of light. When an electron drops from one level to a lower one, energy must be released. This may be in the form of a photon, containing just the amount of energy that has been removed from the atom. What energy can this emitted photon have? Only that lost by the atom when the electron dropped to the lower orbit. According to this theory, each photon that is emitted is caused by an electron dropping from one orbit to another. Each atom, therefore, emits photons that represent differences in energy between two of its own characteristic orbits. Thus the emitted photons can have only certain definite energies. Since the energy of a photon determines its frequency, these photons can have only certain definite frequencies as well. This explains why certain frequencies are emitted from each kind of atom.

If the atom has an energy E_1 when the electron is in one orbit and an energy E_2 after the electron has dropped to a lower orbit, then the energy of the emitted photon must be the difference in energy,

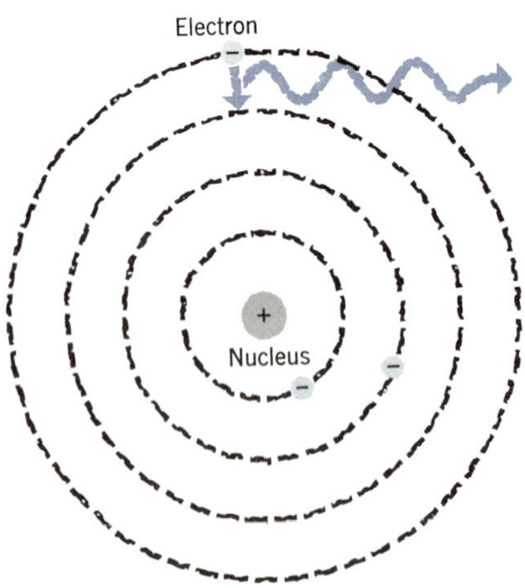

Fig. 11-7 *Photon emitted when electron jumps to lower energy orbit.*

$E_1 - E_2$. This means that we can find the frequency of the photon using $E_1 - E_2$ for the energy in the equation $E = hf$:

$$E_1 - E_2 = hf \quad (3)$$

Since there are only certain characteristic orbits for a given kind of atom, there are only certain energies that can appear on the left side of Equation 3. Therefore, for each kind of atom there will be only certain characteristic frequencies that can appear on the right.

The building shown in Figure 11-8 provides a rough analogy of these electron orbits and their energies. The various electrons live on the various floors, but they cannot live between floors. The floors represent the permitted orbits. Energy is required to hoist an electron from one floor to another. With sufficient energy, one might remove an electron from the building altogether and send it into space. The first floor is right on the ground, representing the lowest-lying orbit in the atom.

Why don't all the electrons in the atom fall down to the lowest orbit? This would correspond to a situation in which all the people in a building lived on the ground floor. The explanation is given by the so-called *exclusion principle* discovered by Wolfgang Pauli (1900–1958). This principle tells us that no two electrons can ever be in exactly the same state or condition. The orbit does not determine the condition of an electron completely, but it does help

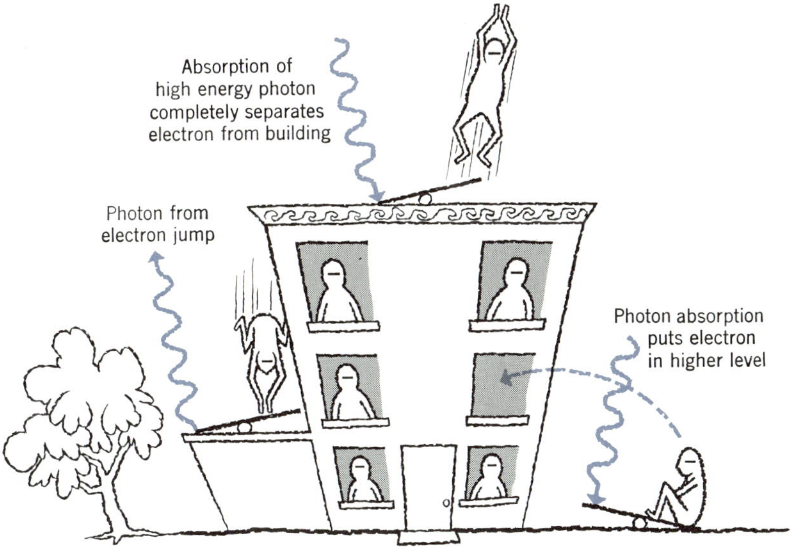

Fig. 11-8

to determine the condition. There can be only a certain number of electrons in each orbit. When an orbit has been filled, it will admit no new electrons unless it gets rid of some of the others.

We have seen that an electron dropped from one orbit to another can cause the atom to emit a photon with an energy equal to the difference in energy between the two orbits. The reverse may also occur if a photon strikes an atom. The photon may have enough energy to remove an electron from the atom completely. A particularly interesting process occurs if the photon energy corresponds exactly to the energy difference between two orbits. If there is an electron in the lower of these two orbits but none in the higher orbit, the atom can absorb the photon, causing the electron to jump to the higher orbit (see Fig. 11-9*a*). The photon, of course, disappears as it is absorbed, just as it appears when it is emitted. If there is already an electron in the higher orbit, however, and none in the lower orbit, then the presence of the photon can actually stimulate the emission of another photon of the same energy, as the electron drops to the lower orbit. This process is called the ***stimulated emission*** of radiation (see Fig. 11-9*b*). Thus, photons of the right energy can actually cause absorption or emission, depending on the state of the atoms they affect.

The Bohr theory of the atom was rather effective in explaining the frequencies emitted by the hydrogen atom, but it was unsuccessful for other atoms.

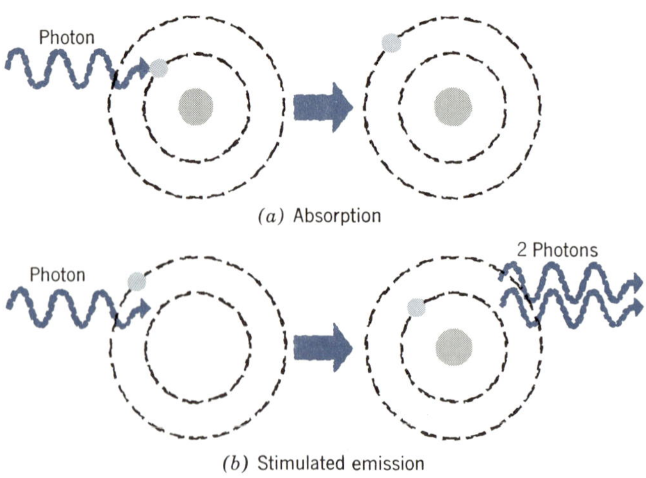

Fig. 11-9

De Broglie Waves

In 1924 Louis de Broglie suggested that particles of matter, such as electrons and protons, should have a wave nature. The wavelength L is given by Planck's constant h divided by the magnitude of the momentum p of the particle:

$$L = \frac{h}{p} = \frac{h}{\text{momentum}} \tag{4}$$

(This same equation is true for photons.) Equation 4 tells us that the greater the momentum of the particle, the smaller its wavelength. In principle the equation applies not only to small particles but also to such things as baseballs and planets. However, the constant h is so small compared to the momentum of these objects at normal velocities that the wavelength would be too small to measure. Once again, as in Planck's equation, we note that the constant h is involved in the wave-particle duality, this time the wave-particle duality of matter.

Several scientists obtained diffraction and interference patterns from beams of electrons and other particles. The wavelengths determined from these patterns proved to be those predicted by de Broglie's formula. Figure 11-10 shows a pattern caused by the diffraction and interference of a beam of electrons passed through a crystal. The pattern is distorted by the presence of a magnetic field. This distortion shows that charged particles are being diffracted, not light or some other electromagnetic wave.

Besides this experimental success, the new concept helped to explain the Bohr orbits. If one calculates the proper wavelengths for the electrons in Bohr orbits, these wavelengths fit into the circular path lengths of the orbits exactly (see Fig. 11-11). The Bohr orbits were just those orbits in which a wave electron would give standing waves, much like the waves on the strings in Figure 9-15 (p. 365).

However, a wave traveling around a central nucleus could not be expected to travel in a narrow circular path without spilling over into the rest of space. Therefore the Bohr orbits were approximations. What was needed was a mathematical treatment of the electron as a space-filling wave, not just as a wave along a curved line. This treatment was provided in 1926 by Erwin Schrodinger (1887–1961), who discovered the equation that describes these waves. Thus the Bohr orbits were replaced by "orbits" or states in which the electron wave was spread through the space around the nucleus in various ways, as illustrated in Figure 11-12. The Schrodinger theory also predicted other states, (or "orbits") which were missing in the Bohr theory.

(Another theory of the atom, formulated in 1925 by Werner Heisenberg, was equivalent to Schrodinger's but far more abstract.) The original de Broglie waves around the Bohr orbit can be likened to the standing waves on a string that we discussed in Chapter 9. Space is not a string, however; it is more like the plate we discussed in Chapter 9 (see Fig. 9-17, p. 368). The more realistic waves of Schrodinger, therefore, are more like the standing waves on the plate. Even the waves on the plate, however, are confined to the plate,

Fig. 11-10. Courtesy of Dr. Darwin W. Smith.

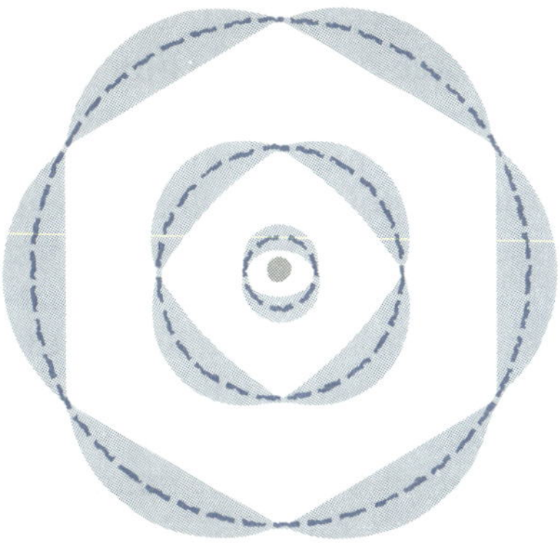

Fig. 11-11

while the Schrodinger electron waves spread out into three-dimensional space.

There are important differences between these waves (introduced by de Broglie and Schrodinger) and the waves on the string or plate that we discussed in Chapter 9. Waves *in* matter are caused by the cooperative motion of many particles. Water waves, for example, arise from the collective motion of many water molecules. A similar statement could be made about the molecules in the strings, plates, etc. The de Broglie waves, as far as we know, are *not* formed in this way. A *single* particle by itself has a de Broglie wave associated with it. These waves *of* matter are evidently part of the nature of matter itself and can pertain to the behavior of a single particle.

The Matter Field

Because of the discoveries discussed above, matter and light presented the same problem. Both material particles (such as electrons)

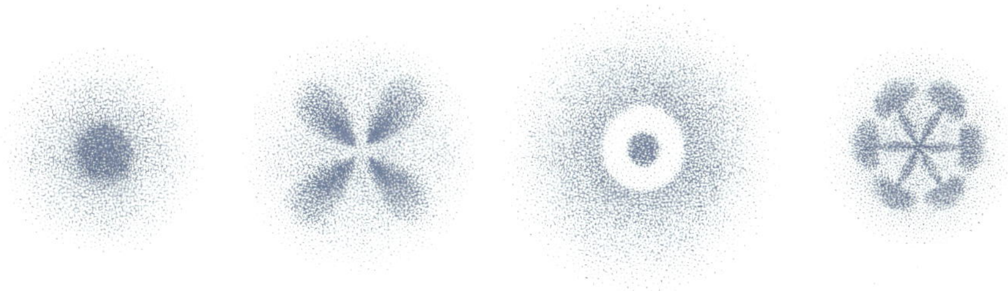

Fig. 11-12

and electromagnetic field waves (such as light) showed wave *and* particle characteristics. There was thus a wave-particle duality problem with both matter and electromagnetic waves. There seemed to be no way to tell which was the more fundamental material reality, the particle or the wave (or field). Is it best to think of a field spread over space with waves traveling in it, or of localized particles moving through space? We have already seen (Chap. 10, p. 409) that both particle and wave can have mass and energy. In fact, because of the wave characteristics of the material particle, one can talk about a *matter field*, in which the matter waves exist. Or, if one is dealing with the field, the field must be lumped into particles, or *quantized* in order to explain some observations.

THE MEANING OF QUANTUM PHYSICS

The Unobserved State

Let us now try to deal with the wave-particle duality problem. We have two apparently necessary but apparently contradictory models for both matter and light—the particle model and the wave model. We need to examine the relationship between these models.

Figure 11-13 diagrams the two-slit interference experiment described in Chapter 9. In this experiment we shall use electrons rather than light, although the main results would be the same for light. To the left of the slits we have a source of electrons. We shall assume that the electrons in the beam from this source all have the same velocity and are going in the same direction. At the right of the screen, some distance off, there is a photographic film. When an electron strikes this film, it exposes a tiny region marking the position at which it strikes.

The result of the experiment is the usual two-slit interference pattern on the film. This means that the electrons must somehow travel as waves through the slits. We also notice, on examining the film carefully, that the interference pattern on the film is caused by the exposure of a large number of individual grains or exposed spots on the film. Individual electrons strike the film at particular spots causing individual grains to be exposed. Thus the electrons must have struck the film as particles, not as waves.

This suggests a way out of the wave-particle dilemma that unfortunately turns out to be incorrect. Perhaps the wave behavior is not characteristic of single electrons but only of electrons in groups. Perhaps the wave behavior is a cooperative behavior, arising from the interactions among the electrons as they come through the slits. After all, the wave behavior is observed in the interference pattern, which cannot be formed unless many electrons strike the film. However, it turns out that this idea will *not* solve our problem, as we can see if we cut down the intensity of the electron beam striking the slits until there is a very low probability for more than one electron at a time to be traveling in the apparatus. Thus the great majority of the electrons go through the apparatus alone. We have to wait a long time for the

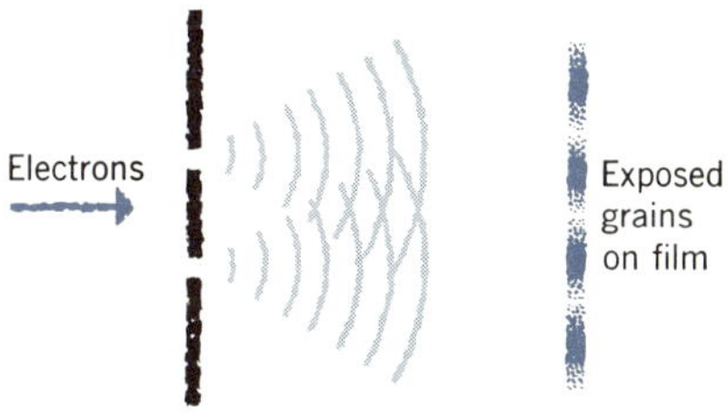

(*a*) Both slits open

(*b*) One slit open

(*c*) Both slits open but with detector at one slit or the other to detect electrons as they pass through slit.

Fig. 11-13

pattern to be built up grain by grain, but when we get it, we see the same two-slit interference pattern we had before. Somehow, the behavior of each individual electron is affected by the wave property.

If each electron participates in the interference pattern, it must be affected by both slits. It is difficult to think of a single electron going through both slits. We can prove, however, that both slits are required to get the pattern. By closing one slit, we force the electron to go through one slit only. But as a result we get not the two-slit pattern but the one-slit pattern. By changing the experiment, we have obtained a different result (see Fig. 11-13*b*).

Does the electron go through one slit or both? We might devise the following experiment to try to answer this question. We set up a device near each slit that signals when an electron goes through that slit. We find from these signals that each electron does indeed pass through only one slit. When we look at the film pattern produced by these observed electrons, however, we find that *we no longer have the two-slit interference pattern*. We have two one-slit patterns instead, one from each slit (see Fig. 11-13*c*). By changing the experiment, we have again obtained a different result.

When we make a measurement to find out where the particle goes, we find that it goes through one slit or the other, never both. We fail, however, to obtain the two-slit pattern. Therefore we cannot answer the question in which we are interested: through which slit does each particle go when we get the two-slit pattern. We want to know where the particle is when we do not observe it, since when we do observe it we change the experiment and get a different result. Let us summarize this unfortunate situation.

(a) When we do not experimentally determine through which slit the electron goes, we must think of it as going through both slits, since we get a two-slit pattern.

(b) When we *do* experimentally determine through which slit the electron goes, we find that it goes through only one slit and we get a one-slit pattern.

This same kind of problem arises in many other situations. Suppose that we wish to investigate a certain physical process. We make all the measurements we can on the system we are dealing with, without actually changing the process we wish to understand. Generally our measurements will not determine all the conditions of the system or the exact way in which the process takes place. In such circumstances we must think of the system as being (to some extent) in all the possible undetermined conditions at once. We must think of the process as occurring in all possible ways at the same time.

Thus, in our two-slit experiment we cannot say that the electron goes through one slit or the other unless it has been found to do so in some experiment or measurement. Furthermore, there is no way to determine through which slit it goes without disturbing the interference process we are studying. Therefore, we must think of the electron as being partially in all possible conditions at the same time, that is, as going through both slits.

When we *do* determine through which slit the electron travels, we actually change the process we are studying. This measurement, therefore, appears to give us information about a situation different from the one that existed before the measurement. The measurement interferes with the process.

THE COPENHAGEN INTERPRETATION OF QUANTUM PHYSICS

The conclusions we shall reach here represent the conventional interpretation of quantum physics today. It is the interpretation suggested by Max Born (1881–) and accepted by Bohr and others at Copenhagen. Some scientists, notably Einstein, Schrodinger, and de Broglie, were or are not happy with this way of looking at things. However, the so-called Copenhagen interpretation seems to agree with all the facts, and other interpretations are not yet required by experiment, as far as we know.

In the two-slit experiment, we found that we could not speak of a definite position for the electron unless we made a measurement that fixed it. If we do not make this measurement, then we must think of the electron as taking all possible paths or being in all possible places and conditions that are consistent with the measurements we *have* made. In the two-slit experiment there are two possible paths through the screen. One path leads through one slit, and another leads through the other slit. Since we have made it possible for the electron to go through either slit by leaving both slits open, we must consider that it does go through both. We must, however, consider it as going through as a wave. The waves from both slits interfere and produce an interference pattern. The pattern, however, cannot be detected by using only one electron. A single electron simply shows up as a single spot on the film. The pattern tells us where it is likely to appear, however, and this brings us to the conventional interpretation of quantum physics.

The waves are regarded as waves of probability. More accurately, the intensity of the wave at a point in space gives us the probability of finding a particle at that point. In the two-slit experiment, the waves must come through the slits. Then the waves interfere, setting up an interference pattern in space. The intensity of the wave at any point in space gives the probability of finding an electron there if a measurement is made. Finally, the waves come to the photographic film. It is very unlikely that the electron will go through the film without exposing it. The interposition of the film causes a measurement—it causes information to be created in an effect that we can observe. Thus the electron wave shows its particle behavior, giving us information in particle form. All we can find out from the wave is the probability of finding the particle at some point on the film. The probability is high where the interference pattern has a maximum. It is low where the interference pattern has a minimum. Apart from this probability the result is completely undetermined. The electron propagates as a wave and is observed as a particle. The wave gives the probability of finding the particle.

Suppose that we only set up a small piece of film that will cover only part of the interference pattern (see Fig. 11-14). Suppose that the electron does hit the film and is observed as a discrete particle. What happens to the rest of the continuous wave pattern that never touched the film? It vanishes instantly. If we know that the electron hit the film, then the probability of its being anywhere else is zero. The waves are waves of probability, so where the probability is zero, they must be zero also. When the particle is observed, the waves must vanish. We start over with a new wave, incorporating the new information that has been gained by observing the particle.

The uncertainty principle

In Newtonian physics, as we pointed out in Chapter 4, the state of a system at one time determined its state at another time. If one knew the position and velocity of all the particles in a system and the forces between them, one could use the laws of motion to predict exactly the future behavior of the system, or even the past. Even if one did not know these positions and velocities, they did exist in principle and would themselves determine the future behavior of the system. There is, of course, some limitation on our knowledge of the position and velocity of a particle—the unavoidable experimental error. However, there was no reason to believe that this error could not be made arbitrarily small by improving our apparatus. The laws of Maxwell for the electromagnetic field were similar in the precision with which they described the field and predicted its behavior.

The principles of quantum physics directly contradict this earlier point of view. According to quantum physics, one *cannot* have precise knowledge of both the position and the momentum (and hence velocity) of a material particle. Therefore the predictions can be made only to a certain restricted degree of accuracy. This limitation arises not from any experimental error but from the laws of quantum physics themselves. We have not given these laws in their mathematical form, but we would like to discuss this consequence. It is called the *uncertainty principle* and was first stated by Werner Heisenberg.

According to the uncertainty principle, it is theoretically possible to know either the position or the momentum of a particle exactly, but we cannot know both exactly. The more we know of one, the less we can know of the other. No exceptions to this principle have been found. For large, ordinary objects, the amount of uncertainty is relatively small and unimportant. We can obtain enough information to predict the motions of the planets and the stars accurately for thousands of years. For electrons, however, there are conditions in which we cannot make predictions that are accurate for even a few seconds.

As an example of the uncertainty principle, we shall describe an attempt to prepare a particle in an accurately defined condition. Suppose that we want a narrow beam of electrons moving in a precisely defined direction. Figure 11-15 illustrates our attempt to obtain such a beam. We first obtain a beam of electrons all moving in the same direction and then try to narrow the beam by making it pass through a single slit. This method works up to a point, but we soon find that as we make the slit narrower, we actually make the beam wider. The wave nature of electrons makes them diffract. We already

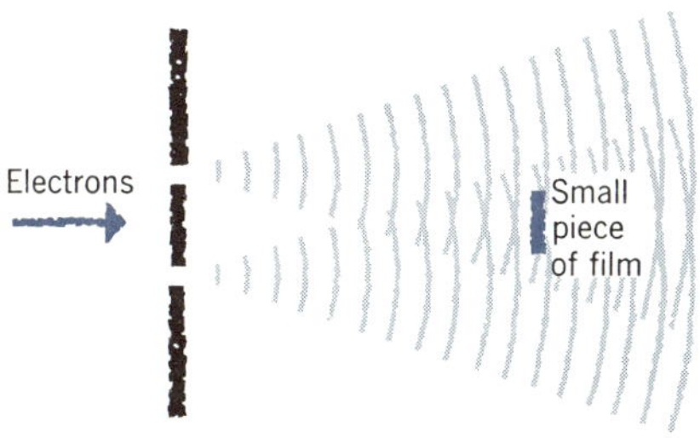

Fig. 11-14

know what to expect from a one-slit diffraction pattern. The narrower the slit, the wider the diffraction pattern (see Fig. 9-8, p. 358). As the slit is made narrower, more precisely defining the position of the electrons as they go through the slit, the direction of their momenta becomes more and more uncertain. The attempt to gain more information about the position of the particles by narrowing the beam thus results in a loss of information about their momenta.

As far as is now known, the uncertainty principle applies to all physical processes. Therefore, exact prediction is physically impossible, according to this principle. Some physicists have been uneasy about this uncertainty. Einstein, in a letter to Max Born, expressed himself in these words.

> You believe in the dice-playing God, and I in the perfect rule of law in a world of something objectively existing which I try to catch in a wildly speculative way.[3]

Some physicists are trying to explain the phenomena of quantum physics by assuming an underlying reality that does not involve an uncertainty principle. Thus far there is no experimental reason for such an assumption.

The Uncertainty Principle and Energy

The uncertainty principle also pertains to time and energy, limiting the amount of information we can have about these two quantities. The more we know of the energy of an object, or system of objects, the less we can know about the exact time at which it had that energy. Also, if a particle or a system remains in a particular condition for only a short period of time, then its energy in that condition is uncertain. The shorter the time interval in which the condition lasts, the more uncertain the energy. If the condition lasts for just an instant, so that it specifies an exact time, then the energy is completely uncertain. If, on the other hand, the condition lasts for an extremely long period of time, the energy may be known with great precision.

According to the uncertainty principle, therefore, a state of definite energy must last for a long time; the longer it lasts, the more certain we can be of the energy. We have reason to be interested in the

[3]Albert Einstein, letter to Max Born, November 7, 1944, published in Max Born, "*Natural Philosophy of Cause and Chance*," New York: Dover, 1964, as quoted in V. Lawrence Parsegian, Alan S. Meltzer, Abraham S. Luchins, and K. Scott Kinerson, *Introduction to Natural Science, Part One: The Physical Sciences*, New York: Academic Press, 1968, p. 437.

states of a system that can exist unchanged for a long period of time. They should be involved in the long-lasting structures observed in the universe. We have already discussed the structure of the atom in terms of its energy states. The orbits in the Bohr theory of the atom, or the standing-wave patterns in the more accurate theory, correspond to states or conditions of definite energy for the atom. When we discuss the long-lasting structure of any system, be it atom, molecule, or nucleus, we are interested in its states of definite energy.

Most frequently, states of definite energy do not correspond to states of definite position and momentum. For example, the standing-wave states of definite energy in the atom do not correspond to states of definite position or momentum for the electron. The position of the electron is uncertain because the electron could be anywhere in the standing-wave pattern.

*Information and Planck's Constant

The amount of uncertainty in the position and momentum of an object is related to Planck's constant h:

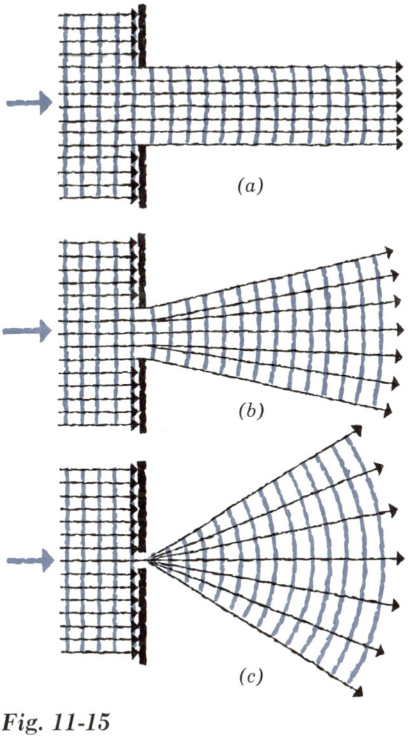

Fig. 11-15

(Uncertainty in position) × (Uncertainty in momentum) $\geq h$ (5)

The uncertainties of energy and time are related by a similar equation:

(Uncertainty in time) × (Uncertainty in energy) $\geq h$ (6)

(The symbol $\geq$ means *greater than* or *equal to*.) Thus the product of the two uncertainties in each equation is greater than or equal to h. This explains why the uncertainty does not bother us in large-scale operations. Planck's constant is a very small quantity. The uncertainty in both the position and the momentum of an object the size of a billiard ball is negligible in comparison with the size and momentum the ball would have under normal conditions. In dealing with very small particles such as electrons, Planck's constant is comparatively large, and the uncertainty may therefore be very important.

Because of its role in the uncertainty principle, Planck's constant determines the amount of information available in a physical situation. Therefore, it controls the maximum accuracy with which predictions of the future can be made on the basis of this information.

*The Zero-Point Energy of an Oscillator

The uncertainty principle allows us to explain why the molecules and other small particles in material objects do not lose all their energy or cease all motion, even at the absolute zero of temperature. Let us consider an oscillator, for example the swing in Figure 11-16. Is it possible for this swing to have an energy state in which it is motionless at the lowest point of its path? If the swing remains at this point, then its position is exactly known. According to the uncertainty principle, therefore, its momentum must be completely uncertain. In such a circumstance, according to the principles of quantum

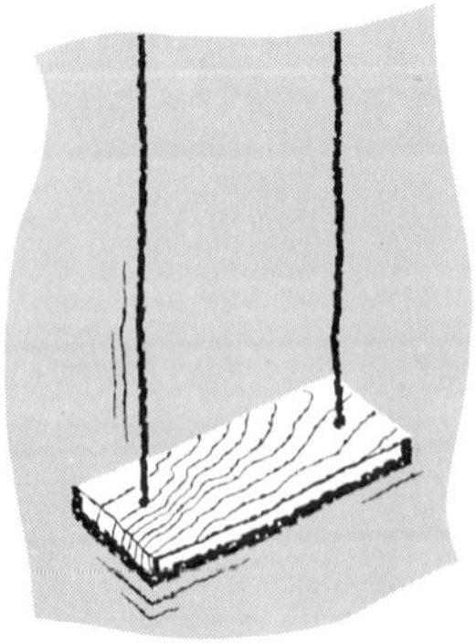

Fig. 11-16

physics, we must think of the swing as being at least partially in all possible momentum states. Most of these states, however, are states of motion, and therefore the swing cannot be considered entirely motionless. Therefore, the swing will be swinging to some minute extent even in its lowest energy state.

Although this zero-point energy is insignificant for an object as large as a swing, it is important for the vibrating molecules, atoms, electrons, etc., in a material object. Even at absolute zero they retain some oscillation because of this zero-point energy.

*The Trapped Electron

Using the uncertainty principle, we can explain why an electron resists being confined to an arbitrarily small volume. Suppose that we try to capture an electron at a certain position and hold it motionless. If we can do this, we will know the position of the electron exactly and also its momentum, since a truly motionless electron should have a momentum that is precisely zero. Figure 11-17 shows the device we will use in trying to perform this feat. One electron is inside the cylinder. The piston is slowly pushed in, decreasing the volume of space that the electron can occupy. The less volume the electron is allowed to occupy, the more information we have about its position. According to the uncertainty principle, however, the more information we have about the position, the less we are allowed to have about the momentum. As we continue to decrease the volume available to the electron, the momentum becomes more and more uncertain. The more uncertain the momentum of the electron

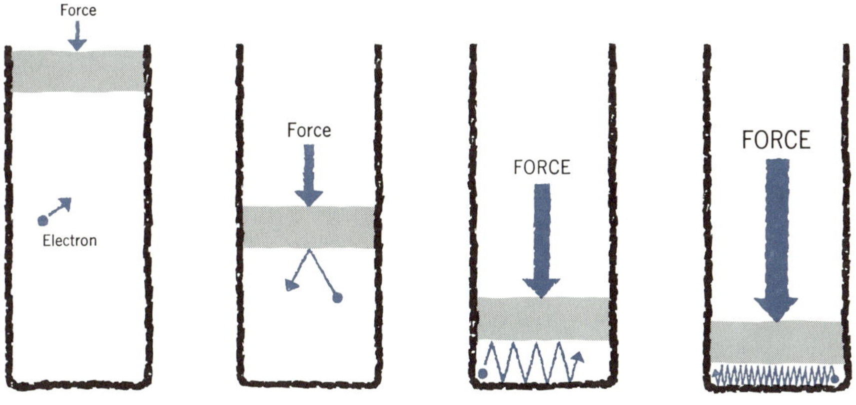

Fig. 11-17

becomes, the greater the likelihood that it is very high. As we have seen, we must consider the electron to be partially in all its possible states, and therefore we must consider the effect of these possible high-momentum states. That effect may be seen in Figure 11-17. The result is a bombardment of the piston by the electron, causing a resisting pressure on the piston. The farther the piston moves, the more accurately we know the position of the electron, since it must be in a smaller and smaller volume. Therefore, the momentum becomes more and more uncertain, and the choice of high momentum for the particle becomes greater and greater. The greater the momentum, the greater the pressure resisting the motion of the piston, because the electron strikes it harder and more often. If the piston could finally trap the electron in one spot, the uncertainty of the momentum would have increased without any limit, and there would be an infinite force pushing back on the piston. No matter how hard the piston is pushed, therefore, the opposing force will eventually stop it.

Of course, we cannot actually observe the effect we have been discussing merely by using an ordinary piston and cylinder. The same principle, however, helps us to account for the size of an atom approximately. As we have seen, before the discovery of quantum physics, an electron moving around the nucleus would have been expected to radiate, lose energy, and finally fall into the nucleus itself. To have the electron located right at the nucleus with no kinetic energy and therefore no momentum, would violate the uncertainty principle, as our examples have indicated. There must be some state of lowest energy below which the electron cannot fall. Therefore, the lowest energy state must be a condition in which the electron is still moving around the nucleus with some momentum, and therefore some energy as well. This state, of course, corresponds to the lowest "orbit," or standing-wave pattern, in which the electron can exist in the atom (see Example 5, p. 486).

*An Illustration of the Uncertainty Principle

In order to clarify how the uncertainty principle arises from the phenomena of quantum physics, we shall outline a thought experiment in which we try to measure both the position and the momentum of an electron precisely. Figure 11-18*a* shows the apparatus we shall use. A single photon of known frequency (and wavelength) leaves the light source, bounces off the electron, passes through the slit, and arrives at the screen. We may imagine a photographic plate on the screen that permanently records the point at which the photon arrives. From a practical standpoint, of course, it would be too much to expect that this single photon would strike our single electron and just happen to pass through the slit. However, it *could* happen, and we are trying to consider the most favorable case.

To simplify our problem, we shall assume that we know the momentum of the electron to begin with. The photon, however, disturbs the electron in striking it and causes it to recoil with a changed momentum. Since we know the frequency and wavelength of the photon, we also know its energy and the magnitude of its momentum (see Equations 1 and 4). Therefore, if we know the direction of the photon before and after it strikes the electron, we can use the laws of conservation of energy and momentum to calculate the magnitude and direction of the recoil momentum of the electron. These two directions of the photon, together with the point at which the photon strikes the screen, will give us information about the position of the electron. We assume that we know the direction in which we have emitted the photon, and that we can easily determine its position when it strikes the screen. In order to determine both the position and the momentum of the electron exactly, we still need to know the direction of the photon after it collides with the electron.

The purpose of the slit is to help us obtain the direction of the incoming photon. In Figure 11-18*a*, we have obviously made this slit too wide. We do not know the direction of the photon very precisely because the photon could have arrived at the point on the screen from any direction within the white cone. Thus the direction of its momentum is very uncertain and, therefore, so is the recoil momentum of the electron. For essentially the same reason, we do not know

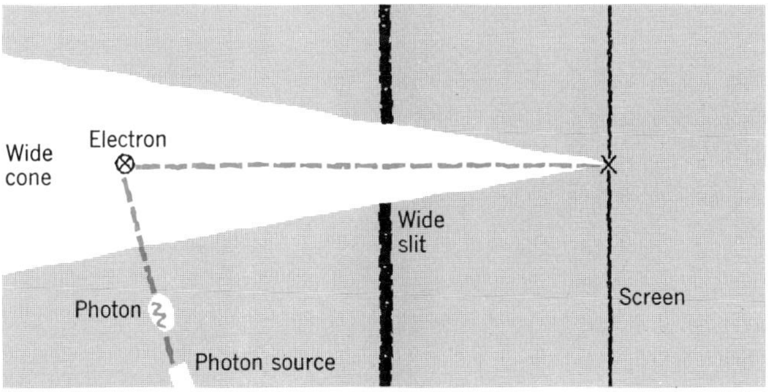

(a)

Fig. 11-18 a

the position of the electron precisely, since we do not know from which direction the photon arrived at the screen.

Obviously, or so it would seem, the way to make our determination more precise is to make the slit narrower, as we have done in Figure 11-18*b*. As the slit is made narrower, however, we become aware of an important factor we neglected in Figure 11-18*a*—that is, the photon passing through the slit is diffracted because of the wave nature of light. The smaller the slit, the larger the diffraction pattern. Therefore, although it is possible that the photon came straight from the electron to the screen, it is more likely that it changed direction as it came through the slit. The bigger the diffraction pattern, the greater will be this likely change of direction. Therefore, after a certain point, decreasing the size of the slit decreases our information about the electron instead of increasing it.

There is an additional change we can make in our experiment in order to reduce the size of the diffraction pattern and thus improve the accuracy of our experiment. We can decrease the size of the diffraction pattern, without increasing the size of the slit, by using photons of shorter wavelength (see Fig. 11-18*c*). In fact, no matter how narrow the slit, we can make the diffraction pattern as small as we please if we use a photon of short enough wavelength. This enables us to determine the position of the electron with great precision. No matter how small the wavelength of the photon, there will still be some uncertainty in its final direction; but we can make that uncertainty theoretically as small as we please by adjusting the slit and the wavelength.

This procedure does not contradict the uncertainty principle be-

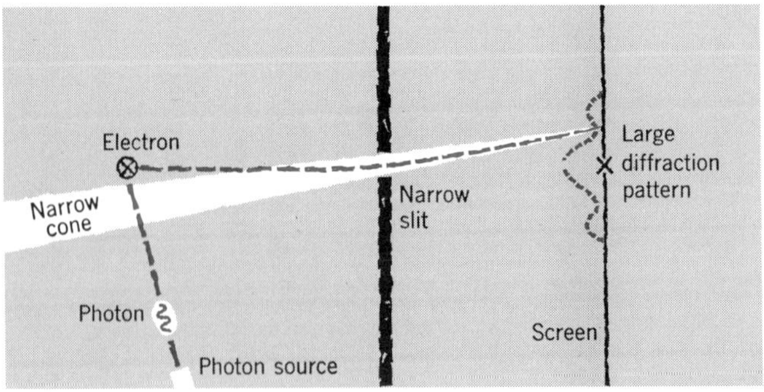

(*b*)

Fig. 11-18 b

cause we must still consider our information about the momentum of the electron after the collision. In order to decrease the size of the diffraction pattern, we had to decrease the wavelength of the photon. This decrease in the wavelength means an increase in the momentum of the photon (Equation 4). Of course, a decrease in the wavelength also means an increase in the frequency, which means an increase in the energy of the photon also (Equation 1). Therefore, the photon greatly disturbs the electron in the collision, and the electron recoils with a much larger momentum in the situation shown in Figure 11-18*c* than it does in that illustrated by Figure 11-18*a* or 11-18*b*.

The more we try to reduce the size of the diffraction pattern by reducing the wavelength of the photon, the greater the momentum of the photon, and therefore the more the photon disturbs the electron when they collide. This disturbance results in a large change in the electron's momentum. The shorter the wavelength of the photon, the greater this change. By trying to reduce the size of the diffraction pattern, therefore, we can determine the position more accurately, but we have caused a greater disturbance in the momentum.

To explain this example precisely, we should add a few remarks about the disturbance of the electron's momentum. The important fact is not that the photon used in our measurement disturbs the momentum of the electron, but that we cannot determine the exact amount of that disturbance. When we reduce the size of the diffraction pattern, we improve our knowledge of the direction of the incoming photon and, therefore, also our knowledge of the direction of the momentum of the recoiling electron as well as our knowledge of the electron's position. However, although we know the direction

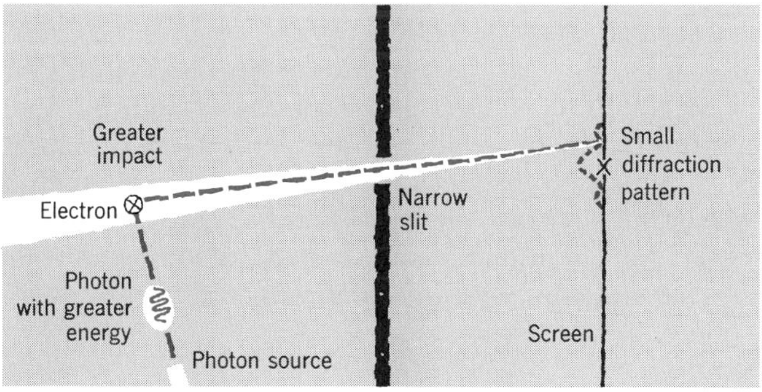

(c)

Fig. 11-18 c

of the momentum more precisely, the uncertainty of the momentum is not ultimately improved because the magnitude of the momentum is so much greater. Figure 11-19 illustrates this effect. Although the direction of the larger arrow is more precisely known, the position of its endpoint is actually more uncertain than that of the smaller arrow. As we use photons of smaller and smaller wavelength to try to measure the position of the electron precisely, we must increase the momentum (and energy) of the photon without limit, causing more uncertainty in our knowledge of the momentum of the electron. Thus we are effectively prevented from determining both the position and the momentum of the electron at the same time.

In our examples of the uncertainty principle, as illustrated in Figures 11-15 and 11-18, we have been concerned with momentum and position in the "up and down" direction, since we have been dealing with a horizontal slit. To measure position and momentum in other directions we might use a small hole instead of a slit, or we might use more than one device, so that we could observe the electron from different positions, with slits oriented in different directions. The uncertainty principle specifies the impossibility of measuring position and momentum along the same direction simultaneously. It would not prevent us from measuring the position along the horizontal direction and the momentum in the vertical direction at the same time. However, we could not measure position along the vertical direction and vertical momentum at the same time, as we tried to do in the last example.

It is not merely the fact that we must disturb a system in order to measure it that is responsible for the uncertainty principle. It was long known that there must be an interaction with a system if some information is to be obtained about the system by measurement or observation. There was no reason to believe, however, that the interaction could not be made as small as you please (in theory), or that the effects of the interaction could not be precisely determined. However, the quantum nature of matter and electromagnetic radiation—that is, the wave-particle aspect of both—does supply such a reason. It was the relation between the wave and the particle aspects of light, as expressed in Equations 1 and 4, that frustrated our attempt to obtain precise and complete information about the electron.

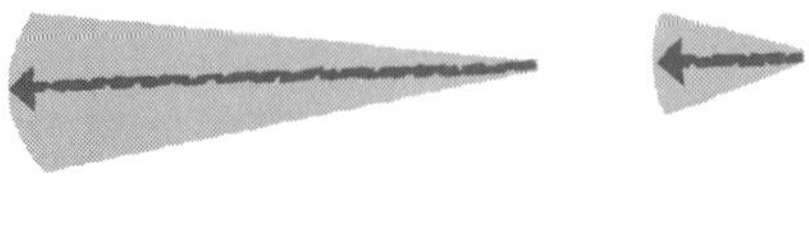

Fig. 11-19

Except for the principles expressed in these two equations, there would be nothing to stop us from using very weak light of very high frequency (short wavelength) to get a very small diffraction pattern without disturbing the electron very much. If, for example, in Figure 11-18*c* we could have used half of a short-wavelength photon, the diffraction pattern, which depends on the wavelength, would still be small, and the "half photon" would have disturbed the electron only about half as much.

The uncertainty principle arises from this unavoidable relation between the momentum (or energy) of the photon, which causes the disturbance, and the wavelength (or frequency) of the wave, which controls our information by determining the size of the diffraction pattern. We have, of course, concentrated on the photon in this illustration, but we could use other particles without changing the results because they also have both a wave and a particle nature.

All experiments designed to outwit the uncertainty principle have failed. The uncertainty principle can be proved from the postulates of the mathematical theory of quantum physics, which we have not given here.

Measurement in Quantum Physics and Time-Reversal Symmetry

One last attempt to outwit the uncertainty principle will allow us to discuss the relations between measurement in quantum physics and the direction of time. We have already mentioned that it is theoretically possible to measure the position of a particle exactly if the momentum is allowed to become totally uncertain. Suppose, therefore, in trying to follow the path of a particle, we measure its position exactly at two different times (see Fig. 11-20). Knowing the time between measurements and the distance between the two measured

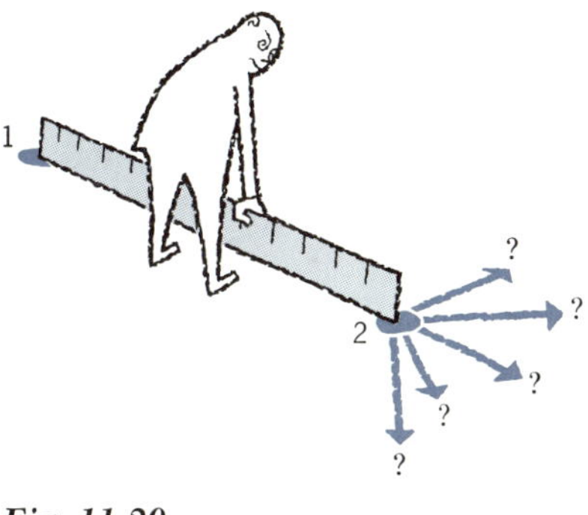

Fig. 11-20

positions of the particle, we ought to be able to calculate the momentum of the particle between the two measurements. Does this violate the uncertainty principle? No.

The uncertainty principle is not concerned with information that refers only to the past, rather, it is concerned with information about the present, information that might be used to predict the future. Information about the momentum of the particle between the two position measurements cannot be used to predict the future because it does not inform us of the present momentum of the particle; almost certainly, the momentum changed when the second measurement was made. A precise measurement of the position of a particle makes the momentum totally uncertain, regardless of how certain or uncertain it might have been before the measurement was made. The uncertainty principle allows us more precise knowledge of the inaccessible (and comparatively useless) past than of the present.

The fact that the uncertainty principle allows us precise knowledge of the past but denies us the same knowledge of the present or future might seem at first to challenge the principle of time-reversal symmetry. That in fact it does not can be shown with the aid of Figure 11-21. Here our observer *A* is making a series of precise measurements of the position of a particle. After he has made a particular position measurement, he will be able to calculate the momentum the particle must have had between the position and the previously measured position. The momentum in the present and, of course, the results of future measurements are unknown to him.

An imaginary observer *B*, moving backward in time and making the same series of measurements in the reverse order, would have exactly the same problem. Knowing only the measurements made in his past (our future), he could calculate the momentum the particle had between those measured positions but could not know the

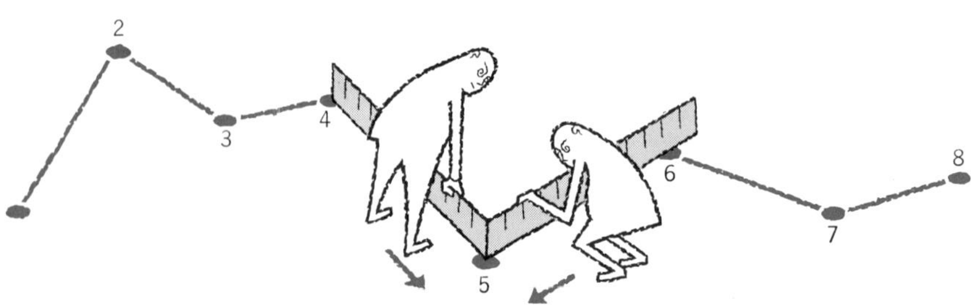

Fig. 11-21

momentum of the particle in his present. Nor, of course, would he know the measurements in his future (our past). Of course, if he had access to our records, matters would be different. (Access to records from the future would make quite a difference, whether quantum physics is valid or not.) However, the laws of quantum physics would work for him just as they do for us, as far as we know, and would present the same problems to him.

*The Philosophical Attitude of Modern Physics

As a result of the success of the theories of relativity and quantum physics, many of the basic ideas of science were questioned, and a new attitude and philosophy spread among physicists. Many physicists became very suspicious of entities that could not be observed, or that had no observable effects. This attitude is characteristic of the philosophy generally known as *positivism.* One early positivistic physicist-philosopher, who even disbelieved in atoms because they were unobservable, was Ernst Mach. A later physicist adopting a positivistic point of view was P. W. Bridgman, who called his philosophy *operationalism.* According to this philosophy, a physical quantity should be defined in terms of the actual physical operations necessary to measure it. Although there are many differences of opinion among those who adhere to this philosophy, it appeals to a great many physicists in one degree or another. It figured very significantly in the development of both relativity theory and quantum physics. Perhaps positivism should be considered to some extent an "antiphilosophy" rather than a philosophy. Many of those who were sympathetic to it held ordinary philosophy in some scorn. One of the prominent theorists of quantum physics is reported to have commented on a certain idea, "That is philosophy, and therefore nonsense."

Albert Einstein was very impressed with Mach's critical treatise on the history of physics. Despite the fact that Mach did not accept Einstein's theory, he probably did have an influence on Einstein's work. One of the results of the theory of relativity is to get rid of the ether, since it could not be observed and apparently had no effect that could be observed. Also, when Einstein examined space and time, he examined them not as absolute abstract entities, but simply as that which is measured by clocks and measuring rods—that is, he concentrated on the instruments and their operation.

In quantum physics we have abandoned the idea that a particle must be in a definite condition when we are not making an observation of that condition. The actual condition of such a particle may be not only unknown but in a sense nonexistent. This seems like an extreme application of positivistic principles, but it is widespread in physics today. This opinion results from the fact that we cannot assign a unique and definite condition to such a particle. The only significance of an unobserved particle is the effect it might have when it is finally observed. Thus it is the effect expected to be observed later, not some hypothetical present condition, that is important.

We have earlier presented the idea that the purpose of science is to make predictions. This view is also consistent with positivistic ideas and expresses a widespread attitude in physics today. Theories, assumed particles, fields, laws, etc., are merely abstract models that enable us to start with known present conditions and predict what will be observed later. Accordingly, the reality is in the observations; the theory merely connects present observations with future observations. Of course, because of the uncertainty principle these predictions cannot be made with absolute precision; we can merely give the probabilities for the events we expect to occur. (Sometimes the probabilities are so great as to amount to certainties, for all practical purposes.) The business of science is to state the probabilities as completely as possible for future events.

From this standpoint, relativity and quantum theory are extremely successful theories, the most successful known today for the very extensive range of phenomena that they cover. Of course, like any new theories, they must yield the same predictions as earlier theories for circumstances in which the earlier theories were correct. The theory of relativity gives the same results as Newton's laws when it deals with ordinary objects moving with speeds much less than that of light. Quantum theory gives the same results as Maxwell's theory of electricity and magnetism in many of the cases in which many electrons and many photons are involved. It also gives the same results as Newton's laws in many situations in which h can be regarded as very small and the wave nature of matter as unimportant.

Both the theory of relativity and the theories of quantum physics are capable of predicting the results of experiments to very great accuracy in many cases. No other theories, at once so comprehensive and so simple, can compete with them at present. Judging from the history of scientific theories, however, we can hardly expect them to be the last word.

WESTERN WASHINGTON STATE COLLEGE

Name Nicholas W. Preboski

Check Pass-Fail courses below	DEPT.	Course No.	Section	Credits 9-wk.	Credits 6-wk.	Credits short courses	Time-Hour M	T	W	Th	F	Instructor	Room
	Gen. Ed	331	33-904	4			8	8	8	8		Krieger	L-4
	Geology	101	39.902	4			10		10		10	Staff	268
	lab	"	"							9/10			BH-428
	Physics	101	61-903	4			11	11	11			Staff	109
	Lab									2/3			BH-437
				12									

FEES PAYMENT:

Students who have paid fees by mail by Friday, June 4: Report directly to your classes. Your Admit to Class Cards have been sent directly to your instructors.

All advance registration class reservations will be canceled on June 22 if fees are not paid, except for those registered in short courses only.

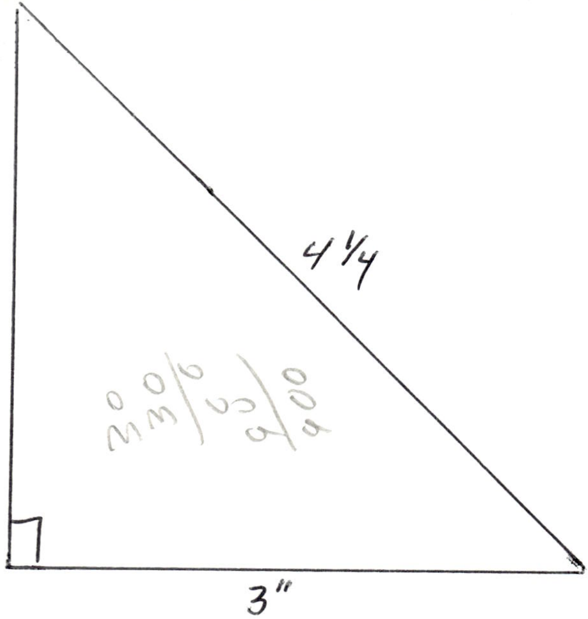

$a = \frac{F}{M}$

$$36\overline{)20^{3}0}\quad 6$$

6

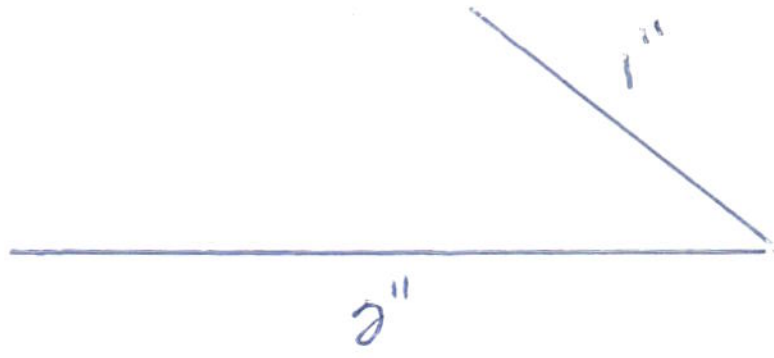

$\frac{20}{36}$

THE LASER: AN APPLICATION OF QUANTUM PHYSICS

An interesting application of quantum physics is the device popularly called a laser (light amplification by the stimulated emission of radiation). The laser produces a very intense beam of light.

To explain the operation of one kind of laser, we recall that electrons are allowed only in states of definite energy in an atom. These states correspond to the "orbits" or standing-wave patterns described earlier. For an electron to jump to a state or "orbit" of higher energy, it must acquire exactly the right amount of energy to make the jump. Once this has happened, the atom can spontaneously emit a photon of light as the electron drops back to a lower-energy "orbit."

Let us consider an atom with at least three allowed electron "orbits" (see Fig. 11-22). Most of the time the electron we are interested in will be found in the lowest of these three "orbits," and the other two "orbits" will be empty. Now suppose that we shine a light on this atom that has photons of just the right energy to cause the electron in the first "orbit" to jump to the third. Suppose then that this electron, now in the third "orbit," immediately drops to the second "orbit." We cannot discuss all the factors that influence the electron in making these changes, but we shall further assume that this drop from the third "orbit" to the second is far more likely than the drop from the second "orbit" back down to the first. Thus, if the light keeps shining on a group of these atoms, electrons will keep rising to the third "orbit" and dropping back to the second. In the second "orbit," however, they will be more or less stuck. Electrons will drop into the second from the third much faster than they fall to the first. Therefore, most of the atoms have electrons in the second "orbit" under these conditions.

If some method could be found to get these electrons to drop back into "orbit" number one all at about the same time, there would be a beam of light from the emitted photons. We have already pointed out that an electron in an upper "orbit" can be stimulated so that it will fall to a lower "orbit" and emit a photon. This stimulation is produced by another photon of the same energy coming from outside the atom (see Fig. 11-9*b*). If we wait long enough, one of the electrons in the second "orbit" of one of the atoms will spontaneously fall back to the first "orbit." This is an unlikely event, but with enough atoms present it is bound to happen. This photon can then stimulate the same electron drop in another atom. Then we have two photons, each of which can cause an electron drop in other atoms, and so on until the atoms have all radiated (see Fig. 11-23). This explains the laser action except for one point.

You will notice that there are mirrors installed at each end of the sample of laser material in Figure 11-23. If a photon is emitted in the direction of one of the mirrors, it can bounce back and forth like the light in a light clock. If a photon is emitted in another direction, so that it does not strike one of the mirrors, it may move out of the apparatus before it stimulates many atoms to emit. The photon bouncing back and forth between the mirrors, however, stays in the apparatus for a while and continues to stimulate the emission of other photons. These emitted photons are in the same direction as the photon that stimulated their emission. Therefore, the beam of photons moving back and forth between the mirrors builds up to a very intense beam in both directions. If one of the mirrors is only half-reflecting, part of this intense beam comes through and can be used outside the apparatus. The other part of the beam is reflected back and continues to stimulate the emission of other atoms.

Problem solving

In this chapter we are interested in problems involving atomic orbits and also the two equations:

$$E = hf$$

and

$$L = \frac{h}{p}$$

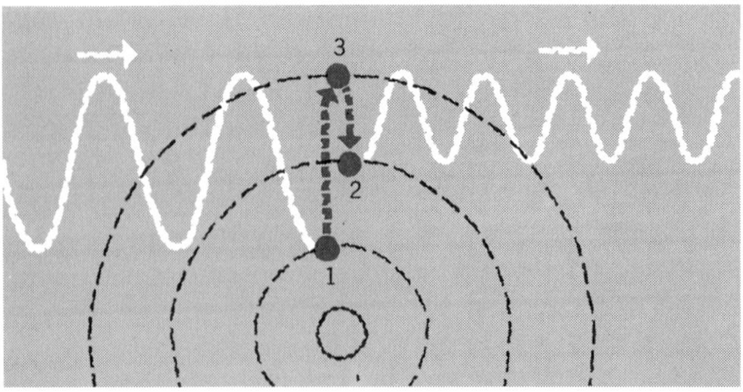

Fig. 11-22

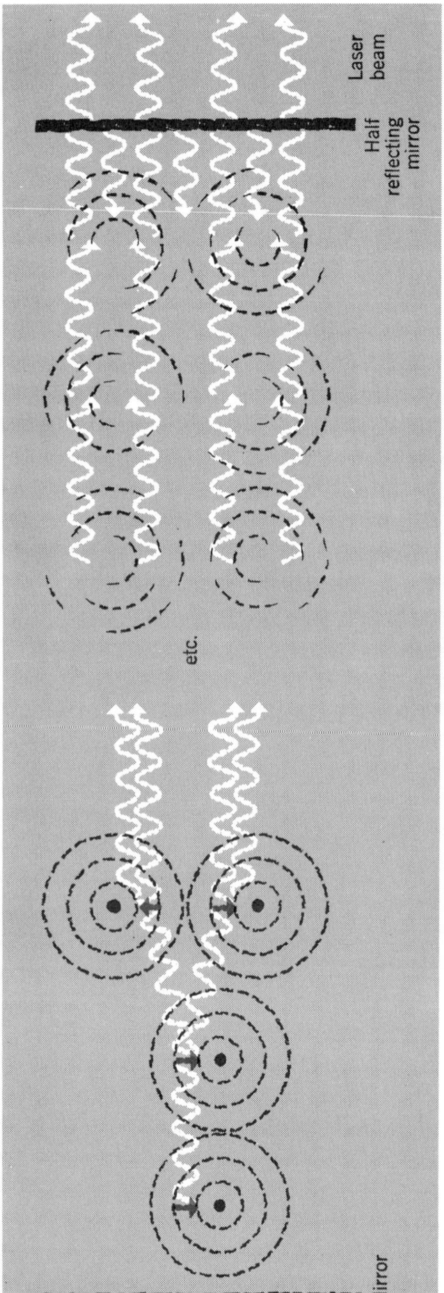

Fig. 11-23

As far as we are concerned here, the first equation applies only to photons. The second applies to all particles and objects.

Example 1

An electron changing from one atomic "orbit" to another causes a photon of frequency 1.5×10^{15} vibrations per second to be emitted. What is the energy difference between the "orbits"?

Solution.

Step 1 Choose the formula for the photon

$$E = hf$$

Step 2 List the knowns and unknowns:

$$\begin{aligned} E &= ? \\ h &= 6.625 \times 10^{-34} \text{ joule second} \\ f &= 1.5 \times 10^{15} \frac{1}{\text{second}} \end{aligned}$$

The frequency of 1.5×10^{15} vibrations per second appears simply as $1.5 \times 10^{15} \times 1/\text{second}$. A "vibration" is not considered a unit.

Step 3 Substitute and solve:

$$\begin{aligned} E &= hf \\ &= 6.625 \times 10^{-34} \times 1.5 \times 10^{15} \quad \frac{\text{joule second}}{\text{second}} \\ E &= 9.94 \times 10^{-19} \text{ joule} \\ E &= 10^{-18} \text{ joule} \end{aligned}$$

Example 2

What is the wavelength of a proton with a speed of 1000 meters/second?

Solution.

Step 1 Choose the equation for the wavelength of a particle:

$$L = \frac{h}{p} = \frac{h}{ms}$$

Step 2 List the known and unknown quantities:

$$\begin{aligned} h &= 6.625 \times 10^{-34} \text{ joule second} \\ m &= 1.67 \times 10^{-27} \text{ kilograms} \\ s &= 1000 \frac{\text{meters}}{\text{second}} \\ L &= ? \end{aligned}$$

STEP 3 Substitute and solve:

$$L = \frac{6.625 \times 10^{-34} \text{ joule second}}{1.67 \times 10^{-27} \times 10^{3} \text{ kilogram} \dfrac{\text{meter}}{\text{second}}}$$

$$= 3.96 \times 10^{-10} \frac{\text{joule second}^2}{\text{kilogram meter}}$$

but

$$\text{joule} = \text{kilogram} \frac{\text{meter}^2}{\text{second}^2}$$

So

$$L = 3.96 \times 10^{-10} \frac{\text{kilogram meter meter second}^2}{\text{kilogram meter second}^2}$$

$$= 3.96 \times 10^{-10} \text{ meter}$$

We can use the rest mass for the mass m since s is much less than c.

EXAMPLE 3

What is the wavelength of a photon with an energy of 16×10^{-19} joules?

Solution.

STEP 1 Choose the equations. We know that

$$E = hf$$

This, however, is in terms of frequency. Therefore we need the equation that relates frequency and wavelength also:

$$c = fL$$

or

$$f = \frac{c}{L}$$

We could now make a list of known and unknown quantities to substitute into these two equations. It is just as easy to eliminate one equation now, however, by substituting $\frac{c}{L}$ for f in Planck's equation.

$$E = hf = \frac{hc}{L}$$

Multiplying both sides by $\frac{L}{E}$, we get

$$L = \frac{hc}{E}$$

STEP 2 List the knowns and unknowns:

$$h = 6.625 \times 10^{-34} \text{ joule second}$$
$$c = 3 \times 10^{8} \frac{\text{meter}}{\text{second}}$$
$$E = 16 \times 10^{-19} \text{ joule}$$
$$L = ?$$

STEP 3 Substitute and solve;

$$L = \frac{6.625 \times 10^{-34} \times 3 \times 10^{8}}{16 \times 10^{-19}} \frac{\text{joule second meter}}{\text{joule second}}$$

So

$$L = \frac{6.625 \times 10^{-34} \times 3 \times 10^{8}}{10 \times 1.6 \times 10^{-19}} \frac{\text{joule meter}}{\text{joule}}$$

$$= \frac{6.625 \times 3}{10 \times 1.6 \times 10^{7}} \text{ meter}$$

$$L = 1.24 \times 10^{-7} \text{ meter}$$

EXAMPLE 4

According to the theory of electricity and magnetism, the equation $E = cp$ gives the relation between the amount of energy and the amount of momentum in a series of electromagnetic waves. From this equation and Planck's equation, $E = hf$, obtain de Broglie's equation (Equation 4) for photons.

Solution. Since $E = cp$, we simply substitute cp for E in the equation

$$E = hf$$
$$cp = hf$$
$$p = hf/c$$

Since $c = fL$, we substitute fL for c in the last equation:

$$p = hf/fL = h/L$$
$$p = h/L$$

*EXAMPLE 5

Using the uncertainty principle, estimate the radius of the smallest "orbit" for the electron in the hydrogen atom.

Solution. The ordinary hydrogen atom consists of one electron and one proton. The first step is to assume a circular orbit, and to equate the electric force of attraction the proton exerts on the electron with the centripetal force necessary to keep the electron in its orbit. The electric force and the centripetal force are, respectively:

$$f = K_e \frac{q_1 q_2}{d^2}$$

$$f = \frac{ms^2}{r}$$

or

$$\frac{ms^2}{r} = K_e \frac{q_1 q_2}{d^2}$$

In this equation, $q_1 = q_2 = e =$ the magnitude of the charge on the electron and the proton; $r = d =$ distance between the electron and the proton in the first Bohr orbit; $m = m_e =$ the mass of the electron; $s =$ the speed of the electron in orbit.
Therefore, we have

$$\frac{m_e s^2}{r} = K_e \frac{e^2}{r^2}$$

Multiplying through by mr, we have

$$\frac{(m_e s)^2}{} = \frac{K_e m_e e^2}{r}$$

The quantity $m_e s = \sqrt{\frac{K_e m_e e^2}{r}}$ is simply the magnitude of the momentum of the electron. This, however, is also a reasonable estimate of the uncertainty of the momentum, since as the electron travels around the proton its momentum is continually changing direction. Along an axis in the same plane as the orbit, the momentum of the electron will vary between plus $m_e s$ and minus $m_e s$. Therefore:

$$\text{Uncertainty of momentum} = m_e s = \sqrt{\frac{K_e m_e e^2}{r}}$$

For the uncertainty of the position, we may take the diameter of the orbit, or twice the radius of the orbit. This assumes that the actual standing-wave pattern that the orbit represents should be about half inside and half outside the orbit:

$$\text{Uncertainty of position} = 2r$$

The minimum of the product of these two uncertainties should be about h, according to the uncertainty-principle equation:

Uncertainty of position × Uncertainty of momentum = h

$$2r\sqrt{\frac{K_e m_e e^2}{r}} = h$$

Squaring both sides:

$$4r^2 \frac{(K_e m_e e^2)}{r} = h^2$$

Solving for r:

$$r\ (4\ K_e m_e e^2) = h^2$$

$$r = \frac{h^2}{4\ K_e m_e e^2}$$

This value, of course, is an estimate. The value used by Bohr was

$$r = \frac{h^2}{4\pi^2\ K_e m_e e^2}$$

Summary

- **I.** FUNDAMENTAL LAWS AND PRINCIPLES
  - **A.** WAVE-PARTICLE DUALITY of both matter and electromagnetic radiation.
    1. Particle nature of electromagnetic radiation, $E = hf$, shown experimentally by:
       - a. Waves in a cavity.
       - b. Photoelectric effect.
       - c. Compton effect.
    2. Wave nature of material particles, $L = h/p$, shown experimentally by:
       - a. Diffraction of electrons.
       - b. Bohr orbits as standing waves.
  - **B.** EXCLUSION PRINCIPLE.
  - **C.** THE UNOBSERVED STATE.
  - **D.** THE UNCERTAINTY PRINCIPLE and its consequences.
- **II.** FUNDAMENTAL CONSTANT
  - **A.** PLANCK'S CONSTANT: $h = 6.63 \times 10^{-34}$ joule second.

III. PARTICLES

A. THE ATOM

1. Composition: positive nucleus and cloud of electrons.
2. Structure: the Bohr atom: electron orbits.
3. Structure: the Schrodinger waves: three dimensional electron waves.

IV. APPLICATIONS

A. THE LASER

V. METHODS AND PHILOSOPHY OF PHYSICS

A. COPENHAGEN INTERPRETATION OF QUANTUM PHYSICS: Waves of Probability

B. OPERATIONALISM and other positivistic influences in (and of) physics.

Questions

1. Outline the photoelectric experiment described on p. 450. What was measured? What result was expected? What was observed? How did Einstein explain his result?
2. Why do we think of light as a wave?
3. Why do we think of light as a particle?
4. Why do we think of an electron as a particle?
5. Why do we think of an electron as a wave?
6. What is the difference between an electron and a photon?
7. Why was it thought that the stability of the atom contradicts the laws of electricity and magnetism?
8. How was the problem posed by the stability of the atom solved?
9. Why does an atom emit certain definite frequencies of light?
10. Does the fact that an atom can emit a photon of light violate the conservation-of-energy law? Explain.
11. What does the uncertainty principle make us uncertain about?
12. What is the importance of Planck's constant?
13. What is the Copenhagen interpretation of quantum physics? Why is it important?
14. Why can't we think of an electron as going through just one slit of the two-slit interference experiment? What happens if we detect the electron going through only one slit, or force it to do so?

15. Why do we think of the waves referred to in Equation 4 as associated with a single particle instead of with a whole group of particles, as is the case for sound or water waves?
16. How can stimulated emission cause a laser to operate?
17. Comment on the following statement. There can be no such thing as free will, because if I know the position and velocity of every particle and the forces between them, I can know everything that will happen.
18. What is an electron like when it is not being observed?
19. If matter propagates as a wave, why does an object such as a billiard ball appear to have a definite path and direction, and thus appear as a particle rather than a wave?
20. If the uncertainty principle holds, why does a particle such as a baseball seem to have a definite and observable position and velocity, while an electron does not?
21. Can you account for the occurrence of the work function W in Equation 2?
22. Describe the changes that have taken place in our understanding of the "field."
23. What are the similarities among waves on a string, waves in a metal plate, and the electron waves in the atom? What are the differences?
24. In what way does quantum physics unify our understanding of the fundamentals of physics? In what way does it destroy a unity in our understanding?
25. Can you think of a physical reason that an observed state of a system should be different from an unobserved state?
26. Why would it require a large force to confine an electron to a very small region of space?
27. Why is the light that comes from a laser different in frequency from the light we shine on the material to make the laser operate?

Problems

1. A certain atom has a separation in energy of 3×10^{-18} joule between two of its orbital levels. What frequency photon is required to make the atom change from the lower state to the higher one?
2. What is the wavelength of a 0.5 kilogram ball going 50 meters per second?
3. What would the formula for the wavelength be in terms of the velocity

if the particle were going so fast that the relativistic effects were important.

4. Find the energies of the photons emitted in the following systems.
   (a) Popular A.M. radio station: frequency = 1110×10^3 vibrations per second.
   (b) Prominent visible radiation from copper: wavelength = 3247×10^{-10} meter.
   (c) Prominent X-ray radiation from copper: wavelength = 1.54×10^{-10} meter.
5. Light of frequency 1.2×10^{17} vibrations per second is falling on a metal of work function 5×10^{-18} joule. What is the maximum energy with which photoelectrons can be emitted from the metal surface?
6. If a metal surface has a work function of 1.0×10^{-17} joule, what is the minimum frequency of light that can cause a photoelectron to be emitted from this surface?

More Difficult Questions and Problems

1. The eighteenth-century philosopher George Berkeley claimed that an object existed only by virtue of someone's perceiving it. How could quantum physics, as illustrated in the two-slit experiment or otherwise, be used to advance, oppose, or modify his argument?
2. The formula $E = hf$ is sometimes used to show that the more accurately we wish to know something, the more energy we must expend to find out about it. How might this type of argument be pursued? This results in information being purchased at the expense of energy. On this principle, how small a length could be measured if one used all the rest energy of the whole earth?
3. Some people claim that we did not have to wait for quantum theory to eliminate exact prediction in physics, since there was always some experimental error. What is the difference between this type of uncertainty and that expressed by the uncertainty principle?
4. A 100-watt light emits energy at the rate of 100 joules per second. If this were all light energy at a wavelength of 3000×10^{-10} meter, how much pressure would it exert on a piece of paper directly facing the light at a distance of one meter?
5. Why must the rest mass of a photon be zero?
6. Show that if Equations 1 and 4 hold for matter waves, then the speed of such waves must be faster than light. (This is explained by the distinc-

tion between the wave and group velocities. Figure 11-24 shows a wave *group*. The individual ripples go along at the *wave velocity*, while the whole group progresses at the so-called *group velocity*. The wave velocity need not be equal to the group velocity. For light in a vacuum, the two velocities are the same. For matter, the wave velocity is greater than the velocity of light, but the group velocity is the velocity of the material particle, which is less than that of light. The group velocity of photons or other particles is the velocity of transfer of information or energy, and is never greater than the velocity of light in vacuum.)

7. Copper atoms from which electrons have been removed can emit radiation as different in energy as the difference between parts b and c of Problem 4 (p. 491). How do you account for this fact? What does it indicate about the "orbits" of the atom?
8. Using the Bohr atom and the de Broglie waves, find the radius of the first Bohr orbit for an atom with Z positive charges on the nucleus.

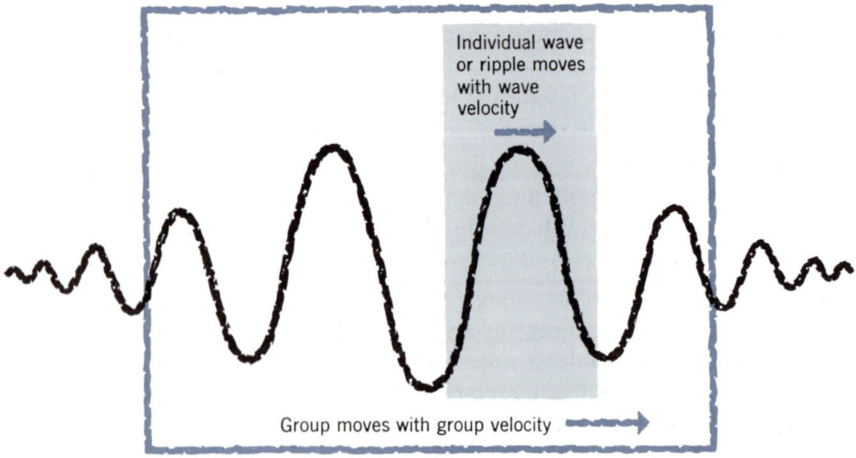

Fig. 11-24

12

THE NUCLEAR FORCES

In this chapter we shall discuss two more fundamental interactions —the so-called *weak* and *strong interactions*. The strong interaction supplies the strongest forces known between particles. The weak interaction is weaker than the electromagnetic and strong interactions but stronger than the gravitational interaction. These two interactions complete the list of fundamental interactions that are at present known to physics.

RADIOACTIVITY

In 1896, the French physicist A. Henri Becquerel discovered the phenomenon of radioactivity. Radioactivity is the emission of high-energy particles (waves) from certain substances. The classic account is that Becquerel chanced to put some samples of the mineral pitchblende into the same drawer with some covered unexposed photographic plates. The radioactivity of the pitchblende exposed the plates. Becquerel investigated this phenomenon (as did such other scientists as Pierre and Marie Curie and Ernest Rutherford). The particles that had exposed the film proved to have come from specific kinds of atoms contained in the pitchblende. Investigation revealed that many other atoms could exist in forms that emit similar types of particles. Such atoms are called radioactive. Some examples are atoms of the metals radium, uranium, and thorium. Three types of particles were observed, called alpha, beta, and gamma particles. The beta particles proved to be high-energy electrons. The gamma particles proved to be photons, much like light, but of much shorter wavelength and much greater penetrating power. The alpha particles were shown to be helium atoms that had lost two electrons—in fact, just what we now call the nuclei of helium atoms.

Probability

The discovery of radioactive emission introduced a new type of process into physics—one that evidently involved probability instead of precisely predictable mechanical behavior. The exact moment at which a particle is emitted from a radioactive atom cannot be predicted; only the probability of emission can be specified. In the last chapter we saw how this probability concept was incorporated into quantum physics.

When an atom emits radioactive particles, a change takes place in its nucleus. The nuclear change is called a *disintegration* or *decay*. These terms are used regardless of whether the change in the nucleus is relatively minor or actually causes the nucleus to break up. In alpha and beta emission the nucleus is changed in such a way that the atom becomes a different kind of atom—an atom of a different chemical element. The great energy of the emitted particle originates in the nucleus, although leading scientists, including Lord Kelvin, once found it difficult to accept this fact.

There does not seem to be any way to predict when a particular nucleus will change by emitting a particle. One nucleus may change immediately while another, apparently just like it, may wait one-hundred years to change, or a million years. If we have a collection of identical nuclei, as for example in a sample of pure radium, we know how many of them *on the average* will disintegrate in a given time period. In the case of radium, we know that half of them will have changed within 1600 years, but we do not know which half, and at present there is no reason to believe that we will ever be able to tell which half. If we wait 1600 years half of the nuclei will have changed. If we wait another 1600 years half of those remaining will have changed, leaving one fourth of the original nuclei unchanged. If we wait still another 1600 years, half of the remainder will have changed (see Fig. 12-1). Thus, for a large sample of such atoms, we can determine the so-called *half-life*. The half-life is the period of time within which half of the nuclei will have disintegrated. Each type of nucleus that can disintegrate radioactively has a definite half-life. These half-lives can vary from billions of years to a few billionths of a second—in fact, there are no known limits as to how long or short a half-life can be. The disintegration rate, and thus the half-life, does not seem to be changed appreciably by the normal forces and chemical changes the atom is subjected to. The disintegration rate can be changed, however, by collisions with radioactive particles, cosmic rays, and high-energy particles from cyclotrons and other particle accelerators.

It should be kept in mind that the half-life refers to average behavior. We can use it to predict the behavior of large groups of radioactive atoms. If we only have one atom, however, we can give odds as to how long it will remain unchanged, but we cannot tell exactly.

The Discovery of the Nucleus

Rutherford's assistants, at his suggestion, allowed a beam of high-energy alpha particles to pass through a very thin gold foil (see Fig. 12-2). The great majority of these particles passed through the foil without much deflection, but a few of them were deflected through

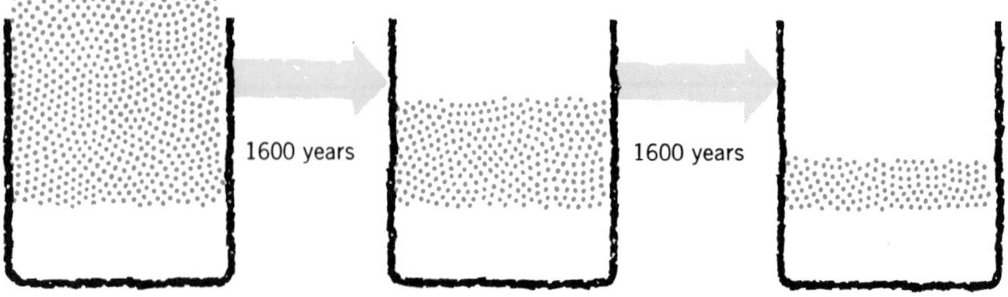

Fig. 12-1 The half-life of radium. Note that the atoms which have decayed have been changed into other atoms; they do not cease to exist.

large angles, some of them even coming straight back. This surprised Rutherford.

> It was quite the most incredible event that has ever happened to me in my life. It was almost as incredible as if you fired a 15-inch shell at a piece of tissue paper and it came back and hit you.[1]

This result meant that most of the cross section of the material target must be empty or covered with light particles, such as electrons. In order to deflect the alpha particles through large angles, however, there must be heavier particles present. (For example in order to bounce a 15-inch shell off a piece of tissue paper, there must be something embedded in the tissue paper that is very massive.) These heavy particles, according to Rutherford, must be small and highly charged. Thus the atomic nucleus was discovered. The atom consists of a surprisingly small, positively charged particle (about 10^{-15} meter in diameter) that contains nearly all the mass of the atom. Around it, out to a distance of about 10^{-10} meter, are the electrons. The alpha, beta, and gamma particles originate from the nuclei of certain kinds of atoms.

Most nuclei have less charge compared to their mass than they would have if they contained only protons. Therefore another kind of particle must also be contained in the nucleus. This fundamental

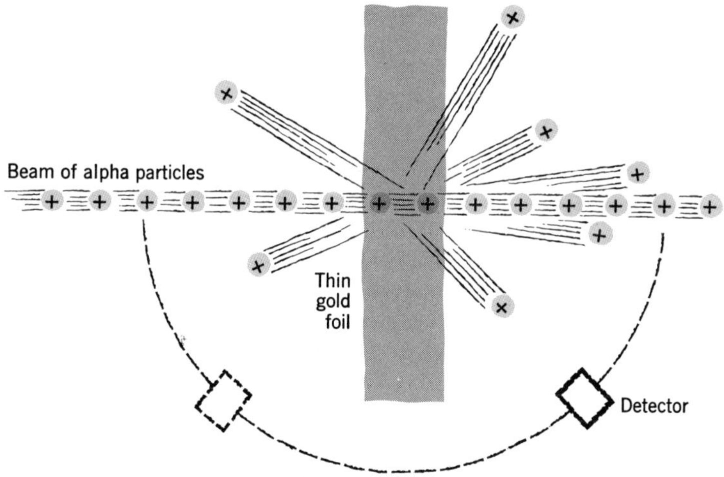

Fig. 12-2

[1]E. N. da C. Andrade, *Rutherford and the Nature of the Atom*, Garden City, N.Y.: Doubleday Anchor Books, 1964, p. 111.

particle, the *neutron*, was identified by J. Chadwick in 1932. The neutron, with no net charge, has a rest mass, m_n, equal to 1.67470×10^{-27} kilogram, which is slightly greater than the mass of the proton. Protons and neutrons are called *nucleons*, because together they make up the nuclei of atoms.

Particle Detectors

When alpha or beta particles pass through a gas, these high-energy particles interact with the molecules of the gas. Since these particles are electrically charged, they can interact with the electrons in the electron cloud outside the nucleus of the atom. They can strike these electrons and remove them from the atom. Thus, as the particle goes through the gas, it leaves gas atoms with missing electrons along its path, and also of course some free electrons that have been knocked out of the atoms. Atoms which have lost (or gained) one or more electrons are called *ions*, and the process of forming ions is called *ionization*. The particle detectors discussed below, as well as others not described here, use these ions to detect the particles that caused them.

THE GEIGER COUNTER. Figure 12-3 shows how a Geiger counter operates. The wire in the center of the gas-filled cylinder is positively charged. The wall of the cylinder is negatively charged. The charges on the wire and the wall are made so large that there is a very large electric field in the cylinder. If a particle comes through

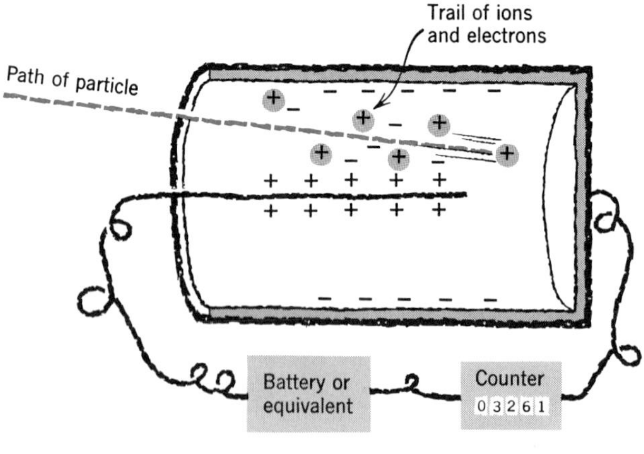

Fig. 12-3

the cylinder and creates ions and free electrons in the gas, the field causes these ions and free electrons to move. The electrons, in particular, are pulled so strongly toward the central positive wire that they strike gas atoms, knocking off other electrons. Thus an avalanche of electrons is formed, which, when it reaches the positive wire, creates a detectible current pulse through the Geiger counter. For each particle passing through the cylinder, a pulse is observed. Thus the Geiger counter is able to count the particles that come through it.

THE CLOUD CHAMBER. The cloud chamber contains a gas (usually air) which itself contains water molecules. When such a gas is cooled, the water tends to condense, forming clouds of small water droplets. This process is what happens in the atmosphere when a cloud is formed. Warm air, with evaporated water in it, expands and cools as it rises in the atmosphere. As it cools below a certain temperature, the water molecules form drops, and a cloud is formed. In order for a cloud to form, however, the water molecules must come together into tiny drops. It is difficult for these drops to form unless there is some dust for them to condense on. The drops will also condense on ions. Thus, if a cloud is formed in clean air, the droplets will first form around the ions present in the air. If a moving particle has left ions along its path, this path will be visible because of the tiny droplets condensing on each of the ions. Each ion will become visible because of the drop that forms on it.

We cool the gas in the cloud chamber by suddenly expanding it (see Fig. 12-4). When it cools, droplets form on the ions left by any particle that has just passed through the chamber. These droplets lie along the path of the particle. Figure 12-4*b* is a photograph of particle paths in a cloud chamber.

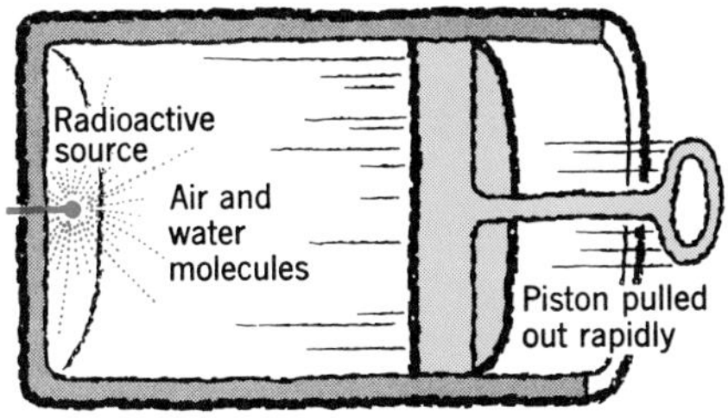

(a)

Fig. 12-4 a

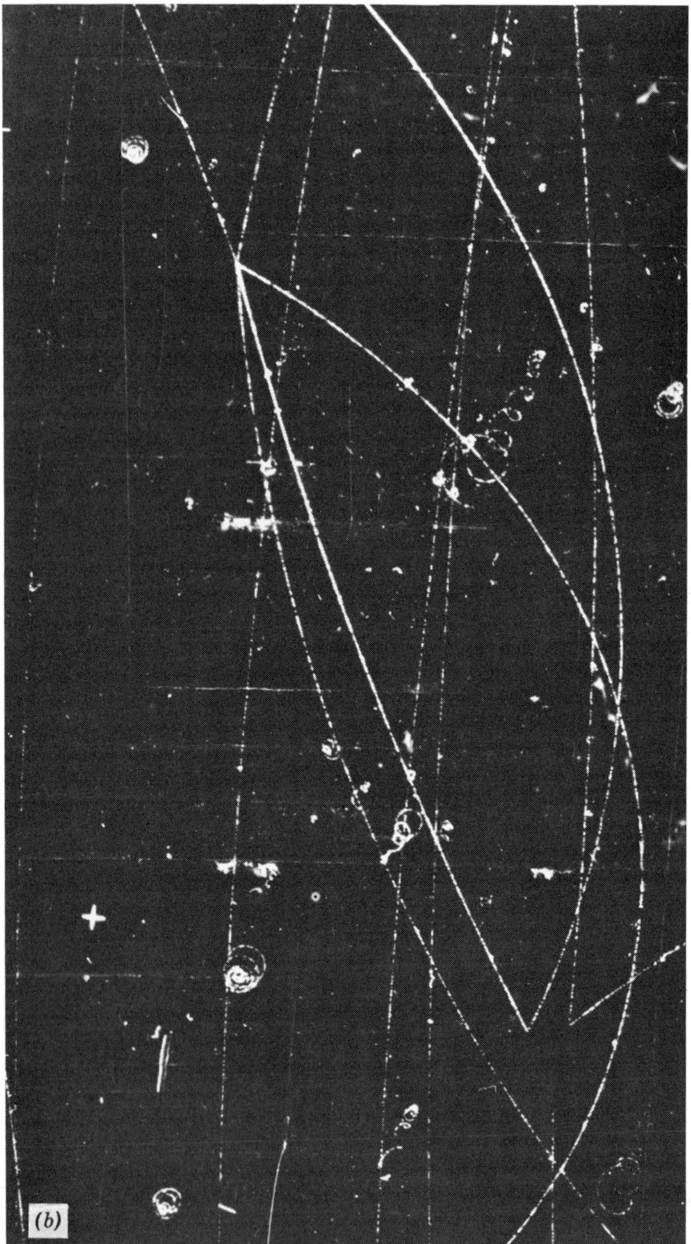

Fig. 12-4 b *Courtesy of Lawrence Radiation Laboratory, University of California.*

NUCLEAR STRUCTURE

We have pointed out that the nuclei of the various atoms are made up of two kinds of fundamental particles, neutrons and protons. The total number of protons in the nucleus determines the total charge on the nucleus. (We recall that the neutron has no net charge.) The electrons in the cloud outside the nucleus are held in the atom because they are attracted to the positively charged protons in the nucleus. In an electrically neutral atom, there are just as many electrons in the cloud outside the nucleus as there are protons in the nucleus.

The number of protons inside the nucleus is called the *atomic number* of that particular atom or nucleus. All atoms of the same chemical element have the same atomic number; in fact, it is the atomic number that determines the chemical element. Thus hydrogen, the simplest element, has an atomic number of one. The common form of hydrogen has one proton (and no neutron) in the nucleus, and one electron in the cloud. The most common form of helium, of atomic number two, has an alpha particle for its nucleus, consisting of two protons and two neutrons.

The *mass number* is the sum of the number of protons and the number of neutrons in the nucleus. Since the neutron and the proton have about the same mass, this number is very nearly proportional to the mass of the nucleus. The mass of the nucleus is very nearly the total mass of the atom, since the electrons are much less massive than the nucleons. The common form of hydrogen has a mass number equal to its atomic number, namely one, since it has only one nucleon in its nucleus (see Fig. 12-5*a*). The common form of helium has a mass number equal to four, because it has two protons and two neutrons in its nucleus (see Fig. 12-5*d*). For all known atoms except the common form of hydrogen, the atomic mass number is greater than the atomic number.

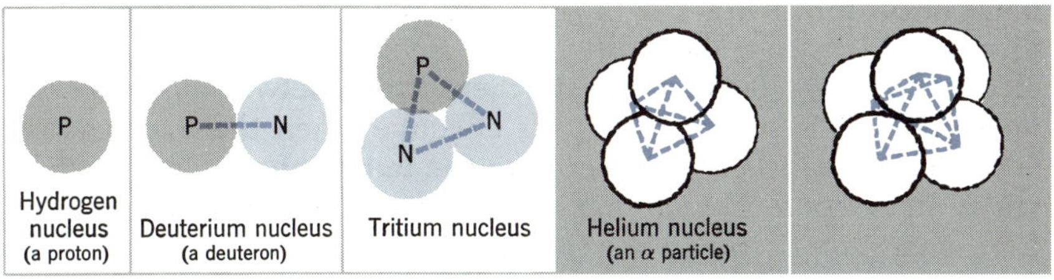

(*a*) (*b*) (*c*) (*d*) (*e*)

Fig. 12-5

We have discussed the common form of hydrogen, which has a nucleus composed of one proton. There are two other forms of hydrogen. One of these, called *deuterium*, has a nucleus composed of one proton and one neutron (see Fig. 12-5*b*). Thus it has a mass number of two and is approximately twice as heavy as the common form of hydrogen. It is still the same chemical element hydrogen, however, because it still has the same number of protons in the nucleus. The third form of hydrogen, called *tritium*, has one proton and two neutrons in the nucleus (see Fig. 12-5*c*). All nuclei that have the same number of protons but different numbers of neutrons are called *isotopes* of each other. Hydrogen, as we have seen, has at least three isotopes. Most elements have several isotopes.

THE STRONG INTERACTION

One characteristic of the strong interaction is that it operates only between certain kinds of particles. We have considered two of the fundamental particles that exert or are subject to the strong interaction—the neutron and the proton. The forces of the strong interaction hold the nucleons together in the nucleus.

Another characteristic of the strong interaction is that it is very strong indeed. It takes considerable energy to break up something that is held together by the strong interaction. This energy is so large that it has an appreciable amount of mass. For example, the mass of the alpha particle is about three quarters of a percent less than the sum of the masses of two protons and two neutrons. One might expect, therefore, to gain or lose a lot of energy in transformations involving the strong interaction. The atomic and hydrogen bombs are evidence that this does occur.

A third characteristic of the strong interaction is the very short range of the forces to which it gives rise. It is most effective at nuclear distances—less than about 10^{-15} meter, roughly the "radius" of a nucleon. Therefore it is strongest when acting between neighboring nucleons and is substantially less effective when the nucleons are separated by a distance greater than the radius of one nucleon. The strong interaction is about one hundred times stronger than the electromagnetic interaction when the nucleons are close together. Therefore we may often ignore electromagnetic forces between neighboring nucleons. The strong forces between protons and protons, protons and neutrons, and neutrons and neutrons are all approximately equal in strength. The electromagnetic force is stronger than the strong force at long ranges, because the strength of the strong interaction falls off much more rapidly with increasing distance than that of the electromagnetic interaction.

At present the exact law governing the strong interaction is unknown.

Nuclear Binding Energies

The strong interaction and the electromagnetic interaction together account for the main features of nuclear binding energies. The binding energy (see Chap. 6, pp. 251–252) is the energy that must be supplied in order to separate particles that are held together by a force. The nucleons that make up a nucleus have less mass when they are combined into a nucleus than when they are separated. Since energy must be supplied to the particles to separate them, mass must also be supplied because energy has mass. Therefore the nucleons will have more mass when separated than when combined in the nucleus.

We shall be concerned here with the total binding energy for the nucleus divided by the total number of particles in it—that is, the binding energy per particle. The actual binding energy of two nucleons is about 3.6×10^{-13} joule, corresponding to a binding mass of 4×10^{-30} kilogram. We shall call this mass B.

Figure 12-5*b* shows a proton and a neutron that have come together to form a *deuteron*, which is the nucleus of the second form of hydrogen, deuterium. The mass of this particle is less than the sum of the masses of the proton and neutron when separated because of the binding energy. The mass loss due to this binding energy is approximately B. The mass loss per particle, then, is $B/2$, since there are two particles making up the deuteron.

Figure 12-5*c* shows a proton and two neutrons that have come together to form the nucleus of tritium, the third isotope of hydrogen. There are now three particles, all attracted to each other so that they bind closely. Thus there are three "bonds," represented by the dotted lines in Figure 12-5*c*, since each particle is in close contact with each of the others. Therefore, we expect that there will be a mass loss of about $3B$. Since there are three particles, the mass loss per particle is $3B/3$, or B. The binding mass per particle is greater in this case than in the last.

Now consider the case of four particles, which will be arranged in a pyramid as shown in Figure 12-5*d*. (This kind of pyramid, with four corners, is called a tetrahedron.) Now there are six bonds and four particles; therefore we expect that the mass loss will be about $6B$ and the mass loss per particle will be $6B/4$, or $(3/2)B$. Thus the mass loss per particle has again increased. (Actually this treatment is much oversimplified for the alpha particle.)

One might think that this could go on forever, with the lowest energy state per particle being achieved when all the neutrons and protons in the world are amassed into one big nucleus. However, there are two other effects to consider. One of these limits the mass loss per particle, and the other actually decreases it.

To see the effect that limits the mass loss per particle consider the nucleus with five particles, as shown in Figure 12-5*e*. There is no way in three-dimensional space by which all of these particles can touch all of the others. Although each of four particles can all touch all of the others, each of five cannot. The fifth particle must be farther away from some of the particles than from others. This is very significant because the strong force is a short-range force. We will assume that it is effective only when the particles are "touching," although of course this is not exactly correct. In Figure 12-5*e*, there are five particles and nine bonds where the particles are touching one another. The mass loss per particle is now $(9/5)B$. This is still larger than, in the last case, but the amount of increase is going down.

What will be the maximum mass loss per particle? Let us consider a particle in the middle of a large nucleus. Even in the closest possible packing, it can have no more than twelve nearest neighbors touching it. (Try this with a stack of oranges—see how many other oranges can touch a single orange.) This means 12 bonds, each of which is shared with another particle. Thus each particle has the equivalent of six bonds for itself. Therefore the mass loss per particle is $6B$ for a particle in the middle of the nucleus. A particle on the edge of the nucleus has fewer nearest neighbors, and therefore the mass loss per particle will be even smaller. The mass loss per particle cannot be greater than $6B$ according to this analysis.

Larger nuclei contain more protons. These protons repel one another because of their electric charge, and thus act against the attractive strong force. The electric force is much weaker than the strong force, but it acts at long range. Therefore each proton is bound by the strong force to its nearest neighbors, but it is repelled by every other proton in the nucleus. Despite the greater strength of the strong force, there is a maximum number of protons that can be housed together in a stable nucleus. These protons, because of their charge, act against the strong force; for large nuclei, the mass loss per particle actually decreases somewhat as the nucleus increases in mass number rather than becoming greater. Therefore, there is a size of nucleus for which the mass loss per particle is greatest. The simplified argument we have given here suggests a graph similar to that shown in Figure 12-6*a*. The actual graph is given in Figure 12-6*b*. The two graphs are somewhat different because of several effects not discussed here.

The Release of Nuclear Energy

An examination of Figure 12-6*b* helps us to understand how nuclear energy can be released. The greatest binding energies per particle occur for nuclei containing about 56 nucleons, such as the nucleus of iron. The fact that the graph has a maximum corresponding to the greatest binding energy per particle means that we could obtain energy in two ways by transforming nuclei.

1. If we could get the light nuclei at the left-hand side of the graph to join together, they would become the bigger nuclei that are farther to the right on the graph. Thus the binding energy per parti-

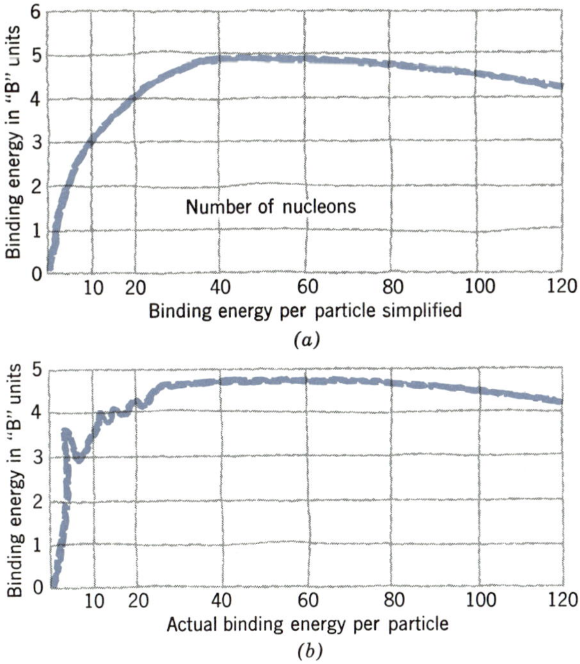

Fig. 12-6

cle in these nucleons would increase. This means that the particles would be more strongly bound together than they were before, and therefore energy would have to be released. This process is known as fusion. Fusion occurs in the explosion of a hydrogen bomb, and in the sun.

2. If we could get the heavy nuclei at the far right of the graph to split approximately in half, the two pieces would be lighter nuclei, found at about the middle of the graph. Here again the binding energy per particle would be increased, although not as much as in the fusion process, as can be seen from the graph. Here also, then, energy would be released, but not as much as in the fusion process. This process, called nuclear fission, occurs in ordinary atomic-bomb explosions and in many nuclear reactions. In both fission and fusion new nuclei are formed, and some of the rest mass of the original nuclei is released as other forms of energy.

Nuclear Fission: The Atomic Bomb

One kind of atomic bomb uses the type of nucleus that has atomic number 92 and mass number 235—the well-known uranium-235. This element appears well to the right on the graph in Figure 12-6*b* and therefore is a candidate for fission. If the uranium-235 nucleus can be made to break up it will indeed release energy. Each

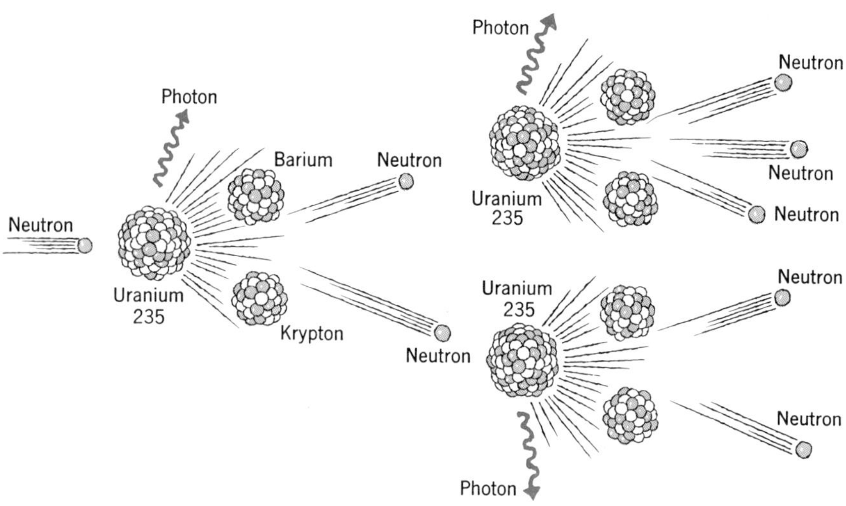

Fig. 12-7 Chain reaction.

uranium-235 atom has a certain probability of breaking up by itself, but we want to stimulate many nuclei to break up at the same time. The fission process can be stimulated by irradiating uranium-235 with slow neutrons. These neutrons have no net charge, and therefore they are not repelled by the nucleus, as a proton would be. They easily fall into the nucleus and become bound to the other nucleons. The binding energy released as the neutron is captured appears as kinetic energy of the neutron and the other particles in the nucleus. This increased energy causes the nucleus to break up, releasing not only energy, but also two or three more neutrons. These neutrons can now help to break up even more uranium-235 nuclei. Thus the process will be self-sustaining. Unfortunately, the neutrons that are released are not slow neutrons but high-energy (fast) neutrons. Before they can be captured easily by other uranium-235 nuclei they must be slowed down to lower kinetic energies. The fast neutrons will become slowed down by repeated collisions and scattering by the atoms of the uranium until they are slow enough to be captured by the nucleus. However, if the piece of uranium-235 is too small, most of the neutrons produced will leave the material before they are slowed down, and the reaction will not build up. If the material is greater than some critical size, however, the neutrons from one fission will be slowed down so that they start two or more other fissions. Then we have a chain reaction—on the average, each fission stimulates more than one more fission, and the reaction rapidly builds up to explosive rates (see Fig. 12-7). This causes an atomic bomb explosion (Fig. 12-8). To explode the bomb it is necessary to take two pieces of uranium-235 below critical size and quickly put them together to make a piece above critical size.

Nuclear Fission: The Nuclear Reactor

If we do not want a fission bomb but wish to obtain nuclear energy from fission at a more controlled rate for less destructive purposes, we use what is called a nuclear reactor. Figure 12-9 is a diagram of one kind of reactor. The "fuel" for this reactor is not pure uranium-235 but uranium that may contain about 20 percent uranium-235. Most of the rest of this uranium fuel is another isotope of uranium, with mass number 238—uranium-238. The uranium-238 will absorb most of the fast neutrons produced by the fission of the uranium-235 before these neutrons can be slowed down enough to be captured by other atoms of uranium-235. To reduce this effect, so that there will be enough slow neutrons to cause fission, the uranium rods are separated by a *moderator*, such as graphite. The moderator slows down the neutrons emerging from a uranium rod without absorbing too many of them. Then these neutrons will have the right speed to cause fission when they again enter a uranium rod.

To make sure that there are not too many slow neutrons, which might cause an uncontrolled chain reaction, control rods of some neutron-absorbing material, such as boron or cadmium, are inserted into the reactor. The deeper these control rods are inserted, the more neutrons they will absorb; thus they afford a method of controlling the process.

Fig. 12-8 *Courtesy of Lookout Mountain Laboratory, U. S. Air Force Photo.*

Fusion

The problem of attaining a practical fusion reaction, in which light nuclei come together to form more massive ones, is more difficult than the similar problem for fission. In a hydrogen-bomb explosion, deuterons combine in processes that lead, for example, to tritium nuclei and alpha particles. The deuterons, however, are positively charged, and therefore repel each other. They must have considerable kinetic energy so that they will collide and get close enough together for the strong force to overcome the electrical repulsion. They will have this high kinetic energy if they are at a high temperature.

In stars, such as the sun, this high temperature does occur. Presumably the sun's energy comes largely from a fusion process. These high temperatures have also been achieved in the hydrogen bomb, which requires an ordinary fission bomb to first obtain the necessary energies for fusion. Controlled release of power, however, presents a formidable engineering problem, since the high temperatures required to sustain the process will destroy any containing structure. At such high temperatures, however, the atoms are nearly all ionized and thus are electrically charged. Such a collection of mostly charged particles is called a *plasma*. In the cyclotron (see Chap. 8, pp. 336–337), a magnetic field is used to keep the charged particles from escaping from the instrument too soon. More complex types of magnetic fields can be used to contain a plasma for very short times. There is some possibility that a plasma can be contained long enough to

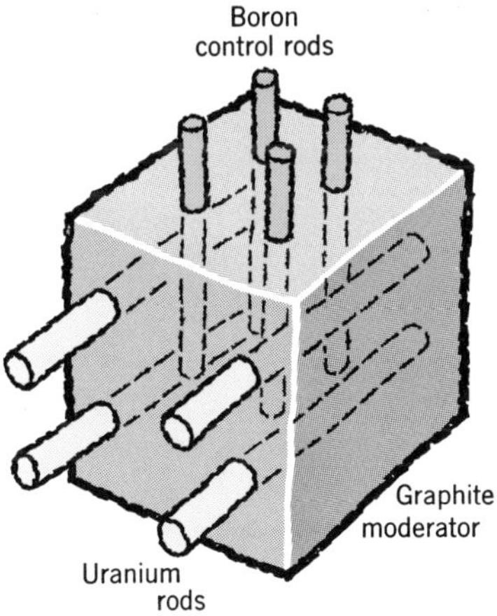

Fig. 12-9

permit a controlled fusion reaction. Research is still being done on this problem.

ANTIMATTER

The Discovery of Antimatter

In 1930, the American physicist C. D. Anderson announced the discovery of a new particle—the positive electron, also known as the *positron*. This particle has the same mass as the electron but it has a positive instead of a negative charge. The charge has precisely the same magnitude as that of the electron but it is opposite in sign. The positron appears to behave like an electron except for this difference in charge. Cloud-chamber tracks of an electron are curved in the presence of a magnetic field. The tracks of the positron are also curved, but in the opposite direction (see Fig. 12-10).

Today it is believed that for every particle there is an antiparticle, of the same mass as the particle itself but of opposite charge. Most of these antiparticles have been discovered. Thus there is the antiproton, with the same mass as the proton, but of negative charge. There is the antineutron, of the same mass as the neutron. Since the neutron has no net charge, the antineutron also has no net charge but it differs from the neutron in properties we have not yet discussed. The antiphoton is itself a photon; hence we say that the photon is its own

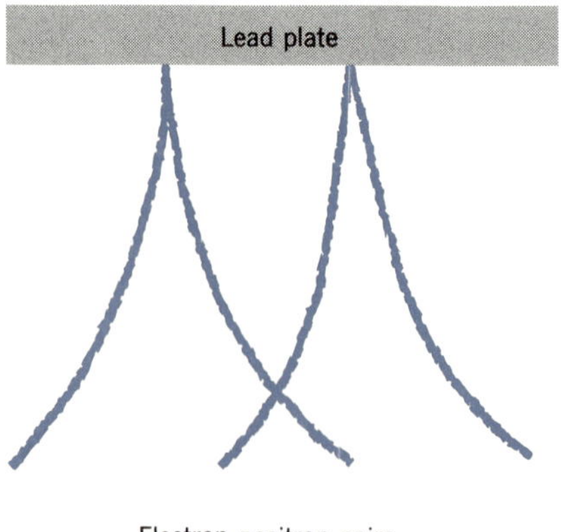

(*a*)

Fig. 12-10a

antiparticle. There is also an antimuon, as is to be expected, since the muon behaves like an electron but has a larger mass. It is possible to have antiatoms made up of antiprotons, antineutrons, and positrons, rather than protons, neutrons, and electrons. We can even speculate about antiuniverses, made of antimatter. Perhaps some of the galaxies we see through the telescope are actually made up of antimatter rather than matter. Perhaps it is a coincidence that we ourselves are made of matter rather than antimatter.

The Annihilation of Matter

When a particle and its antiparticle come together they may annihilate each other; that is, they may both disappear, leaving two or

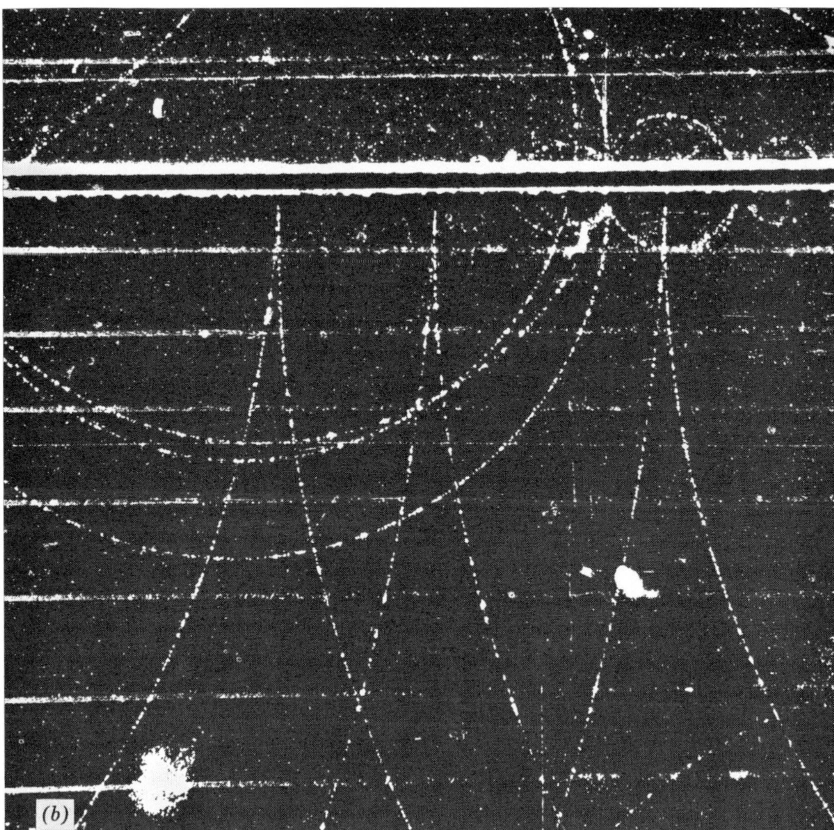

12-10b. *Courtesy of Lawrence Radiation Laboratory, University of California.*

more photons or perhaps other particles (see Fig. 12-11). The remaining photons ensure that energy and momentum will be conserved; they have the energy and momentum that were previously contained in the particles that were annihilated. In our part of the universe, when antimatter is produced, it soon finds some ordinary matter and is annihilated. Similarly, in certain processes of sufficient energy, particles and antiparticles are created in pairs. This process is called *pair production.* Of course, there must be at least enough energy available to make up the rest mass of the pair of particles before this can take place.

The annihilation of matter is an excellent example of the conversion of rest mass into other forms of mass-energy. When an electron and a positron annihilate each other to form photons, their rest mass completely disappears, since photons have no rest mass. The rest mass of the electron and positron is completely transformed into the kinetic energy (mass) of the photons.

Particle-Antiparticle Symmetry

The existence of antimatter appears to introduce another symmetry into the physical world. For every possible configuration of matter, we could get another by changing all the particles into antiparticles. This suggests that the laws of physics should be the same even if we were to change all the particles into antiparticles—that every possible process in the physical world would still be possible if the particles were antiparticles. This symmetry principle holds for a wide variety of phenomena, but it is not universally valid, as we shall see later.

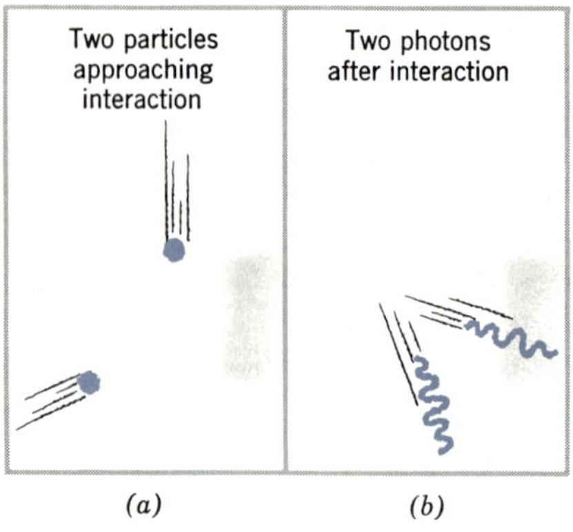

Fig. 12-11

THE WEAK INTERACTION

The Neutrino

When the neutron is separated from other particles, it is unstable. It emits two particles and changes into a proton, according to the following process:

$$n \rightarrow \mathrm{p}^{+} + \mathrm{e}^{-} + \bar{\nu}_e \tag{1}$$

This "equation" describes a particle reaction. The two sides of the reaction are separated by an arrow. To the left of the arrow are represented the particles that start the reaction, in this case the neutron, n, alone. To the right are represented the particles that are formed by or result from the reaction. In this case they are the proton, p^+, the electron, e^-, and a new particle, $\bar{\nu}_e$, the *antineutrino*. The antineutrino is the antiparticle of another new particle, the neutrino, which is represented by the symbol ν_e. (We often represent an antiparticle by putting a bar over the symbol representing the particle.)

The neutrinos and antineutrinos are particles with no charge and no rest mass. These particles apparently travel at the speed of light. They are very hard to detect, since they can pass through solid matter a tenth the size of the galaxy before being absorbed.

Besides describing the decay of an isolated neutron, "equation" 1 also describes what happens to a neutron in the nucleus in the ordinary radioactive process of beta emission. The emitted beta particle is the electron that appears on the right side of the reaction. The neutron emitting the electron changes into a proton. This process changes the atomic number of the nucleus, but not its atomic mass number. When this process was first studied, it appeared that energy and momentum were not conserved because neutrinos and antineutrinos were unknown. In fact, these particles were first assumed to exist as a way of accounting for the apparent loss of energy and momentum in the process of beta emission. Since then the detection of neutrinos has been reported in some very difficult experiments.

The reaction described in "equation" 1 involves a kind of interaction which we have not yet discussed. The forces of this interaction are responsible for the emission of the electron and the antineutrino from the neutron, as it changes to a proton. This interaction is called the weak interaction, because its forces are weaker than those of either the strong interaction or the electromagnetic interaction. (They are, however, stronger than the forces of the gravitational interaction.)

In some nuclei another kind of beta emission can occur in which a proton changes to a neutron while emitting a positron (represented by the symbol e^+) and a neutrino:

$$p^+ \rightarrow n + e^+ + \nu_e \tag{2}$$

This reaction also involves the weak interaction. It cannot occur when the proton is alone, separated from other particles, because the rest mass of the proton is smaller than that of the neutron. In such a case the reaction would violate the law of conservation of mass-energy if it occurred spontaneously. In certain nuclei, however, the difference between the binding energy of the neutron and that of the proton may be great enough to permit release of the necessary energy when the proton changes to a neutron.

A stable nucleus—that is, one that does not disintegrate spontaneously by any process—is characterized by a certain number of neutrons and a certain number of protons. For a stable nucleus of a given mass number there are only a certain number of possibilities for the number of neutrons and protons. (There may be only one possibility, or even none.) If there are too many neutrons, beta emission of electrons can occur, changing the neutrons to protons. On the other hand, if there are too many protons, beta emission of positrons can occur, changing the protons to neutrons. These stabilities and instabilities result from the small differences in the binding energies of neutrons and protons in the various nuclei. Stable nuclei, especially the lighter ones, usually have about as many neutrons as protons. The only nucleus that contains no neutrons is that of the common isotope of hydrogen, which consists of a single proton. The more massive nuclei, such as uranium, have up to 62 percent more neutrons than protons. The large forces of electric repulsion due to the presence of the charged protons in large nuclei tend to disrupt the nucleus and limit the total possible size of a stable nucleus. Since the neutron has no net charge, more neutrons can be present in these larger nuclei without having this disruptive effect. There are no stable nuclei containing more than 83 protons or a total of more than 209 nucleons.

The Electron and Muon Family Numbers

The symbol for the antineutrino in "equation" 1 has "*e*" as a subscript. This subscript is used to differentiate the neutrino and antineutrino associated with the electron (the so-called electron neutrino and electron antineutrino) from another neutrino and antineutrino associated with the muon (called the muon neutrino and muon antineutrino). Thus there are two kinds of neutrinos and antineutrinos. The muon neutrino is sometimes called a neutretto.

We can obtain new conservation laws by assigning a "family number" to each of the members of certain groups of particles. One such group is comprised of the electron, the positron, and the electron neutrino and antineutrino. The electron and electron neutrino are each assigned the electron family number of one. The positron and the electron antineutrino are assigned the electron family number of minus one. All other particles are assigned the electron family number zero. The electron family number of a collection of particles is the sum of the electron family numbers of the particles in the collection. This family number is conserved in all particle reactions. For example, it is conserved in the reaction described by "equation" 1. On the right side of this reaction we have an electron (electron family number one) and an electron antineutrino (electron family number minus one). The total electron family number on the right is therefore zero. On the left size of "equation" 1 we have only a neutron, which has electron family number zero. Therefore the electron family number is not changed in the reaction. The electron family number is conserved in all physical processes.

The muon family number is assigned in exactly the same way as the electron family number except that the members of the group are different. We assign the negative muon and the muon neutrino the muon family number of one. We assign the positive muon and the muon antineutrino the muon family number of minus one. All other particles are assigned the muon family number zero. The muon family number is also conserved. As an example of these laws let us consider the process by which the muon changes into an electron and other particles. This is another reaction that involves the weak interaction:

$$\mu^- \rightarrow e^- + \bar{\nu}_e + \nu_\mu \tag{3}$$

In this reaction the negative muon is represented by the symbol μ^-. Its antiparticle, the positive muon would be represented by the symbol μ^+. In this reaction the electron and muon family numbers are conserved separately. On the left the muon family number is one. On the right it is also one, since the muon neutrino (ν_μ) appears. On the left the electron family number is zero, since neither electrons nor electron neutrinos appear. On the right there is an electron (electron family number one) and an electron antineutrino (electron family number minus one). Thus the total electron family number on the right is zero, just as it is on the left. There are no known exceptions to these two conservation laws.

*Violation of Reflection Symmetry and Particle-Antiparticle Symmetry

Processes that involve the weak interaction do not necessarily obey the principle of reflection symmetry discussed in Chapter 1 (p. 34). This means that processes are known to occur, such as beta emission, which when viewed in a mirror give processes that do *not* occur. The beta emission from cobalt-60 is a famous example. (The cobalt-60 nucleus has an atomic number of 27 and a mass number of 60.) Figure 12-12 shows how the spinning nucleus emits an electron in beta emission. The electron comes out in the opposite direction from the antineutrino, but the directions are usually in the same relation to the spin of the nucleus. If one looks at the mirror image, as indicated in the figure, the process does not look the same. The mirror image is not the same as the original process, just as the reflection of a right hand in a mirror looks like a left hand, not another right hand. This would be all right, except that the process that is observed in the mirror is not observed in nature—in reality the particles generally come out in the way indicated in the diagram and not as indicated in the mirror. For true reflection symmetry to hold, however, the reflected process should be just as likely as the unreflected process.

The weak interactions also violate the principle of particle-antiparticle symmetry. However, weak interactions may obey a combined symmetry principle. If we not only reflect the process illustrated in Figure 12-12 in a mirror but also exchange all particles for their antiparticles, we obtain what is believed to be a possible

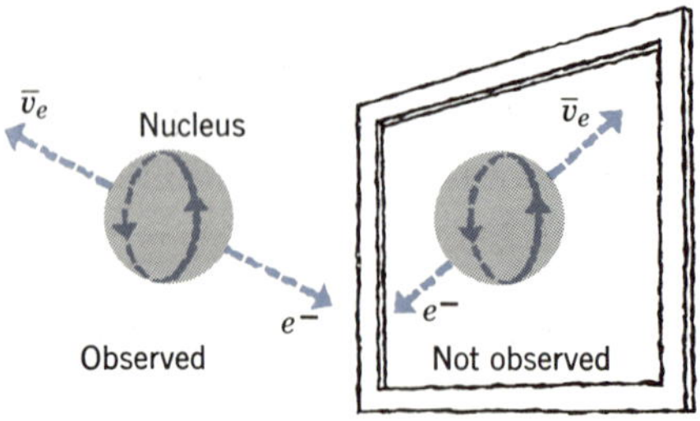

Fig. 12-12

process. As far as is known, reactions involving the weak interaction always obey this combined symmetry principle of reflection plus particle-antiparticle exchange. The failure of reflection symmetry in the weak interactions was a surprise to scientists when it was discovered in 1956. Just which symmetry principles are valid for weak interactions is still under investigation.

FURTHER CONSIDERATIONS OF THE STRONG FORCE

Conservation of Baryon Family Number

The proton by itself has enough rest mass to change into many electrons and positrons. Why does it not do so? We do not know, but we can invent a principle to account for it. Both the neutron and proton are heavy particles. (We shall discuss some other heavy particles later.) Heavy particles are called *baryons*. Both protons and neutrons are assigned the baryon family number (or simply baryon number) of one, and their antiparticles are assigned the baryon family number of minus one. All the other particles we have discussed so far are assigned the baryon family number of zero. The baryon family number is conserved in all physical processes. This conservation law explains why the free proton is stable – there is no other heavy particle that has less rest mass. The proton cannot change into lighter particles only, since the baryon family number must be conserved. Similarly, a nucleon in the nucleus cannot change completely into lighter particles but must change into other heavy particles, preserving the heavy nucleus.

**Alpha Emission*

Since nuclei are positively charged, they repel other positively charged particles. When the positively charged alpha particles are directed toward the nucleus, as in Rutherford's experiment (p. 496), it is this electric repulsion that deflects the alpha particles. If an alpha particle has sufficient energy, it may be able to overcome this repulsion and reach the nucleus.

The electric repulsion acts like a barrier, keeping the alpha particles out of the nucleus. Figure 12-13 shows this barrier schematically. The height of the barrier at any point represents the energy that an alpha particle must have to reach that point. (The height of the barrier is the potential energy that an alpha particle would have at that point because of two fields of force acting on it, one caused by the strong force of attraction and the other by the repulsive electric force.) The top of the barrier is the point at which the attraction due to the strong force is equal in magnitude to the repulsion due to the electric force. At points closer to the nucleus than the top of the barrier, the strong force of attraction dominates the electric force. Farther away from the nucleus than the top of the barrier, the repulsive electric force dominates.

In the radioactive process of alpha emission, alpha particles are coming out of the nucleus rather than being directed into it. One would expect that such alpha particles would have to have at least as much energy as that represented by the top of the barrier in order to get out. However, alpha particles that do not have this much energy are often emitted from nuclei, changing the nuclei in the process. Somehow the alpha particle gets through the barrier without having enough energy (apparently) to do so. This is one example of a process known as "barrier penetration," which is possible for a material particle according to quantum physics but not according to the earlier Newtonian physics.

According to Newtonian physics, the top of the barrier is inaccessible to the alpha particle because the particle does not have enough energy to reach it. It does have enough energy to reach the regions outside the barrier, however, if some way could be found for it to get across the barrier. According to the laws of quantum physics, however, a particle can "leak" through an energy barrier between one region and another. This can occur if the particle has enough energy

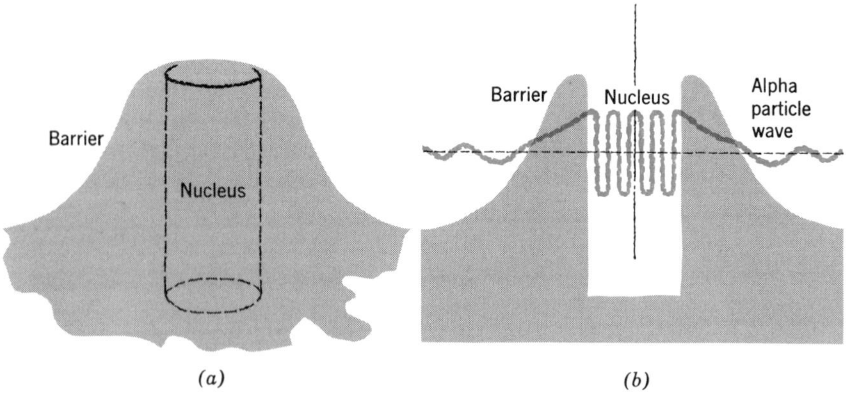

Fig. 12-13

so that both regions would be accessible to it except for the barrier. The higher and wider the barrier, the lower the probability for such leakage. If the barrier is infinitely high or infinitely wide, leakage is impossible.

If we picture the particle as a wave, we can illustrate the process of barrier penetration as in Figure 12-13*b*. The alpha particle inside the nucleus is represented as a standing wave. Such a standing wave can be thought of as a sum of waves moving in different directions, being reflected from the barrier. At the barrier, however, the standing wave does not have a node but penetrates with decreasing amplitude into the barrier. When this decreasing wave reaches the outside of the barrier, it still has some small amplitude, which then represents the particle outside the barrier. (This discussion, of course, is not a proof of barrier penetration but merely an illustration of how it occurs.)

We can use the uncertainty principle that relates time and energy to explain why it is possible for some of the alpha particles to penetrate into and through the barrier region. Over a long time the energy of the alpha particles corresponds to a certain precise energy state within the nucleus. For a very short time period, however, their energies can be very uncertain. This uncertainty, for very short times, allows them to have higher energies, and with these higher energies they can move into the barrier region. The distance through which they penetrate the barrier depends on how fast they can move with their additional energy during the short time the extra energy is allowed to them by the uncertainty principle.

APPLICATIONS: RADIOACTIVE DATING

Lord Kelvin had estimated the age of the earth as no more than twenty or forty million years by calculating the time it must have taken for the earth to cool to its present temperature. This time span was an embarrassment to Darwin and other proponents of organic evolution, for it did not appear to be long enough for evolutionary processes. The apparent conflict of evolutionary theory with the laws of physics hampered the acceptance and development of the theory for over fifty years.[2]

Rutherford, in a paper presented to the Royal Institution, pointed

[2]Loren Eiseley, *Darwin's Century: Evolution and the Men Who Discovered It*, Garden City, N.Y.: Doubleday Anchor Books, 1961, Chap. 9.

out that radioactivity in the earth was a source of heat, and that Kelvin's argument was therefore no longer valid.

> I came into the room, which was half dark, and presently spotted Lord Kelvin in the audience and realized that I was in for trouble at the last part of my speech dealing with the age of the earth, where my views conflicted with him. To my relief, Kelvin fell asleep, but as I came to the important point, I saw the old bird sit up, open an eye and cock a baleful glance at me! Then a sudden inspiration came, and I said Lord Kelvin had limited the age of the earth, *provided no new source* [*of heat*] *was discovered.* That prophetic utterance refers to what we are now considering tonight, radium! Behold! the old boy beamed upon me:[3]

It was, however, a few years before Kelvin accepted radium as a new chemical element.

The discovery of radioactivity not only made Kelvin's argument irrelevant but also provided a means of making meaningful estimates of the age of the earth and certain structures and strata in it. These methods depend on the fact that the rate of radioactive disintegrations is not appreciably influenced by normal changes of temperature and pressure, or by normal chemical processes. Therefore, if we can measure or estimate how much of a sample of radioactive material has already disintegrated, we can find out how long the sample has been in its present condition. We must, of course, know the half-life of the radioactive material in order to do this. We shall discuss two of these methods of radioactive dating here—the fission-track method, and the carbon-14 method.

Fission Track Dating

Uranium-238 naturally undergoes fission, with a half-life of 10^{16} years.[4] When the fission occurs, the large fragments of the nucleus acquire kinetic energy. If the fission occurs in a sample of a glassy substance, these fragments cause tracks (see Fig. 12-14) which can be seen in a microscope if a smooth surface of the sample is etched. The original tracks can be made to disappear if the material is heated to a high temperature (annealed). Natural glassy substances, such as volcanic glass, sometimes contain small amounts of uranium. If

[3]From *Rutherford and the Nature of the Atom*, by E. N. da C. Andrade. Copyright ©, 1964 by Educational Services Inc. (Science Studies Series). Reprinted by permission of Doubleday & Company, Inc.

[4]See R. L. Fleischer, P. B. Price, and R. M. Walker, *Application of Fission Tracks and Fission-Track Dating to Anthropology*, General Electric Research Laboratory Report No. 65-RL-3878M, Schenectady, N.Y.: General Electric Company, February 1965. This is a readable account of some of the applications of fission-track dating.

fission occurs among these uranium atoms, the fragments leave tracks. By counting these tracks one can find out how many uranium nuclei have undergone fission since the glass was last annealed.

The number of remaining uranium-238 atoms may be found by irradiating the sample with slow neutrons. These slow neutrons will cause the uranium-235 nuclei in the sample to undergo fission. The fission fragments from this process are then counted, which gives the number of uranium-235 atoms in the sample. If, in addition, the ratio of uranium-235 to uranium-238 is known, the number of remaining uranium-238 atoms can be calculated. Knowing the number of uranium-238 nuclei that have undergone fission, the number remaining, and the half-life for the fission process, we can calculate how long the process has been going on in the sample since it was last annealed. The fission-track method of dating can be used to date glassy objects from thirty to a billion years old, depending on the amount of uranium in the object.

Several precautions must be observed in using this method. For example, one must be sure that there are no other kinds of nuclei in the sample that will also undergo fission when the sample is irradi-

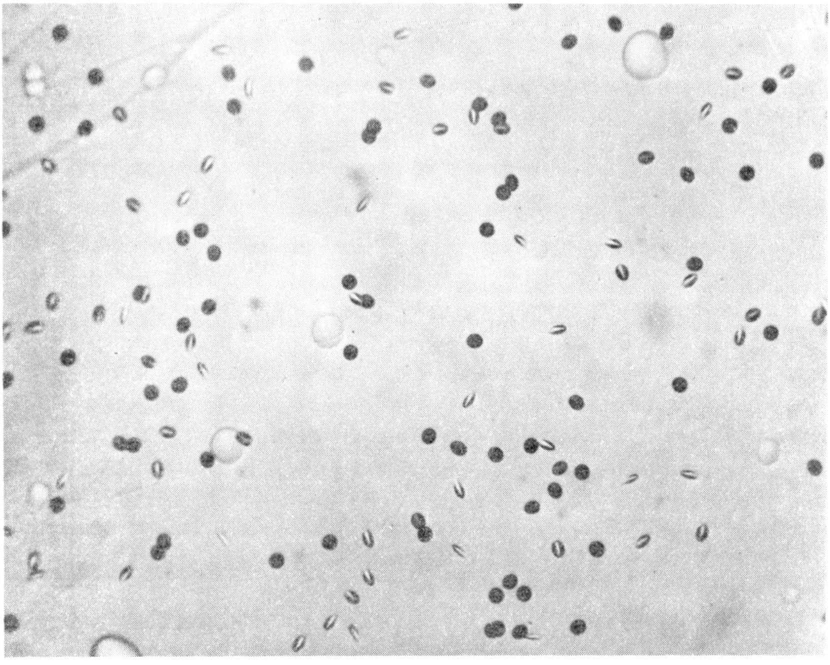

Fig. 12-14. Courtesy of General Electric, Research Division.

ated with neutrons. Also, the age assigned using this method applies to the last time the sample was annealed, but the sample may have been in existence for a long time before this annealing.

There are, of course, possible sources of error in all radioactive dating methods, arising from contamination, diffusion of material in or out of the sample, outside radiation, etc. However, there is good correlation, according to experts in the field, among the different methods. This helps to account for the confidence placed in the results.

Carbon-14 Dating

Carbon-14 is produced from nitrogen-14 in the atmosphere when the nitrogen-14 is struck by neutrons, themselves produced by cosmic rays. The reaction that produces the carbon-14 is

$$N^{14} + n \rightarrow C^{14} + p^{+} \tag{4}$$

(N^{14} and C^{14} represent nitrogen-14 and carbon-14, respectively. Nitrogen-14 has a mass number of 14 and an atomic number of 7. Carbon-14 has a mass number of 14 and an atomic number of 6.)

The carbon-14 thus formed becomes mixed with the ordinary carbon in the air and water and thus finds its way into living organisms. Carbon-14, however, is itself radioactive. Thus it is continually disintegrating, with a half-life of about 5730 years.[5] The reaction by which the carbon-14 disintegrates is

$$C^{14} \rightarrow N^{14} + e^{-} + \bar{\nu}_e \tag{5}$$

The net result of this process is that there is a certain amount of carbon-14 in the air and water. If the rates of creation and disintegration are the same, then the amount of carbon-14 in the air and water will come to some constant value and remain there.

Living plants and animals take in carbon-14 from the air and water as part of their total carbon intake. When a plant or animal dies, however, it no longer takes in any more carbon. The carbon-14 it has already taken in, however, continues to disintegrate. As time goes on, therefore, there is less and less carbon-14 in the remains of the plant or animal. From the fraction of carbon-14 remaining, we can calculate how long the plant or animal has been dead. Wood, for example, can be dated in this way. Because of its relatively short half

[5] D. T. Goldman and J. R. Roesser, *Chart of the Nuclides*, 9th Ed., Schenectady, N.Y.: General Electric Company, July 1966.

-life, carbon-14 is very useful in dating relatively recent artifacts, up to about 30,000 years old.

The use of the C^{14} method requires an estimate of the fraction of carbon-14 in the environment at the time the plant or animal died. The usual assumption is that this fraction remained the same until about 1850, when industrial processes began adding large quantities of carbon to the atmosphere. Recent evidence, however, indicates that the rate of formation of carbon-14 exceeds the rate of disintegration, and that therefore the amount of carbon-14 is at present increasing. Tree-ring data indicate that there have been changes in the fraction of carbon-14 in the air and water.

PROBLEM SOLVING

In this chapter we are interested in problems dealing with radioactive behavior and the application of old and new conservation laws. In Example 3 we introduce and discuss the use of atomic mass units.

EXAMPLE 1

A radioactive sample containing one kind of nuclei emits beta particles at the rate of 60 particles per second on the average. The emission process has a half-life of 30 days. How long will it be before the same sample emits only $7\frac{1}{2}$ beta particles per second on the average?

Solution. The number of beta particles emitted per second is a measure of how much radioactive material there is in the sample. (We are assuming that it is not transformed in the process into something of comparable or greater beta emission.)

Every 30 days the amount of radioactive material is reduced by half. Therefore, every 30 days the number of emitted particles per second is cut in half also. Thus we merely ask how many times we must halve 60 to get $7\frac{1}{2}$:

$$60 \times \tfrac{1}{2} \times \tfrac{1}{2} \times \tfrac{1}{2} = 7\tfrac{1}{2}$$

Thus the total would be three half-lives, or 90 days.

EXAMPLE 2

What is the missing particle in the following reaction:

$$\mu^+ \rightarrow e^+ + \nu_e + ? \qquad (6)$$

Solution. We know that both charge and family number must be conserved in the reaction. We observe that charge is already conserved. Therefore, the missing particle must be electrically neutral. Next we check the various family numbers. We notice that we have a muon family number of −1 on the left, but of zero on the right so far. Since we cannot use a charged particle, we must have a neutral particle of muon family number −1. The only possibility is the muon antineutrino:

$$\mu^+ \rightarrow e^+ + \nu_e + \nu_\mu \tag{6}$$

Example 3

The atomic mass unit (amu) is approximately (but not exactly) the mass of the proton and neutron:

$$1 \text{ amu} = 1 \text{ atomic mass unit} = 1.66 \times 10^{-27} \text{ kilogram}$$

The *amu* is so defined that the isotope of oxygen which has an atomic mass number of 16 has a mass of 16 atomic mass units. The masses of the proton, neutron, and deuteron in atomic mass units are as follows:

$$\begin{aligned} m_p &= \text{proton mass} = 1.007593 \text{ amu} \\ m_n &= \text{neutron mass} = 1.008983 \text{ amu} \\ m_d &= \text{deuteron mass} = 2.01420 \text{ amu} \end{aligned}$$

What is the binding energy of the deuteron? How much energy will be released if a kilogram of deuterons is formed from neutrons and protons?

Solution. We can write an equation for the conservation of mass energy when the deuteron is formed:

$$m_p + m_n = m_d + m_B$$

(m_B is the binding mass of the deuteron.)

Thus

$$m_B = m_p + m_n - m_d$$

$$\begin{aligned} m_p &= 1.007593 \\ m_n &= 1.008983 \end{aligned}$$

$$m_B = \begin{cases} m_p + m_n = 2.016576 \\ \quad - m_d = -2.01420 \end{cases}$$

$$m_B = 0.002376$$

$$
\begin{aligned}
m_B &= 2.38 \times 10^{-3} \text{ amu} \\
&= 2.38 \times 10^{-3} \, (1.66 \times 10^{-27} \text{ kilogram}) \\
m_B &= 3.95 \times 10^{-30} \text{ kilogram.}
\end{aligned}
$$

We convert this to joules, using

$$
\begin{aligned}
E_B &= m_B c^2 \\
c &= 3 \times 10^8 \text{ meters/second} \qquad \frac{\text{kilogram meter}^2}{\text{second}^2} \\
E_B &= 3.95 \times 10^{-30} \, (3)^2 \times 10^{16} \\
&= 3.56 \times 10^{-13} \text{ joule}
\end{aligned}
$$

This does not look like very much energy. Now, however, let us calculate how much energy is released when a kilogram of deuterons is formed.

We obtained 3.56×10^{-13} joule from one deuteron. One deuteron has a mass of about 2 amu:

$$
\begin{aligned}
2 \text{ amu} &= 2 \times 1.66 \times 10^{-27} \text{ kilogram} \\
&= 3.32 \times 10^{-27} \text{ kilogram}
\end{aligned}
$$

Thus one deuteron has a mass of 3.32×10^{-27} kilogram. The number of deuterons in one kilogram of deuterons, therefore, must be

$$
\frac{1 \text{ kilogram}}{3.32 \times 10^{-27} \text{ kilogram}} = 3.01 \times 10^{26}
$$

If each deuteron releases 3.56×10^{-13} joule, then 3.01×10^{26} deuterons will release

$$
\begin{aligned}
&3.01 \times 10^{26} \times 3.56 \times 10^{-13} \text{ joule} \\
&= 1.1 \times 10^{14} \text{ joules}
\end{aligned}
$$

Summary

I. FUNDAMENTAL LAWS AND PRINCIPLES

A. FUNDAMENTAL INTERACTIONS

1. The strong interaction—supplies strong, short-range forces between nucleons (and some other particles). Exact force law unknown.
2. The weak interaction—responsible for beta emission and some other reactions.

B. CONSERVATION LAWS

1. Electron family number.
2. Muon family number.
3. Baryon family number.

C. SYMMETRY PRINCIPLES
1. Particle-antiparticle symmetry.
*2. Combined particle-antiparticle symmetry and mirror reflection symmetry.
*3. Breakdown of particle-antiparticle symmetry and mirror reflection symmetry for weak interactions.

II. PARTICLES
A. FUNDAMENTAL PARTICLES
1. The neutron: much like a proton, but with no net charge.
2. Antiparticles; creation and annihilation of particles.
3. Electron neutrino and antineutrino.
4. Muon neutrino and antineutrino.
B. THE NUCLEUS
1. Radioactivity.
a. Alpha emission– alpha particle is a helium nucleus.
b. Beta emission—beta particle is an electron, positive or negative.
c. Gamma emission—gamma particle is a high-energy photon.
d. Half-life and probability.
2. Nuclear structure.
a. Atomic Number.
b. Mass Number.
c. Isotopes; the deuteron and deuterium.
d. Nuclear binding energies.
3. Nuclear fission.
4. Nuclear fusion.

III. APPLICATIONS AND INSTRUMENTS
A. INSTRUMENTS
1. Geiger counter.
2. Cloud chamber.
B. APPLICATIONS
1. Radioactive dating techniques.
a. Fission-track dating.
b. Carbon-14 dating.
2. Nuclear explosives using fission and fusion; nuclear reactor.

Questions

1. What are the fundamental forces or interactions of nature?
2. Under what circumstances do the fundamental forces and interactions occur? What structures do they hold together?

3. What is radioactivity?
4. What was surprising about the experiment that sent alpha particles into a gold foil? What conclusions were drawn from it and why?
5. What is meant by the term "half-life"?
6. How does radioactive emission differ from the processes governed by Newton's laws of motion?
7. What are the particles that make up an atom? How are they arranged?
8. What might make you think that the charge on the electron is precisely the same in magnitude as the charge on the proton?
9. What are the main properties of protons and neutrons?
10. What is ionization?
11. Explain the operation of a cloud chamber.
12. Explain the operation of a Geiger counter.
13. Of what use are cloud chambers and Geiger counters?
14. What are isotopes?
15. A proton is a hydrogen nucleus. So is a deuteron. Explain.
16. Define and distinguish between mass number and atomic number.
17. What is a nucleon?
18. Why is the proton stable?
19. Explain the main features of the curve in Figure 12-6*a*.
20. What does the binding energy per particle have to do with the energy release in the atomic and hydrogen bombs?
21. What are the characteristics of an antiparticle?
22. Why would it be difficult for matter and antimatter to exist in close contact?
23. What difference is there between the cloud-chamber tracks of an electron and those of a positron? Explain this difference.
24. Can electrons be created or destroyed? Explain.
25. Can matter be created or destroyed? Explain.
26. Why was the neutrino invented?
27. How many kinds of neutrinos are there?
28. Give examples of reactions between particles that theoretically could take place if electron and muon family numbers were not conserved.
29. What determines whether or not a particle will disintegrate into other particles?
*30. Does reflection symmetry really hold? Explain. What other symmetry principle does not hold?
31. Describe the various family-number conservation laws.
32. What are some processes that are possible according to quantum phys-

ics but impossible according to Newtonian physics? What are some processes that are possible according to Newtonian physics but are impossible according to quantum physics?

33. What did the new discoveries in nuclear physics have to do with our knowledge of the age of the earth?

34. How do you suppose one could tell that alpha particles are positively charged, beta particles negatively charged, and gamma particles uncharged?

35. A particle track in a cloud chamber appears to be very straight. How do you know that the path shown by the droplets is not straight "microscopically"?

36. Why does the gas in a cloud chamber cool down when it is forcibly expanded?

37. In the annihilation of an electron with a positron, two charged particles disappear, and uncharged photons appear. Why does this not violate the law of conservation of charge?

38. The rest mass of the alpha particle is smaller than the sum of the rest masses of the two protons and two neutrons of which it is made. Explain how this is possible.

39. How can nuclear fusion and fission both be used to produce energy when they are opposite processes?

40. Why do we believe in neutrinos?

41. Why might it be easier to shoot neutrons into a nucleus than to shoot protons into a nucleus?

42. Sometimes it is possible to break up deuterons by shooting them into a nucleus. The proton is stripped off the deuteron. Why?

43. Complete the following reactions. (The small Greek letter gamma, γ, represents a photon.)
(a) $\bar{n} + p^+ \rightarrow e^+ + ?$
(b) $\gamma + e^- \rightarrow \gamma + ?$
(c) $p^+ + e^- \rightarrow \nu_e + ?$
(d) $p^+ + \bar{\nu}_\mu \rightarrow ? + ?$

Problems

1. A certain sample of a chemical element emits alpha particles at the rate of 144 particles per second. Eight hours later it is only emitting 9 alpha particles per second. What is the half-life of this material?

2. X rays are high-energy photons given off when high-energy electrons are stopped or slowed down in matter. If the electrons in the beam which causes the x rays are going at a speed v when they strike the metal plate in the x-ray tube, what is the maximum frequency that can be emitted in the x rays? Why don't all the x rays have the same frequency when this beam strikes matter, even though all the electrons are going the same speed? (The electrons are going at a speed that is not very small compared to the speed of light.)
3. A certain isotope of platinum emits beta particles with a half-life of one-half hour. How long will it be before only 1/32 of the original platinum is left?
4. An electron and a positron at rest annihilate each other to form two gamma particles of the same energy. What is the frequency of the two photons?

*5. In the decay of the neutron:
$n \rightarrow p^+ + e^- + \bar{\nu}_e$
What is the maximum kinetic energy that the electron can have? What is the maximum momentum?

6. In the beta decay of C^{14}, how much energy is released from rest mass?

C^{14}	Mass = 14.00762 amu
N^{14}	Mass = 14.00764 amu
Electron	Mass = .000545 amu

The masses for the atoms include the nucleus and the atomic electrons.

7. How much energy would be released if two neutrons and one proton came together to form a tritium nucleus? (The tritium atom has a mass of 3.001686 atomic mass units.)
8. How much energy would be released if four hydrogen atoms came together to form an alpha particle, releasing, of course, some electrons?
9. An alpha particle is emitted from a thorium-229 nucleus to form a radium-225 nucleus. What is the maximum energy you would expect for the alpha particle?

	Atomic Number	Atomic Mass	
Radium	88	225.0931	includes atomic electrons
Thorium	90	229.1022	
Alpha	2	4.00218	

10. Suppose that one particle has a mass of 45 amu, another particle has a mass of 79 amu, and the binding mass is 20 amu. What is the mass of a composite particle made up of the two particles?

More Difficult Questions and Problems

1. One could argue that because of the uncertainty principle, in order to restrict an electron to the size of the nucleus, the electron's velocity must be so high that it would fly away. Explain.
2. When a nucleus emits a particle, the particle generally does not have as much energy as would be necessary to put it back into the nucleus. Why not?
3. Because of the phenomenon noted in Question 2, the emitted gamma photon in Figure 12-15 cannot be absorbed by the similar nucleus at the right, since it does not have enough energy. This can be remedied by setting the nucleus at the right in motion toward the incoming photon. Why?
4. A nucleus behaves in many ways like a liquid drop. What does this fact suggest about the forces between molecules in a liquid?
5. The particle reaction

$$\mu^+ \rightarrow \pi^+ + \bar{\nu}_\mu$$

cannot take place with the muon at rest, because the rest mass of the first particle on the right, the so-called pion, is larger than that of the muon. If we try to give the muon extra mass by increasing its kinetic energy, we can indeed get enough mass, but still the process will not occur. Why not?

6. Why cannot the following reaction occur:

$$e^+ + e^- \rightarrow \gamma$$

7. If an electron and a positron at rest annihilate each other to form two gamma photons, why must the two photons have the same energy?

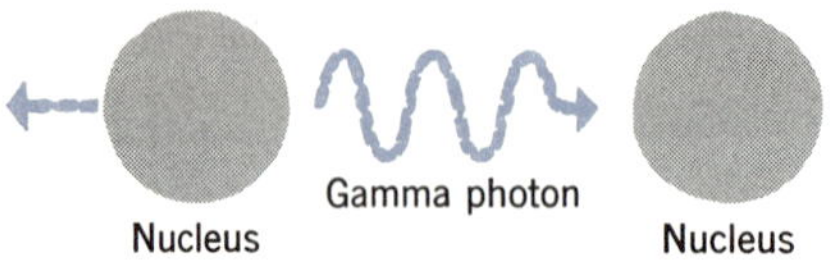

Fig. 12-15

13

☼SYSTEM OF THE WORLD, MODERN VIEW

> You boil it in sawdust: you salt it in glue:
> You condense it with locusts and tape:
> Still keeping one principle object in view—
> To preserve its symmetrical shape.[1]

IN THIS CHAPTER WE SHALL DISCUSS THE PICTURE MODERN PHYSICS GIVES US OF THE UNIVERSE in which we live. The new insights that modern physics has contributed to the world center mainly in our ideas of the very small and the very large. We begin with the very small.

THE MICROCOSM

What has modern physics contributed to the ancient philosophical attempt to reduce everything in our experience to matter and motion, atoms and the void? The constituents of matter and light, as we now understand them, are the fundamental particles, which interact according to four fundamental interaction laws.

The Fundamental Particles, or Fields

Table 13-1 lists some of the fundamental particles that have been discovered. Each of these particles, usually called elementary parti-

[1]Lewis Carroll, *The Hunting of the Snark*, as quoted in Martin Gardner, *The Annotated Snark*, New York: Bramhall House, 1962, p. 73.

cles, has its antiparticle. The photon is its own antiparticle, as is the graviton and the pion of zero charge. The other antiparticles are different from their particles. There are some recently discovered particles which we have not included. There seem to be too many fundamental particles for them all to be fundamental. It is hoped that there will some day be some deeper understanding which will enable us to reduce the number of particles we think of as fundamental. It may be, for example, that several particles will turn out to be just different states of the same particle, or combinations of other particles.

FAMILIES OF PARTICLES. The fundamental particles are grouped into families. The first family listed in the table includes only one member, the photon, which has no rest mass. The second particle, the graviton, may also be put in a family of its own. The graviton has never been detected, but it is believed to be associated with the as yet undetected gravitational waves, much as the photon is associated with electromagnetic waves. The next two families have already been discussed—the electron and muon families, each with its characteristic neutrino. Each of these families has a family number that must be conserved in interactions, as we have noted. Since there is no conserved family number for photons, they may be produced or destroyed without regard to such a conservation law, although, of course, other conservation laws must be observed. The next family on the list, the meson family, also has no family number. It is divided into two groups, the pi and *K* mesons. The baryon family comprises the heavy particles, such as the proton and the neutron, and others. We have already seen that the baryon family number is conserved in particle reactions.

Table 13-1 Some Properties of Some Fundamental Particles

Particle Family	Particle	Symbol	Rest Mass[b]	Spin	Charge	Antiparticle[a]	Strangeness	Family Numbers: Electron No.	Family Numbers: Muon No.	Family Numbers: Baryon No.	Average Lifetime in Seconds
	Photon	γ	0	1	0	γ Itself		0	0	0	Stable
	Graviton	g	0	2	0	g Itself		0	0	0	Stable
Electron	Electron	e^-	1	½	−1	e^+ Positron		1	0	0	Stable
	Electron neutrino	ν_e	0	½	0	$\bar{\nu}_e$ Electron antineutrino		1	0	0	Stable
Muon	Muon	μ^-	206.77	½	−1	μ^+ Positive muon		0	1	0	2.20×10^{-6}
	neutrino	ν_μ	0	½	0	$\bar{\nu}_\mu$ Muon antineutrino		0	1	0	Stable
Mesons	Pions	π^+	273.2	0	+1	π	0	0	0	0	2.55×10^{-8}
		π^0	264.2	0	0	π^0 Itself	0	0	0	0	1.8×10^{-16}
		π^-	273.2	0	−1	π^+	0	0	0	0	2.55×10^{-8}
	Kaons	K^+	966.3	0	+1	K^-	1	0	0	0	1.23×10^{-8}
		K^0	974.6	0	0	K^0 Anti K^0 (no charge)	1	0	0	0	Double
		K^-	966.3	0	−1	K^+	−1	0	0	0	1.23×10^{-8}
Baryons	Proton	p^+	1836.12	½	+1	p^- Antiproton	0	0	0	1	Stable
	Neutron	n	1838.65	½	0	$\bar{n}$ Antineutron (no charge)	0	0	0	1	1.013×10^3
	Lamda	Λ^0	2182.8	½	0	$\bar{\Lambda}^0$ Anti Λ^0	−1	0	0	1	2.62×10^{-10}
	Sigma	Σ^+	2327.6	½	+1	$\bar{\Sigma}^+$ Anti Σ^+	−1	0	0	1	7.9×10^{-11}
		Σ^0	2333.5	½	0	$\bar{\Sigma}^0$ Anti Σ^0 (no charge)	−1	0	0	1	10^{-20}
		Σ^-	2342.6	½	−1	$\bar{\Sigma}^-$ Anti Σ^-	−1	0	0	1	1.6×10^{-10}
	Xi	Ξ^-	2584.7	½	−1	$\bar{\Xi}^-$ Anti Ξ^-	−2	0	0	1	1.7×10^{-10}
		Ξ^0	2572	½	0	$\bar{\Xi}^0$ Anti Ξ^0	−2	0	0	1	3×10^{-10}

[a]The antiparticle has the same rest mass, spin, and average lifetime as the corresponding particle, and the opposite charge, strangeness, and family number as the corresponding particle.

[b]The rest mass is given in terms of the electron rest mass.

We speak of these particles as fundamental particles, but we must remember that they also display wave characteristics. We might as well speak of fundamental waves, or fundamental fields.

Each fundamental particle has certain characteristic properties. Each particle has a definite rest mass, charge, and spin. The masses of the particles vary widely, as the table demonstrates. So far we do not know how to account for these large differences in mass. However, if a fundamental particle has a charge, that charge is equal in magnitude to the charge on the electron. It may be either positive or negative, depending on the type of particle. The antiparticle has a charge opposite to that on the particle. If we call the charge on the proton one, then the charge on the electron is minus one. Fundamental particles have charges of plus one, minus one, or zero. Table 13-1 gives the charges of the fundamental particles in these terms.

The *spin* of a particle is a curious property. We cannot, of course, look at a particle and see it spin. However, certain particles behave like tiny magnets. We have seen that magnets can be formed by electric currents moving around in a circle. A charged particle would form such a circulating electric current if it were spinning. Charged particles are not the only particles that act as tiny magnets, the neutron does this as well. Although the neutron has no total charge, it might be thought of as having distributions of opposite charge that cause magnetic effects as the charge moves. A spinning particle should also have angular momentum. We find that it does; we must include this spin angular momentum in order to maintain conservation of angular momentum in interactions.

A particle can have only certain amounts of spin angular momentum around an axis. The photon has what we will call one unit of spin; we say that it has spin one. The electron has spin one-half. The amounts of angular momentum corresponding to these spins are very small—of the order of magnitude of Planck's constant. Most fundamental particles have spins of zero, one half, or one. We can classify particles according to their spin. If a particle has a spin equal to zero, one, two, or any whole number, it is called a *boson*. If it has a spin equal to 1/2, 3/2, 5/2, etc., it is called a *fermion*. Thus the photon, the graviton, and all the mesons are bosons. The electron, muon, neutrinos, and all the baryons are fermions.

PARTICLE IDENTITY AND EXCHANGE SYMMETRY. All electrons are like all other electrons. All positrons are like all other positrons. All photons are like all other photons. This does not mean that they all have the same energy, momentum, or frequency, but that they are intrinsically alike—they have the same rest mass, spin, and charge.

Indeed, we may say that all fundamental particles of the same kind are not merely alike but identical. This fact has some curious consequences. For example, if two particles are really identical, then they could switch places without changing anything, for all we know. We can, in fact, define a new symmetry principle.

The laws of physics are not changed if identical particles are exchanged for each other.

If, in fact, one electron should exchange its identity with that of another electron, there would be no way of our knowing about it. We cannot put a mark or a tag on a fundamental particle to identify it. We *can* put a mark on a billiard ball, or some other large object, to distinguish it from a very similar object, but this trick will not work for fundamental particles. Ordinarily we might not even bother to mark two balls in order to distinguish between them. We can follow them with our eyes and observe each moving in its separate path. This simple technique does not work for electrons (or other fundamental particles), as we can see by referring to Figure 13-1. We are trying to follow the paths of two electrons by making measurements of their positions from time to time. The circles on the figure represent positions at which an electron has been observed. We cannot "see" the electrons all the time but can only make measurements of their positions from time to time. When we make a measurement of an electron's position, however, we have no way of being sure which electron we are observing. We cannot tell, for example, whether the

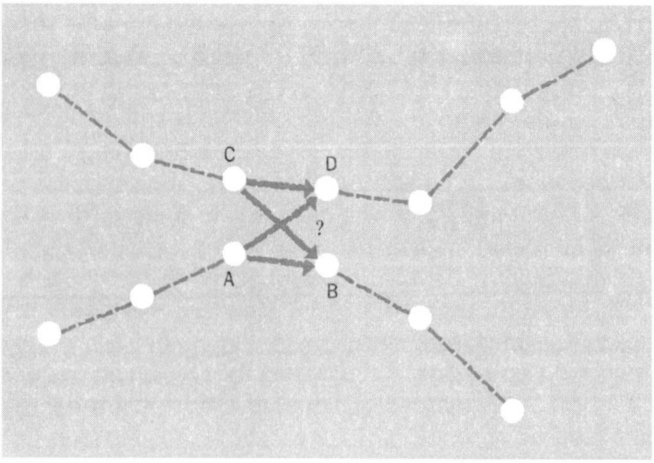

Fig. 13-1

electron originally observed at position A on the diagram then went to position B or to position D. The electron carries no mark to distinguish it from any other electron. If we knew the momentum the electron had when it left point A, we would know whether it was heading for point B or point D. However, according to the uncertainty principle, when we measure the position at A, we introduce uncertainty into the momentum, and we do not know exactly where the particle is going. Because we cannot mark electrons and because of the uncertainty principle, we cannot follow the path of an electron with certainty.

The fundamental particles are identical in a way far different from the way in which two billiard balls or two coins are identical. In order to show the difference between these two kinds of identity, we shall discuss an example involving two identical coins. We shall first treat these two coins as ordinary coins. Then we shall assume that the two coins represent bosons. Finally, we shall assume that the two coins represent fermions.

If the coins are tossed, they can land in several conditions. Each coin by itself can land in one of two conditions: it may come up heads or it may come up tails with equal probability. The pair of coins may therefore come up in four equally likely conditions.

	First Coin	*Second Coin*
First condition	Heads	Heads
Second condition	Heads	Tails
Third condition	Tails	Heads
Fourth condition	Tails	Tails

This table lists all the possible conditions for two coins. We can see that there are four of them. We may briefly write this table as follows, using H to stand for heads and T for tails:

H H
H T
T H
T T

That this listing means something can be found by determining the odds for throwing the coins so that they come up in a certain condition. We might ask for the probability that the coins will both come up heads. Since this condition is one of four equally likely possibilities, there is a probability of one fourth for this to occur; there is one chance in four that both coins will come up heads. We can also ask for the probability that the two coins will come up in different conditions—that is, either with the first one heads and the second one tails, or with the first one tails and the second one heads. This situation holds for two of the four possible conditions, so the probability is two out of four, or one half. Thus, if we throw two coins together, we expect that they will come up different half the time. These probabilities can be verified experimentally for two ordinary coins by tossing them many times.

Now let us assume that the two coins represent two identical bosons, and that "heads" and "tails" represent two different specific states or conditions in which the particles can exist. Under these circumstances there are only three possible conditions for the two "coins":

H H
H T
T T

You may wonder what happened to the T H condition. Because the two "coins" are identical, the T H condition is the *same condition* as the H T condition. Thus, for identical "coins," these two conditions are not merely similar or alike but are the *same identical condition*. There being no way to distinguish between them, even in principle, they are regarded as the same condition. This means that there are only three possible conditions instead of two. If one were to find the probability that one coin would come up heads and the other tails, the result would be one third instead of one half as before. This is true because only one of the three possible conditions has the two particles in different conditions. The fact that the particles, or "coins," are identical *reduces* the number of possible conditions. Bosons actually show this reduction of the total number of possible conditions in experiments.

Now we shall assume that the two coins represent fermions, and that "heads" and "tails" represent two different specific conditions or states in which the fermion particles exist. For example, the coins may represent electrons. We already know that the exclusion principle holds for electrons; in fact, it holds for any identical fermions. The exclusion principle states that no two identical fermion particles can exist in exactly the same condition. This means that if our coins represent fermions, there is (in this example) only one condition that is possible:

H T

Now the probability for this condition is exactly one, since it is the only condition possible. Experiments verify this kind of reduction in total number of possible conditions or states for fermions.

We have presented this example in order to show, in principle, what identity means for fundamental particles, and how it differs from our ordinary notions of identity. This reduction of the total number of states is only one of the ways in which the possible exchange of identical particles affects quantum physics. In processes involving two or more identical particles we consider the possible exchange of these particles in order to obtain the correct result. The situation prompted Erwin Schrodinger to remark, "The elementary particle is not an individual."[2] All other fundamental particles of the same kind are identical to it.

Fundamental Interactions

The fundamental particles interact with one another in four fundamental interactions. As far as is known at present, these are the only kinds of interaction that occur. Therefore, it is possible that all physical phenomena can be explained in terms of numbers of fundamental particles interacting according to four laws of interaction. We shall discuss the interactions here in the order of their strength, starting with the weakest.

THE GRAVITATIONAL INTERACTION. The gravitational interaction is so weak that it is negligible, as far as we know, in experiments dealing with interactions between small numbers of fundamental particles. Any two particles that have mass should have a gravitational interaction with each other. The mass may have to be very large for the forces of interaction to be important. This interaction is a long-range interaction as well as a very weak interaction. Thus it acts at large distances between large masses and helps to hold the solar system, the galaxy, and, as far as we know, all the large-scale systems of the universe together. The gravitational interaction between massive objects at a distance from each other presumably occurs because small particle-waves, called gravitons, are transmitted from each object to the other. These particles carry energy and momentum, and thus cause the influence that each body exerts on the other. This enables us to think of the gravitational interaction from a particle point of view rather than from the field point of view we used before. As we have mentioned, however, neither gravitons nor gravitational waves have been observed. We do not know what the law of gravita-

[2]Erwin Schrodinger, *Science, Theory, and Man*, New York: Dover, 1957, p. 194.

tional interaction is between any two fundamental particles. The laws of gravitational force we have discussed in earlier chapters apply to comparatively large masses, and presumably large numbers of gravitons. They do not involve the ideas introduced in quantum physics.

THE WEAK INTERACTION. The weak interaction is the interaction involved in beta emission, and it can occur in reactions involving any particle listed in Table I except, presumably, the graviton. The weak interaction often involves the various kinds of neutrinos. Much progress has been made in understanding the interaction law by using a theory similar to that of the electromagnetic interaction, but much still remains to be done. We have seen that the weak interactions obey neither the principle of reflection symmetry nor the principle of particle-antiparticle symmetry, although they do obey the combination symmetry principle which involves reflection and particle-antiparticle exchange at the same time. (However, see below)

We can add to these two symmetries the principle of time-reversal symmetry and obtain a triple symmetry: we can reflect a process in a mirror, change all particles to antiparticles and vice versa, and consider the process reversed in time. If we do this we have a combined symmetry principle which seems to hold for all interactions. There have been some recent developments in this general subject of symmetry principles, however, which cast these matters again into doubt. The combination reflection and particle-antiparticle symmetry is apparently violated in a reaction involving K° mesons. This has opened the possibility that this combination symmetry principle fails for weak interactions, or perhaps for some other interactions. If the triple combined symmetry principle holds but the combined reflection and particle-antiparticle symmetry does not, then time-reversal symmetry should also fail.

THE ELECTROMAGNETIC INTERACTION. The electromagnetic interaction supplies a long-range force that falls off with distance in the same manner as the gravitational interaction does. The electromagnetic interaction is much stronger than either the gravitational or the weak interaction. It is, roughly, 10^{10} times stronger than the weak interaction. The relative strengths of the electromagnetic and gravitational interactions are such that the electric repulsion between two protons is about 10^{36} times greater than the gravitational attraction between them. The great strength of the electromagnetic interaction is not always appreciated, however, because large imbalances of charge are seldom encountered in our environment. (Even lightning involves only a very small percentage of the total number of charged particles in the air and water that make up a cloud.) The electromagnetic interaction is strong enough to keep the positive and negative charges held together so that most matter in our environment is very nearly electrically neutral.

The electromagnetic interaction occurs in reactions which can involve any of the particles listed in Table I except neutrinos (and antineutrinos) and gravitons. It is responsible for the forces of attraction between the nucleus and the electrons of an atom, and it accounts for the forces between molecules in solids, liquids, and gases.

As indicated in Chapter 8, the electromagnetic interaction involves the electromagnetic field, with electromagnetic waves such as light being radiated from one charge to another. We can also consider the interaction from a particle point of view. The photon is the particle representing the electromagnetic field. The electromagnetic interaction occurs when particles exchange photons with each other. The photons carry energy and momentum from one particle to the other, and thus cause each particle to change its motion.

The electromagnetic interaction is the best understood of the fundamental interactions. Using the laws of this interaction, we can make very accurate predictions about the behavior of fundamental particles. Here we shall discuss some of the results of the modern theory of this interaction.

The equation that Schrodinger developed for quantum physics (see Chap. 11, p. 459) was very successful, but it did not take into account either the special theory of relativity or the creation and annihilation of particles. P. A. M. Dirac wrote an equation in 1925 which was eventually successful in including these matters, especially for electrons and photons. The equation was designed to deal with electrons, but it predicted that electrons could exist with negative masses and negative kinetic energies. Electrons, however, appear to have positive masses and positive kinetic energies. The equation could not be dismissed out of hand, however, because it accounted very accurately for some observed properties of the hydrogen atom. It seemed worthwhile, therefore, to find out just what the equation was saying and take it seriously.

The real problem of having possible negative masses and negative energies for electrons was that ordinary electrons could lose energy by giving off photons, much as electrons in the Bohr orbits could drop to a lower orbit. In the Bohr atom, however, there was a lowest orbit, and an electron having dropped into it could fall no further. If electrons could have negative kinetic energies, however, there would be no limit to how far they could fall in energy. In fact, all electrons could lose energy until they were all in states of negative energy. This theoretical catastrophe can be avoided, however, if we assume that there are an infinite number of electrons filling up all these negative energy states. Then the exclusion principle would prevent any other electrons from falling into these same states. This infinite sea of negative-energy electrons is not directly observable.

What happens when one of these negative-energy electrons is somehow knocked up into an ordinary positive-energy state and appears as a normal electron? This leaves a "hole" or unoccupied negative-energy state in the sea of negative-energy electrons. Another electron in the invisible "sea" can enter this unoccupied state, leaving another state unoccupied. Thus this unoccupied state, or hole, will appear to move around like a bubble in the sea. If we calculate the behavior of this "bubble," we find that it behaves like a positron. Therefore we can describe the process as follows: an electron, leaving a negative-energy state, suddenly appears. At the same time a "hole" or "bubble" appears that behaves like a positron. We have, therefore, pair creation. If the electron once again enters the sea, both the hole and the electron disappear together. Therefore, we have electron-positron annihilation. Thus the Dirac theory accounts for both creation and annihilation of particles.

To discuss the interactions of photons and electrons we shall use diagrams devised by the American physicist R. P. Feynman (called Feynman diagrams). One such diagram is given in Figure 13-2*a*, which shows a collision between a photon and an electron. The two particles are drawn as they come together, collide, and then separate. (In the diagram, time increases upward.) The diagram shows something about the collision that is different from our normal expectation. Instead of merely bouncing off the electron, the photon is actually absorbed by the electron, and then, very shortly afterward, another photon is emitted by the electron. Because of this the wavy line representing the photon ends when it reaches the line representing the electron. The electron line does not end at this point. Later another photon line appears and leads away from the electron line. It turns out that we must consider the process in this way in order to obtain a correct result. Actually, of course, an observer would not see this stepwise process. He might send a photon and an electron into a region, and then observe an electron and a photon leaving. As far as the observer knows, therefore, the electron may have emitted the second photon before it ever absorbed the first one. This process is illustrated in Figure 13-2*b*. Therefore there are actually two diagrams to be considered. There is no way to tell which one represents the actual process. We learned in our discussion of quantum physics that whenever we cannot distinguish between two ways in which a process can occur, we must consider that the process occurs in both ways at once to some extent. (This is much like the situation described in Chapter 11, when we could not tell through which slit the electron passed. We had to consider that it went through both, in some sense, in order to get the correct result. Here we have to consider that the two series of events represented by these diagrams contribute to the final result.) Actually there are more diagrams that should be considered than just these two, but they are less important, and we shall ignore them for the time being. It should be stated that although we are merely discussing the diagrams, there is a mathematical theory that goes with them that enables one to make calculations.

Let us take Figure 13-2*a* and change it a little. We move the line representing the incoming photon around until it looks like an outgoing photon. We move the outgoing electron line around until the arrow points backward into the past. Then we get the diagram in Fig-

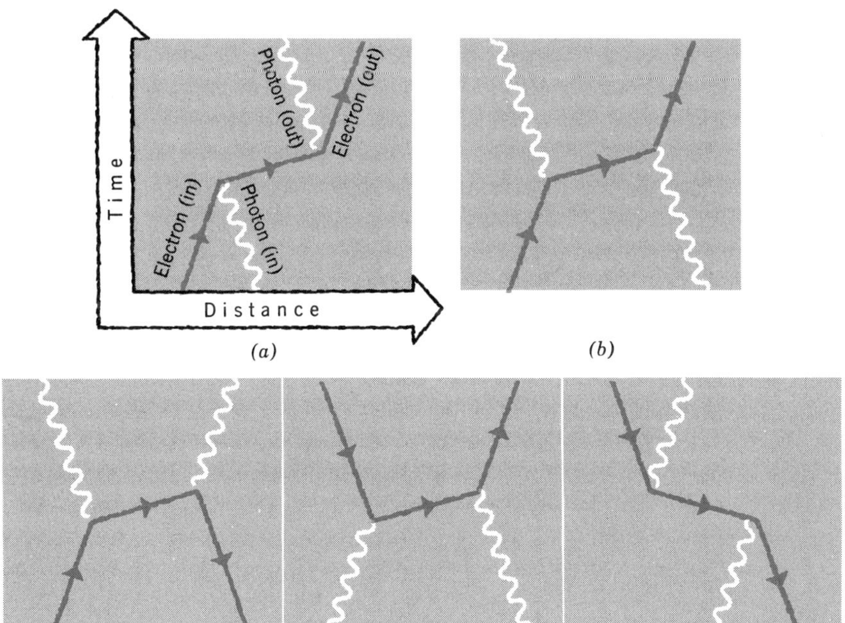

(*a*) (*b*)

(*c*) (*d*) (*e*)

Fig. 13-2

ure 13-2*c*. This at first looks like an impossible process. We have an electron moving along, emitting first one photon and then another. As it emits the second photon, it receives such recoil that it moves off *backward in time*. It turns out, however, that this process is not so "impossible" as it first appears. Mathematically, the electron moving backward in time looks just like a positron moving forward in time. What we really have, therefore, is an electron and a positron being annihilated, leaving two photons, which conserve energy and momentum. If we make the mathematical computation, we find that this interpretation of the positron as an electron going backward in time gives the right answers for the process of electron-positron annihilation. Thus, for all we know, the positron actually is an electron going backward in time. To our crude, unsophisticated perceptions, it appears as a positron. This theory does not upset our notions of cause and effect, however, since the electron going backward in time cannot carry a message with it. The theory does help us to predict the future results of experiments.

We can twist the diagram around some more, as shown in the succeeding parts of Figure 13-2, obtaining diagrams for pair creation (Fig. 13-2*d*) and the interaction of a photon with a positron (Fig. 13-2*e*). In Figure 13-3 we see a diagram of an electron going backward and forward in time several times. To us it looks like many electrons and positrons at once. No wonder all electrons are alike; as John A. Wheeler remarked facetiously, they all have the same charge and mass because "they are all the *same* electron."[3] This interpretation of

[3]Quoted in R. P. Feynman, "The Development of the Space-Time View of Quantum Electrodynamics," *Physics Today*, **19**, 8 (August 1966), pp. 31–44.

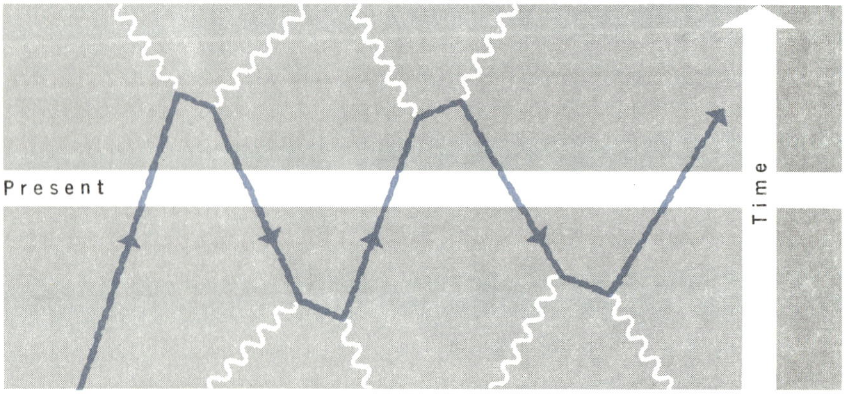

Fig. 13-3

the positron as an electron going backward in time may seem strange, but it is probably no stranger than the alternate theory about "holes," which we considered earlier. (Pairs are commonly created when high-energy photons interact with strong electromagnetic fields near the nuclei of atoms. It is of only theoretical interest that pair creation could be obtained by directing two beams of photons at each other.)

The first diagram in Figure 13-4 shows an electron that emits a photon and then later absorbs it back again. We might question whether or not such an event is possible. For an electron to emit a photon all by itself and remain an ordinary electron would cause a violation of the law of conservation of energy. There is, however a way in which this process can occur temporarily, even if the law of conservation of energy is temporarily "violated." This possibility can be shown using the uncertainty principle which involves time and energy (see Chap. 11, pp. 468–469).

We can apply this uncertainty principle to the event shown in Figure 13-4*a*. If the time interval between emission and absorption is short enough, an energetic photon can exist even though it seems to have received its energy from nowhere. During the short time that the photon exists, the energy is very uncertain for the system consisting of the electron and the photon. This means that over this short time the system is allowed by the uncertainty principle to have possible energies that are different from the energy the electron has over a long time period. In quantum physics we must consider the effects of these possible energies, and one of these effects is the possible existence of a short-lived photon. The shorter the time that the photon exists, the more energy it can have on the average. The longer it

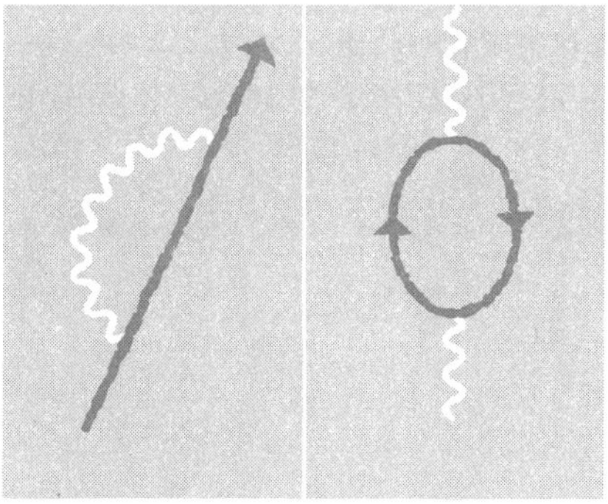

(a) (b)

Fig. 13-4

exists, the less energy it can have. Such photons are called *virtual* photons, since they are not directly observed. If they are absorbed by the same particle that emitted them, they do not cause an interaction with any other particle.

This concept of virtual photons suggests that the electromagnetic field around the electron (and other charged particles) is made up of a swarm of virtual photons, continually being emitted and absorbed. Virtual photons that exist for a long time can travel a long way from the electron that emitted them. Most of these photons have comparatively low energy, however, since the uncertainty principle requires the long-lived virtual photons to have low energies on the average. Close in to the electron we can have more of the short-lived, high-energy virtual photons. Two charged particles can interact by absorbing each other's virtual photons. If they are distant from each other, they will receive low-energy virtual photons from each other, and the effect of the interaction will be small. If they are closer together, they will probably "exchange" the high-energy virtual photons that remain closer to the emitting particle, and the effect of the interaction will be greater. Thus the electro-magnetic field around a charged particle is strong close to the particle where the high-energy virtual photons are and falls off with distance. Figure 13-5 shows one virtual photon in transit between two interacting electrons.

Other particles may also occur in virtual states. Figure 13-4*b* shows a virtual electron-positron pair. A photon suddenly produces a pair of particles, which exist for only a short time before recombining to give the "original" photon. Once again, if they last for only a short time, the law of conservation of energy may be partly violated, because the energy itself is uncertain. Electrons and photons, as normally existing, should have all possible virtual processes occurring all the time. We have shown some of the possibilities for the electron in Figure 13-6. The electron considered with all the virtual processes included is called a *dressed* electron, while the theoretical electron

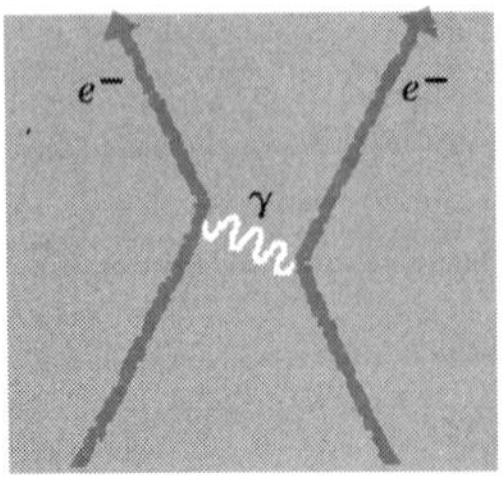

Fig. 13-5

without them is called a *bare* electron. Real electrons, of course, are dressed. Thus the mathematical theory of the electron and its interactions must take these virtual particles into account. This requirement caused great problems, because certain calculated results came out infinite. A systematic mathematical trick was developed to overcome this defect and obtain the right result. The corrected theory is amazingly accurate although it is not too satisfying from an aesthetic point of view. This theory of the electro-magnetic interaction is still the most successful theory that combines the concepts of both quantum physics and modern special relativity. When one applies this theory to situations involving large charges and many electrons and photons, one obtains an extremely close approximation to the familiar laws of electricity and magnetism discussed in Chapter 8.

Nevertheless, this technique actually involves an infinite number of diagrams. For the interaction of an electron and a photon, for example, there are more diagrams than the first two in Figure 13-2.

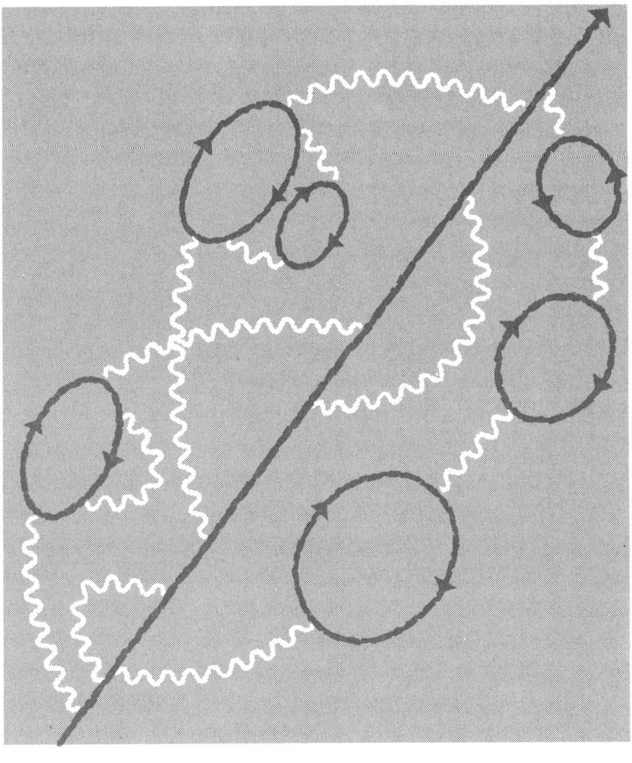

Fig. 13-6

This process also involves, for example, the diagrams illustrated in Figure 13-7. These diagrams have one photon and one electron going in, and one electron and one photon going out. They correspond to ways in which the interaction between an electron and a photon can occur, and therefore they must be accounted for if we wish to calculate the effects of the interaction. Fortunately, from a mathematical point of view, these more complex diagrams are less important; the more complex a diagram is, in general, the less important it is. Therefore, as it turns out, we can get an accurate result using just the first two diagrams in Figure 13-2. More complex diagrams are used for small corrections.

If we try the same kind of calculation using similar diagrams for the strong interaction, however, it turns out that the more complex diagrams are more important than the simple ones. Thus this diagram method fails for strong interactions.

THE STRONG INTERACTION. The strong interaction is responsible for the strongest known forces between particles. In the nucleus the strong interaction between neighboring nucleons is about a hundred times stronger than the electric repulsion between protons the same distance apart. The strong interaction is responsible for holding the

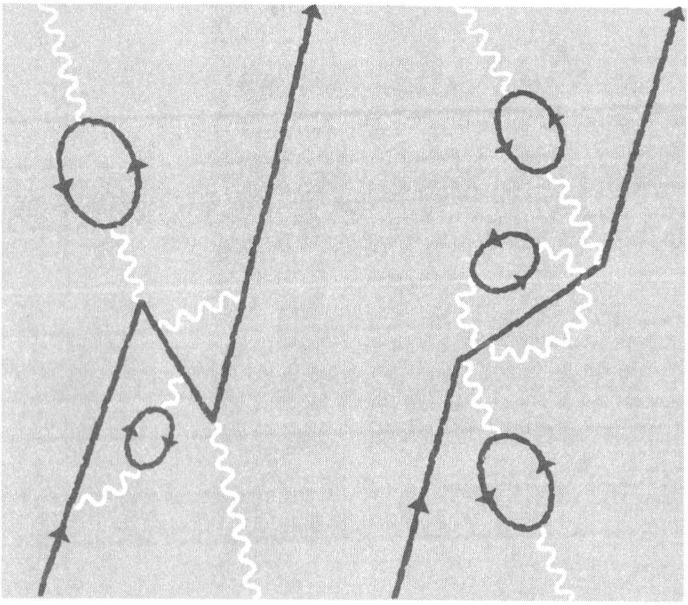

Fig. 13-7

nucleus together and is of very short range. The particles that are subject to the strong interaction are the mesons and the baryons (see Table 13-1).

The strong interactions are imperfectly understood. We do not know the law of interaction that would enable us to calculate the results of experiments in which the particles are subject to the strong interaction. Several techniques have been used to overcome this disadvantage and to explain the strong interactions as well as possible. One such technique is to use the conservation laws and symmetry principles to show what behavior is allowable and what is not. The strong interactions obey more conservation laws and symmetry principles than either electromagnetic or weak interactions—indeed, as far as we know at present they obey all the conservation laws and symmetry principles discussed in this book. (The electromagnetic interaction also obeys all the symmetry principles and conservation laws discussed in this book, but there is one symmetry principle and a related conservation law that it does not obey. We shall not discuss this matter further.)

In the course of investigations of reactions involving the strong interaction it was noticed that there were certain reactions that never occurred. There appeared to be no reason why these reactions should not occur, since there was no known conservation law or symmetry principle that prevented them. Therefore the quantity *strangeness* was introduced to explain why these reactions did not occur. Each of the baryons and the mesons is assigned a certain amount of strangeness (see Table 13-1). The antiparticle is assigned a strangeness equal to the negative of the strangeness of the corresponding particle. The reactions that were not observed would have failed to conserve strangeness. As far as is known at present, strangeness is conserved by the strong and electromagnetic interactions. The weak interaction, however, need not obey the law of conservation of strangeness.

The strong interaction between nucleons, which holds the nucleus together, may be explained as an exchange of pions between nucleons. Earlier in this chapter we explained how an electron may be thought of as surrounded by a cloud of virtual photons, and how electrons could interact by absorbing virtual photons. The nucleon, be it a proton or a neutron, can be thought of as surrounded by a cloud of virtual pions. Pions, as Table 13-1 indicates, may be positive, negative, or uncharged. In Figure 13-8 we see several different possibilities represented in diagrams. When a charged pion is exchanged, the proton involved in the process is transformed into a neutron, and vice versa. When the uncharged pion is exchanged, this change of nucleon does not occur. In each case, however, a strong interaction takes place. The fact that the pions have rest mass helps to explain why this nuclear force is of short range. In discussing virtual photons we said that the virtual particles furthest away from the electrons had the smallest energies, on the average. In order to get that far away, they had to last a comparatively long time, and the uncertainty principle did not allow them much energy. For the pion, some of this energy is tied up in rest mass and is not available as kinetic energy to increase its speed. Besides, the speed of the pion cannot be so great as that of the photon anyway. Thus, in a given time interval the pion cannot get so far away from the nucleon which emitted it. The great majority of the virtual pions are therefore restricted to a region close to the nucleon (see Example 1, p. 577).

Since there are many strongly interacting particles (more than are listed in Table 13-1), various schemes have been proposed to classify them so that they can be considered as merely representing different conditions of a few *really* fundamental particles. This, of course, would be in accordance with the long tradition of physics that searches for the most elementary building blocks of matter. Matter is made up of molecules, molecules of atoms, atoms of nuclei and electrons, nuclei of protons and neutrons. Where does this decomposition end? However, an alternate view has been suggested which puts all these strongly interacting particles on about the same footing, arguing that the particles are as likely to be made up of each other as of some few more fundamental particles. This point of view stems partly from the fact that the strong interaction is capable of large binding energies.

Let us consider the deuteron as a simple example. The binding

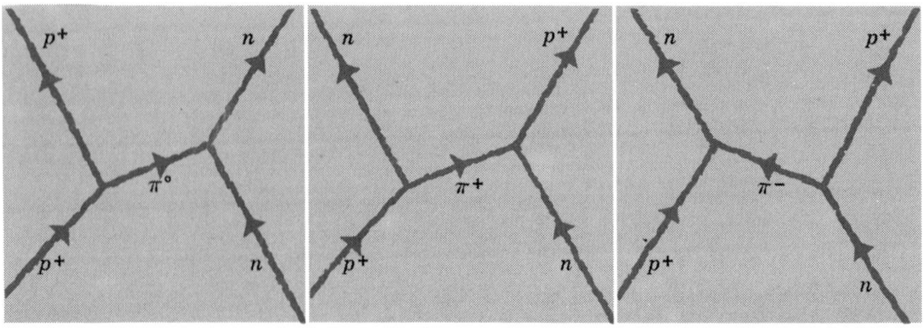

Fig. 13-8

energy of the deuteron is comparatively large, as we have seen, but it is still only a small fraction of the combined mass of the proton and the neutron. Consequently, the total mass of the deuteron is only a little less than the sum of the mass of the proton and the mass of the neutron. Therefore we recognize it easily as being made up of these two nucleons. Suppose, however, that the binding energy were about the same size as the mass of one of the nucleons. Then the combination particle would not be so easily recognized as a combination of two particles. In fact it would appear to be another particle of about the same mass as a nucleon. We would probably not know that it was indeed a combination of two nucleons. The two nucleons could react to give the combination particle all right, but the extra energy represented by the binding energy would have to escape in the form of one or more other particles. How would we know which of all the resulting particles was really the combination particle, and which of the particles were merely removing extra energy? The process would appear to be a particle reaction and not a particle combination. Even if we were to recover the original nucleons by breaking up the combination, we would have to add the binding energy in the form of another particle. How would we know which particle really broke up? Mass-energy can take different forms. Perhaps the strongly interacting particles are just different forms of mass energy, none of them more fundamental than the others in the sense of being unique building blocks from which the others are made.

Fundamental Constants

The fundamental constants of nature determine the scale or size of various phenomena. We have already seen that the velocity of light in vacuum, c, determines the maximum rate of communication or transfer of information and the minimum time delay between cause and effect. The constant h determines the uncertainty of our knowledge and prediction of events. Wavelengths and frequencies of waves are related to the energy and momentum of particles through this constant. It determines the wavelength of the photon for a given energy, and also helps to determine the size of the atom. The fundamental charge, e, and the masses of the fundamental particles determine the scale of the particle nature of both charge and matter. The size and compactness of the various structures in the universe are determined by the strengths of the fundamental interactions. These strengths are controlled by certain fundamental constants, such as the gravitational constant G and the electromagnetic constant K_e.

A theory that would relate these fundamental constants to each other and thus reduce the number of really fundamental constants would be very welcome. Some attempts to relate them have been made.

*SIC Physics: Taboos

The three letters SIC refer to Symmetry, Invariance, and Conservation. We have already mentioned that we do not know all the exact laws of interaction for fundamental particles. Even without these exact laws, however, we can understand a great deal about the behavior of fundamental particles by using the conservation laws and the symmetry principles. The picture that emerges is one of particles undergoing all possible kinds of reaction and transformations. Mass-energy, in the form of particles and kinetic energy, is constantly being transformed from one kind of particle to another, now bound up in the rest mass of particles, now appearing as photons or neutrinos with no rest mass whatever. This underlying chaotic behavior in all matter fully exploits the uncertainties of quantum physics. Through this chaos, limiting and regulating it, are the conservation laws and symmetry principles. According to these principles, some conceivable reactions and processes are forbidden. We have already seen that the disintegration of the proton is forbidden by the law of conservation of baryon number. The disintegration of the electron into less massive particles is forbidden by the law of conservation of charge. (All fundamental particles that have less mass than the electron are uncharged.) Indeed, whenever we find that a reaction or process does not occur, we think that there may be a conservation law or symmetry principle that forbids it. It was in this way, as we have seen, that the conservation of strangeness was discovered.

Many of the conservation laws and symmetry principles are related to each other. In fact, one would like to say that every conservation law is equivalent to some symmetry principle, and vice versa. Unfortunately, this is not quite true. We know of no conservation law that would correspond to time-reversal symmetry. However, it is true for many such principles. For example, according to our present understanding of the laws of physics, the law of conservation of energy follows from the fact of time symmetry. The law of conservation of momentum follows from position symmetry. The law of conservation of angular momentum follows from directional symmetry. We cannot give the reasoning here that demonstrates the connection between these symmetries and conservation laws. However, we can suggest why such connections exist by using the relation between conservation of energy and time symmetry as an example.

We recall that the principle of time symmetry states that the laws of physics do not depend on when we take our measurements. That is, it should be possible to state the laws of physics in such a way that they remain the same from minute to minute, from year to year, from age to age. To see what happens if this is not so, let us again consider the example of a simple pendulum swinging without friction. Each successive swing will take the same time. Suppose, however, that we arbitrarily decide to change the way in which we measure time. We can, after all, measure time any way we choose, as long as we are willing to face the consequences. Let us then agree that every successive swing of the pendulum will be defined to be twice as long as the one before it. This will never lead to a contradiction by itself, since there is no way to resurrect a past swing to compare it with a present or future swing. The past is inaccessible, and there is no direct way to bring back a past hour or second to compare with a present hour or second. If we measure time in this way, however, we will have to conclude that the pendulum is losing speed. If the bob is taking twice as long to travel the same path, it must be moving more slowly. The speed at the bottom, therefore, must be less on each successive swing. Thus the kinetic energy, $(1/2)ms^2$, would also be less on each successive swing. Energy as we know it therefore would not be conserved. This would not be the only change, however. The physical laws relating to the pendulum would also change with time, since the same conditions at the beginning of each swing would systematically give different results for the motion of the pendulum on successive swings. Our redefinition of time measurement would eliminate both time symmetry and conservation of energy.

As a second example, let us consider an object moving without any forces on it. According to the first law of motion, the object should move in a straight line with a constant speed. If, however, we arbitrarily change our way of measuring time, this law need no longer hold. We could agree to measure time so that the particle travels twice as far in each successive second. (Someone would undoubtedly accuse us of making each second twice as long as the last—but we are free to choose this way of defining time if we so desire, and it is just as precise and valid as any other way of defining time.) If we do this, we find that the kinetic energy of the particle is not conserved, since it is moving faster and faster. The laws of physics have also been changed, since the first law no longer holds. Consider, however, what happens if the particle hits something. The dent it makes is a physical reality and does not depend on our definition of time. As time goes on, the object is moving faster and faster according to our definition of time. This means that the physical law that describes the impact of objects as a result of speed must change—the apparently greater speed would cause no greater dent. Again, we have disturbed both time symmetry and conservation of energy by changing our definition of time measurement.

We can see from these examples that the validity of the time symmetry of physical laws is at least partly a consequence of the way in which we decide to measure time. Therefore we can arbitrarily decide to measure time in such a way that time symmetry does hold. This makes this symmetry principle seem to be merely a consequence of our decision to choose a given method of time measurement. However, the matter is not that simple. A definition of time that causes one law to obey time symmetry might cause another law to violate it. Therefore it hardly seems trivial that there is a way of measuring time that makes *all* the laws obey time symmetry. Furthermore, it seems remarkable that the conservation, over long periods of time, of one quantity, energy, is equivalent to the fact that all the laws obey time symmetry.

The Space-time Continuum

At the beginning of this book we mentioned the distinction between continuous and discrete quantities. Matter, it appears, has both discrete and continuous aspects. The discreteness appears in the particle aspects, as when information is obtained by the detection or measurement of a particle. The continuity appears in the waves of probability. In this book we have treated space and time as continuous entities. It is difficult to conceive of them as made up of particles. Nevertheless, as certain scientists have pointed out, we have no direct experimental evidence for the *infinite* divisibility of space and time. No matter how small the distances and time durations that we measure experimentally, there should theoretically be even smaller distances and time durations in a continuous space and time. Therefore the concept of really continuous space and time cannot be tested fully by our experiments and, according to some scientists, should be rejected. For a long time scientists have been thinking about a possible fundamental distance, a shortest possible distance that might do for space what the constant e, the smallest possible charge, does for electric charge. Thus far, however, the existence of this distance remains a matter of speculation.

LIFE, MIND, INFORMATION; CHEMISTRY

We have just discussed the physicist's picture of the very small —the fundamental particles. In the next section we shall discuss the physicist's picture of the very large—the astronomical universe. Between these two extremes lies the world of familiar "normal-sized" objects. For "ordinary" sizes and "ordinary" velocities, the "old" laws of Newton and Maxwell still hold very accurately. However, living beings also fall into this class of "normal-sized" objects. Many scientists believe that when we finally understand the nature of life and mind we will be able to understand them in terms of matter in motion—that is, in terms of the laws of interaction between fundamental particles. They feel that life processes are merely very complex instances of processes which involve the known laws of physics. On the other hand, several excellent physicists, such as Schrodinger and Eugene Wigner, have felt that we should expect to discover new fundamental laws as we increase our knowledge of life processes. In this section we shall outline briefly some of the ways in which life processes can now be understood in terms of particles obeying the laws of physics. We remind the reader that the bodies of living plants and animals (with some exceptions, such as viruses) are made up of cells, visible in the microscope. First we shall discuss the way in which atoms combine to form molecules, and then some of the ways in which molecules are important in the living cell.[4]

Atoms Combine to Form Molecules: Chemistry

One of the important facts that must be explained by any theory of molecular formation is the fact that atoms combine in definite proportions to form certain definite molecules. This phenomenon is more obvious for the smaller molecules. For example, one atom of oxygen combines with exactly two atoms of hydrogen to form a molecule of water. The chemical properties of pure water are fixed by the fact that the constitution of water molecules is fixed—exactly one atom of oxygen joined to exactly two atoms of hydrogen, no more, no less. If the atoms could combine in any indiscriminate way, these fixed properties of chemically pure substances would not nec-

[4]As general references for this section, see J. D. Watson, *The Molecular Biology of the Gene*, New York: Benjamin, 1965. p. 277; E. Schrodinger, *What Is Life? The Physical Aspects of the Living Cell*, Cambridge, England: Cambridge University Press, 1955, pp. 71, 72, 76; and Louis B. Flexner, *The American Scientist*, **56**, 1 (Spring 1968), pp. 52–57. See also recent issues of *The Worm Runner's Digest.*

essarily exist. There is another type of substance that can also be formed from oxygen and hydrogen atoms. If two atoms of oxygen combine with two atoms of hydrogen, a molecule of hydrogen peroxide is formed. This combination has different properties than water and is rather unstable. It readily breaks up to form one molecule of water and a free atom of oxygen which is available for such useful purposes as bleaching. The stable molecules thus do not include all the possible combinations that one might imagine, but only certain combinations.

According to the uncertainty principle, the position and momentum of a particle or system cannot both be known exactly at the same time. We might wonder how a molecule, or even an atom, could be considered in a definite state or condition if one of these quantities, the position or the momentum, were unknown or uncertain. However, as we have discussed earlier, there is also an uncertainty relation between time and energy. We can consider the energy of a system as very certain if the system lasts for a long time. Thus, when we are thinking in terms of a permanent system lasting for long periods of time, we are naturally led to think of the system as in a state of definite energy. We have already thought of atoms in this way—the Bohr "orbits" correspond to states of definite energy for the atom. When an atom exists with its electrons in definite Bohr orbits, the atom is in a state of definite energy. Actually, as we pointed out in Chapter 11, the concept of the Bohr orbit is too simple to describe the atom accurately. Each orbit should be thought of as a possible electron wave in space—a standing wave around the center of the atom. Both the precise position and the precise momentum of the electron represented by this wave are uncertain, but the whole atom is in a state of definite energy.

Now let us consider a simple molecule—the hydrogen molecule, which is made of two hydrogen atoms. Each atom of the common isotope of hydrogen is made up of one proton (the nucleus) and one electron. The two protons are too far apart in the hydrogen molecule for the strong interaction to hold them together against the electric repulsion. However, an electron between these two protons can attract both the protons to itself and thus act as an electric binding agent between them. The electron wave is concentrated between the two protons, so that the electron is more likely to be at a position between the two protons than at other positions (see Fig. 13-9). This electric attraction holds the protons together. They cannot get too close together, however, or their electric repulsion will push them apart. Thus the two protons are held a certain distance apart (or they oscillate back and forth in a certain energy state at about that distance). The molecule as we have described it thus far has two protons in it and only one electron. Thus it is electrically charged. We normally call this structure the hydrogen molecule ion.

We need another electron to form the neutral hydrogen molecule. Because of the exclusion principle, we might think that there could not be another electron in the same standing-wave pattern as the first electron. However, electrons have spin, and therefore two electrons can be in the same standing-wave pattern if their spins are in different directions. The laws of quantum physics determine how many completely different possible states there are for the spin of a particle. Without discussing this matter more deeply here, we shall simply say that there are two completely different states for the spin of an electron, or any particle of spin one-half. This means that there can be no more than two electrons in any standing-wave pattern. The electrically neutral hydrogen molecule consists of two protons with a pair of electrons (in a standing-wave pattern) acting as a bond between them. Even if there were a third electron, it could not have the same standing-wave pattern at the same place as the first two electrons, because of the exclusion principle. A third electron in the same standing-wave pattern would have to have the same spin state as one of the other electrons, since there are only two spin states possible. Thus it would be in exactly the same state as one of the other electrons, and this is impossible according to the exclusion principle. Therefore a pair of electrons acts as a natural electric bond to hold atoms together in molecules.

We often express this fact by saying that atoms can combine together by sharing pairs of electrons. It is common to think of one of

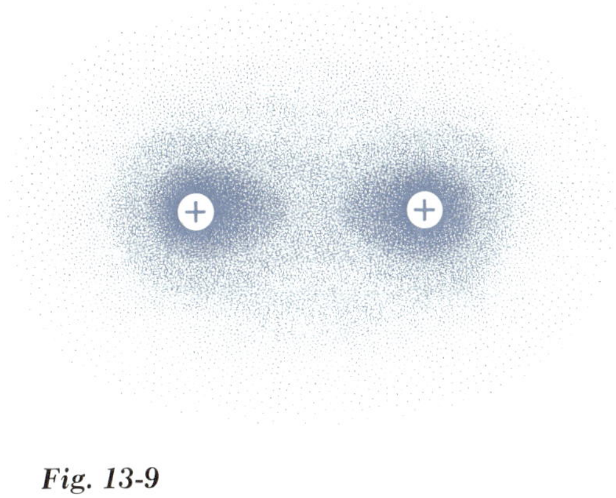

Fig. 13-9

the pair of electrons as coming from one atom, and the other from another, although there are cases in which both electrons of the pair can be thought of as originating from the same atom. Electrically neutral atoms other than hydrogen have more than one electron. Because of the exclusion principle, not more than two of the electrons can be in the lowest energy state, or standing-wave pattern, of the atom. In fact, not more than two electrons can be in any standing-wave pattern at the same time. Therefore some of the electrons will generally be in "outer orbits" or states of greater energy than others. It is these outer electrons, in the outer orbits, which are generally involved in electron-pair bonds between the atoms of a molecule.

There is another kind of bonding between atoms and groups of atoms in molecules besides the electron pairs. Sometimes one atom may acquire an electron from another atom, thus acquiring also an overall negative charge and leaving the other atom with a positive charge. The two atoms thus become oppositely charged ions held together by the electric attraction between their opposite charges. Various intermediate kinds of bonding are also possible, in which an electron is transferred part way from one atom to another.

The exact theory of the energy states of the atom, and the relation of these states to the electrons that can be used to bind atoms together is very complex. It is deeply involved in the directional symmetry of space, but we cannot pursue the subject further here. Here we emphasize that because of the wave nature of the electron, etc., there are definite energy states for the atom, in which the electrons are found in definite orbits, or standing-wave patterns. Because of the exclusion principle, these electrons are generally not all in the same orbit; some of them are in the outer orbits and can form pairs with other electrons to bind atoms together. Only these outer electrons are generally available for a single bond with another atom. Therefore, since there is a restricted number of these electrons available for binding, there is a restricted number of ways in which atoms can join together into molecules. Thus there are only certain molecules that can be formed. The amount of binding energy will help to determine the stability of these molecules.

Sometimes the same kinds of atoms can be arranged in different ways to form different kinds of stable molecules. Usually these arrangements differ in energy. Even if the energy difference is not large, however, considerable energy may be required to change one arrangement into another. The situation may be likened to the problem illustrated in Figure 13-10. Although the loads of lumber have the same gravitational potential energy in village *A* as in village *B*, it still takes considerable extra energy to get a load over the mountain from one village to another. Similarly, it may require considerable extra energy either to change one arrangement of atoms in a molecule into another arrangement, or to break up the molecule. This requirement of extra energy helps to keep the molecule from being broken up or altered by the constant bombardment it receives from other molecules.

The fact that atoms form molecules in certain discrete arrangements, and that it requires considerable energy to change from one stable arrangement to another, helps to explain the stability of molecules. We can illustrate this with two more examples. Consider the dart player in Figure 13-11*a*. He is attempting to hit the center of the bullseye with a dart. Any small disturbance of his hand or eye will cause the dart to strike off center. If he throws one-hundred darts, they will be grouped around the center, but very few will be very close to the center. On the other hand, if a funnel is installed as shown in Figure 13-11*b*, many more precise bullseyes will be obtained. There is now only one possible state in the center region. If the dart hits anywhere in the center region it will be funnelled into the bullseye. Only a comparatively large disturbance, causing the dart to miss the funnel entirely, will put the dart outside the bullseye.

Figure 13-12 illustrates a second example. The marbles in the egg box stay in their compartments even when the box is shaken. The marbles will roll around in their individual compartments, but they will not move from one compartment to another or out of the box

Fig. 13-10

unless the shaking becomes particularly vigorous. Enough energy must be added to raise a marble high enough to cross over the barrier wall between one cell and another. Similarly, in a molecule, a given arrangement is preserved until enough energy is supplied to one or more of the atoms to cross the energy barrier that holds them in a given arrangement and allow them to break up or fall into another arrangement. Thus the arrangement remains the same or is changed in one of a number of definite ways.

The energy needed to change a molecule may be supplied in many ways. The molecule may be struck very forcefully by other molecules in their random motion. Therefore heating a collection of molecules, and so increasing their kinetic energy, increases the likelihood of a molecular change. Some molecular changes can be caused by photons of light; for example, the exposure of a photographic film. Alpha, beta, and gamma particles as well as X rays and cosmic rays can also cause molecular changes.

Structure: DNA and RNA

If one were to create a human being artificially, assuming that it were possible to do so, one would need a great deal of information. One would need not only enough information to describe the overall human form and anatomy but also enough to describe the structure and operation of all the cells, including the molecules and molecular processes that occur within them. Yet we know that the human body develops from the union of microscopically small egg and sperm cells. In these cells must be stored sufficient information to direct the formation and operation of the whole human body.

(a) (b)

Fig. 13-11

Today we believe that most of this stored information is contained in long slender molecules of desoxyribonucleic acid—the famous DNA. These molecules may be as long as a thousandth of a meter. They may have a mass number of about 2×10^9; that is to say, they may have a mass equal to about 2×10^9 protons. They may close on themselves to form loops. Figure 13-13 shows the structure of one of these DNA chains. It is made up of molecules of a kind of sugar connected together by phosphate groups. (A phosphate group is made up of atoms of phosphorus and oxygen.) Each sugar group has attached to it another group of atoms called a base. (This is *not* the kind of base discussed in chemistry when talking about acids and bases.) There are four main kinds of bases; adenine, cytosine, thymine, and guanine, symbolized by the letters A, C, T, and G. The order of these bases along the DNA chain determines the "message" or information contained in the DNA molecule. The information is contained in the sequence of these bases much as the information in a sentence is contained in the sequence of letters and words of the sentence. The situation is actually more complex than a sentence, however, since the DNA molecule must not only store the information but act physically in the cell to use this information to direct the operation of the cell, and indeed to control the development of the fertilized egg cell into the adult organism. The DNA molecule must also be able to help create another DNA molecule similar to itself when the cell divides, so that each resulting cell will have its controlling DNA with the right information to direct the cell. In its direction of the living cell, DNA is aided by several types of a similar but less stable molecule called ribonucleic acid, or RNA. RNA also consists of a chain of sugar and phosphate, with bases attached to the sugar.

One might wonder how information can be safely stored in so small a structure as DNA. The same information must be preserved for ages so that the offspring resemble the parents for generation aft-

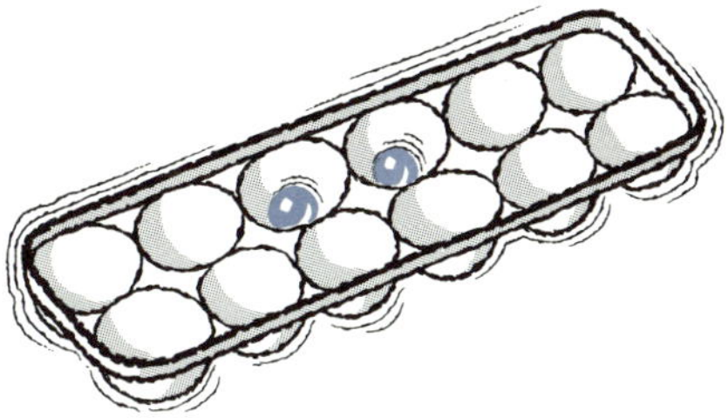

Fig. 13-12

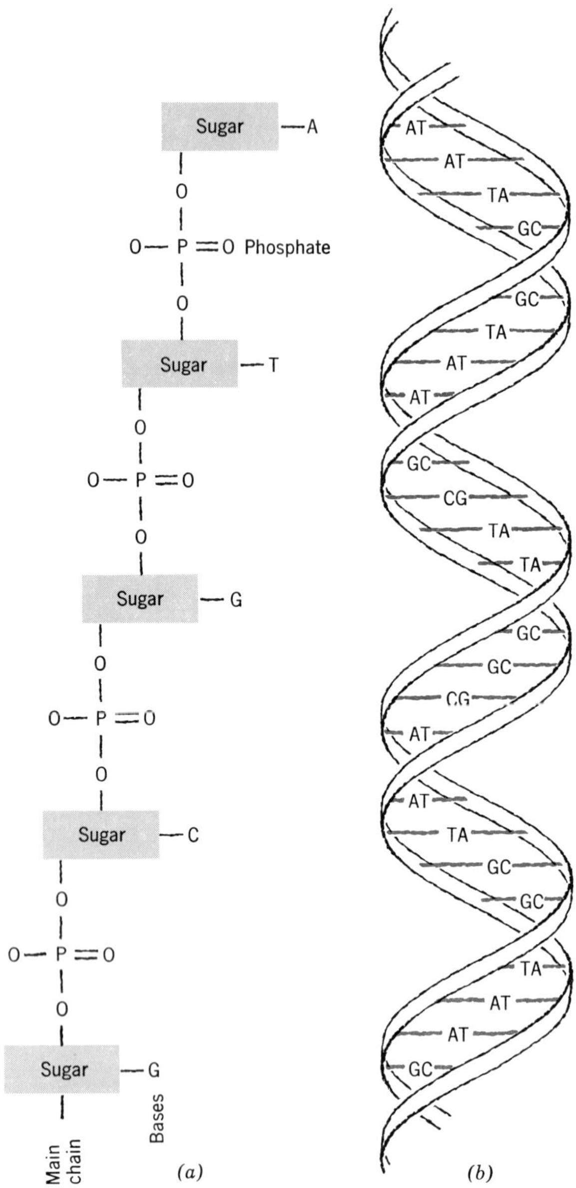
Sugar
A
O
O
P
O
Phosphate
O
Sugar
T
O
O
P
O
O
Sugar
G
O
O
P
O
O
Sugar
C
O
O
P
O
O
Sugar
G
Main chain
Bases
(a)
AT
AT
TA
GC
GC
TA
AT
AT
GC
CG
TA
TA
GC
GC
CG
AT
AT
TA
GC
GC
TA
AT
AT
GC
(b)

Fig. 13-13

er generation. We have here not the sudden discharge in a Geiger counter, nor a spot on a photographic film that gives information to be observed by man. Rather, we have the information comprised in the stable standing-wave structures of the atoms combined into molecules. We have already indicated why molecules can be very stable structures in space under the right conditions, and thus capable of storing information. There is one other property of most DNA that increases its stability. In its resting state it is coiled into a double helix, as shown in Figure 13-13*b*. The helix is formed by two chains of DNA twisted together, so that the bases are in the interior. Thus the "message" contained in the sequence of bases is protected.

Notice that the order of the bases is not the same in each chain of the helix—each base pairs with another base. The same bases always occur in pairs together, however, unless some "error" has occurred. Therefore either of the two chains of the helix could be determined if the other were known. The discovery of the double-helical structure of DNA resulted from the efforts of F. Crick and J. D. Watson.

Mutation

Sometimes a plant or animal will exhibit a characteristic that has not been seen in its ancestors. Such a change is called a mutation. If this characteristic is inherited by offspring of the individual showing the mutation, then there has been some change in the information that determines the nature of the plant or animal. We would suspect that there has been a change of some kind in the DNA of the individual. Perhaps the wrong base was added into the chain at a certain point when the molecule was duplicated. Perhaps some disturbance occurred when the DNA was struck by high-energy photons or fundamental particles. A great deal of evidence shows that mutations can occur because of alterations of DNA.

Anything that changes the molecular structure of DNA bases can conceivably cause mutations. Thus we would expect that mutations would continually occur throughout the generations of a given species of animal or plant. Most of these mutations would undoubtedly be "harmful." According to the present theory of natural selection, however, these "harmful" mutations would put the organism at a disadvantage. Such an organism would be less likely to survive and reproduce. "Helpful" mutations, although rare, would be those that helped individuals to survive and reproduce. These mutations therefore would become more common in the population and thus would change the nature of the species. Thus DNA helps to account for both the stability of the species found in nature and the mutations required by the theory of evolution by natural selection.

Memory

The fact that a human being or an animal has a memory suggests that there must be some kind of information storage system in the body, presumably in the brain. The exact manner and location in which this information is stored are long-standing puzzles which are yet unsolved. The problem is complicated by the fact that there appear to be different kinds of memory. Apparently the processes responsible for immediate recall are different from those responsible for long-term memory.

The electrical activity of the brain has been known for many years. Nerve signals are associated with electrical impulses traveling along nerve fibers. The brain also exhibits electrical activity in the so-called brain waves, whose characteristics depend on the animal's state—sleep, arousal, etc. The vast network of nerve fibers in the brain suggests that the brain is a complex electrical system, in which information is stored by making certain pathways less resistant than others to the passage of electrical impulses. There is evidence that extensive regions of the brain cooperate in the storage of each item of information. Certain areas appear to be used for the processing of information, since electrical stimulation of these areas seems to arouse memories. The model of the brain as an electrical computer arises naturally from some of these considerations.

With the discovery of the function of DNA and RNA in the living cell, however, an additional idea is suggested. These large molecules store and transfer information in the cell. Memory is essentially an information transfer and storage process. Might not DNA and RNA be involved in the storage, transfer, and processing of information in the brain? Recent evidence suggests strongly that RNA in the brain is a very important factor in memory and learning processes. Treatment with RNA has been reported to reduce memory loss in older people. Changes in RNA in the brain have been reported to accompany learning in experimental animals. RNA is important in the construction of certain large molecules (proteins) in the living cell. If this activity is obstructed in the brain by treatment with antibiotics such as puromycin, certain memory processes are also obstructed. Several researchers have reported the transfer of learned behavior from one group of experimental animals to another as a result of the transfer of RNA from the animals of one group to the other. Thus

RNA is believed to be important in memory and learning processes, but its precise function in these processes is still unknown.

In this section we have illustrated the usefulness of physical and chemical concepts in the study of biology. There are, of course, many other areas of biology in which such concepts are useful. The living organism acquires and uses energy. Also, since the processes in living cells are irreversible, they result in an increase of entropy, or disorder. The organism takes in as food, etc., material with a certain amount of molecular order, and rejects some of it as waste in a more disordered state, thus getting rid of the disorder it creates. Such facts indicate the importance of the principles of physics in living processes.

THE MACROCOSM

The picture of the universe we sketched at the end of Chapter 4 is still correct in its main features: satellites revolve around the planets, planets rotate on their axes and revolve around the sun, the sun moves in its orbit around the center of the galaxy, and the galaxy, like countless other galaxies, moves through space. However, new observations obtained from outside the earth's atmosphere by satellites and rocket-borne instruments, as well as new techniques such as radioastronomy, have increased our knowledge of matter moving in space. Much of this knowledge concerns fundamental particles. It is the particles of light and matter coming to us through space that supply our information about the universe.

Particles in Space

THE SOLAR WIND. In 1744 it was observed that the somewhat awesome but beautiful displays of northern lights were accompanied by disturbances of compass needles. (These disturbances in the earth's magnetic field are called magnetic storms.) It was found that these storms often occurred two or three days after such disturbances as sunspots appeared centrally located on the sun. It was assumed that jets of ionized gas were emitted from these disturbed regions of the sun and streamed outward into the nearly empty regions of interplanetary space. At times, the earth would pass through such a stream of gas. The earth's magnetic field would deflect the particles in the gas toward the poles, and the particles would there crash down into the atmosphere, ionizing and energizing (or "exciting") the molecules. The glow would occur as the electrons in the excited molecules returned to orbits of lower energy. At the same time, currents were supposedly set up in the atmosphere that traveled round the earth, producing the magnetic disturbances observed at the surface. These ideas were not quickly accepted. Not only were observational data sparse, but very noted physicists were opposed to the proposed explanations. For example, Lord Kelvin said, "The supposed connection between magnetic storms and sunspots is unreal, and the seeming agreement between the periods has been mere coincidence."[5]

During the 1950's it was suspected that isolated jets of charged gas or plasma could not account for all of the observed phenomena and that a more continuous gas in the entire interplanetary region might exist. For example, light pressure alone could not account for the characteristics of comet tails. (These tails always point away from the sun, whichever way the comet is going.) Furthermore, there is some disturbance to compass needles and some display of northern lights almost every night. This is hard to explain if the jets of gas are supposedly few and far between. In 1958 E. N. Parker proposed a model that predicted a continuous high-speed solar wind streaming out from the luminous outer atmosphere (corona) of the sun (see Fig. 13-14).

Observations by spacecraft announced in the early 1960's led to widespread acceptance of the notion of a continuous, high-speed plasma blowing out from the sun (see Fig. 13-15). Scientists have found that protons, alpha particles, and electrons move away from the sun at speeds ranging from about 3×10^5 to more than 8×10^5 meters/second. The number of particles in this plasma varies from about 10 to 200 particles per cubic centimer. Furthermore, there is a weak but important magnetic field that moves outward with the solar wind.

The emerging view is that interplanetary space is not only filled with light (and other electromagnetic waves) but also contains particles of matter in the rapidly expanding outer part of the solar atmosphere that extends far beyond the earth. Embedded in the general outward flow of this solar wind are streams of higher speed particles. These streams in the solar wind are characterized by large variations in density and in the magnetic field. Some of these streams are rather

[5]Lord Kelvin (William Thompson), Address to the Royal Society at Their Anniversary Meeting, November 30, 1892, *Proc. Roy. Soc.* London, **52** (1892), pp. 300–310.

long-lived and persist for several months. As the sun rotates, the stream itself also rotates and can be observed again and again as it sweeps past a spacecraft. Frequently, however, the "wind" pattern changes rapidly, and the streams form and dissipate in a complex way.

The magnetic field of the earth presents an obstacle to the solar wind. In fact, spacecraft observations show that the solar wind is excluded from a large region of space around the earth by the earth's magnetic field, as shown in Figure 13-16. In the process the solar wind distorts the earth's magnetic field. Instead of a simple field resembling that of a bar magnet as observed near the earth's surface, the outer part of the earth's field is swept backward away from the

Fig. 13-14 Courtesy of High Altitude Observatory, Boulder, Colorado.

sun into a long tail-shaped region. It is now thought that this magnetic "tail" of the earth provides the conditions necessary for the energy in the solar wind to be converted into the effects seen from the earth. Many scientists now believe that the northern (and southern) lights and magnetic storms are caused by particles from the solar wind that enter the tail and approach the earth from the rear. More recently, correlations have been found between magnetic storms and fluctuations in the magnetic field associated with the solar wind. The precise mechanism that produces the northern and southern lights and magnetic storms is not yet known.

INTERGALACTIC PARTICLES. Like the interplanetary region, the interstellar region also contains particles. In fact, there is about as much matter in the form of dust and gas spread through space as there is collected in the stars. Dust and gas in space should scatter the blue wavelengths of the light passing through it more efficiently than the red. This would produce a "reddening" of the light passing

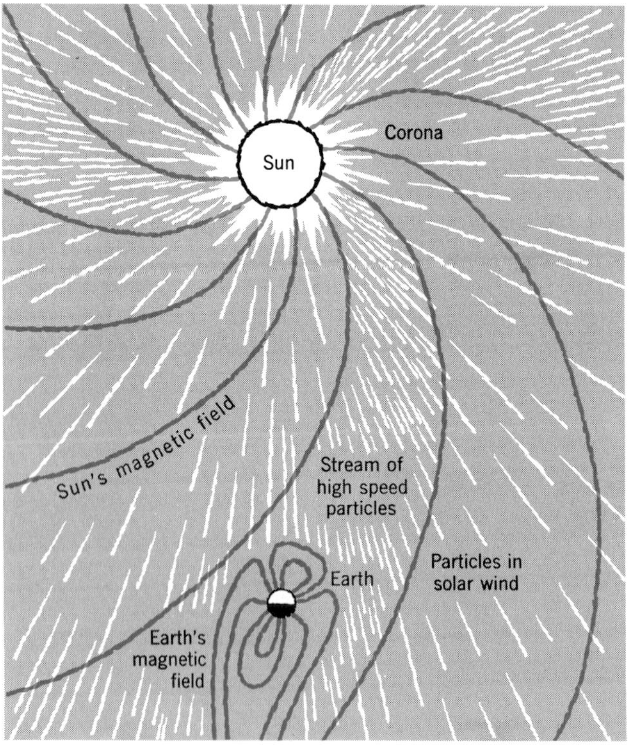

Fig. 13-15

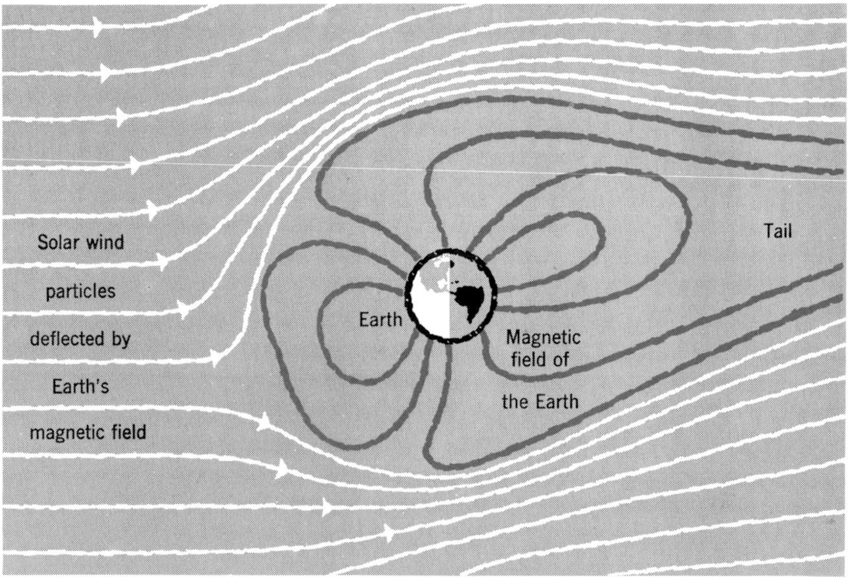

Fig. 13-16 Solar wind particles deflected by earth's magnetic field.

through the dust and gas. Such a reddening is actually observed.

The light coming through this gas and dust on its way to the earth has certain frequencies missing or weakened (see Fig. 13-17). These frequencies are absorbed by the atoms in space. Energy from the photons of this light raises an electron to a higher orbit when the photon is absorbed. Therefore the missing frequencies of light are characteristic of the kinds of atoms in space and enable us to find out which atoms are there. Lines of calcium, potassium, iron, and sodium are frequently observed. (Hydrogen is present in far greater quantities than other elements but is detected by a different technique.) If

Fig. 13-17

the light travels through a gas cloud that is moving relative to the source of light, the Doppler effect causes a small change in the frequency absorbed. In such a situation the dark lines in Figure 13-17 would appear in slightly shifted positions. From this shift the velocity of the dust cloud can be determined. If the light passes through two clouds moving with different velocities, this shift will be different for each cloud. The lines will then appear to be doubled, or at least broadened. In this way as many as seven different clouds of gas have been detected at once.

COSMIC RAYS. Besides the particles we have already mentioned, there are very high-energy particles, called *cosmic rays*, moving through space. Cosmic rays consist of atomic nuclei, mostly protons, but with many alpha particles and heavier nuclei. Some of these particles have more kinetic energy than any other fundamental particles yet detected or produced in man-made particle accelerators. When these particles collide with nuclei in the earth's atmosphere, particle transformations and creations occur, resulting in a descending shower of high-energy particles. Because of the extremely high kinetic energy of the cosmic-ray particles, other particles of very high rest mass are created in these transformations. In fact, many of the elementary particles were discovered in cosmic-ray experiments.

We have seen that the earth's magnetic field shields the earth from the solar wind. The earth's field also helps to shield the earth from cosmic rays. Charged particles moving in a magnetic field are deflected. The higher the energy of the particles, the less effective this shielding is. The atmosphere itself also helps to shield the earth by absorbing some of the cosmic rays. This shielding is very important in protecting life from high-energy radiation.

Some cosmic rays come from the sun, but these rays are unimportant and infrequent compared to the more energetic and more numerous cosmic rays from outer space. The cosmic rays from space appear to come in nearly equal numbers from all directions. No one knows where they come from, but several possibilities have been suggested. One possibility is that they are accelerated in electric and magnetic fields which are caused by exploding stars—each one acting as a sort of cyclotron in the sky. Another possibility is that the particles are accelerated in collisions with the magnetic fields associated with moving dust clouds.

Objects in Space.

THE EXPLORATION OF THE SOLAR SYSTEM. The great era of the exploration of the land surface of the earth appears to be ended. Virtually all of the earth's surface has been explored, studied, and mapped. The direct exploration of the solar system, however, is just beginning, with the advent of remote-controlled and manned spacecraft. It is unnecessary here to do more than mention that men have been put in orbit around the earth and the moon; and that spacecraft have flown by Mars and Venus, into Venus; and that spacecraft have landed on the moon (see Figs. 13-18 to 13-22). Scientists have finally obtained photographs of the hidden side of the moon, the side that always faces away from the earth (Fig. 13-18). Much information

Fig. 13-18 *A large lunar crater the floor nearly covered with a layer of dark material, is prominent in this Lunar Orbiter II photograph of the moon's far side.* *Courtesy of NASA.*

has been gathered by instruments on these spacecrafts and by other new techniques. Scientists have discovered that Mars and Venus have very weak magnetic fields, if any. The Mars photographs—for example, Figure 13-22—show craters much like those on the moon. The results so far generally make the earth appear to be unique in its fitness for life as we know it. For example, the atmosphere of Mars now appears to be even thinner, according to spacecraft obser-

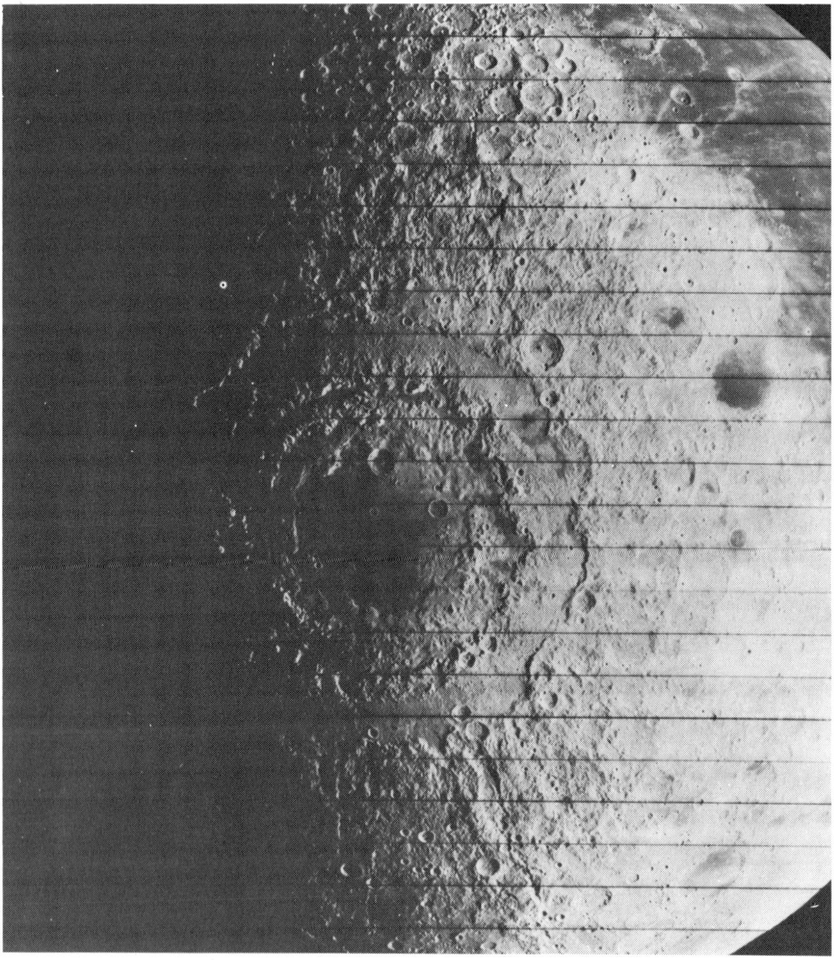

***Fig. 13-19** The Cordillera Mountain range of the moon forms a concave curve leading off the southwest part of the moon's visible face. Sweeping around this area, Lunar Orbiter IV took a photograph showing that the mountains formed a vast ring around the 600-mile-diameter Oriental Basin. Within this ring are several smaller rings, giving the effect of a giant bullseye.* Courtesy of NASA.

vations, than according to earlier estimates. We now think that the pressure of Mars' atmosphere on the planet's surface is about 0.5 percent of the atmospheric pressure at the surface of the earth.

STRANGE OBJECTS IN SPACE. In 1929 the astronomer Edwin Hubble noticed that the frequencies of characteristic lines in the light from distant galaxies were shifted toward the red. That is, these frequencies were actually lower than they would be if they came from a source on earth. If this red shift is caused by the Doppler effect, then these galaxies are moving away from us. In fact, the red shift is greater for the more distant galaxies. Therefore, the more distant the galaxy, the faster it is apparently moving away from us. Thus the whole universe appears to be expanding. This naturally suggests that the universe may have begun in a large explosion—the so-called *big bang* theory.

The fact that all these galaxies appear to be moving away from us does not necessarily mean that we are at the center from which the expansion began. We would expect to see about the same effect from any galaxy, not merely from our own. Suppose that we have an expanding balloon, as shown in Figure 13-23. If we let each dot on the surface of the balloon represent a galaxy, or a cluster of associated

Fig. 13-20 *Close-up of part of the moon Crater Copernicus taken by the camera of Lunar Orbiter II. Mountains rise as high as 1000 feet from the crater floor. Horizontal distance: about 17 miles. Distance from base of photograph to horizon: about 150 miles. Courtesy of NASA.*

galaxies, we have a useful model of the expanding universe. No matter which dot we stand on, we would observe all the other dots moving away from us.

Recently scientists have developed other means of looking at the stars and galaxies besides light telescopes. X rays from the sun and certain stars and other objects in the sky cannot penetrate the atmosphere of the earth, but they can be detected by instruments carried in rockets. Objects have been found that emit far more energy as X rays than as visible light. One source of X rays is the Crab nebula, which is believed to be the remains of an exploding star.

Electromagnetic waves of some wavelengths much longer than light waves can also penetrate the earth's atmosphere. Radio telescopes, which are large, carefully shaped antennas designed to receive radio waves, have been used to scan the heavens for sources of radio waves. One such source is the center of our own galaxy. Another

Fig. 13-21 *Lunar Orbiter I photographed earth when all but a crescent was in darkness. To an observer on the moon, the earth goes through phases, as does the moon viewed from the earth. The phases in both cases are due to the positions of the earth and moon relative to the sun as the moon revolves around the earth. Courtesy of NASA.*

is the sun. Other sources are thought to be exploding galaxies. Some of the most interesting radio sources are the so-called quasars. These are radio sources that have been identified with visible stars having curious properties. They have surprisingly large red shifts, which correspond to speeds of recession of up to 80 percent of the speed of light. According to the theory of the expanding universe, objects of this speed would be among the most distant objects ever observed. If they are so distant, however, they must be brighter than a trillion suns in order to appear to us as bright as they do. The light of these objects can change in brightness by as much as 50 percent in a few years. This change would be unlikely for a galaxy, consisting of many stars spread out over large regions of space. If quasars are really

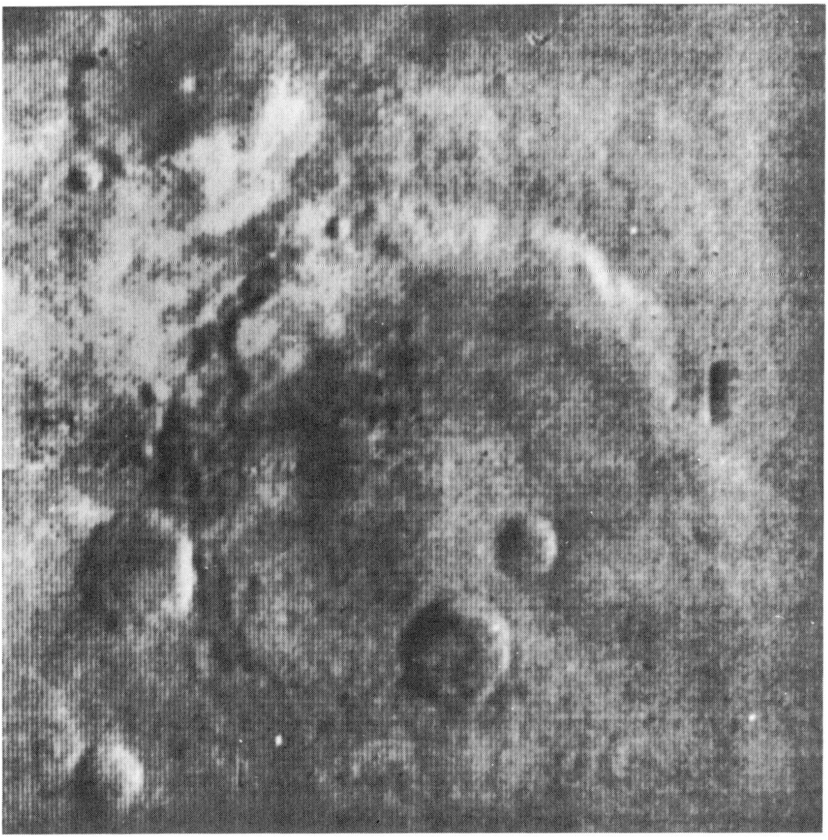

Fig. 13.22 A Mariner IV photograph of Mars. Notice the similarity between the craters shown here and those observed on the moon. Courtesy of NASA.

not so far away, then they would not have to be so bright. Then, however, their large red shifts, with no other apparent motion, are hard to explain. Thus these objects are still considered mysterious.

THE PHILOSOPHY AND METHODOLOGY OF SCIENCE.

We have mentioned both method and philosophy several times in this book, but we have not tried to identify any specific philosophy and method as *the* philosophy of science, or *the* scientific method. Instead we have tried to discuss specific philosophical attitudes and methods that are or have been influential in science, and to illustrate their usefulness in specific cases. Most scientists take a rather practical view of the matter, considering any method that works as a valid one.

There are some things, however, that can be tentatively included within the world view shared by many physicists, at least in their scientific work. One of these is the attempt to explain physical phenomena in terms of particles of matter in motion. We have seen the curious concepts to which this attempt has led. Both matter and motion turned out to be rather different from what the earlier, more naive philosophers expected. It would probably be even more naive for us to think that our present understanding of matter in motion will not be drastically changed by new developments.

Another thing that can be included within the world view of most

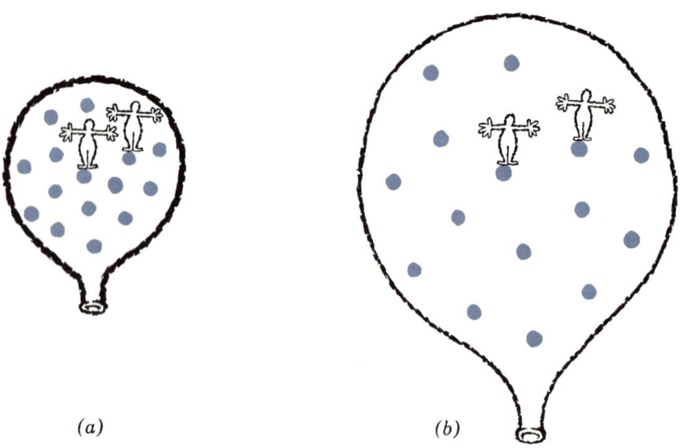

(a) (b)

Fig. 13-23

scientists in their scientific work is an emphasis on experiment and observation, not reason alone. Reason and logic are very important in science. However, no matter how logical a theory may be, it must still meet the test of experiment if it is to be considered valid or true. We have seen how theories and concepts have been abandoned or modified when they failed to predict correct results. Philosophers of science have discussed this basic attitude and expressed it in many ways, trying to arrive at a satisfactory notion of scientific truth.

The necessity for agreement with experiment exerts a strong constraint on possible scientific theories. An artist, painting on a flat surface, or a sculptor, working with clay or stone, is of course also subject to the limitations of his medium. He is subject to the constraints imposed by the nature of the paints and canvas, or the chisels and stone, with which he works, and his achievement must be evaluated with these constraints in mind. One who practices the art of pure mathematics is somewhat more constrained—logical consistency is desired. The scientist operates under the constraint that his theories must be true—that is, they must agree with experiment. In a practical sense, they should be as simple as possible (and thus as convenient as possible) for the purpose at hand. For pure science, they should be as simple as possible and still agree with experiment. The ideal is to find theories that are simple, comprehensive, and in agreement with all correctly performed experiments, past, present, and future. Needless to say, we have not yet achieved this ideal.

Despite the elegance of some theories, and the ingenuity required to construct them, they often appear crude and inadequate when we consider what we do not yet understand. When we consider, however, what can be done with the knowledge we now have, and how this knowledge affects our technology and culture, our scientific information appears of great importance. Even the little scientific knowledge we now have has made a great difference to our society.

Problem solving

We here include one example to show the use of the uncertainty principle in estimating the range of an interaction.

Example 1

Using the uncertainty principle relating time and energy, and also the mass of the pion, estimate the range of the strong force exerted by a nucleon.

Solution. The uncertainty principle relating time and energy is

$$(\text{Uncertainty in time}) \times (\text{Uncertainty in energy}) \geqq h \qquad (1)$$

In this problem we are interested in the minimum possible uncertainty—the uncertainty that we cannot remove by making further measurement. We must consider the effects of the conditions allowed by this minimum uncertainty. The minimum uncertainty is obtained from Equation 1 by using an equal sign instead of the symbol $\geqq$ (greater than or equal to):

$$(\text{Uncertainty in time}) \times (\text{Uncertainty in energy}) = h \qquad (2)$$

The strong force exerted by a nucleon is a result of the emission of virtual pions. The total time t, between the emission of these virtual pions and their reabsorption corresponds to the uncertainty of time allowed by Equation 2. Their energy corresponds to the uncertainty of energy allowed by Equation 2. The total energy of the pion is just its mass m multiplied by c^2, the square of the speed of light. Therefore we have

$$t\,(mc^2) = h \qquad (3)$$

or,

$$t = \frac{h}{mc^2} \qquad (4)$$

The range of the force is estimated by the maximum distance that the pion can travel during the time t. If the pion has a speed s, then this distance will be

$$d = st = \frac{hs}{mc^2} \qquad (5)$$

The maximum speed that the pion can have is c, the speed of light. Therefore the numerator of the fraction on the right side of Equation 5 cannot be greater than hc. The minimum mass that the pion can have is its rest mass m_0. Therefore the denominator of the fraction on the right of Equation 5 cannot be less than m_0c^2, and the fraction itself should be no greater than

$$\mathrm{d} = h/m_0\mathrm{c} \qquad (6)$$

Since the uncertainty principle used in this way gives only an estimate, this distance is not the "maximum" range for the strong force. The strong force does have appreciable effects beyond this range. The distance h/m_0c is, however, a good estimate of the range, as we can see by finding its numerical value. Using the rest mass of one of the pions:

$$\begin{aligned} m_0 &= \text{rest mass of neutral pion} = 264.2 \times m_e \\ m_e &= \text{electron rest mass} = 9.1 \times 10^{-31} \text{ kilogram} \\ h &= 6.63 \times 10^{-34} \text{ joule second} \\ c &= 3.0 \times 10^{8} \text{ meters/second} \end{aligned} \tag{7}$$

$$\begin{aligned} d = h/m_0c &= \frac{6.63 \times 10^{-34}}{264.2 \times 9.1 \times 10^{-31} \times 3.0 \times 10^{8}} \text{ meter} \\ &= 0.93 \times 10^{-15} \text{ meter} \end{aligned} \tag{8}$$

This is approximately the value cited in Chapter 12 as the "radius" of a nucleon.

Summary

In general, this summary omits material introduced in earlier chapters and concentrates on the principles and topics covered in this chapter for the first time.

I. FUNDAMENTAL LAWS AND PRINCIPLES

- **A.** PARTICLE IDENTITY; the distinction between fermions and bosons.
- **B.** SYMMETRY UNDER THE EXCHANGE OF IDENTICAL PARTICLES
- **C.** CONSERVATION OF STRANGENESS
- **D.** THE RELATION BETWEEN SYMMETRY PRINCIPLES AND CONSERVATION LAWS

II. FUNDAMENTAL PARTICLES

- **A.** THE PIONS (for others see Table 13-1, p. 534).

III. FUNDAMENTAL CONSTANTS

- **A.** MASSES OF FUNDAMENTAL PARTICLES (see Table 13-1, p. 534).

IV. FUNDAMENTAL INTERACTIONS

- **A.** ELECTROMAGNETIC
  1. The photon as the carrier of the electromagnetic interaction.
  2. Discussion of the relativistic theory, involving Feynman diagrams, "backward in time," etc.
- **B.** THE PIONS: as carriers of the strong interactions.

V. APPLICATIONS
   A. VIRTUAL PARTICLES; the uncertainty principle.
   B. COMBINING OF ATOMS TO FORM MOLECULES
   C. STABILITY, especially of molecules.
   D. HEREDITY, MUTATIONS, MEMORY: DNA and RNA.
   E. PARTICLES IN SPACE: the solar wind and cosmic rays.
   F. DARK-LINE SPECTRA
   G. THE EXPANDING UNIVERSE
   H. X RAY AND RADIOASTRONOMY: Quasars.

Questions

1. How do the various fundamental constants determine the scale of the world?
2. What is strangeness?
3. From Table 13-1, find some characteristics that might be used to distinguish between the different families of particles.
4. What different amounts of spin can fundamental particles have, and what important distinction can be made on this basis?
5. What are fermions and bosons? Why do we distinguish between them?
6. What properties can a particle have?
7. What is a virtual particle?
8. Why would violation of the double symmetry mentioned for weak interactions but the keeping of the triple symmetry mean that time reversal should fail?
9. Why would the keeping of the double symmetry mentioned for weak interactions but failure of the reflection symmetry mean that particle-antiparticle symmetry would also fail?
10. What is the "hole" theory?
11. To what extent can we take the idea seriously that a positron is an electron going backward in time?
12. How does the exclusion principle help the "hole" theory?
13. What do you know about Feynman diagrams?
14. Why is it necessary to consider Figure 13-2*b* as well as Figure 13-2*a*?
15. One could claim that a Feynman diagram represents an interaction

between photons and electrons that is independent of direction in time. What do you know that would support this point of view?

16. Why do the virtual photons that contribute to the outermost parts of the electron's electric field have, on the average, less energy than those closer to the electron?
17. What is the uncertainty principle that relates time and energy?
18. A photon is emitted and absorbed very quickly by an electron, as shown in Figure 13-4*a*. What alteration of conservation laws is allowed in this process, and why?
19. List the fundamental interactions and describe either systems that are held together by each of them or reactions that take place because of them.
20. Which interaction obeys the greatest number of symmetry principles and conservation laws?
21. Assuming that the strong interactions occur by the exchange of virtual pions, explain why the force has such a short range.
22. Suppose that we have three states and two electrons. How many different combination states are possible?
23. Suppose that we have three states and two photons. How many different combination states are possible?
24. Suppose that we have three states and two coins. How many different combination states are possible?
25. What do we mean when we say that all electrons are the same? How does their identity differ from the similarity of two coins, for example?
26. What do you think Schrodinger meant by saying that a fundamental particle is not an individual?
27. Would you expect a virtual photon of wavelength 3000×10^{-10} meter to have influence in the electric field of the electron at a greater or lesser distance than 1000×10^{-10} meter? Explain.
28. What is SIC physics?
29. What determines how big everything is?
30. What is the distribution of matter in space, as far as you know?
31. What is the possible significance of the fact that the light from distant galaxies appears redder than would normally be expected?
32. What is the solar wind?
33. Could we, despite the uncertainty principle, know the energy of a system in a certain condition exactly?

34. What would prevent the following reactions from occurring? (See Table 13-1 for the meaning of the symbols.)

$$e^- + \mu^+ \rightarrow \nu_\mu + \nu_e$$
$$\Sigma^0 \rightarrow p^- + \pi^+$$
$$\Sigma^+ + \Sigma^- \rightarrow K^+ + p^- + \gamma + \gamma$$
$$p^+ + K^+ \rightarrow n + e^+ + \overline{\nu}_e + \pi^+$$
$$p^+ + \Sigma^0 \rightarrow \pi^+ + \pi^- + K^0 + n$$
$$\Lambda^0 \rightarrow p^+ + \Sigma^-$$

35. How do atoms combine to form molecules?

36. State two possible attitudes toward the relation of life to the laws of physics.

37. How can an electron hold two protons together?

38. Why do chemically pure substances exhibit definite properties?

39. Why is it that the uncertainty principle does not prevent a small system from existing in a definite state? What characterizes the state?

40. Discuss the relation among the following for atoms: orbits, standing-wave patterns, energy states.

41. Distinguish between the hydrogen molecule ion and the hydrogen molecule.

42. Why does a pair of electrons rather than a trio of electrons make up a bond between two atoms?

43. Why don't all the electrons of a many-electron atom just fall into the lowest orbit?

44. Why can atoms form molecules in certain ways and not in other ways?

45. Why does it sometimes take a comparatively large amount of energy to change the state of a system even though the first and second states of the system differ very little in energy? How does this fact increase the stability of the system?

46. What can cause a molecule to break up or rearrange its atoms?

47. Why could heating a solution cause a chemical reaction taking place within it to proceed more rapidly?

48. How is information stored in the living cell?

49. How can the information stored in the living cell be preserved from generation to generation?

50. How do we know that not all the information stored in the living cell is preserved without change from generation to generation?

51. Justify the phrase "like father like son."

52. Why don't snakes hatch out of birds' eggs?

53. A scientist has remarked that the information of life is written in four-letter words. Explain. (Actually, it appears that the four letters are used in groups of three.)

54. What does the term "double helix" have to do with life?

55. What does DNA have to do with evolution?

56. What does physics have to do with the nature of the mind?

57. What reasons are there for suspecting that RNA might have something to do with memory?

58. Describe the magnetic fields of the earth and the sun.

59. What is the tail of the earth?

60. Explain how the light spectrum coming from distant objects in space could show

    (a) Dark lines on a bright band.
    (b) Light lines on a darker bright band.
    (c) Light lines alone.
    (d) A bright band with no lines whatsoever.
    (e) Dark lines split in two on a light band.
    (f) Light lines split in two.
    (g) Dark lines split into three-line groups on a light band.

61. What are cosmic rays?

62. How are cosmic rays formed?

63. Suppose that a star were formed by dust particles in space being pulled together by their gravitational attraction. Why would the star get hot?

64. Why might an expanding universe appear to be expanding from each galaxy, just as if that galaxy were the center of expansion?

65. What good are artificial satelites?

66. What limitations are there to earth-bound astronomy?

67. What is a quasar?

68. What has been discovered by radioastronomy?

69. What sorts of philosophical problems and attitudes are important to scientists as scientists, and why?

70. What similarities and differences are there between scientific and artistic endeavors?

71. Why should a practical man be interested in science?

*72. Would barrier penetration alter our discussion of the eggbox in Figure 13-12?

73. What would lead you to suspect that the strong interaction supplies a repulsive force at very short distances?

More Difficult Questions and Problems

1. Is it of any practical consequence that the nucleus of an atom is much more massive than the electrons?
2. Figure 13-24 diagrams one possible interaction between an electron and a positron. Find another diagram for the same interaction which also uses only one photon.
3. Justify the uncertainty principle involving energy and time in terms of the wave nature of a particle. (Use the formula $E = hf$, and see what happens if a wave is confined to a short existence in time.)
4. If you were ordered to discover a new conservation law or get rid of an old one, what would you do?
5. What would the rest mass of a deuteron be if the binding mass were equal to the mass of a neutron? If it were equal to the sum of the masses of the proton and the neutron? If it were greater than the sum of the masses of the proton and the neutron?
6. What relationships exist between conservation laws and symmetry principles?
7. Why is it impossible to prove the continuous nature of space experimentally?
8. What is the scientific method?
9. What is physics?
10. What are the limitations of science, if any?
11. What problems remain to be solved in physics? If you were a physicist, which one would you like to work on?
12. Suppose that A, B, C, D, and E represent particles. If only the following reactions are possible, find a conserved quantity:

$$\mathrm{A} \rightarrow B + C + D$$

$$A + B \rightarrow D + D + \gamma + \gamma$$

$$A + C \rightarrow \gamma + \gamma$$

$$A + D \rightarrow C + \gamma + \gamma$$

13. Estimate the amount of energy that would have to be available to measure a distance of 10^{-45} meter. How much mass would this correspond to?
14. Is it possible for an electron by itself to simply emit or absorb a photon? Explain. Is it possible for an electron by itself to simply emit or absorb one photon and then emit another, as in Figure 13-2*a*?

15. Given that the following reaction occurs:

$$\mu^- \rightarrow e^- + \bar{\nu}_e + \nu_\mu$$

explain why the following should also occur:

$$e^- + \bar{\nu}_e + \bar{\nu}_\mu \rightarrow \mu^-$$

Why would this second reaction be difficult to obtain in practice?

16. The light emitted by an atom when an electron drops from one orbit to another can be slightly different in frequency from the light emitted by another atom of the same kind and the same orbits. Explain.

17. Why are men about six feet tall, more or less?

18. Is there any way in which the earth could affect the speed of revolution or rotation of the moon, or vice versa?

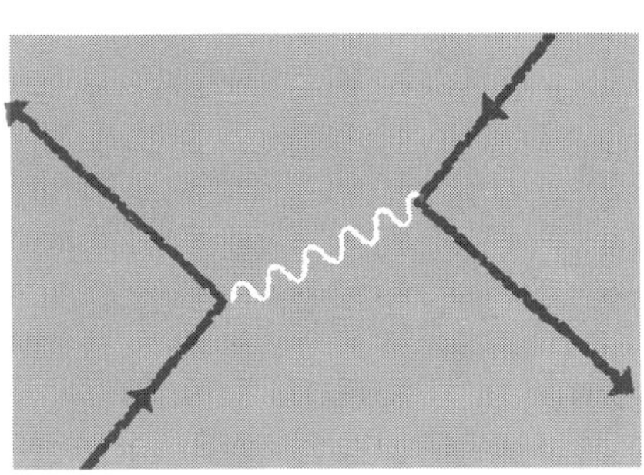

Fig. 13-24

Appendix I Systems of Units

	Meter-Kilogram-Second[a]	Centimeter-Gram-Second	Foot-Slug-Second
Length	meter	centimeter = 10^{-2} meter	foot = .305 meter
Time	second	second	second
Mass	kilogram	gram = 10^{-3} kilogram	slug = 14.59 kilogram
Charge	coulomb	statcoulomb = 3.34×10^{-10} coulomb	coulomb
Force	newton = $\frac{\text{kilogram meter}}{\text{second}^2}$	dyne = 10^{-5} newton	pound = 4.45 newton
Energy	joule = $\frac{\text{kilogram meter}^2}{\text{second}^2}$	erg = 10^{-7} joule	foot-pound = 1.36 joule

[a]The meter-kilogram-second system is used throughout this book.

APPENDIX II. SOME MATHEMATICAL CONCEPTS

In this appendix we shall consider several mathematical subjects. Some of these, on significant figures, scientific notation, and proportionality, are useful for the elementary sections of the book. The others are important only for some of the more difficult problems.

A. *Significant Figures*

The quantities that we obtain from scientific experiments are not regarded as absolutely accurate. For example, there is always some margin of error in the measurement of a distance. If several people were to try to measure the same distance as accurately as their rulers would allow, they would probably obtain slightly different results. Even if the same person, using the same ruler, were to try to make three very accurate measurements of the same distance, he would probably obtain three slightly different values. If he were measuring a distance of about a yard, for example, it would not be too surprising to find his values differing by about 1/64 inch. With more care and a better instrument the uncertainty could be reduced, but there would still be some "experimental error" of measurement. The same type of small error would also occur if several people using stopwatches were timing a 100-yard dash. Each might obtain a slightly different value. Usually in scientific work we would average the several slightly differing values obtained by measuring a quantity, unless there were some reason for believing that one or more of the values was especially good or bad.

Let us suppose that three different individuals actually did time a runner in the 100-yard dash, and obtained the following times for the event:

First timer	9.891 seconds
Second timer	9.872 seconds
Third timer	9.923 seconds

Assuming that the timers are competent, there is no particular reason to prefer one of these slightly differing times to the others. The differences are due to experimental errors. We could obtain the best value for the time by averaging these three values. If we do this we obtain 9.891 seconds. But how good is this value? The 9.89 would appear to be significant, judging from the variation in the three values obtained. The 1 at the end, however, would seem to be almost totally meaningless. The experimental error, or, one might say, the difference between any two of the three measured values, is much greater than the quantity represented by the 1 at the end of 9.881. Therefore this 1 cannot be considered a significant figure. We would say that the quantity 9.891 is good to three significant figures. The nine, the eight, and the other nine are meaningful; the one is not. For this reason we would write this quantity as 9.89 seconds, not as 9.891 seconds.

When we give quantities in this book we usually give only significant figures. The quantity 6.87 meters, for example, would be considered to have three significant figures. The quantity 2.2 seconds would have two significant figures. The quantity 2.20 seconds, however, would have three significant figures. When we write 2.20 seconds, we are implying that we have enough information to justify the zero in the third figure. If we write 2.2 seconds we suggest either that we do not have enough information to put in a third figure or that for some reason we do not care what this third figure is.

When we multiply numbers together or divide them, the results of our calculations cannot be more accurate than the numbers we put in. This means that the answers will have the same numbers of significant figures as that quantity in the calculation which has the least number of significant figures. For example, suppose we were to perform the following computation:

$$\frac{6.1 \times 9.26 \times 3.14159}{2.77 \times 0.188 \times 4.00}$$

The proper answer is 85, not 85.1908. The quantity with the fewest significant figures is the number 6.1, which has two significant figures. Therefore, the answer can have only two significant figures.

The use of significant figures in addition and subtraction can be shown by examples. Let us suppose that we have the following addition to perform. We have written only significant figures for the numbers above the line but have included all the figures in the number below the line.

$$\begin{array}{r} 807.25 \\ 22.1 \\ 3.304 \\ \hline 832.654 \end{array}$$

We find the number of significant figures in the answer by the following rule.

We look at the right side of each of the numbers to be added (or subtracted). Some of the numbers extend farther to the right than others. The number that has the *fewest* places following the decimal is the one we use to determine the last significant figure in the answer. The last significant figure in the answer will be directly beneath the last significant figure in the number that extends to the right the least. For example, in the addition above, the number with the fewest places after the decimal is 22.1. The last significant figure in this number is 1. The last significant figure in the answer is 6, which is directly below the 1. Therefore the 5 and the 4 are not significant figures, but the other figures in the answer are.

Notice that the last significant figure in the answer lies at the bottom of a column that contains a significant figure from each of the numbers to be added. Therefore it has a complete column of significant figures above it. In an addition or subtraction, the last significant figure in the answer will be the one that lies beneath the last such column on the right if such a column exists.

We write the answer to the above addition as 832.7, not as 832.6, since the whole answer, 832.654, is nearer to 832.7. In general, in a calculation, if the first nonsignificant figure in an answer is five or more, we raise the value of the last significant figure by one. If the first nonsignificant figure is less than five, we keep the last significant figure as it is. This process is called "rounding off."

Two other examples will illustrate the procedure in slightly different circumstances. First suppose that we wish to add the numbers 2.3×10^3, 1.390×10^4, and 3.30×10^2. We may find the sum as follows:

$$\begin{array}{r} 2300. \\ 13900. \\ 330. \\ \hline 16530. \end{array}$$

The last complete column of significant figures lies above the 5 in the resulting sum. Therefore, 5 is the last significant figure in the sum and the sum should be given as 1.65×10^4.

As another example let us find the sum of 37.3, 2.1, and 5.2×10^{-2}:

$$\begin{array}{r} 37.3 \\ 2.1 \\ 0.052 \\ \hline 39.452 \end{array}$$

In this case there is no complete column of significant figures. However, the zeros to the left of the 5 in the number 0.052 cannot cause any inaccuracy in the sum. Therefore, the 4 is last significant figure in the resulting sum. Rounding off, we have 39.5.

The procedure we have illustrated here for finding the significant figures in an addition process is also valid for subtraction.

B. Scientific Notation for Numbers

Often in science we must use numbers that are very large or very small. For example, the average distance from the earth to the sun is

$$149{,}680{,}000{,}000 \text{ meters}$$

Only the first five of these twelve figures are significant, because only these first five are accurately known. Another distance that is important in atomic physics is the approximate distance between the particles in the hydrogen atom:

$$0.000\ 000\ 000\ 052\ 917\ 2 \text{ meters}$$

Only the last six of these figures are significant. (The zeros to the left of the five do not contribute to the accuracy of this length and are not counted as significant figures.) Each of these quantities may be expressed in a more compact and convenient form that does not use so many zeros.

The method for representing very large and very small numbers compactly and conveniently depends on the use of powers of ten. We shall explain the use of powers of ten by a few examples. The number "ten squared" is just ten multiplied by itself:

$$\text{ten squared} = 10 \times 10 = 10^2 = 100$$

The number "ten cubed" is 10^2 multiplied by ten:

$$\text{ten cubed} = 10 \times 10 \times 10 = 10^3 = 1000$$

We often call 10^2, ten raised to the second power. Similarly, 10^3 is called, ten raised to the third power. Then,

$$\text{ten raised to the fourth power} = 10 \times 10 \times 10 \times 10 = 10^4 = 10{,}000$$

and

$$\text{ten raised to the seventh power} = 10 \times 10 \times 10 \times 10 \times 10 \times 10 \times 10$$
$$= 10^7 = 10{,}000{,}000.$$

We notice that the number 7 written above and to the right of the number 10 in the last example corresponds to the seven zeros found to the right of the 1 in the number 10,000,000. This relation is correct for positive powers. For example, 10^4 is equal to 10,000 which is one followed by four zeros. By this general rule, 10^1 is equal to 10, and 10^0 is equal to 1:

$$10^1 = 10$$
$$10^0 = 1$$

We can use a similar technique to express numbers of the form 0.01, 0.0001, etc. We do this by defining what is meant by negative powers of ten. Ten raised to a negative power is just one over ten raised to the same positive power. In other words,

$$0.1 = 1/10 = 1/10^1 = 10^{-1}$$
$$0.01 = 1/100 = 1/10^2 = 10^{-2}$$
$$0.001 = 1/1000 = 1/10^3 = 10^{-3}$$
$$0.0001 = 1/10{,}000 = 1/10^4 = 10^{-4}$$
$$\cdots\cdots$$
$$0.000\,000\,1 = 1/10{,}000{,}000 = 1/10^7 = 10^{-7}$$

The negative seventh power of ten is thus just one over ten to the positive seventh power. It represents the number with the figure one in the seventh decimal place, and is thus preceded by six zeros to the right of the decimal point.

Using these techniques we can represent numbers in compact form. We shall start with the following equation:

$$12 = 1.2 \times 10$$

This is very simple, and we can follow it with

$$120 = 1.2 \times 100 = 1.2 \times 10^2$$
$$1200 = 1.2 \times 1000 = 1.2 \times 10^3$$

or, to use higher numbers,

$$1{,}200{,}000 = 1.2 \times 1{,}000{,}000 = 1.2 \times 10^6$$
$$42{,}300{,}000 = 4.23 \times 10{,}000{,}000 = 4.23 \times 10^7$$

For very small numbers, we can use a similar technique:

$$0.12 = 1.2 \times 0.1 = 1.2 \times 1/10 = 1.2 \times 1/10^1 \times 1.2 \times 10^{-1}$$
$$0.012 = 1.2 \times 0.01 = 1.2 \times 1/100 = 1.2 \times 1/10^2 = 1.2 \times 10^{-2}$$

or, for even smaller numbers,

$$0.000\,000\,12 = 1.2 \times 0.000\,000\,1 = 1.2 \times 10^{-7}$$
$$0.000\,000\,000\,375 = 3.75 \times 0.000\,000\,000\,1 = 3.75 \times 10^{-10}$$

Thus we can represent any number by a number in which there is only one figure to the left of the decimal point, multiplied by some power of ten. If the original number is less than one, then the power of ten will be negative. Otherwise the power of ten will be zero or positive. The number one itself can be represented as

$$1 = 1 \times 10^0$$

This method of representing numbers is very convenient because it displays the significant figures accurately. For example, we might not know exactly how accurately a number is known if it were represented as

$$3{,}703{,}000$$

However, we know immediately how many significant figures the number has if it is represented as

$$3.7030 \times 10^6$$

It is clear that this number has five significant figures. The zero on the end is put there because it is significant, and not, as before, because it is necessary to express the number itself.

In order to use powers of ten in calculations, it is necessary to know how to multiply and divide exponential figures. The basic rules are as follows:

$$10^m \times 10^n = 10^{m+n} \tag{1}$$

$$\frac{10^m}{10^n} = 10^{m-n} \tag{2}$$

$$\frac{1}{10^n} = 10^{-n} \tag{3}$$

In these equations, n and m are two numbers, either positive or negative. Equations 1-3 give the usual rules for the use of powers and apply to the powers of any non-zero number, and not just to powers of ten.

Example 1

Find the ratio of the distance between the earth and the sun to the distance between the particles in the hydrogen atom.

Solution. The ratio we wish to obtain is

$$\frac{149{,}680{,}000{,}000 \text{ meters}}{0.000\ 000\ 000\ 052\ 917\ 2 \text{ meter}}$$

First we express the numbers using powers of ten:

$$\frac{1.4968 \times 10^{11} \text{ meter}}{5.29172 \times 10^{-11} \text{ meter}} = \frac{1.4968}{5.29172} \times \frac{10^{11}}{10^{-11}}$$

Then we perform the division of the two numbers, retaining five significant figures since the number with the fewest significant figures had five.

$$0.28286 \times \frac{10^{11}}{10^{-11}}$$

Now we must evaluate the exponent, using equations 1-3:

$$\frac{10^{11}}{10^{-11}} = 10^{11-(-11)} = 10^{22}$$

therefore the ratio is

$$0.28286 \times 10^{22}$$

which is equal to

$$2.8286 \times 10^{21}$$

EXAMPLE 2

Perform the computation indicated:

$$\frac{2.1 \times 10^{13} \times 653}{0.0000372169 \times 53}$$

Solution. We first express each of the numbers using powers of ten:

$$\frac{2.1 \times 10^{13} \times 6.53 \times 10^{2}}{3.72169 \times 10^{-5} \times 5.3 \times 10}$$
$$= \frac{2.1 \times 6.53}{3.72169 \times 5.3} \times \frac{10^{13} \times 10^{2}}{10^{-5} \times 10}$$

Performing the numerical computation, we obtain an answer good to two significant figures:

$$0.70 \times \frac{10^{13} \times 10^{2}}{10^{-5} \times 10}$$

Computing the power of ten we have

$$0.70 \times \frac{10^{15}}{10^{-4}}$$
$$= 070 \times 10^{15-(-4)}$$
$$= 0.70 \times 10^{19}$$

which is equal to

$$= 7.0 \times 10^{18}$$

C. Proportion

The concept of proportion is important in physics because many laws of physics state that certain quantities are proportional to other quantities. For example, the behavior of an object traveling at a constant speed in a given direction can be expressed as a simple proportion. If an object is traveling at three meters per second, it will travel three meters in one second, six meters in two seconds, nine meters in three seconds, etc. In fact, we can make a table of these values (Table A-1).

Table A-1

Distance (in meters)	Time (in seconds)
0	0
3	1
6	2
9	3
12	4
15	5
18	6
21	7
24	8
etc.	etc.

The distance traveled during any period of time can be found if the period of time is known. The distance and the time are therefore related to each other. If the time is doubled, the distance is doubled also. If the time is tripled, the distance traveled is tripled also. If the time is multiplied by four, the distance is multiplied by four also. This behavior is characteristic of *proportion.* Any two quantities that change together so that if one is doubled (or tripled, etc.) the other is doubled (or tripled, etc.) are said to be *proportional* to each other. Thus the distance is proportional to time for an object moving at a constant speed in a given direction. As another example, the number of hours in a period of time is proportional to the number of minutes in that same period of time. If we decide to spend three times as many hours as usual at the beach, we spend three times as many minutes also.

The concept of proportion can be precisely expressed in an equation in two ways. First, suppose that there are two quantities, x and

y, which can take on different values. If x and y are proportional to each other, we can write an equation expressing this relation as follows:

$$y = kx \tag{4}$$

The letter k represents a constant quantity—a quantity that remains the same whatever values x and y may have. For example, if we once again consider an object that is moving at a speed of three meters per second, we may write

$$d = kt \tag{5}$$

In this case d represents the distance traveled, t the time, and k the constant of proportionality. In this equation the constant k is just the constant speed s of the object. Thus we can also write

$$d = st \tag{6}$$

In the example chosen, the speed s is equal to three meters per second. We may write, therefore

$$d = (3 \text{ meter/second}) \cdot t \tag{7}$$

Now if we put in two seconds for t, we obtain six meters for d. If we put in three seconds for t, we obtain nine meters for d. This gives us two of the lines in Table A-1. We may obtain the other lines in a similar way. Thus the equation expresses the same relation of proportionality as the table in a more complete, compact, and convenient way. This type of proportion is called *direct proportion*, to distinguish it from other proportional relations. When the word *proportion* is used without any qualifications, however, it refers to direct proportion, the type of relation shown by Equation 4.

There is another way to express this direct proportion in an equation. Suppose we choose two pairs of corresponding values of x and y. We shall use the numbers 1 and 2 as subscripts on the letters x and y to represent these values. We may write the proportional relationship between these quantities as

$$y_1 = k\,x_1 \tag{8}$$

$$y_2 = k\,x_2 \tag{9}$$

If we divide the sides of Equation 8 by the corresponding sides of Equation 9, we obtain

$$\frac{y_1}{y_2} = \frac{k\,x_1}{k\,x_2}$$

The k's cancel out, giving us

$$\frac{y_1}{y_2} = \frac{x_1}{x_2} \tag{10}$$

This equation is another way of representing the direct proportion.

Proportionality may also be represented by means of a graph. Figure A-1 is a graph made using Table A-1. We have marked off equal distances representing equal periods of time on the horizontal line (or horizontal axis). We have also marked off equal distances on the vertical line (or vertical axis) representing equal distances traveled by the object. The point at which these two lines cross is called the origin. We have marked a point on the graph for each of the first five lines of Table A-1. The distance of a point from the vertical line represents a time, taken from a line of Table A-1. The distance of the same point from the horizontal line represents a distance traveled, taken from the same line of Table A-1. The important thing to notice about the graph is that the points all fall on a straight line that goes through the origin. This is characteristic of direct proportion. Any direct proportion can be graphed as a straight line going through the origin, and any such graph is the graph of a direct proportion. If any two quantities are related to each other so that their graph is a straight line through the origin, then those two quantities are proportional. The constant k, in Equation 4, determines the slope of the straight-line graph. This may be most easily seen in Figure A-2, which gives two different graphs of Equation 1 for two different values of k. The larger the value of k, the steeper the graph.

If two quantities are in direct proportion, then either both quantities increase together or both quantities decrease together. In another type of proportion, called *inverse proportion*, one quantity increases as the other decreases. For example, suppose that we are adding water to a sugar solution. The more water we add, the weaker the

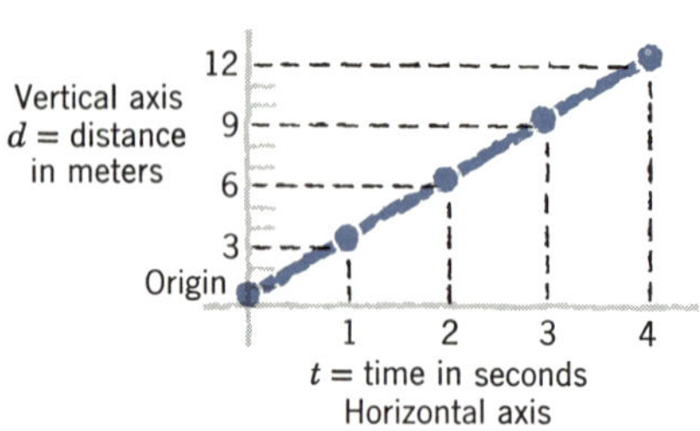

Fig. A-1

solution becomes. In fact, the concentration (or strength) of the solution is inversely proportional to the amount of water in it. (It is also directly proportional to the amount of sugar.)

There are two ways of expressing an inverse proportion mathematically. They are, for two quantities x and y,

$$y = k/x \tag{11}$$

$$\frac{y_1}{y_2} = \frac{x_2}{x_1} \tag{12}$$

Notice that the x appears in the denominator for the inverse proportion (Equation 11) but not for the direct proportion (Equation 4). Notice also that Equation 12 is different from Equation 10 in that one side has been turned upside down. The graph of Equation 11 for inverse proportion is as shown in Figure A-3. The graph shows y decreasing as x increases.

There are two other kinds of proportion that we shall consider briefly here because they are involved in physical applications. Often we speak of one quantity as proportional to the square of another quantity. The equations for this kind of relationship are as follows:

$$y = kx^2 \tag{13}$$

$$\frac{y_1}{y_2} = \frac{x_1^{\,2}}{x_2^{\,2}} \tag{14}$$

This relationship is similar to the direct proportion, except that the x is squared. We express this relationship by saying that y is proportional to the square of x, or proportional to x squared. The quantity y increases as the quantity x increases, but it increases at an ever-increasing rate. An example of a graph for this kind of relationship is

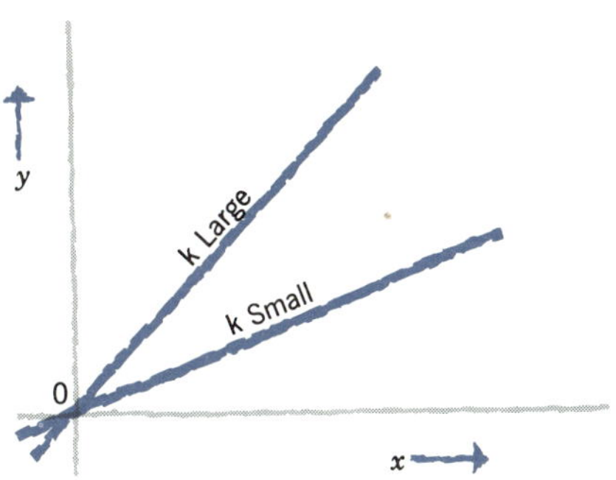

Fig. A-2

shown in Figure A-4. (Compare this graph with the graph for direct proportion, Fig. A-1.)

The other type of proportion is expressed by saying that a quantity y is inversely proportional to the square of another quantity, x. The equations for this relationship are as follows:

$$y = k/x^2 \tag{15}$$

$$\frac{y_1}{y_2} = \frac{x_2^{\,2}}{x_1^{\,2}} \tag{16}$$

This relationship is similar to inverse proportion, except that x appears squared. An example of a graph for this kind of relationship is shown in Figure A-5. (Compare Fig. A-5 with the graph for inverse proportion, Fig. A-3.)

D. The Pythagorean Theorem

The Pythagorean theorem, a very famous and important law of geometry, is useful in solving problems that involve *right triangles.* Right triangles are triangles that have two perpendicular sides. The angle found where these two perpendicular sides join is called a *right angle.*

The letters in Figure A-6 represent the lengths of the sides and the angles of a right triangle. It is customary to label the right angle with the capital letter C. The side opposite the right angle is called the *hypotenuse*, and its length is represented by the small letter c. The sides of the triangle next to the right angle are called the legs, and their lengths are represented by the small letters a and b. With these symbols the Pythagorean theorem can be stated as follows:

$$a^2 + b^2 = c^2 \tag{17}$$

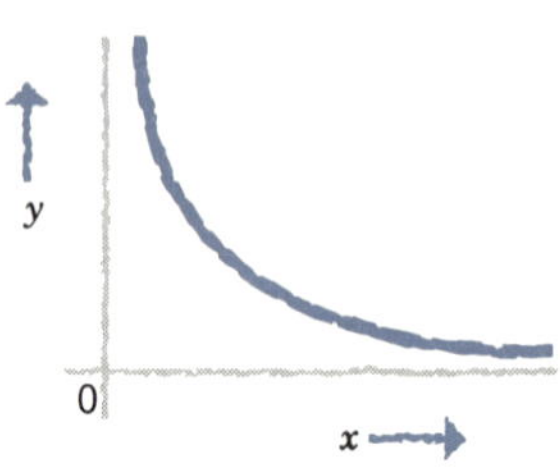

Fig. A-3

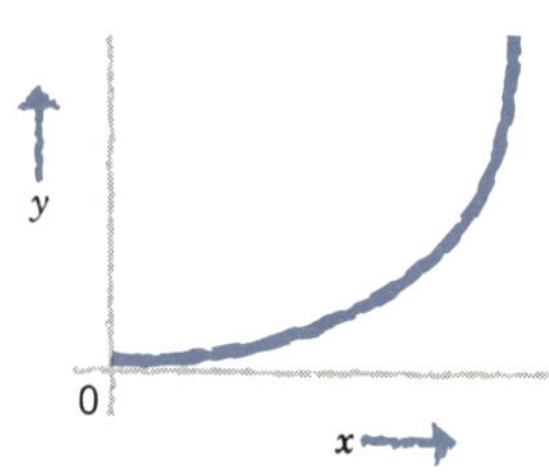

Fig. A-4

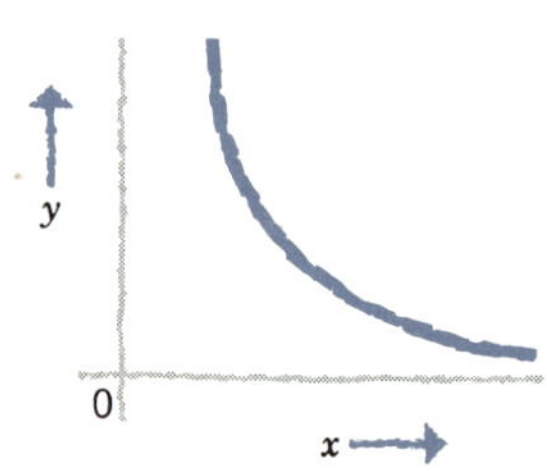

Fig. A-5

Thus the theorem states that the square of the length of the hypotenuse is equal to the sum of the squares of the lengths of the other two sides. If we know any two sides of a right triangle, we can find the other one by using Equation 17. Note that this equation holds only for *right* triangles, and that c must be the length of the hypotenuse, *not* of one of the legs.

There are two simple kinds of right triangles that we shall describe here. The first of these is the right triangle in which the two legs are equal in length (see Fig. A-7). This means that the two angles at A and B are also equal to each other; in fact each is equal to 45 degrees. There is a simple relation between the length of the hypotenuse of such a triangle and the length of its legs. We can easily find this relationship using the Pythagorean theorem. Let us use c to represent the length of the hypotenuse and a to represent the length of the legs. (Since the legs are of equal length, $b = a$.) According to the Pythagorean theorem:

$$a^2 + a^2 = c^2 \tag{18}$$

$$2a^2 = c^2 \tag{19}$$

Taking the square root of both sides of Equation 19, we obtain:

$$a\sqrt{2} = c \tag{20}$$

This relationship between the legs and the hypotenuse can be used to find either one of them if the other is known. It holds for all right triangles, regardless of size, which contain a 45-degree angle. Thus we can use it for all right triangles in which one leg is equal to the other.

Figure A-8 shows another very useful type of right triangle, the so-called 30 degree-60 degree right triangle. One of the angles is a 30-degree angle and another is a 60-degree angle. (The right angle is a 90-degree angle.) This triangle has a hypotenuse whose length is just twice the length of the shortest leg (see Fig. A-8). If we call the length of the short leg a, then the hypotenuse must be of length $2a$.

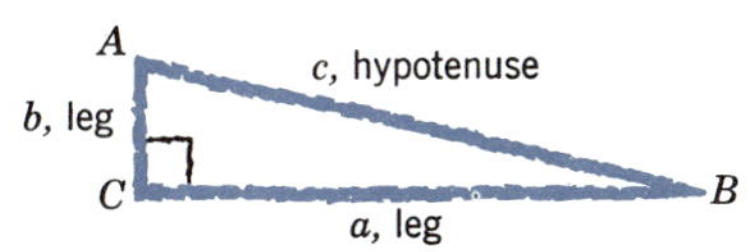

Fig. A-6

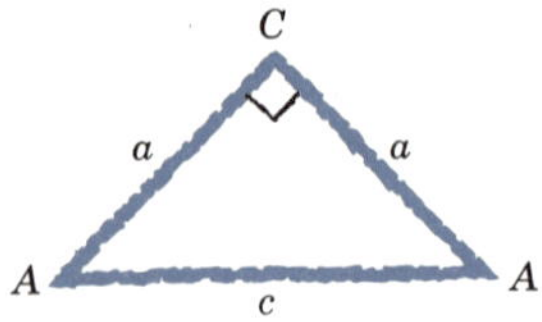

Fig. A-7

To find the length of the side b, we can use the Pythagorean theorem as follows:

$$a^2 + b^2 = (2a)^2 \tag{21}$$

$$a^2 + b^2 = 4a^2 \tag{22}$$

$$b^2 = 3a^2 \tag{23}$$

Taking the square root of each side, we obtain:

$$b = a\sqrt{3} \tag{24}$$

This relation holds for any 30 degree-60 degree right triangle, regardless of size.

E. Similar Triangles

The relations that hold between the sides of similar triangles are also useful in solving physical problems. The basic relation may be stated as follows: if the angles of one triangle are equal, respectively, to the angles of another triangle, then the sides of the one triangle are proportional, respectively, to the sides of the other triangle. The converse is also true; if the sides are respectively proportional, then the angles are respectively equal. These facts are illustrated in Figure A-9, which shows three triangles which are all similar to one another. We have used the same letters, A, B, and C, to represent the vertices of these three triangles, since the angles at these points are

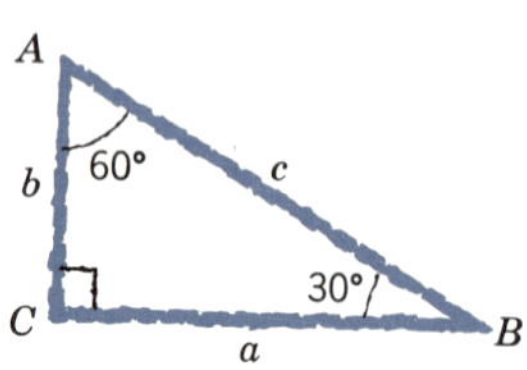

Fig. A-8

equal. Thus all the angles at vertices labeled A are equal, etc. We have labeled the sides opposite these vertices using the small letters a_1, a_2, a_3, etc., depending on which triangle is intended. These sides are not equal, but they are proportional according to the following equations:

$$\frac{a_1}{a_2} = \frac{b_1}{b_2} = \frac{c_1}{c_2} \tag{25}$$

$$\frac{a_1}{a_3} = \frac{b_1}{b_3} = \frac{c_1}{c_3} \tag{26}$$

$$\frac{a_2}{a_3} = \frac{b_2}{b_3} = \frac{c_2}{c_3} \tag{27}$$

**APPENDIX III. THE MATHEMATICAL DERIVATION OF THE FORMULAS OF THE SPECIAL THEORY OF RELATIVITY

In this appendix we shall obtain the equations that relate the description of an event in one reference frame to the description of an event in another reference frame moving at a constant velocity with respect to the first. These equations taken together are called the Lorentz transformation, after H. A. Lorentz, who first obtained them. We shall also obtain other relations arising in the special theory of relativity.

We shall assume Einstein's two postulates (see Chap. 10, pp. 397–398) and some symmetry principles. In order to obtain the mass increase with velocity, we shall assume conservation of momentum as well.

An event can be specified by three items of information that fix its location in space, and one item that fixes its position in time. We shall describe the position of a point in space by using the three let-

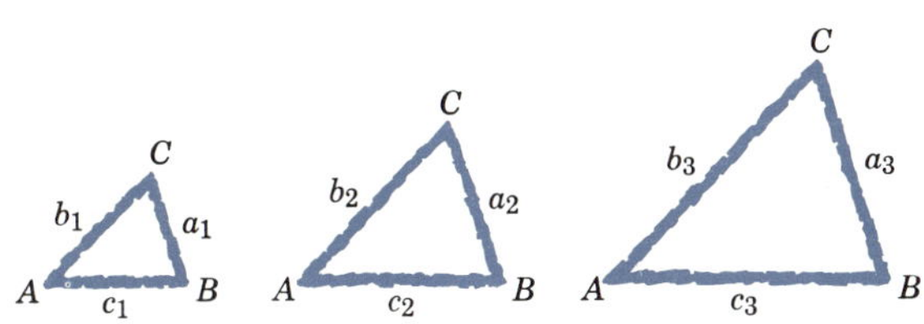

Fig. A-9

ters x, y, and z to represent the distances to the point along the three axes, called the X, Y, and Z axes, as shown in Figure A-10. For simplicity we shall use the two reference frames in relative motion shown in Figure A-11. The frame designated S is called the unprimed reference frame, and the unprimed letters x, y, and z locate points in it. The frame designated S' is called the primed reference frame, and the primed letters x', y', and z' locate points in it. The letters t and t' represent the time of events as observed in the S and S' frames, respectively.

The primed frame is moving to the right along the x axis of the unprimed frame, at a speed v. (Notice that v refers here to a speed, and that v^2 is the square of this speed.) When the times, t and t', both equal zero, the two frames have the same origin, and in fact fall exactly on top of each other, so that all their axes are the same. As time passes, the primed frame moves out to the right. The Z and Z' axes extend perpendicular to the plane of the paper, in the direction of the reader.

The S' frame is moving with a speed v to the right as viewed from the S frame. This means that the S frame is moving in the opposite direction, with a speed v, (that is, a speed v to the left) as viewed from the S' frame. If this were not the case, then there would be a difference between the two directions in space, which would violate one of our symmetry principles.

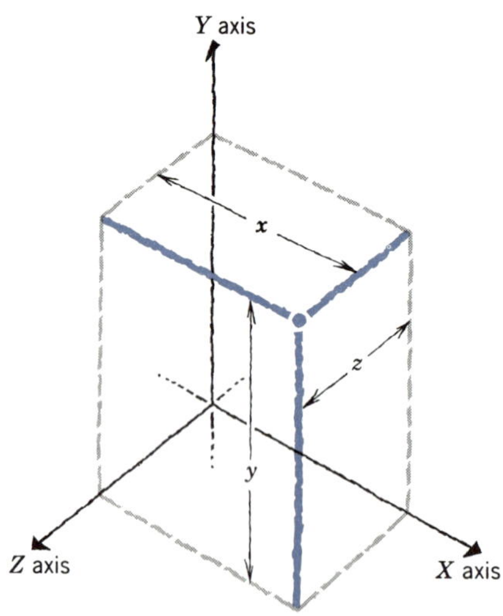

Fig. A-10

We shall find it convenient to define a new symbol, the Greek letter gamma, as follows:

$$\gamma = \frac{1}{\sqrt{1 - v^2/c^2}} \tag{1}$$

In Chapter 10, we obtained the time dilation by analyzing the behavior of a light clock (see Example 4, p. 437). We obtained the Lorentz contraction from the time dilation and the fact that the relative speed of the two reference frames is the same as viewed from either frame (see Chap. T10, p. 405). Thus we have already derived the following formulas:

$$T' = \gamma T \tag{2}$$

$$L = \gamma L' \tag{3}$$

(Note that we have used the primed and unprimed quantities in equations 2 and 3, in such a way that the ruler or clock is at rest in the unprimed system.)

The Lorentz Transformation

Suppose that we have an event that is observed from both of our reference frames. It will have a specification x, y, z, and t in the S frame, and a specification x', y', z', and t' in the S' frame. We wish to find the equations that relate these two sets of quantities.

First of all, we state that the distances perpendicular to the direction of relative motion are observed to be the same in each reference frame. We can see that this must be true by considering what would happen if they were not. In Figure A-12 we have two cylinders, one at rest in each frame. The radii of the two cylinders, as measured for each in its own frame, are the same. If the radius of one of the cylinders appeared contracted as viewed from the other frame, then as the two systems went by each other, one of the cylinders would be seen

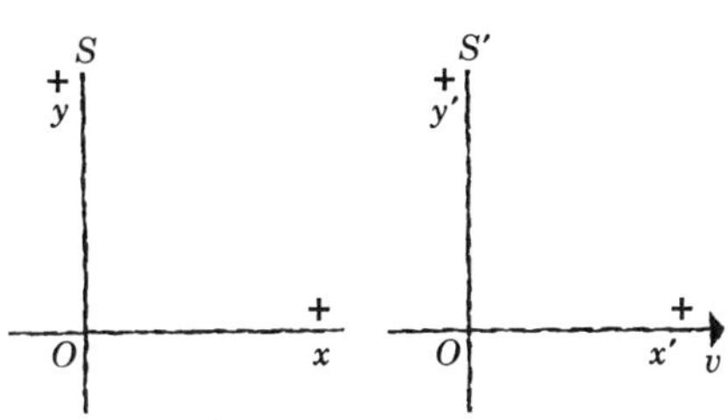

Fig. A-11

to pass inside the other. Let us suppose that cylinder C' passed inside cylinder C, as observed from the S frame. What will we see from the S' frame? If we also see cylinder C' passing through cylinder C, then we can define a special direction in space along the X axes of the two frames—a cylinder coming from one direction passes over a cylinder coming from the other. This distinction would violate one of our symmetry principles. If, on the other hand, we saw cylinder C pass inside cylinder C' as observed from the S' frame, while still seeing the opposite from the S frame, we would have a physical impossibility. Suppose, for example, that cylinder C was solid while cylinder C' was hollow. Then the occurrence would be perfectly feasible as observed from one frame, but impossible as viewed from the other. Therefore, the assumption that one cylinder passes over the other must be false. This leaves us with only one conclusion—the radii of the cylinders are the same as observed from either frame. Thus distances perpendicular to the X axis are unchanged, and

$$y = y' \tag{4}$$

$$z = z' \tag{5}$$

Now we put a measuring rod in the unprimed system, which reaches along the X axis from the origin to a point at a distance x from the origin (see Fig. A-13). An observer in the S' frame touches the right end of this rod at a certain time t', thus specifying an event. Because of the Lorentz contraction, the observer in the S' frame thinks that the length of the measuring rod is $\frac{1}{\gamma}x$. He also thinks that the separation of the two origins after the time t' will be vt'. At the time t' the end of the ruler is at the point we shall call x' in the S' system.

From Figure A-13 we can see that the following relation must hold:

$$x' + vt' = \frac{1}{\gamma}x \tag{6}$$

$$x = \gamma(x' + vt') \tag{7}$$

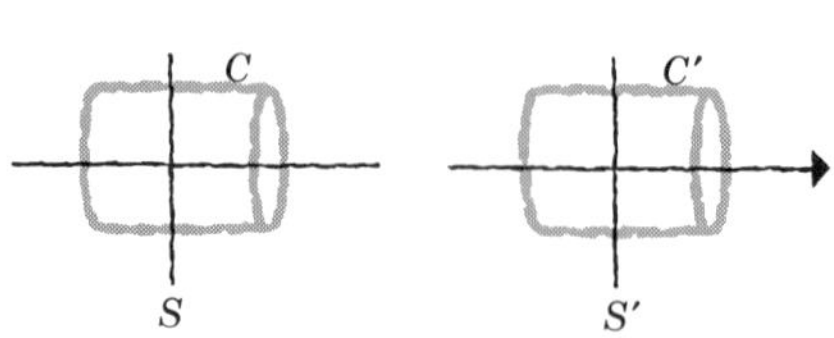

Fig. A-12

Now let us try the same scheme in the primed system. We put a measuring rod in the primed system, so that it reaches from the origin out to a distance x' along the X' axis (see Fig. A-14). Then the observer in S thinks that this length must be $\frac{1}{\gamma}x'$, and

$$\frac{1}{\gamma}x' + vt = x \tag{8}$$

$$x' = \gamma(x - vt) \tag{9}$$

Thus we have found the two equations:

$$x = y\,(x' + vt') \tag{10}$$

$$x' = \gamma\,(x - vt) \tag{11}$$

We may obtain two other relations by using algebra to eliminate first x and then x' from Equations 10 and 11:

$$\begin{aligned}
x' &= \gamma(x - vt)\\
x' &= \gamma x - \gamma vt\\
x &= \gamma x' + \gamma vt'\\
\frac{1}{\gamma}x' &= x - vt\\
\frac{1}{\gamma}x' + x &= \gamma x' + x + \gamma vt' - vt\\
vt &= \gamma vt' + \gamma x' - \frac{1}{\gamma}x'\\
&= \gamma\left(vt' + \left[1 - \frac{1}{\gamma^2}\right]x'\right)\\
vt &= \gamma(vt' + \left(1 - 1 + \frac{v^2}{c^2}\right)x')\\
t &= \gamma\left(t' + \frac{v}{c^2}x'\right) \qquad (12)
\end{aligned}$$

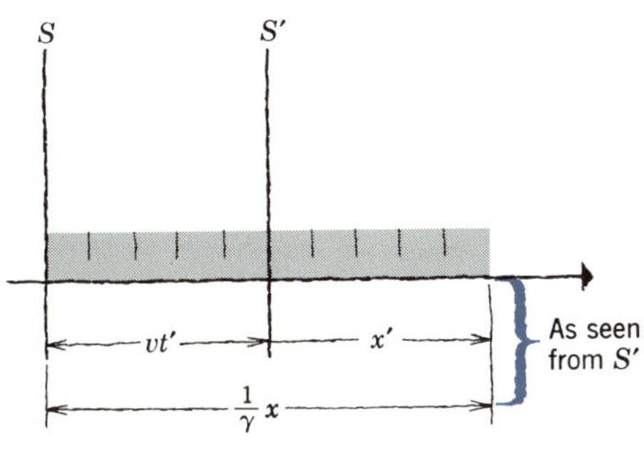

Fig. A-13

Similarly,

$$\frac{1}{\gamma}x = x' + vt'$$
$$x' = \gamma x - \gamma vt$$
$$\frac{1}{\gamma}x = vt' + \gamma x - \gamma vt$$
$$vt' = \gamma vt + \left(-\gamma + \frac{1}{\gamma}\right)x$$
$$t' = \gamma\left(t + \frac{1}{v}\left[-1 + \frac{1}{\gamma^2}\right]x\right)$$
$$t' = \gamma\left(t - \frac{v}{c^2}x\right) \tag{13}$$

The equations we have found give us the relations we are seeking:

$$\begin{aligned} x' &= \gamma(x - vt) \\ y' &= y \\ z' &= z \\ t' &= \gamma\left(t - \frac{v}{c^2}x\right) \end{aligned} \qquad \text{or} \qquad \begin{aligned} x &= \gamma(x' + vt') \\ y &= y' \\ z &= z' \\ t &= \gamma\left(t' + \frac{v}{c^2}x'\right) \end{aligned} \tag{14}$$

These equations relate the events in one system to the events in the other.

The Transformation of Velocities

We ought to be able to use the equations obtained above to answer questions relating to distance and time in the two frames. Let us suppose that a given object is moving to the right in the primed frame, S', with a velocity u' (see Fig. A-15). We would like to know u, the velocity of the same object in the unprimed frame, S.

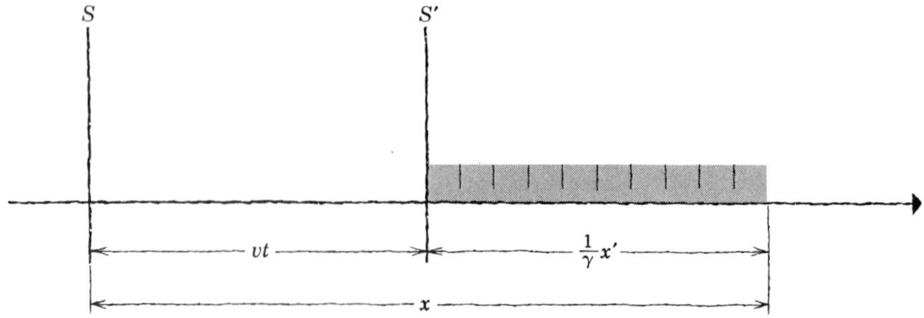

Fig. A-14

The velocity u' represents a distance d' divided by a time t' in the primed system. Let us suppose that the object left the y axis when the two systems were together, at time $t = t' = 0$, and that it has proceeded to a point x' in the primed system at time t'. Then we must have

$$u' = \frac{x'}{t'} \tag{15}$$

The velocity u in the unprimed system must also be the distance divided by the time, so

$$u = \frac{x}{t} = \frac{\gamma\,(x' + vt')}{\gamma\left(t' + \dfrac{v}{c^2}\,x'\right)} = \frac{\left(\dfrac{x'}{t'} + v\right)}{\left(1 + \dfrac{v}{c^2}\dfrac{x'}{t'}\right)} \tag{16}$$

$$u = \frac{u' + v}{1 + \dfrac{u'v}{c^2}} \tag{17}$$

We note that u is not equal to $u' + v$. If $u' = c$, the speed of light, then

$$u = \frac{c + v}{1 + \dfrac{cv}{c^2}} = \frac{c + v}{\dfrac{1}{c}\,(c + v)} = c \tag{18}$$

We find that the velocity is also c in the unprimed reference frame.

The Formula for Increase of Mass with Velocity[1]

We assume that we have an object of mass m_0 at rest at the origin in each frame; that is, each mass is measured in its own frame and

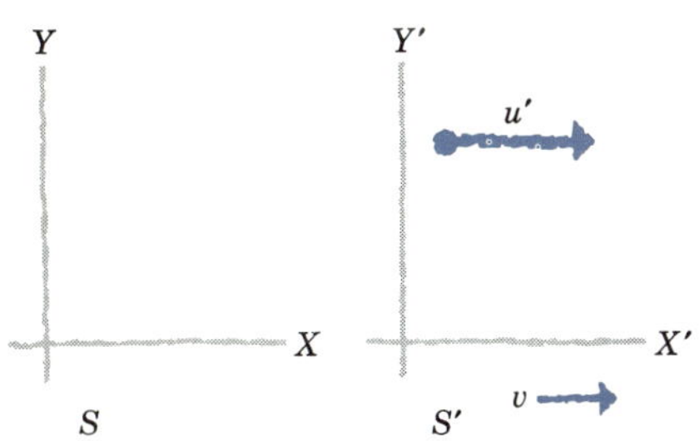

Fig. A-15

[1]See Eric Rogers, *Physics for the Inquiring Mind,* Princeton, N. J.: Princeton University Press, 1960, pp. 485–486. Our thought experiment is similar to that of Rogers, who mentions that he is following a suggestion by Tolman.

found to have rest mass m_0. As the masses go by each other, an interaction takes place between them. This interaction may arise from the momentary force of a string or spring connecting them, or it may be caused by identical particles fired by each into the other. Each mass will acquire a momentum, and therefore a velocity. For convenience, suppose that this resulting velocity is up the Y axis for the object originally at rest in the unprimed frame. Then it turns out that the velocity of the other mass in the primed system will be down the Y' axis, with the same speed that the other has in going up the Y axis in the S frame.

We can show that this must be true by considering a third reference frame in which both masses are originally approaching each other at equal speeds (see Fig. A-16). (We have not drawn in the axes of this third reference frame but indicate the other two reference frames, S and S', both in motion in opposite directions. Before the interaction, the two objects have zero total momentum, since they are approaching each other at equal speeds. Therefore, they must have zero total momentum after they interact also. Thus the momenta of the two objects after the interaction must be equal in magnitude and opposite in direction. This also must be true of the components of momentum along any axis. Since we know the object in the S frame is traveling up the Y axis, we must have the object in the S' frame traveling down the Y' axis at an equal speed, as seen from our third system. Now we wish to look at the object moving along the Y axis from the point of view of the S frame, and the object moving along the Y' axis from the point of view of the S' frame. This involves changes of reference frame to each of the two frames, S and S'. Because of the symmetry of the situation, the new speed, s, of each object as observed in its respective frame must be the same.

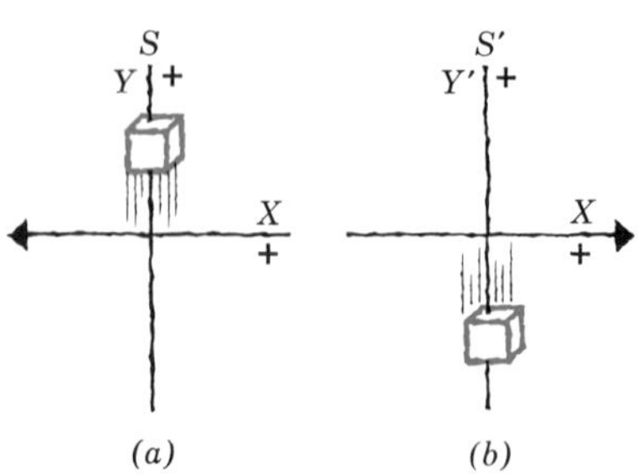

Fig. A-16

Let us now look at things from the S frame. The observer in the S frame writes the law of conservation of momentum (y-component):

$$m_0 s = m\ u \tag{19}$$

The symbols m_0 and s represent the mass and speed of the object moving up the Y axis, and the letters m and u represent the mass and y-speed of the other object as observed from the S frame. According to the observer in the S frame, the speed u is obtained from the distance y traveled along the Y' axis during the time t since the interaction. Thus

$$m_0 s = m\ y/t \tag{20}$$

The observer in the S' frame, however, will calculate one of these values differently. According to Equation 14

$$y = y' \tag{21}$$

$$t = \gamma\ (t' + vx'/c^2) = \gamma\ t', \text{ since } x' = 0 \text{ along the } Y' \text{ axis} \tag{22}$$

Thus we have

$$m_0 s = m\ y/t = m\ y'/\gamma t' = \frac{1}{\gamma} m\ s \tag{23}$$

This last step follows from the fact that y'/t' is the speed of the object traveling along the Y' axis as observed from the S' frame, and therefore $\frac{y'}{t'}$ is also equal to s. Dividing both sides by s, we have

$$m_0 = \frac{1}{\gamma} m \tag{24}$$

$$m = \gamma m_0 = \frac{m_0}{\sqrt{1 - \frac{v^2}{c^2}}} \tag{25}$$

We have used m_0 for the mass of the object moving along the Y axis as seen from the S frame. This is not really the rest mass, since the object is actually moving. Nothing about our reasoning depended on the speed s, however, so we can take s extremely small. In the limiting case we obtain Equation 25 for the equation relating the mass of a moving object to its rest mass.

Relativistic Energy

The expression for the mass, Equation 25, has an interesting relation to energy.

From Equation 25 we obtain

$$m\sqrt{1 - \frac{v^2}{c^2}} = m_o \tag{26}$$

$$m^2\left(1 - \frac{v^2}{c^2}\right) = m_o^2$$

$$m^2\left(1 - \frac{v^2}{c^2}\right)\left(1 + \frac{v^2}{c^2}\right) = m_o^2\left(1 + \frac{v^2}{c^2}\right) \tag{27}$$

$$m^2\left(1 - \frac{v^4}{c^4}\right) = m_o^2\left(1 + \frac{v^2}{c^2}\right)$$

We shall assume that the speed, v, is much smaller than the speed of light, c, so that $\frac{v^2}{c^2}$ is small and therefore such terms as $\frac{v^4}{c^4}$ are very small. Therefore we may add or delete these terms as follows and still have approximately correct equations:

$$m^2 \approx m_o^2\left(1 + \frac{v^2}{c^2} + \frac{1}{4}\frac{v^4}{c^4}\right) \tag{28}$$

$$m^2 \approx m_o^2\left(1 + \frac{v^2}{2c^2}\right)^2 \tag{29}$$

$$m \approx m_0\left(1 + \frac{v^2}{2c^2}\right) = m_0 + \frac{1}{c^2}\frac{1}{2}m_0v^2 \tag{30}$$

or,

$$mc^2 = m_0c^2 + 1/2\, m_0v^2 \tag{31}$$

or,

$$mc^2 - m_0c^2 = 1/2m_0v^2 = \text{Kinetic Energy} \tag{32}$$

This result strongly suggests two things.

1. $mc^2 - m_0c^2$ is also the kinetic energy even when v/c is large. Then the formula $1/2\ mv^2$ is no longer correct for the kinetic energy.
2. The term m_0c^2 itself also represents some form of energy. In other words, perhaps there are processes in which this quantity, called the rest energy, can change to release more kinetic energy.

Both these possibilities turn out to be true experimentally. In general, then, we can say

$$E = mc^2 = \text{Total Energy} \tag{33}$$

$$mc^2 - m_0c^2 = \text{Kinetic Energy} \tag{34}$$

We should like here to list the conserved quantities of momentum and energy as they appear in special relativity.

$$p_x = mv_x = \frac{m_0 v_x}{\sqrt{1 - v^2/c^2}} = m_0 \gamma v_x \tag{35}$$

$$p_y = mv_y = m_0 \gamma v_y \tag{36}$$

$$p_z = mv_z = m_0 \gamma v_z \tag{37}$$

$$E = mc^2 = m_0 \gamma c^2 \tag{38}$$

The terms v_x, v_y, and v_z represent the three components of velocity; p_x, p_y, and p_z represent the three components of momentum; and E represents the energy of an object. An interesting relation between these quantities is given by the expression

$$E^2 - p^2 c^2 = m_0{}^2 c^4 \tag{39}$$

$$p^2 = p_x{}^2 + p_y{}^2 + p_z{}^2 \tag{40}$$

in which p^2 is just the square of the magnitude of the momentum vector. Equation 39 can be proved simply by substituting the values for the energy and the components of the momentum found in Equations 35 to 38 and 40 into Equation 39.

Energy and momentum are very closely associated in relativity theory. Actually, we could define

$$p_t = \frac{E}{c} \tag{41}$$

Then we have four components of momentum, the energy furnishing us with the component in the "time direction." Thus the three-dimensional momentum vector becomes expanded into a four-dimensional vector with four components. The three-dimensional vector conservation of momentum law and the scalar conservation of energy law are thus combined into one four-dimensional vector conservation law for momentum and energy. Because of the mass-energy equivalence, this contains the law of conservation of mass also. We can also form another four-dimensional vector from the four items of information needed to specify an event—the quantities x, y, z, and t, or rather ct. The use of these four-dimensional vectors has enabled modern physicists to put the special theory of relativity in a very compact mathematical form.

**APPENDIX IV. THE SPACETIME CONTINUUM

The matters discussed here are sometimes regarded as mathematics, although from our point of view they are a fundamental form of physics. A pure mathematician would probably agree with us, relegating to mathematics the task of obtaining formal structures by reasoning from some of the concepts we shall discuss. Although our discussion follows a chronological order, we are more concerned with logical than with historical development and make no claim to precise historical accuracy; indeed, the actual events of ancient times are often difficult to reconstruct. Because of the speculative nature of much of this material, we have not tried to make the discussion as precise as it might be.

A. The Pythagorean Hope

The philosophy of Pythagoras, who lived in the sixth century B.C. and reportedly studied in Egypt, in Babylon, or under the Magi, was said to combine mathematics with mystical doctrines. His mysterious "brotherhood" was at once a religious order and a scientific research society. Pythagoras' opinion that "all things are numbers" was supported by his alleged discovery of musical relations in harmony. We have seen, in Chapter 9 that the wavelengths of the oscillations on a stretched string are numerically related in a simple way. Pythagoras could have discovered this simple relation by studying stretched strings. By "number," of course, he meant the simple whole numbers that we call the positive integers, or their ratios. Much mystical-seeming numerology has been ascribed to Pythagoras and his brotherhood; all things in heaven and earth were numbers to the Pythagoreans.

A seeming triumph of this point of view was Pythagoras' possible discovery of the first proof of the theorem in geometry which bears his name: the Pythagorean theorem. This theorem, set forth in Appendix II-D, states that the sum of the squares of the legs of a right triangle is equal to the square of the hypotenuse. Using the letters a and b to represent the legs and c to represent the hypotenuse, we can state the Pythagorean theorem as follows:

$$a^2 + b^2 = c^2 \tag{1}$$

This theorem appeared to show a deep relationship between numbers and space and thus to support the Pythagorean point of view. In fact, however, it can be used to apparently disprove that very thesis.

If we choose a right triangle in which the two legs are equal, then we have, by the Pythagorean theorem,

$$a^2 + a^2 = c^2 \tag{2}$$

or

$$2a^2 = c^2 \tag{3}$$

(The letters a and c represent the lengths as measured in terms of some unit.) This is all we need to destroy the Pythagorean philosophy. According to that philosophy, everything must be represented in whole numbers, or as ratios of whole numbers—that is, simple fractions. Actually, if we use a small enough unit of length, we can forget about the fractions and consider a and c as whole numbers. We shall assume, then, that we have done this and that both a and c are whole numbers. This would mean that the hypotenuse and a leg of the triangle are commensurate—they can both be measured exactly in terms of the same small unit of length. But this is impossible, as we can easily show. The number on the left of Equation 3 is even, since it is divisible by two. Therefore, the number on the right must be even also, since it is equal to the number on the left. But if c^2 is divisible by two, then c itself is divisible by two. This means that c^2 is divisible by two twice. But if c^2 is divisible by two twice, then the equal number $2a^2$ must also be divisible by two twice. Hence, a^2 must be divisible by two once. But, by the same reasoning as for c^2, if a^2 is divisible by two once, it must also be divisible by two twice. But if a^2 is divisible by two twice, then $2a^2$ is divisible by two three times. The point of all this dividing by two, therefore, is that the left side of Equation 2 can be evenly divided by two one more time than the right side can be divided by two. This is impossible, because both sides of the equation are equal—they represent the same number.

Thus the Pythagorean hope failed. The story is that the Pythagoreans tried to conceal this catastrophe. The two sides of this particular right triangle cannot both be measured in terms of the same unit of length if we want to express both lengths as whole numbers. This result apparently disproves the discrete nature of space. Suppose that there were some small fundamental length; all other lengths would be some whole number times that length. We have already seen a similar situation with respect to charge. Electric charge comes in small bundles, equal to the charge on the electron. Any larger charge could in principle be expressed as a whole number of these fundamental charges. The same trick does not work for length, however, according to the reasoning here. The length a and the length c in the example above cannot both be represented as whole numbers times a certain fundamental length. We could, of course, reject the "self-evident" assumptions of ordinary Euclidean geometry, on which the proof of the Pythagorean theorem is based, or we could reject the existence of precise lengths, but let us not do that at present.) Then the discrete nature of space is disproved because it leads to a contradiction. Length and number, therefore, are different types of quantity. Modern mathematics "solves" this problem by inventing new numbers by means of the notions of limits and infinite series. The result than is that one can represent every length by a number if one uses the new set of numbers.

B. A Paradox Of Zeno

The Pythagorean fiasco was part of what Fraenkel and Bar-Hillel have called the first of three "crises" in mathematics[1]. Another part of the same crisis involves the paradoxes of Zeno (495–435 B.C.). We are not sure what Zeno meant by his famous paradoxes, which are still discussed today. We shall therefore describe (following Russell, more or less)[2] what we think he *should* have meant.

The first paradox is that of the racecourse, illustrated in Figure A-17. Before the runner can traverse the whole length of the racetrack, he must first traverse half of it. Suppose that he has indeed traversed half of it, and is at point A in Figure A-17. Then, in order to make it across the remaining half, he must first travel half of that remaining half. Suppose he does so and gets to point B. Then in order to go the remaining distance, he must first travel half of that remaining distance, to point C. But we can keep dividing the remaining distance in half, ad infinitum. In order to get to the end of the track, the runner must cover an infinite number of halves of halves of halves, etc. Since he must travel an infinite number of distances, he can never get to the end. His total trip has been divided into an infinite number

[1]A. A. Fraenkel and Y. Bar-Hillel, *Foundations of Set Theory*, Amsterdam: North Holland Publishing Co., 1958, pp. 14–15, Chap. I. For antinomies, see also W. V. Quine, "Paradox," *Scientific American*, April 1962, p. 84.

[2]Bertrand Russell, *Our Knowledge of the External World*, New York: Mentor Books, New American Library of World Literature, 1956, Chap. VI.

of smaller trips. Each of these trips must take some time; therefore, according to the paradox, the whole time must be infinite.

It is true, of course, that the trips get shorter and shorter as the runner gets closer to the goal. But we have already seen that in order to allow the existence of right triangles and conventional geometry we had to allow space to be divisible without limit. Therefore we can keep taking halves of halves of halves, etc. However, this infinite number of ever-decreasing distances must add up to the total length of the racetrack. Thus, because space is infinitely divisible, this infinite number of distances can add up to a finite distance. If time is also infinitely divisible, in the same way that space is, then these small trips over ever-decreasing distances could have ever-decreasing time durations. Then all the time durations could add up to a finite time duration and resolve the paradox.

We know, of course, that it is possible for runners to finish races, and therefore we elect to regard the time duration as infinitely divisible, as well as the space. Thus we have the spacetime continuum. At least, the naive discrete theory of space and time seems unsatisfactory. Instead of reviewing Zeno's other paradoxes, therefore, we shall move forward to the time of Newton.

C. *Instantaneous Velocity: The Differential Calculus*

In the foregoing discussion, we were led to accept a continuous nature for space and time. Instead of a finite number of instants in each time duration, or a finite number of small lengths in each distance, we think instead of an infinite number of instants within a given

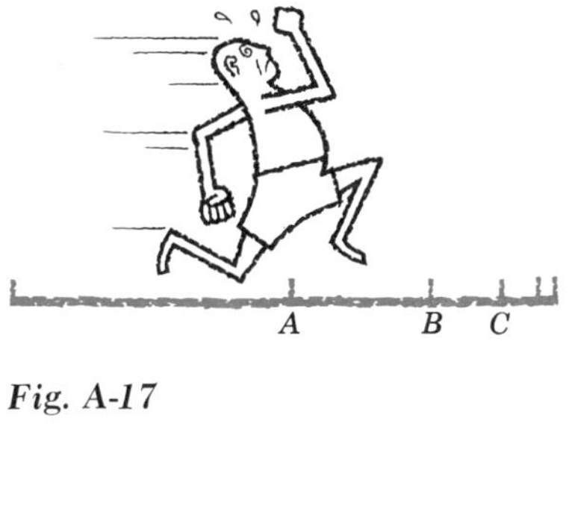

Fig. A-17

time duration, or of points along a line. Thus, infinity, even for short distances and time durations, has entered mathematics. Not surprisingly, it has caused trouble ever since.

The paradoxes of Zeno can all be discussed with reference to uniform motion. When the motion is not uniform, however, other problems arise. Let us then look at problems that might have arisen in discussing Newtonian physics. We have seen how Galileo, in studying motion with constant acceleration, obtained the distance-time equation, $d = 1/2\ at^2$, from the velocity-time equation, $v = at$. There is a more general way to obtain these equations from each other. (We shall use a small l to represent distance instead of d, which will be needed for another purpose.)

If an object is moving at a constant velocity v along a line, then the distance l and the time t are related by the following equation:

$$l = vt \tag{4}$$

The graph of this equation is shown in Figure A-18 The slope of this curve gives the velocity:

$$\text{slope of curve} \rightarrow v = l/t \tag{5}$$

We can easily see that the slope must represent the velocity, since it is simply the rise in the curve, in distance units, divided by the horizontal distance over which the rise occurs, measured in time units.

Suppose, however, that the velocity is not uniform. Then we get a graph something like Figure A-19. If the velocity changes from instant to instant, how are we to discover the object's velocity at a given time? For a small period of time, we can find an *average velocity*. We simply take the distance traveled during a period of time and divide it by the duration of that period. In Figure A-19 we have indicated a small time duration between t_1 and t_2. If we divide the distance traveled by this time duration, we have

$$\text{average velocity} = \frac{l_2 - l_1}{t_2 - t_1} \tag{6}$$

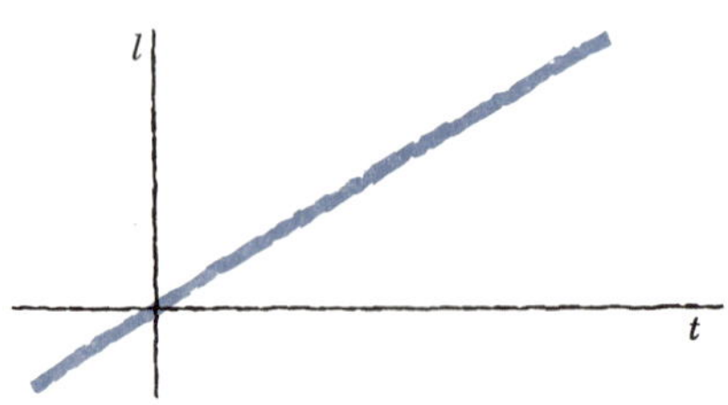

Fig. A-18

Thus we have obtained an average velocity by taking the small change in the distance and dividing it by the small change in the time. We usually represent changes in a quantity by the Greek letter delta. For example:

$$l_2 - l_1 = \Delta l \tag{7}$$

$$t_2 - t_1 = \Delta t \tag{8}$$

Thus, the average velocity may be written

$$\text{average velocity} = \frac{\Delta l}{\Delta t} \tag{9}$$

Let us find the average velocity in a simple example. We start with the equation $l = 1/2\, at^2$, which may be graphed as in Figure A-20. In this case,

$$l_1 = \tfrac{1}{2}\, a t_1^{\,2} \tag{10}$$

$$l_2 = \tfrac{1}{2}\, a t_2^{\,2} \tag{11}$$

Therefore, we may write the average velocity as follows:

$$\begin{aligned} \text{average velocity} &= \frac{\Delta l}{\Delta t} = \frac{l_2 - l_1}{t_2 - t_1} = \frac{\tfrac{1}{2}\, a t_2^{\,2} - \tfrac{1}{2}\, a t_1^{\,2}}{t_2 - t_1} \\ &= \tfrac{1}{2} a \;\frac{(t_2^{\,2} - t_1^{\,2})}{t_2 - t_1} \end{aligned} \tag{12}$$

The average velocity seems rather inexact, however. We would like to define the precise velocity at a definite time, as for example time t_2. We shall call this the *instantaneous velocity.* We might try to define it by merely setting t_1 equal to t_2 in Equation 12. If we do this, however, the following catastrophe occurs:

$$\text{average velocity} = \frac{\tfrac{1}{2}a(t_2^{\,2} - t_1^{\,2})}{(t_2 - t_1)} = \frac{0}{0} \tag{13}$$

The expression 0/0 is meaningless. Algebra alone will not help us here, and we must use a new form of mathematics. To save this situa-

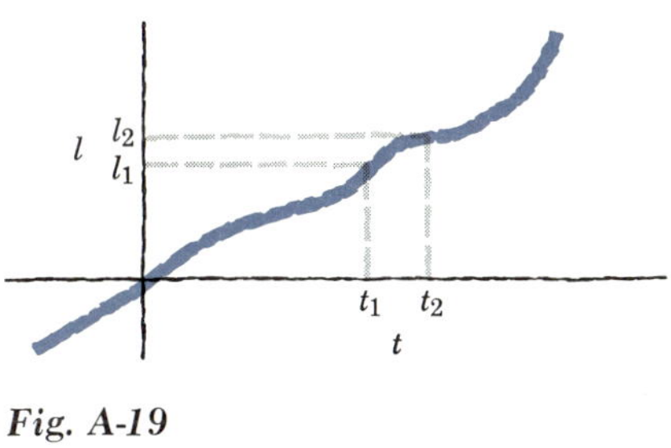

Fig. A-19

tion, we define the mathematical notion of a limit. We notice that it is only when t_1 is exactly equal to t_2 that the catastrophe occurs. We therefore allow t_1 to approach t_2 but never to be exactly equal to it. Then, as we let t_1 get closer and closer to t_2, we ask whether our average velocity, as expressed in Equation 12, also gets closer and closer to some value. As long as t_1 is not allowed to be quite equal to t_2, the following algebra is valid:

$$\text{average velocity} = \frac{\frac{1}{2}a(t_2^2 - t_1^2)}{t_2 - t_1} = \frac{\frac{1}{2}(a)(t_2 - t_1)(t_2 + t_1)}{(t_2 - t_1)} = \tfrac{1}{2}a(t_2 + t_1) \tag{14}$$

Now we ask for the value that this last quantity approaches as t_1 goes to t_2. This value, called the limit, gives us our instantaneous velocity. We can find it quite easily now that the troublesome quantities have been removed from the equation:

$$v = \text{instantaneous velocity} = \text{limit as } t_1 \text{ goes to } t_2 \text{ of } [\tfrac{1}{2}a\,(t_2 + t_1)]$$

$$v = \tfrac{1}{2}a(t_2 + t_2) = at_2 \tag{15}$$

This gives us the formula for the velocity-time relation. We drop the subscript in t_2 and have

$$v = at \tag{16}$$

In order for this limiting process to be exact, time and space must be infinitely divisible so that we can approach as closely as we please to the time t_2. For convenience we define the symbol dl/dt as follows:

$$\frac{dl}{dt} = \text{limit as } t_1 \text{ goes to } t_2 \text{ of } \left(\frac{\Delta l}{\Delta t}\right) \tag{17}$$

This is the instantaneous rate of change of distance, with respect to

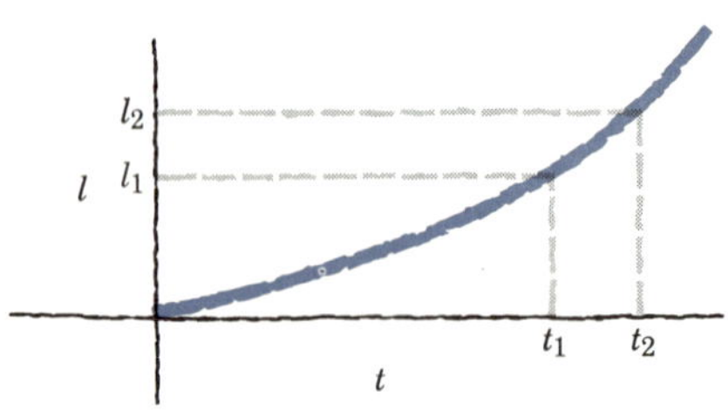

Fig. A-20

time. It is, of course, the instantaneous velocity, usually simply called the velocity:

$$v = \frac{dl}{dt} \tag{18}$$

We can define the instantaneous acceleration in a similar manner, since it is the rate of change of the instantaneous velocity:

$$a = dv/dt \tag{19}$$

If we look at the graph of distance and time again (Fig. A-21), we notice that the quantity $(l_2 - l_1)/(t_2 - t_1)$ is just the slope of the solid straight line shown on the graph. When we let t_1 approach t_2, this line becomes the line tangent to the curve at t_2. The instantaneous velocity thus is given by the slope of the tangent to the distance-time graph. In similar fashion, the acceleration of an object is given by the slope of the tangent line to the velocity-time graph. From the distance-time graph, therefore, we can obtain both the velocity-time graph and the acceleration-time graph, either by mathematically taking limits or by drawing tangents to curves.

D. The Integral Calculus

The analysis above, which outlines some of the ideas of differential calculus, enables one to find the rate of change of a quantity when the quantity itself is known over a period of time. There is another process by which the total change in the quantity itself can be found if its rate of change is known. This is, of course, the opposite process to the one we have just discussed. We can illustrate this by starting with the velocity-time equation for uniform acceleration and finding the distance-time equation.

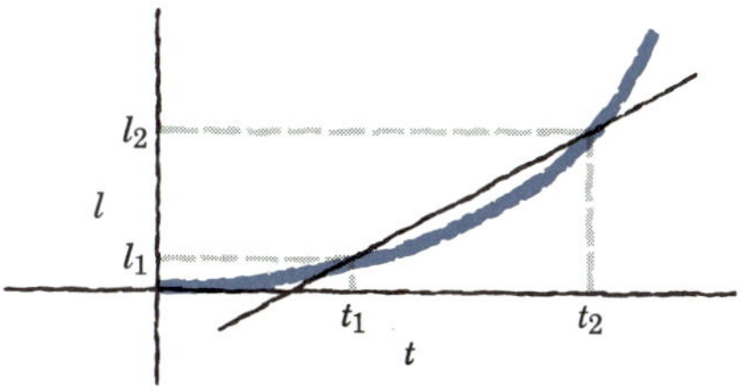

Fig. A-21

The velocity time equation is

$$v = at \tag{20}$$

The graph for this equation is given in Figure A-22. Let us consider a small interval about the time t_1, as shown in the figure. In this small interval of time, the velocity does not change much. Therefore we can use the equation for uniform motion to find the approximate distance traveled by the object during the small interval of time, Δt_1. This small distance Δl_1, will be

$$\Delta l_1 \approx v_1\ \Delta t_1 \tag{21}$$

This, however, is just the area of the small rectangle in Figure A-22. The quantity v_1 gives the altitude of the rectangle, in units appropriate to the graph, and the Δt_1 gives the base of the rectangle. We can now fill the horizontal axis between the origin and the final time, t_f, with similar small rectangles (see Fig. A-23). Then the total distance traveled is approximately

$$l = \Delta l_1 + \Delta l_2 + \Delta l_3 + \Delta l_4 + \text{etc.} \tag{22}$$

$$\approx v_1\ \Delta t_1 + v_2\ \Delta t_2 + v_3\ \Delta t_3 + v_4\ \Delta t_4 + \text{etc.} \tag{23}$$

There is nothing to stop us, however, from taking smaller and smaller rectangles. (More rectangles are then needed.) The smaller the rectangle, the less the velocity changes during the small interval of time, and the closer the sum in Equation 23 is to the true distance. As the width of the rectangles goes to zero, the sum will be, in the limit, exactly the distance traveled. At the same time, however, the combined areas of the rectangles will become closer and closer to the area underneath the line on the graph. The total distance traveled, therefore, is simply the area underneath the curve in the velocity-time graph. We define a new symbol to represent this area:

$$\int_{t_1}^{t_2} v\, dt = \text{area under } v,\, t \text{ graph between } t = t_1 \text{ and } t = t_2 \tag{24}$$

In this case we are interested in the distance traveled from time

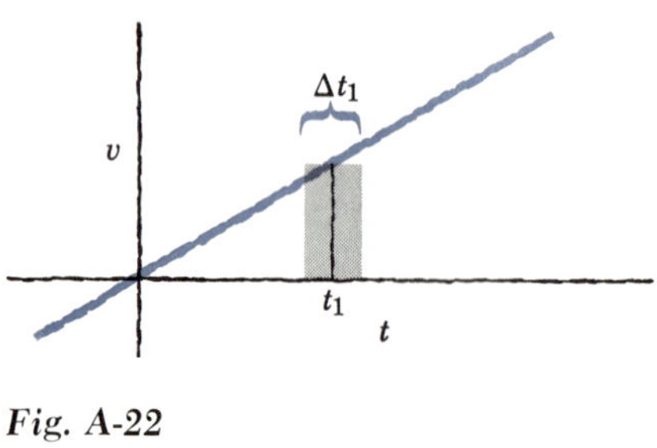

Fig. A-22

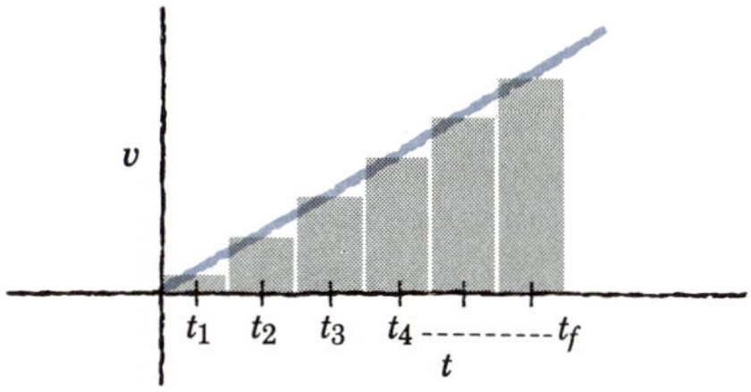

Fig. A-23

$t = 0$ to time $t = t_f$. The total area under the curve between these two times is the area of a triangle. We can find the area quite easily—it is one half the base of the triangle multiplied by its altitude (see Fig. A-24).

$$l = \int_0^{t_f} v\,dt = \tfrac{1}{2}\,t_f\,at_f = \tfrac{1}{2}\,at_f^2 \tag{25}$$

$$l = \tfrac{1}{2}\,at_f^2 \tag{26}$$

We have now accomplished the reverse procedure we desired. Notice that this procedure gives the distance traveled between two times. Actually this is the difference in the displacement of the object from the origin of the reference frame at the two times. (In adding up the area, any area below the horizontal axis counts as negative and must be subtracted from the positive area.) In similar fashion we can find the velocity from the acceleration-time curve:

$$\text{change in velocity between } t_1 \text{ and } t_2 = \int_{t_1}^{t_2} a\,dt \tag{27}$$

The procedure that we have just outlined uses some of the concepts of integral calculus. The differential and integral calculus furnish us with the mathematics necessary to deal with varying velocities and accelerations. According to Newton's laws, the velocity changes when force is applied to an object. Thus the calculus is very important in applications of Newton's laws of motion and, in fact, in nearly all physical laws discovered since Newton's time.

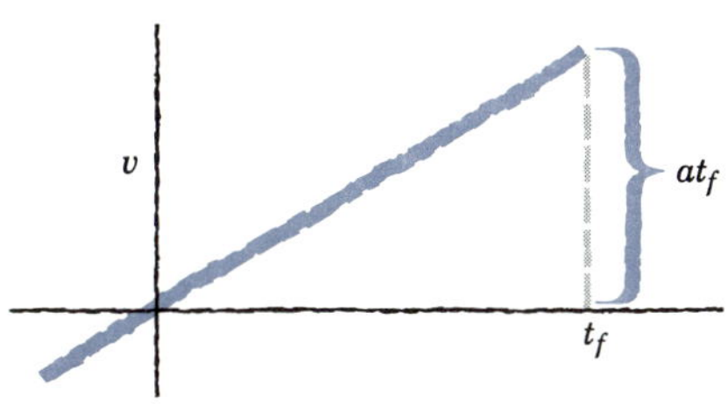

Fig. A-24

E. Problems Raised By The Continuum

At the time that the limiting processes of the calculus were first developed, the apparent ratio dl/dt was actually thought of as the ratio of two infinitesimally small quantities and was called the fluxion of l. The philosopher Bishop Berkelely, annoyed by certain antireligious mathematicians, correctly pointed out certain logical fallacies and ambiguities in the theory of the calculus. He called fluxions the "ghosts of departed quantities" and said that "anyone who can digest a second or third fluxion need not be too squeamish over any point of divinity."[3]

This controversy and attempts to resolve it constitute the second of the crises in mathematics referred to earlier. D'Alembert voiced his confidence in the new mathematics with the advice that one should go on using it, and faith would follow. A more rigorous analysis of the concepts of limits and derivatives removed many of the objections that had been raised. Currently, however, the foundation of this analysis involves the use of infinite sets, which have problems of their own. Certain paradoxes of set theory raise questions as to the validity of even this attempt at a logical understanding of the continuum. (Thus arose the third crisis in mathematics.) Yet if the continuum is rejected, we must face Zeno's paradoxes again and also reevaluate Pythagoras' theorem.

We have here presented the method of defining instantaneous velocity by the use of the concept of a limit. We are used to thinking of instantaneous velocities because they seem to appear on our speedometers. Actually our speedometers continually average the speed as our automobiles move along. The concept of instantaneous velocity is actually more difficult than it first appears. Its precise definition, as we have pointed out involves continuous space and time, and also the possibility, in principle at least, of precise measurements.

F. The Present Status Of The Continuum

From time to time physicists raise the question of the validity of the spacetime continuum, not because of mathematical or logical difficulties, but because a continuum is intrinsically unobservable. How can one measure an arbitrarily small length? To discern a very

[3]George Berkeley, "The Analyst," included in James R. Newman, *The World of Mathematics*, New York: Simon and Schuster, 1956, Vol. I, pp. 286–293.

small region, one would have to use a very small wavelength of light. That small wavelength means a large frequency, and therefore a large energy. The smaller the length, the larger the energy needed to see it. All the energy in the world would still leave us with some small lengths unmeasurable. In the modern spirit of rejecting things that are in principle unobservable, some physicists would like to get rid of the spacetime continuum. Perhaps this would help to solve some of the theoretical problems in modern high-energy particle physics. Other physicists believe that the spacetime continuum should be retained as long as it is useful and does not lead to incorrect results. The present status of this continuum is that it is very useful in physics and mathematics. However, it is deeply involved in some paradoxes and problems in mathematics and is unobservable in physics. How it could be replaced with a discrete space and time is unknown.

Appendix V. Suggested supplementary reading

Many of the following books and articles can be easily understood by the general reader. Most of them are available in inexpensive paperback editions. We list them, not because we agree in every respect with the authors, but because they discuss the issues in an interesting, valuable, and usually clear way. There are, of course, many other books and articles which could be included in such a list.

Andrade, E. N. da C., *Rutherford and the Nature of the Atom,* Anchor Books, Doubleday & Co., Garden City, New York, 1964.

Aristotle's Physics, translated by Richard Hope, University of Nebraska Press, Lincoln, Nebraska, 1961.

Arons, Arnold B., and Alfred M. Bork, eds., *Science and Ideas, Selected Readings,* Prentice-Hall, Inc., Englewood Cliffs, New Jersey, 1964.

Barger, Vernon D., and David B. Cline, "High-Energy Scattering," *Scientific American,* **217** (December 1967), pp. 76–91.

Barzun, Jacques, *Darwin, Marx, and Wagner, Critique of a Heritage,* Rev. Second Edition, Anchor Books, Doubleday & Co., Garden City, New York, 1958. A critical attack on scientistic or mechanistic ideas. The original edition was published in 1941, and the comments on evolution should be read in the light of recent developments.

Boorse, Henry A., and Lloyd Motz, eds., *The World of the Atom,* Basic Books, Inc., New York, 1966. Two volumes of readings containing excerpts from the writings of famous physicists.

Bronowski, J., *The Common Sense of Science*, Modern Library Paperbacks, Random House, New York, n.d.

———, *Science and Human Values*, Harper Torchbooks, Harper & Row, New York, Rev. Edition, 1965.

Bronowski, J., and Bruce Mazlish, *The Western Intellectual Tradition from Leonardo to Hegel*, Harper Torchbooks, Harper & Row, New York, 1960.

Burtt, E. A., *The Metaphysical Foundations of Modern Physical Science*, Anchor Books, Doubleday & Co., Garden City, New York, Rev. Edition, 1932. This book is more interesting than its title may suggest to those suspicious of metaphysics.

Chew, Geoffrey F., Murray Gell-Mann, and Arthur H. Rosenfeld, "Strongly Interacting Particles," *Scientific American*, **210** (February 1964), pp. 74–93.

Cohen, I. Bernard, *The Birth of a New Physics*, Anchor Books, Doubleday & Co., Garden City, New York, 1960.

Dantzig, Tobias, *Number, the Language of Science*, Anchor Books, Doubleday & Co., Garden City, New York, Fourth Ed. rev. and augmented, 1956.

de Santillana, Giorgio, *The Crime of Galileo*, The University of Chicago Press, Chicago, 1955.

———, *The Origins of Scientific Thought*, Mentor, The New American Library of World Literature, Inc., New York, 1961.

Drake, Stillman, *Discoveries and Opinions of Galileo*, Anchor Books, Doubleday & Co., Garden City, New York, 1957. This very entertaining book contains translations of Galileo's writings, including the letter to the Grand Duchess Christina and parts of *The Assayer*, in which Galileo replied to "Sarsi."

———, and C. D. O'Malley, *The Controversy on the Comets of 1618*, University of Pennsylvania Press, Philadelphia, 1960. This book contains the full translated text of Galileo's *The Assayer* and writings of others in the controversy.

Eddington, Arthur S., *Space, Time, and Gravitation*, Harper Torchbooks, Harper & Row, New York, 1959. This book, originally published in 1920, begins with an interesting dialogue on the nature and measurement of space.

Einstein, Albert, and Leopold Infeld, *The Evolution of Physics*, Simon & Schuster, New York, 1938.

Fermi, Laura, and Gilberto Bernardini, *Galileo and the Scientific Revolution*, Fawcett Publications, Inc., Greenwich Conn., 1965. (Originally published by Basic Books in 1961.)

Feynman, Richard, *The Character of Physical Law*, M. I. T. Press, Cambridge, Mass., 1967. (Originally published in 1965.)

Ford, Kenneth W., *The World of Elementary Particles*, Blaisdell Publishing Company, New York, 1965. Very understandable account of elementary particles.

Gamow, George, *Biography of Physics*, Harper Torchbooks, Harper & Row, New York, 1964. (Originally published in 1961.)

———, *One, Two, Three, . . . Infinity, Facts and Speculations of Science*, Mentor, The New American Library of World Literature, Inc., New York, 1955. We would like to call special attention to Chapter VIII, The Law of Disorder.

Galilei, Galileo, *Dialogue Concerning the Two Chief World Systems, Ptolemaic and Copernican*, trans. by Stillman Drake, University of California Press, Berkeley and Los Angeles, 1953.

———, *Dialogues Concerning Two New Sciences*, trans. by Henry Crew and Alfonso de Salvio, Dover Publications, Inc., New York. (Translation originally published in 1914.)

Gardner, Martin, "Can Time Go Backward?" *Scientific American*, **316**, 1 (January 1967), pp. 98–108.

———, *Relativity for the Millions*, Pocket Books, New York, 1965. (Originally published by Macmillan in 1962.) A well-illustrated and interesting account of the special and general theories of relativity.

Gillispie, C. C., *The Edge of Objectivity*, Princeton University Press, Princeton, New Jersey, 1960. The development of science, with attention to its cultural and intellectual importance.

Helm, E. Eugene, "The Vibrating String of the Pythagoreans," *Scientific American*, **217** (December 1967), pp. 92–103.

Hoffman, Banesh, *The Strange Story of the Quantum*, Second Ed., Dover Publications, Inc., New York, 1959.

Jeans, James, *The Growth of Physical Science*, Fawcett Publications, Inc., Greenwich, Conn., 1961. (Originally published in 1947.)

———, *Physics and Philosophy*, Ann Arbor Paperbacks, The University of Michigan Press.

Kearney, Hugh, *Origins of the Scientific Revolution*, Longmans, Green and Co., Ltd., London, 1964.

Kemeny, John, *A Philosopher Looks at Science*, Van Nostrand, Princeton, New Jersey, 1959.

Koestler, Arthur, *The Sleepwalkers*, Macmillan, New York, 1959. Sympathetic in his treatment of Kepler, less so in his treatment of Copernicus and Galileo.

Mach, Ernst, *The Science of Mechanics*, Open Court Publishing Co., La-Salle, Illinois, Sixth American Ed., 1960.

Minnaert, M., *The Nature of Light and Color in the Open Air*, Dover Publications, Inc., New York, 1954.

Newman, James R., *The World of Mathematics*, Simon & Schuster, New York, 1956. Four volumes of very interesting readings, many relevant to physics.

Newton, Isaac, *The Mathematical Principles of Natural Philosophy and His System of the World*, trans. by Andrew Motte in 1729, rev. by Florian Cajori, University of California Press, Berkeley, California, 1934.

Reichenback, Hans, *Experience and Prediction*, Phoenix Books, University of Chicago Press, Chicago, Illinois, 1961. (First published in 1938.)

"An Analysis of the Foundations and Structure of Knowledge," the last section, discusses the problem of induction.

———, *From Copernicus to Einstein*, trans. by Ralph B. Winn, The Wisdom Library, The Philosophical Library, New York, 1942.

Rogers, Eric M., *Physics for the Inquiring Mind*, Princeton University Press, Princeton, New Jersey, 1960.

Russell, Bertrand, *The ABC of Relativity*, Mentor, The New American Library of Literature, Inc., New York, 1959. (Revised edition first published by Allen and Unwin, Ltd., 1958.)

Schneer, Cecil J., *The Evolution of Physical Science*, Grove Press, New York, 1960.

Schrodinger, Erwin C., *Science, Theory, and Man*, Dover Publications, Inc., New York, 1957. (Originally published in 1935.) See especially Chapter IX on the nature of an elementary particle.

Thayer, H. S., ed., *Newton's Philosophy of Nature, Selections from His Writings*, Hafner Publishing Company, New York, 1953.

Watson, J. D., *The Molecular Biology of the Gene*, Benjamin, New York, 1965.

Weisskopf, Victor F., *Knowledge and Wonder, The Natural World as Man Knows It*, Anchor Books, Doubleday & Co., Garden City, New York, 1963.

Weisskopf, Victor F., "Quantum Theory and Elementary Particles," *Science*, **149** (September 10, 1965), pp. 1181–1189.

White, A. D., *A History of the Warfare of Science with Theology in Christendom*, reprinted by Dover Publications, Inc., New York, 1960. Two volumes.

Wigner, Eugene P., "Violations of Symmetry in Physics," *Scientific American*, **213** (December 1965), pp. 28–36.

Young, Louise B., ed., *Exploring the Universe*, McGraw-Hill Book Co., New York, 1963. Selected readings.

———, *The Mystery of Matter*, Oxford University Press, New York, 1965. Selected readings.

INDEX